MW01628865

GLOBAL CHANGE BIOLOGY

GLOBAL CHANGE BIOLOGY

The Study of Life on a Rapidly Changing Planet

ERICA BREE ROSENBLUM
UNIVERSITY OF CALIFORNIA, BERKELEY

NEW YORK OXFORD
OXFORD UNIVERSITY PRESS

Oxford University Press is a department of the University of Oxford.
It furthers the University's objective of excellence in research, scholarship,
and education by publishing worldwide. Oxford is a registered trademark
of Oxford University Press in the UK and certain other countries.

Published in the United States of America by Oxford University Press
198 Madison Avenue, New York, NY 10016, United States of America.

© 2021 by Oxford University Press

For titles covered by Section 112 of the US Higher Education Opportunity Act,
please visit www.oup.com/us/he for the latest information about pricing and
alternate formats.

All rights reserved. No part of this publication may be reproduced, stored
in a retrieval system, or transmitted, in any form or by any means, without
the prior permission in writing of Oxford University Press, or as expressly
permitted by law, by license, or under terms agreed with the appropriate
reproduction rights organization. Inquiries concerning reproduction outside
the scope of the above should be sent to the Rights Department, Oxford
University Press, at the address above.

You must not circulate this work in any other form
and you must impose this same condition on any acquirer.

Printed by Bridgeport National Bindery, Inc., United States of America

This book is dedicated to my family, the millions of species on our planet, and you. May our efforts as a learning community benefit ourselves, each other, and all living things.

CONTENTS

Preface xvii

UNIT I **SETTING THE STAGE**

Learning Objective: Understand the approaches used by global change biologists, analyze historical patterns of biodiversity, evaluate impacts on the environment across human history, and distinguish among different global change stressors.

CHAPTER 1 Approaches in Global Change Biology *1*

Learning Outcomes 1

The Blank Page 1

Introduction 1

How Did the Field of Global Change Biology Develop? *2*

Early History *2*

Contemporary Stressors *3*

Ways of Knowing *4*

Honoring Diversity and New Horizons *6*

How Are Global Change Biology Studies Designed? *7*

Integrative Approaches *7*

Key Elements of Study Design *8*

Personal and Societal Relevance *10*

What Key Research Approaches Are Used in Global Change Biology? *12*

Observational Approaches *13*

Experimental Approaches *13*

Modeling Approaches *13*

Synthesis Approaches *14*

Participatory Approaches *15*

In Sum *15*

What Key Tools Are Used in Global Change Biology? 17
Environmental Monitoring Tools 17
Organismal Monitoring Tools 17
Molecular Tools 18
Computational Tools 19
In Sum 20
Core Concepts: How Are Data Displayed? 20
Conclusion 22
Meet the Data: The Economic Value of Nature 23
Taking a Closer Look: The Value of Biological Diversity 26
Key Concepts 27
Consolidate Your Knowledge 28
Literature Cited 29

CHAPTER 2 Brief History of Life on Earth 32
Learning Outcomes 32
The Blank Page 32
Introduction 32
What Key Transitions Led to the Emergence of Life on Earth? 33
Origin of the Universe 33
Formation of the Earth 34
Development of a Crust and Hydrosphere 34
Prebiotic Chemistry 34
Origin of Heredity 35
How Did Cellular Life Evolve and Diversify? 36
Evolution of Bacteria and Archaea 36
Evolution of Eukaryotes 37
Evolution of Multicellularity 37
What Evolutionary Processes Shape Biological Diversity? 39
Natural Selection 39
Genetic Drift 39
Speciation 39
Extinction 40
When Have Speciation and Extinction Rates Been Particularly High? 41
Evolutionary Radiations 42
Core Concepts: What Is a Phylogenetic Tree? 42
Mass Extinctions 45
Conclusion 46
Meet The Data: The Ring of Life 46
Taking a Closer Look: Biological Levels of Change 48

Key Concepts 52
Consolidate Your Knowledge 52
Literature Cited 53

CHAPTER 3 Rise of the Humans 56
Learning Outcomes 56
The Blank Page 56
Introduction 56
When and How Did Early Hominids Evolve? 56
Divergence from Great Ape Common Ancestor 56
Overlapping Early Human Lineages 57
Core Concepts: What Is in a Name? 58
Evolution of the Homo Group 60
When and How Did Modern Humans Spread Around the World? 61
How Did Early Human Civilizations Impact the Environment? 63
Hunting and Megafaunal Extinctions 63
Agriculture and Domestication 65
Urbanization and Population Growth 67
Conclusion 68
Meet the Data: Ice Age Genetics 68
Taking a Closer Look: The Evolutionary Success of Humans 71
Key Concepts 74
Consolidate Your Knowledge 75
Literature Cited 76

CHAPTER 4 The Anthropocene 80
Learning Outcomes 80
The Blank Page 80
Introduction 80
What Is the Anthropocene and When Did It Begin? 80
Core Concepts: What Is Climate and How Is It Measured? 82
What Are Patterns of Contemporary Population Growth? 85
How Are Contemporary Human Civilizations Impacting the Environment? 86
Land-Use Change 86
Pollution 88
Globalization 90
Additional Stressors 91
How Do Anthropogenic Stressors Interact with Each Other? 91

What Influences Overall Vulnerability to Global Change Pressures? *92*
Exposure *93*
Sensitivity *93*
Capacity to Respond *93*
Conclusion *93*
Meet the Data: Pollinators and Pesticides *94*
Taking a Closer Look: Historical and Contemporary Climate Change *97*
Key Concepts *102*
Consolidate Your Knowledge *103*
Literature Cited *103*

UNIT II CORE RESPONSES TO GLOBAL CHANGE STRESSORS

Learning Objective: Analyze the four core responses of organisms and species to global change stressors (move, adjust, adapt, die) and interpret results from contemporary global change biology studies.

CHAPTER 5 Core Responses: Move *106*
Learning Outcomes *106*
The Blank Page *106*
Introduction *106*
How and Why Do Organisms Move? *107*
What Is a Geographic Range? *109*
What Factors Determine a Species' Geographic Range? *110*
Evolutionary History *110*
Species Characteristics *110*
Ecological Requirements *111*
In Sum *111*
Core Concepts: What Is a Niche? *112*
Do Range Changes Occur Even Without Anthropogenic Influence? *113*
What Types of Range Changes Occur in Response to Anthropogenic Pressures? *115*
Range Contractions *116*
Range Expansions *116*
Range Marches *117*
In Sum *119*
How Do Scientists Predict Range Changes? *120*
Conclusion *124*
Meet the Data: A Century of Change in Yosemite *125*
Taking a Closer Look: Globalization and Invasive Species *128*
Key Concepts *133*
Consolidate Your Knowledge *134*
Literature Cited *135*

CHAPTER 6 Core Responses: Adjust *139*

Learning Outcomes *139*

The Blank Page *139*

Introduction *139*

What Is Phenotypic Plasticity? *140*

Core Concepts: What Are the Mechanisms of Heredity? *140*

Is the Capacity for Plasticity Consistent Across Traits and Species? *143*

What Types of Plasticity Occur in Response to Global Change Pressures? *144*

Shifts in Development *145*

Shifts in Physiology *146*

Shifts in Behavior *146*

Shifts in Morphology *148*

In Sum *148*

What Mechanisms Underlie Phenotypic Plasticity? *148*

Instantaneous Responses *149*

Gene Regulation Responses *149*

Epigenetic Responses *150*

Induction and Reversal *151*

How Do Scientists Assess and Predict Phenotypic Plasticity? *152*

Experimental Approaches *152*

Molecular Approaches *154*

Can Plasticity Facilitate Long-Term Persistence? *155*

Conclusion *156*

Meet the Data: Phenology and Global Warming *157*

Taking a Closer Look: Urbanization *160*

Key Concepts *164*

Consolidate Your Knowledge *165*

Literature Cited *166*

CHAPTER 7 Core Responses: Adapt *171*

Learning Outcomes *171*

The Blank Page *171*

Introduction *171*

What Conditions Are Required for Adaptation? *172*

Heredity *172*

Variation *172*

Differential Survival or Reproduction *173*

Core Concepts: Where Does Genetic Variation Come from? *174*

What Is an Example of Evolution by Natural Selection? *176*

What Types of Adaptation Occur in Response to Global Change Pressures? *177*

Adaptation to Environmental Contaminants *177*

Adaptation to Introduced Species *179*

Adaptation to a Changing Climate *179*

In Sum *180*

How Do Scientists Identify Adaptations and Predict Adaptive Potential? *181*

Field Studies *182*

Lab Studies *183*

Molecular Approaches *185*

In Sum *187*

Can Adaptation Prevent Extinction? *187*

Conclusion *190*

Meet the Data: The *Daphnia* Time Machine *191*

Taking a Closer Look: Coral Reefs *193*

Key Concepts *197*

Consolidate Your Knowledge *198*

Literature Cited *199*

CHAPTER 8 Core Responses: Die *205*

Learning Outcomes *205*

The Blank Page *205*

Introduction *205*

How Is the Survival of Individuals, Populations, and Species Connected? *206*

Loss of Individuals *206*

Loss of Populations *208*

Loss of Species *209*

What Are Examples of Extinction in Response to Global Change Pressures? *209*

Pinta Island Tortoise *209*

Passenger Pigeon *210*

Polynesian Tree Snail *210*

Superb Cyanea *210*

Yangtze River Dolphin *211*

Smallpox Virus *211*

In Sum *211*

How Do Scientists Estimate Extinction Risk? *211*

Species Distribution Modeling *212*

Population Viability Analysis *212*

In Sum *216*

How Do Scientists Summarize Global Patterns of Extinction Risk? *216*

Biodiversity Databases *216*

Meta-analyses *219*

In Sum *220*

What Is the Sixth Mass Extinction? *220*

Core Concepts: What Is Extinction Debt? *220*

Conclusion *222*

Meet the Data: The Sixth Mass Extinction *222*

Taking a Closer Look: Amphibian Declines *226*

Key Concepts *231*

Consolidate Your Knowledge *232*

Literature Cited *233*

UNIT III COMPLEX RESPONSES TO GLOBAL CHANGE PRESSURES

Learning Objective: Analyze higher level responses of communities, ecosystems, and the biosphere to global change stressors and interpret results from contemporary global change biology studies.

CHAPTER 9 Community-Level Responses *237*

Learning Outcomes *237*

The Blank Page *237*

Introduction *237*

What Are Key Types of Biological Interactions? *238*

Facultative Versus Obligate Interactions *238*

Multispecies Interactions *239*

Direct Versus Indirect Interactions *240*

In Sum *240*

How Do Global Change Pressures Affect Biological Interactions? *240*

Effects of Species Loss *240*

Effects of Species Gain *241*

Effects of Species Change *242*

In Sum *245*

How Does Extinction Affect Communities? *245*

Coextinction of Mutualists *245*

Coextinction of Parasites *246*

Coextinction of Predators and Herbivores *247*

Core Concepts: What Are Above- and Below-Ground Food Webs? *248*

In Sum *250*

What Are Cascading Effects? *251*

Conclusion *253*

Meet the Data: The Collapse of Mutualisms *253*

Taking a Closer Look: Kelp Forests and Trophic Cascades *257*

Key Concepts *261*

Consolidate Your Knowledge *261*

Literature Cited *262*

CHAPTER 10 Ecosystem-Level Responses *266*

Learning Outcomes *266*
The Blank Page *266*
Introduction *266*
What Are Biogeochemical Cycles? *267*
Water Cycle *267*
Carbon Cycle *267*
Nitrogen Cycle *267*
In Sum *268*
How Do Global Change Pressures Impact Ecosystems? *268*
Ecosystem Structure *268*
Ecosystem Functions *268*
Ecosystem Services *270*
Links Between Ecosystem Properties *270*
How Do Global Change Pressures Impact Large-Scale Earth Systems? *271*
Terrestrial Systems *274*
Atmospheric Systems *274*
Aquatic Systems *277*
Cryospheric Systems *277*
In Sum *279*
Core Concepts: What Is a Biodiversity Hotspot? *279*
What Is a Feedback? *281*
What Is Ecosystem Collapse? *283*
What Is Ecosystem Resilience? *286*
Conclusion *287*
Meet the Data: Greenhouse Gases in the Soil *287*
Taking a Closer Look: Factors Influencing Response to Global Change *292*
Key Concepts *295*
Consolidate Your Knowledge *296*
Literature Cited *297*

UNIT IV NEW HORIZONS

Learning Objective: Critique different conservation and management approaches and evaluate issues at the intersection of science and society.

CHAPTER 11 Conservation in an Era of Global Change *301*

Learning Outcomes *301*
The Blank Page *301*
Introduction *301*
Why Is It Important to Explicitly Define Conservation Priorities? *302*
Core Concepts: What Is Climate Mitigation? *304*

Why Is It Important to Match Conservation Actions to Particular Biological Levels? *306*

What Are Examples of Fine-Filter Conservation Strategies? *306*

Reducing Overharvest *306*

Captive Breeding *308*

Translocations *309*

What Are Examples of Coarse-Filter Conservation Strategies? *311*

Creating Reserves *311*

Re-establishing Corridors *313*

Re-establishing Natural Disturbance Regimes *315*

Enhancing Habitat in Highly Modified Settings *315*

In Sum *318*

What Is Adaptive Management? *318*

Conclusion *320*

Meet the Data: Maximizing Evolutionary Diversity *321*

Taking a Closer Look: Emerging Technologies and Conservation Ethics *325*

Key Concepts *328*

Consolidate Your Knowledge *329*

Literature Cited *329*

CHAPTER 12 Aligning the Interests of Biodiversity and Human Society *334*

Learning Outcomes *334*

The Blank Page *334*

Introduction *334*

What Are Coupled Human–Natural Systems? *335*

What Societal Levers Can Be Used to Support Biodiversity Conservation? *336*

Core Concepts: What Is I=PAT? *336*

How Can Individuals Support Biodiversity Conservation? *339*

As Consumers *339*

As Funders *340*

As Practitioners and Activists *341*

In Sum *342*

How Can Collectives Support Biodiversity Conservation? *343*

Ecotourism *343*

Sustainable Food Production *345*

Cradle-to-Cradle Manufacturing *345*

In Sum *348*

How Can Policy Action Support Biodiversity Conservation? *348*

Fisheries Governance *348*

Endangered Species Legislation *349*

International Climate Treaties *350*
International Debt-for-Nature Swaps *352*
In Sum *352*
What Is the Forecast for the Future? *353*
Conclusion *353*
Meet the Data: Financial Incentives for Dynamic Conservation *354*
Taking a Closer Look: Environmental Worldviews *359*
Key Concepts *361*
Consolidate Your Knowledge *362*
Literature Cited *363*

Glossary *367*

Index *375*

PREFACE

THE BIG PICTURE

We are living during a profoundly important period of our planet's history. Life on Earth—and the conditions that support it—have been evolving and changing for billions of years. But today we are in an age of acceleration: the magnitude and velocity of contemporary environmental change are staggering. Never before has a single species precipitated such radical and rapid changes on Earth. Humans have modified Earth systems from the depths of the oceans to the upper reaches of the atmosphere. Scientists have long warned that we are approaching a planetary tipping point, where our impacts on the biosphere will have dramatic and long-lasting consequences. Moreover, there are deep and inextricable links between environmental, economic, social, and cultural systems. Many pressing societal issues around equity, sustainability, and public health cannot be separated from the broader biological and planetary context. It is therefore urgent to understand how humans are changing the conditions for life on Earth, how different species respond to these changes, and how we can better conserve the biological heritage of our planet for the future.

GLOBAL CHANGE BIOLOGY

The goal of this textbook is to provide a dynamic and integrative exploration of the emerging field of Global Change Biology. Global Change Biology addresses time-critical questions about how environmental change impacts life on Earth. Global change biologists address questions over many time periods (from past to present), across many spatial scales (from local to global), and throughout the tree of life (from microbes to mammals). They use novel tools and approaches to develop a cohesive understanding of how complex biological systems respond to rapid environmental change. This textbook offers a comprehensive introduction to the field of Global Change Biology, focusing on cutting-edge developments in studying and conserving life on a changing planet.

THIS PARTICULAR MOMENT IN HISTORY

Perhaps not surprisingly, it takes years to write a textbook. Periods of intensive writing are interspersed with rounds of extensive review and subsequent revision and refinement. During these years, not only did the field of Global Change Biology continue to evolve, but our world and our lives changed dramatically. As a society, we were confronted by global disease outbreaks (including COVID-19), increased frequency and severity of climate-related calamities (fires, floods, hurricanes), and the pervasive inequities of our social and economic systems, including structural racism and discrimination against many identity groups. Disease, displacement, death, and violence have amplified a sense of fear and uncertainty for many about our future on this planet.

What has become increasingly clear is that these issues are not separate from each other, nor are they separate from our core topic of Global Change Biology. At its most dysfunctional, modern society has treated people, other species, and natural resources as objects to be used and discarded. Thus, the ideologies that have led us to a social justice crisis are the same that have led us to an ecological crossroads. If we perpetuate systems of exploitation, all living things suffer. However, periods of change—and even upheaval—are also times of great opportunity. Many people across sectors are working with renewed urgency to divest from systems that perpetuate alienation and instead invest in creative and life-affirming initiatives. We can all begin envisioning new ways of being that honor all living things and their deep interconnection.

Global Change Biology is inherently focused on these connections. All life on Earth is interdependent, and we cannot consider planetary change without evaluating our place in it. The journey through this textbook will focus on the science of Global Change Biology, but this moment in history reminds us of the deep connections between biological and social phenomena. As you engage with the textbook, consider ways that the content is relevant to your own experience and how you can explore the question of what it means to be human at this time point in history.

Ultimately, this book is intended to inspire you about a new field of study. As a university professor, I hear many students express how environmental science literature often leaves them feeling angry about the past, hopeless about the future, or guilty about being human. My hope is that this book offers a different perspective: one aligned with new ways of relating to ourselves, each other, and the entirety of the natural world. By studying the impact humans have on the biosphere, we can begin to see our individual and collective power and potential. We each have so much inspiration, creativity, and presence to contribute.

INTENDED AUDIENCE

When I was hired as a faculty member at the University of California, Berkeley, I was asked to develop a Global Change Biology course. The opportunity was exciting, but I quickly discovered that there was no definitive textbook synthesizing the research in this emerging field. Textbooks in Ecology, Evolution, Conservation, and Environmental Science introduced relevant themes but did not provide a sophisticated treatment of Global Change Biology as a cutting-edge, interdisciplinary field. Texts in the Global Change arena focused more narrowly on specific stressors or ecosystems. Many were edited volumes and did not present a unified voice with a clear conceptual arc. I wanted a text that would provide an integrative perspective on the complex scientific opportunities and challenges in Global Change Biology. The result is this book.

I wrote this textbook with upper division undergraduate biology and environmental science majors in mind. Juniors and seniors often use my course as a capstone to apply key concepts in ecology and evolution to the Global Change theme area and develop a more analytical and integrative skill set as scientists. However, Global Change Biology has crossover power to engage other audiences, and this textbook is intended to be flexible. Lower division students or nonmajors can use this book to understand contemporary issues confronting the biosphere. Graduate students beginning their professional journey as independent researchers can use this book as an introduction to a new field and a roadmap for identifying important areas for future research. Throughout the textbook, there are features that both review fundamental concepts and expose students to leading-edge topics in research, providing opportunities to adapt the text for different audiences.

APPROACH

This book provides a comprehensive introduction to the field of Global Change Biology and a roadmap for structuring Global Change Biology courses. The approach encourages students to think across spatial and temporal scales, grapple with real-world questions, and integrate ecological, evolutionary, and conservation perspectives. The conceptual arc of the textbook is organized around four fundamental learning objectives corresponding to four units.

Unit I: Setting the Stage

The first section of the book creates context by taking a journey into the past. We first evaluate the approaches and tools used in the science of Global Change Biology (Chapter 1). We then analyze patterns of biodiversity and environmental change throughout Earth's history (Chapter 2). This leads us to explore the history of human evolution (Chapter 3) and the Anthropocene, the geological period where human activities have drastically altered our planet (Chapter 4). Finally, we evaluate global change stressors such as climate change, habitat alteration, and globalization (Chapter 4).

Unit II: Core Responses to Global Change Stressors

The second section of the book explores how contemporary changes to the global environment impact living things, with an emphasis on molecular, individual, population, and species levels. We evaluate four core responses organisms exhibit in changing environments: move (Chapter 5), adjust (Chapter 6), adapt (Chapter 7), and die (Chapter 8). Case studies across scales, across the tree of life, and across Earth's major biomes are used to bring concepts alive.

Unit III: Complex Responses to Global Change Stressors

The third section of the book continues to investigate how changes to the environment impact life on Earth, with an emphasis on community, ecosystem, and biosphere levels. We evaluate complex and interacting responses and feedbacks across ecological communities (Chapter 9) and large-scale Earth systems (Chapter 10). Again, case studies across scales, across the tree of life, and across Earth's major biomes are used to develop analytical skills and address pressing contemporary issues.

Unit IV: New Horizons

The final section of the book focuses on evaluating what actions can be taken to maintain resilient ecosystems in a changing world. We critique conservation and management options through a Global Change Biology framework (Chapter 11), with a focus on emerging approaches and complex trade-offs. We also address issues at the intersection of science and society (Chapter 12) because conserving biological diversity will depend on aligning the interests of individuals, societies, and the ecosystems on which humanity depends.

Enrichment and Assessment Features

The core text is enhanced by features that provide opportunities for reviewing foundational concepts, interpreting scientific data, developing systems-oriented thinking, and reflecting on the learning process. These features also provide explicit pre- and post-assessment opportunities to promote active engagement with the text and integrative learning.

The Blank Page is an opening feature of each chapter to provide an opportunity for reflection and self-assessment. Each *Blank Page* provides a guided drawing or writing exercise that draws on the student's own experience about the topic in the upcoming chapter. The *Blank Page* exercises are inspired by the National Science Foundation "Picturing to Learn" initiative that recognizes how drawing scientific concepts can reveal misconceptions and deepen student understanding. This feature also provides students with a more personal entry point into material.

Learning Outcomes are presented at the beginning of each chapter to specify the knowledge and skills students will gain in each chapter.

Core Concepts boxes are included to review foundational concepts from ecology, evolution, and genetics that are necessary to understand Global Change Biology. The *Core Concepts* feature helps level the field for students approaching this field with different prior coursework.

Meet the Data features provide direct experience interpreting Global Change Biology data. By generating predictions, interpreting data figures, and articulating key findings, students learn how data are collected, analyzed, and interpreted. The *Meet the Data* features support students to go beyond learning content and practice being Global Change Biologists.

Taking a Closer Look features provide an opportunity to evaluate multifaceted responses in complex systems. These features encourage students to integrate material across chapters and explore the relevance of Global Change Biology themes in their own lives.

Reflection questions are presented with each figure and table throughout the text so students can engage directly with the material.

Key Concepts are summarized at the end of each chapter to help students review important concepts and draw conceptual links.

Consolidate Your Knowledge questions are presented at the end of each chapter to help students assess their progress toward meeting the learning objectives.

In Sum

Not only is this a pivotal time in history to be a human, but it is a fascinating time to be a scientist. Never before have humans had such complex and dramatic impacts on the biosphere. But simultaneously, never before have scientists had such innovative tools for studying and conserving life on a rapidly changing planet. While there are few one-size-fits-all solutions, there are many ways to contribute to science and society in this period of change. Whether or not you pursue a career in Global Change Biology, the goal of this book is to assist in developing a conceptual base and analytical skills that will serve you well in whatever ways you may contribute to sustainable solutions on our planet.

ACKNOWLEDGMENTS

Dozens of reviewers and hundreds of students have helped refine the approach, the arc, and the content of this book. My students, past and present, have all served as an incredible inspiration. Thanks to students who provided feedback through my Global Change Biology course and my Environment and the Self course. Thanks especially to those students and colleagues that pitched in at crucial times to assist with the manuscript directly, including Rainbow DeSilva, Shannon O'Hara, Allison Byrne, and Elizabeth McAlpine-Bellis. Thanks also to the undergraduate students, graduate students, and postdocs in my research lab, who demonstrate again and again that with dedication and caring, professional communities can feel like family.

Numerous reviewers (some listed here, and some anonymous) provided helpful feedback at many junctures:

David Allard, Texas A&M University, Texarkana
Jennifer Boyd, University of Tennessee, Chattanooga
Colin Carlson, University of California, Berkeley
Sarah Fitzpatrick, Michigan State University
Richard A. Gill, Brigham Young University
Alex R. Gunderson, Tulane University
Michelle Hersh, Sarah Lawrence College
Luke M. Jacobus, Indiana University–Purdue University, Columbus
Thomas Jenkinson, University of California, Berkeley
Jason Knouft, Saint Louis University
Jason J. Kolbe, University of Rhode Island
Kristy Kroeker, University of California, Santa Cruz
Jay Lunden, Haverford College
Bruce Robertson, Bard College
John Skillman, California State University, San Bernardino
Madhu Srinivasan, University of Kentucky
Robert Warren, State University of New York, Buffalo State

In addition, several colleagues went above and beyond in providing additional input when I felt stuck, including Thomas Jenkinson, Sarah Fitzpatrick, and Sarah Evans. I am also grateful to those scientists highlighted in the *Meet the Data* features who graciously provided photographs and checked my portrayal of their work, including

Gerardo Ceballos, Qiaomei Fu, David Reich, Terry Root, Camille Parmesan, Maria Rivera, Craig Moritz, James Patton, Carolina Voigt, Maj Rundlöf, Eric Hallstein, and Mariano Rodriguez-Cabal.

I am deeply grateful to my own teachers, mentors, and colleagues who supported me to pursue a career in science and who continue to inspire. David Miles, David Rand, Craig Moritz, David Wake, and Michael Eisen played particularly pivotal roles as scientific mentors. Jeanne Robertson, Luke Harmon, Kristen Ruegg, Jamie Voyles, Seema Bhangar, Noah Whitman, Ari Makridakis, and Jordan Rosenblum provided key moments of support as I was working on this project. I also thank various mentors who provided guidance as I was considering this project, including Harry Greene, James Collins, Jonathan Losos, Dolph Schluter, and Justin Brashares. I also appreciate those who discouraged me from undertaking a textbook project. They helped me refine my goals, clarify my purpose, and develop a system to complete this book. It is a great pleasure to have trusted mentors on this path.

My gratitude goes to those at Oxford University Press who helped shepherd this project, including Jason Noe, senior acquisitions editor; Sarah D'Arienzo, editorial assistant; Katie Tunkavige, assistant editor; Louise Karam, senior production editor; Brad Rau, project manager, SPi Global; Michele Laseau, art director; Joan Lewis-Milne, marketing manager; Chris Bowers, director of marketing; Bill Marting, national sales manager; and Petra Recter, director of content and digital strategy.

Finally, it is impossible to fully acknowledge the incredible ongoing support provided by my family. Four generations of my family were living together during the time this book was written, including my grandmother (who has witnessed planetary change for over 100 years and counting!), my mother, and my children. The patience, love, and support from family—and extended family—provided a solid foundation during this time. I extend especial heartfelt appreciation to Sybil, Wendy, Mike, Chloe, and Bodhi. You are so loved. Gratitude also to the "endless forms most beautiful" that have kept my passion for studying biodiversity alive. Even at home, various cats, frogs, and lizards served as guardians of my writing space (thank you Sumo, Apollo, Mina-bird, Sticky, and Crescents). And beyond all, thanks to the great mystery that is constantly revealing itself and keeping us curious about—and inspired by—life itself.

Approaches in Global Change Biology

Learning Outcomes

After working with this chapter, you will be able to:

- Describe the development of Global Change Biology as a field.
- Classify and apply different key approaches and tools in Global Change Biology.
- Compare frameworks for assessing the value of the natural world.
- Apply your knowledge to real-world case studies and interpret data from recent scientific studies.

THE BLANK PAGE

Every chapter in this text book begins with a "blank page" exercise. The purpose of the blank page is for you to pause and reflect on the topics in the upcoming chapter. It is an opportunity for you to make your learning process more active by reflecting on what you know, what you don't know, and what you are curious about. For each of these exercises, sit with a blank sheet of paper and simply explore the topic without assuming you should know anything specific in advance. Feel free to be creative with your responses to the blank page exercises—you can write stream of consciousness, make lists or flowcharts, or draw. Because you do the blank page exercises before reading the chapter, they serve as a *preassessment* tool, but you can also revisit them after you complete the chapter to see how your response could be refined based on new knowledge.

For this blank page exercise, let's explore the following question: What can we learn from studying life on our planet? Consider how new knowledge about biological diversity could be of value to human society and also to the millions of other species on our planet. Make a list of possible motivations for studying global biodiversity. Which of the motivations you listed resonates most with your own personal interests?

INTRODUCTION

Studying the causes and consequences of environmental change has become a key focus of modern science. Global Change Biology is a recently formalized field of biology that focuses on understanding environmental change and its effects on—and interactions with—life on Earth.

As we will explore together in this textbook, we live in an unprecedented era in which a single species—*our* species—has become a planetary force. Global Change Biology thus centers increasingly—but not exclusively—on anthropogenic (i.e., human-caused) impacts on biological diversity. Contemporary environmental change has urgent implications for all species on our planet. Thus, global change biologists work on a diversity of species (from microbes to mammals) in a variety of ecosystems (from marine to montane).

Understanding dynamics of past environmental change, evaluating current anthropogenic pressures, and predicting future ecological trajectories are all intimately linked. Thus, global change biologists conduct studies across different spatial and temporal scales. They often employ cutting-edge tools and technologies to address pressing questions about how living things respond to rapid environmental change.

The goal of this chapter is evaluate the history of Global Change Biology, the research methods global change biologists employ, and cross-cutting themes in the field. This introduction will provide a foundation for the rest of our exploration together.

HOW DID THE FIELD OF GLOBAL CHANGE BIOLOGY DEVELOP?

Early History

The scientific enterprise dates back millennia. Detailed knowledge of the natural world predates the emergence of the written word, and early societies had oral traditions to pass along observations of nature (like astronomical and agricultural knowledge). The development of early writing systems ~2,000–5,000 years ago provides the first evidence that early cultures recorded meticulous observations about the natural world, as shown in Figure 1.1.

By 500 BC, the Greek philosophers were actively pursuing empirical studies—and developing scientific theories—to explain natural phenomena. However, advances in astronomy, mathematics, medicine, and many other branches of science were occurring around the world well before the Common Era. Whether ancient mathematical artifacts from Africa, Ayurvedic medicine practices from India, or metallurgy in China, numerous cultures display evidence of early and sophisticated scientific practices (e.g., Selin 2008).

Nearly 2,000 years later, the core scientific disciplines—such as physics, chemistry, geology, astronomy, and biology—were formalized during the Scientific Revolution of the 1600s and 1700s. By the late 1800s the first broadly distributed weekly scientific journals were in print (the first issue of *Nature* in 1869 and the first issue of *Science* in 1880). The establishment of a field-specific journal is one way to chart the formalization of modern areas of scientific inquiry. Many of the fields of biology from which Global Change Biology draws were delineated in the 20th century (the journal *Ecology* was established in 1920, *Evolution* in 1946, and *Conservation Biology* in 1987). The journal *Global Change Biology*, which aims "to promote new understanding of the interface between biological systems and all aspects of environmental change that affects a substantial part of the globe," was established in 1995, making this one of the youngest branches of biology.

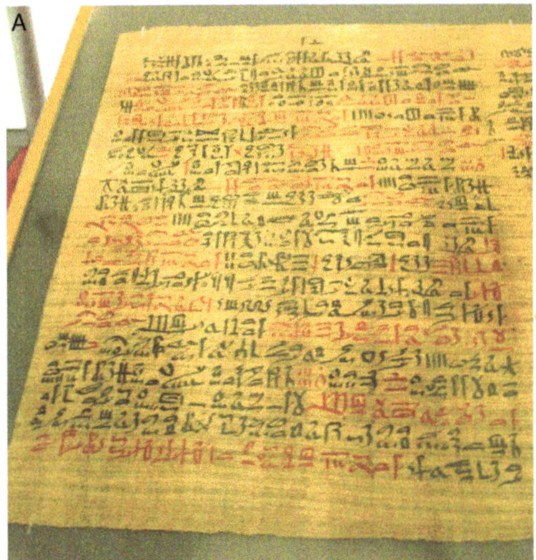

FIGURE 1.1 Science in antiquity. (A) Ancient Egyptian texts (like the "Ebers Papyrus" shown) are among the oldest preserved scientific records. Some papyri detailing mathematical, astronomical, and medical knowledge date back to 2000 BC. (B) Greek philosophers—like Thales of Miletus (624–546 BC; sometimes referred to as the "father of science")—developed philosophical and scientific approaches to explain natural phenomenon.

Reflection: What kinds of events or observations can spark the creation of new scientific fields?

Source: (A) https://www.sciencesource.com/archive/Ancient-Egypt--Ebers-Medical-Papyrus-1550-BC-SS2501699.html; (B) Naci Yavuz/Shutterstock

Contemporary Stressors

What precipitated the formalization of Global Change Biology as a stand-alone field in biology? Simply put, it resulted from the increasing awareness that humans were rapidly altering the conditions for life on Earth. Studies of Global Change Biology do not need to focus on anthropogenic impacts. In fact, studies of any environmental trend that affects the biosphere—past, present, or projected—are encompassed by this field. However, it was the recognition of the dramatic and globalized impact of human activities that imbued this nascent field of study with a sense of urgency.

Take climate change, for example. In 1896, the Swedish scientist Svante Arrhenius published calculations showing that the Earth's temperature is affected by atmospheric carbon dioxide (CO_2) and arguing that burning fuel (coal at the time) could warm the surface temperature of the planet (Arrhenius 1897). Decades later, the British engineer Guy Callendar published the first data showing that increasing global temperatures were correlated with increasing CO_2 (Callendar 1938). Over the next 50 years, the link between anthropogenic emissions and global warming would be firmly established, and the impacts of global warming on biological diversity would become irrefutable. The Intergovernmental Panel on Climate Change was established in 1998, and its first assessment report in 1990 explicitly warned that anthropogenic climate change could have catastrophic effects on the biosphere. A flurry of scientific activity followed, and climate impacts on biodiversity became a central theme in the emerging field of Global Change Biology.

However, Global Change Biology addresses myriad stressors on the biosphere, not only climate change. Take the impacts of land-use change and chemical contaminants as additional examples. As we will discuss in detail in upcoming chapters, the last several centuries have been characterized by exponential population growth and dramatic land-use change with increased urbanization, high-intensity agriculture, and displacement of indigenous people. Awareness slowly emerged that these activities were leading to the destruction of forests, degradation of soils, decreases in biodiversity, and changes to biogeochemical cycles. Increased awareness precipitated the emergence of many nature conservation societies and preserves (e.g., the United States National Park System) in the early 1900s. However, it was not until the mid-1900s that popular and scientific attention was galvanized. Often cited as a turning point in Western environmental history, Rachel Carson's book *Silent Spring* (1962) brought the effects of pesticides on biological systems to the fore. The first Earth Day followed soon after in 1970 with an estimated participation of 20 million people (an event that now reaches nearly 1 billion people per year). Coincident with this, scientific research on the impacts of land-use change and chemical contamination on biological diversity began to accelerate, as shown in Figure 1.2.

Ways of Knowing

Scientific publications are important as formalized contributions to Global Change Biology, and we will draw from the primary literature throughout this book. However, other ways of understanding and communicating changes in the natural world are also essential. For instance, indigenous ecological knowledge has played—and continues to play—a key role in the development and progress of the field of Global Change Biology.

One specific example comes from indigenous communities like the Inuit in northern latitudes, who provide narrative evidence of climate change and its impacts on the Arctic. Traditional knowledge and scientific measurement can corroborate each other—as they did in the early days of climate change research—increasing the total weight of evidence. However, these approaches can also complement each other if they provide information across different spatial or temporal scales (Alexander et al. 2011). Thus, greater partnership and dialogue between formalized scientific study and traditional knowledge will continue to enrich the field of Global Change Biology (e.g., Berkes 2009; Makondo & Thomas 2018). This has been termed "bi-directional" learning by one prominent global change researcher (Middleton et al. 2019).

An exemplar of this approach is the collaboration between Dr. Middleton (shown in Figure 1.2) and the Yorok Tribe in northern California. Using a community-based participatory research framework, university scientists and members of the indigenous community worked together to study toxic contaminants in a key watershed that were impacting human and ecosystem health. The partnership allowed traditional ecological knowledge to be integrated with high-resolution mass spectrometry data to quantify—and identify the source—of toxins in the environment. This project serves as a model not only for how indigenous knowledge can be explicitly prioritized in global

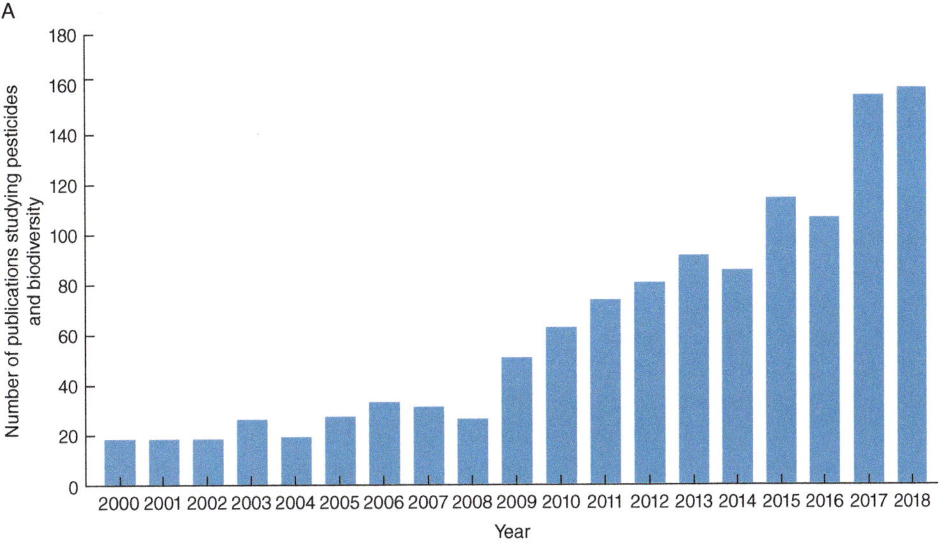

FIGURE 1.2 Understanding human impacts on the environment. (A) Scientific research on human impacts has accelerated dramatically over the last decades as illustrated by the total number of scientific publications per year on the effects of pesticides on biodiversity. (B) Indigenous knowledge has helped chronicle rapid environmental change, as illustrated by Inuit elders who have seen the impacts of climate change—such as melting of polar ice—within their lifetimes. (C) Dr. Beth Rose Middleton, a professor in the Department of Native American Studies at University of California, Davis. Dr. Middleton's work focuses on indigenous analysis of climate change, rural environmental justice, intergenerational trauma and healing, and participatory research approaches.

Reflection: What interests, experiences, perspectives, or identities do you personally hold that could contribute to the study of Global Change Biology and to creating an environment where diverse voices feel welcomed, heard, and amplified?

Source: (A) Web of Science; (B) Ciril Jazbec/National Geographic; (C) College of Letters and Sciences, University of California, Davis

change studies but how partnerships built on shared learning can begin to address the very painful environmental and social legacy of exploitation of native lands.

Honoring Diversity and New Horizons

Although there are many studies that explicitly incorporate diverse ways of knowing, it also essential to acknowledge that many perspectives, voices, and ways of knowing have long been marginalized, oppressed, and ignored in the mainstream scientific enterprise. The social, economic, and cultural systems built over the last millennia rely on exploitation of myriad people, species, ecosystems, and landscapes. Scientific systems are no different. Systemic biases in science have directly excluded many identity groups and have discounted many indigenous ways of knowing. Many practitioners have long called for diversifying the sciences (e.g., Puritty et al. 2017; Vaughan et al., 2019), but 2020 provided a watershed year to honestly and soberly evaluate the racism and exclusion that are still embedded in STEM fields (e.g., Chen 2020). Authentically addressing inequality and racism in science will require sustained and multi-faced efforts by individuals and institutions (e.g., Schell et al. 2020a).

It is important to recognize that this textbook draws mainly on the primary literature in the form of published scientific studies. Therefore, while the text provides a solid grounding in the history, practice, and key findings of Global Change Biology, it is also subject to the same inherent biases that permeate academic research and publishing. This textbook therefore presents Global Change Biology as it is today, but fields are constantly changing, often for the better. There is now an opportunity to create a new way of viewing, studying, and engaging with Global Change Biology that honors all voices and the deep and inextricable links between scientific, ecological, economic, social, and cultural systems.

A number of inspiring initiatives have recently launched to do the meaningful work of integrating global change research with values of equity, justice, and societal change. One example is the Critical Ecology Lab (www.criticalecology.space), which works to understand the relationship between extractive and oppressive social systems and key ecological and biogeochemical processes. Another is the Civic Laboratory for Environmental Action Research (CLEAR; www.civiclaboratory.nl), which integrates community-based and citizen science approaches to studying marine plastic pollution. Both initiatives conduct scientific research while simultaneously and directly challenging the underlying systems of oppression that have pervaded Western science. Both also recognize that any system of alienation will simultaneously lead to human inequality and environmental degradation. The directors of these two efforts are shown in Figure 1.3.

The hope is that in the coming years, integrative and equity-based approaches will become the norm in Global Change Biology. As scientists become more self-aware of their own biases and more willing to embody their deeper values, this field will continue to evolve. This book can evolve with them. Perhaps most importantly, *you* are invited to be part of this change. This textbook will ask you to engage directly with the material and bring your own unique perspective to bear on the content. In this way, we will enter a new era where different perspectives, identities, and ways of knowing are integrated into this critically important field of study.

FIGURE 1.3 (A) Dr. Suzanne Pierre is the director of the Critical Ecology Lab. Dr. Pierre is a global change ecologist and biogeochemist who studies processes at the interface between plants, microbes, and the abiotic environment. Dr. Pierre's research at the Critical Ecology Lab focuses on understanding the connections between environmental change and human pursuits of wealth and power to create a more engaged culture of science and a more just world. (B) Dr. Max Liboiron is the director of the Civic Laboratory for Environmental Action Research (CLEAR). Dr. Liboiron is a science and technology studies scholar, environmental scientist, and activist who works to understand and address plastic pollution in marine food webs. CLEAR is an anti-colonial, marine science laboratory that specializes in grassroots and community-based environmental monitoring of plastic pollution. The photo shows Dr. Liboiron (right) with lab member Emily Wells (left) using surface trawls to study plastic pollution in Newfoundland.

Reflection: If you could interview these scientists about their research, career path, and personal experience as scientists, what are three questions you would ask?

Source: (A) photo: Amy Snyder, text: http://www.criticalecology.space; (B) photo: David Howells (https://www.davehowellsphoto.com/), text: www.civiclaboratory.nl

HOW ARE GLOBAL CHANGE BIOLOGY STUDIES DESIGNED?

Global Change Biology has become a thriving area of study focused on understanding the impacts of many different facets of environmental change and integrating across many different spatial and temporal scales. There are challenges inherent in studying dynamically changing systems at a planetary scale. Therefore, it is important to evaluate the key methodological approaches used by global change biologists, a topic we will turn to now. It is also essential to recognize that *all* of the scientific approaches described below can be applied in ways that honor and include diverse voices and perspectives.

Integrative Approaches

Studies in Global Change Biology address biological responses to environmental change across different spatial scales (local to global), temporal scales (minutes to millennia), and biological scales (genes to ecosystems). While some studies may evaluate the response of a single population to a single stressor in a single year, others may measure the response of an entire ecosystem to myriad aspects of environmental change over thousands of years. Effective studies are explicit about the spatial, temporal, and biological scale they address.

Studies in Global Change Biology also address a wide variety of environmental issues, both anthropogenic and nonanthropogenic. While some environmental pressures have localized effects on particular ecosystems, others have global impacts. Similarly, some stressors have consequences that are immediately measurable (e.g., a volcano eruption or a large oil spill), while others reveal their impacts over longer timescales (e.g., changes in ocean circulation patterns). We will evaluate key contemporary stressors in more detail in Chapter 4.

Ultimately, Global Change Biology is a highly integrative field because it addresses diverse stressors and responses across scales. As such, Global Change Biology uses data and methods from many branches of biology (such as ecology, evolution, conservation biology, paleontology, and physiology) and many other fields of science (including geology, chemistry, atmospheric science, and computer science). These fields are all unified by their application of the scientific method (whereby systematic observation leads to the development, testing, and modification of hypotheses or predictions).

Key Elements of Study Design

Our purpose here is not to review the basics of the scientific method but to summarize some best practices in designing Global Change Biology studies. We will revisit these elements of study design throughout the textbook as we analyze approaches and results from real-world studies in the field. Table 1.1 defines key study design elements,

TABLE 1.1 Key Elements of Scientific Study Design

Key Term	Definition
Independent variable	A factor whose value does not depend on the other factor(s) studied. There can be one or more independent variables in a study. Sometimes called *explanatory variable*.
Dependent variable	A factor whose value depends on the independent variable. There can be one or many dependent variables in a study. Sometimes called *response variable*.
Treatment group	A set of individuals or samples that is exposed to the independent variable.
Control group	A set of individuals or samples that is not exposed to the independent variable and is used as a standard of comparison for the treatment group.
Confounding variable	An extraneous or hidden factor that influences both the dependent and independent variables and can lead to biased or misinterpreted results.
Main effect	The individual effect of one independent variable, ignoring any effects of other independent variables.
Interaction effect	The joint effect of two or more independent variables, which can be greater than, less than, or equal to the sum of their individual effects.
Replication	The repetition of all experimental conditions in multiple individuals, samples, groups, and/or locations.
Randomization	The use of chance methods to choose samples or assign samples to experimental groups to avoid systematic bias.

Reflection: In your own words, explain the difference between (a) independent and dependent variables, (b) control and treatment groups, and (c) main and interaction effects.

FIGURE 1.4 Abandoned mines are found around the world, illustrated here by an example from the Ural Mountains of Russia.

Reflection: Imagine that you were hired to assess the effect of heavy metal pollution from a former mining operation on local biodiversity. Where would you begin? Make a list of three specific things you could measure and incorporate a few key elements of study design from Table 1.1.

Source: https://mishainik.livejournal.com

and the *Core Concepts* feature of this chapter reviews how data from scientific studies are typically displayed.

Let us use a hypothetical example to practice applying these concepts to a Global Change Biology scenario. Imagine that you grew up near an abandoned copper mining operation, like that illustrated in Figure 1.4. Although no active mining occurred during your lifetime, your community always wondered whether there might be heavy metal leaching into the nearby soils. Heavy metals (like lead, cadmium, iron, copper, and arsenic) naturally occur, and many heavy metals are essential as micronutrients for life on Earth. However, heavy metals are highly toxic in excess quantities, and human activities such as mining, agriculture, and industrial production release excess heavy metals into the environment, where they can have adverse effects on living things (Nagajyoti et al. 2010).

Your community decided to crowdsource funding for a scientific study to test whether there are lingering effects from the mining, and they chose you as the lead scientist. You are charged with studying whether there have been long-term effects of the copper mine on local biodiversity.

You predict that wastes produced by the mining operation contained high concentrations of heavy metals, which leached into the soil and impacted local biodiversity. You expect that the heavy metal pollutants had effects on soil microbes, plants, and animals, but you decide to focus your first study on one local plant species that is particularly important in the food chain. In this case, heavy metal concentration would be your **independent variable**, and you could explicitly measure concentrations of different heavy metals in the soil and/or taken up by the plant tissue. You could then measure plant density, biomass, and/or growth rate as your **dependent variable(s)**.

You would want to carefully select your study sites. Obviously, you want to study the plants close to the mine. But you might want to measure plant growth in multiple independent populations thought to be affected by heavy metals. Conducting measurements in multiple independent populations provides **replication** so you can see if your results are generalizable. You would also want to find comparison populations that are not affected by heavy metal contamination. The comparison with unpolluted sites provides a **control**. The unpolluted sites should be as similar in environmental conditions (other than heavy metal concentration) as possible. When you choose which specific plants to measure in your study sites, you would want to use a method of **randomization** so as not to introduce bias into your study (e.g., if you accidentally choose the largest plants to measure in one site but the smallest in another).

In addition to monitoring key variables such as heavy metal concentration and plant biomass, you could measure other possibly confounding environmental factors (for example, if there were also differences across sites in temperature) to increase your confidence that heavy metal contamination alone explains the observed effects. It is possible during your research that you might discover that your study sites had been exposed to other stressors besides heavy metal contamination. If this is the case, you might want to conduct additional studies to parse the **main effects**—and the **interaction effects**—of these different stressors.

Of course, you could extend your study in countless ways. For example, you could evaluate changes over time if you had access to historical data from before the mining project began; you could study the effects of metal pollution across multiple biological levels; or you could use mathematical models to predict future changes in the system. In the remainder of this chapter, we will look at some key approaches and cutting-edge tools that can be applied to this or other questions in Global Change Biology.

Personal and Societal Relevance

Before we move on, let us briefly look at the relevance of the copper mining case study presented. You may not have been exposed to the impacts of mining where you grew up, but you may have experienced other environmental stressors. Think back to your own community and consider what impacts of global change have affected you personally. You may have experienced intense transient impacts like flooding due to increased storm severity or poor air quality due to increased fire severity. Or your community may have experienced the long-term impacts of soil or water pollution. Whether you grew up in a rural or urban setting, environmental change impacts us all.

Although global change has many universal effects, not all stressors have equal impact. Much of our exploration together will focus on understanding the diverse impact of global change stressors on the millions of nonhuman species on our planet. But environmental threats also have uneven impacts across human society. Numerous factors—including geography, race, and class—are correlated with vulnerability to the detrimental effects of climate change, pollution, and other global change stressors (e.g., Kelly-Reif & Wing 2016).

Inequities in the distribution of resources, wealth, and power mean that some nations—and some regions or identity groups within nations—are disproportionately impacted by global change. For example, one recent synthesis study demonstrates how myriad forms of structural racism and classism lead to a cascade of impacts on ecological and evolutionary processes in urban environments (Schell et al. 2020b). Figure 1.5 illustrates how social issues like immigration policy, residential segregation, and political representation have cascading impacts on both abiotic and biotic elements of the environment. For example, not only does the distribution of green space and chemical

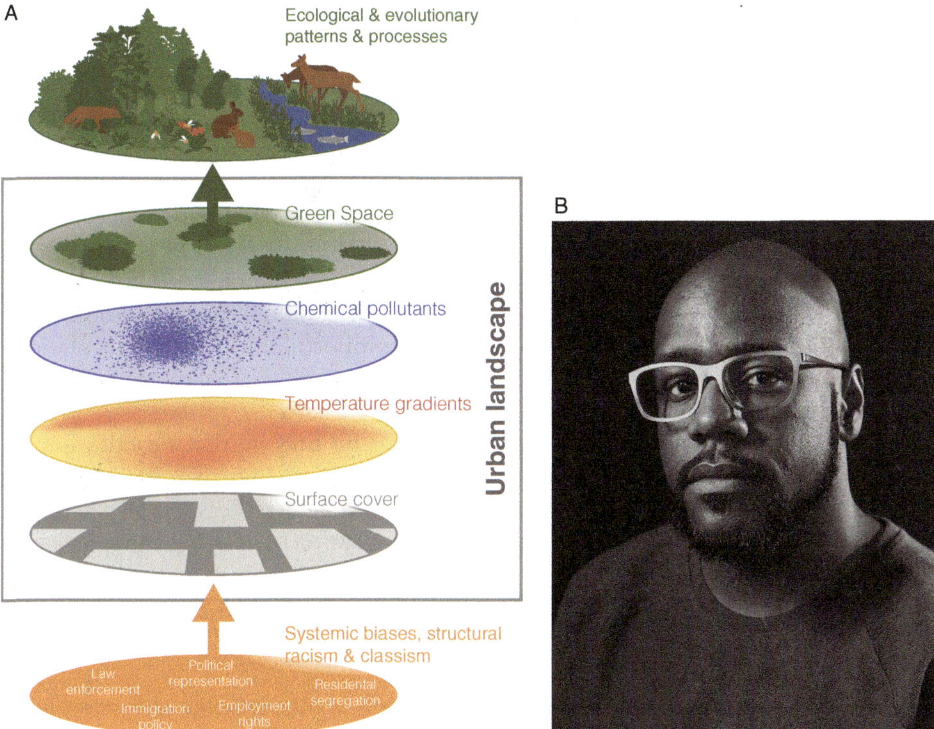

FIGURE 1.5 (A) A representation of how human social processes influence myriad components of the urban abiotic and biotic environment. Systemic biases and structural racism and classism shape policies that exacerbate inequalities in cities. These inequalities then impact biotic and abiotic components of urban landscapes, which in turn affect the ecology and evolution of both human and nonhuman species. (B) Dr. Chris Schell, the lead author of the study that produced this framework. Dr. Schell is an urban ecologist who studies human–wildlife conflict and how social and ecological systems interact. Much of Dr. Schell's work involves collaboration with underrepresented communities, cultural institutions, government agencies, and wildlife managers. In this way, Dr. Schell's research applies integrative and inclusive approaches to linking global change biology and environmental justice.

Reflection: What questions interest *you* about the interaction between global change and social justice?

Source: (A) Simone Des Roches, modified from Schell et al. 2020b; (B) Quinn Russell Brown (https://quinnrussellbrown.com/)

pollutants impact human health but also the abundance of wildlife. Therefore, ecosystem health and social justice are inextricably linked when we evaluate the causes and consequences of global change (Drescher 2019).

The study and practice of **environmental justice** is one way of addressing these inequities (e.g., Brulle & Pellow 2006; Ramirez 2019). The Environmental Protection Agency defines environmental justice as "the fair treatment and meaningful involvement of all people regardless of race, color, national origin, or income, with respect to the development, implementation, and enforcement of environmental laws, regulations, and policies. This goal will be achieved when everyone enjoys the same degree of protection from environmental and health hazards, and equal access to the decision-making process to have a healthy environment in which to live, learn, and work."

Not only can environmental justice and social justice be a focus in the policy arena, but it can directly inform research in Global Change Biology. A number of recent studies have proposed meaningful ways of doing this (e.g., Leong et al. 2018; Kuras et al. 2020; Schell et al. 2020b). For example, researchers can explicitly ask whether metrics of social equity correlate with metrics of ecosystem health. Does biodiversity vary with affluence within and between cities? Do countries with more economic equality also exhibit a deeper commitment to biodiversity conservation? Do populations with vested economic interest in extractive practices disproportionately oppose regulations that would protect biodiversity? These examples only scratch the surface of what is possible when a holistic approach is taken.

Although most of our work together will focus on how human societies impact other living things, continue to draw connections back to your own life, your own experiences, and your own interests. Global Change Biology is a diverse and dynamic field, and you may find a way to make a unique contribution by offering your particular perspective. The *Meet the Data* and the *Taking a Closer Look* features at the end of this chapter further explore different ways to value our natural world.

WHAT KEY RESEARCH APPROACHES ARE USED IN GLOBAL CHANGE BIOLOGY?

Here we will briefly compare four core research approaches that can be applied to questions in Global Change Biology (observational, experimental, modeling, and synthesis approaches). Each of these research approaches has strengths and limitations, so it is important to choose a research approach most appropriate to the scientific question under investigation. Of course, these approaches are not mutually exclusive, and multiple approaches can be complementary when brought to bear on a research question.

In this chapter, we focus simply on defining key approaches. Throughout the remaining chapters of this book, we will draw from real-world examples of these approaches to inform our understanding of how global change stressors impact life on Earth. We will also explore nuances about how data from these different approaches are collected, interpreted, and communicated in *Meet the Data* features throughout the textbook.

Observational Approaches

Observational approaches are those where a researcher measures biotic and/or abiotic factors without manipulating the study system. Observational studies can be powerful, especially when researchers employ a comparative approach. For example, comparisons can be made among different points in time, different geographic regions, different environmental conditions, and different species. The hypothetical study we just designed to measure the effects of heavy metal contamination on local plants is an example of an observational study.

Observational studies are powerful ways to understand changes in the biosphere. They also can employ an incredible variety of sophisticated tools, a topic we will explore later in this chapter. However, observational studies can also be limited by their power to directly test cause and effect and to disentangle the effects of multiple interacting stressors in natural systems. For example, nearly all 20th-century population declines are correlated with a background pattern of global warming, but this does not mean that climate change was the causal factor in all population declines. Thus, observational studies can be paired with other approaches for more explicit hypothesis testing.

Experimental Approaches

Experimental manipulation is an effective tool for addressing mechanistic questions in Global Change Biology. Experiments in controlled conditions allow variables to be isolated and provide increased confidence in the causal links between organismal response and particular global change stressors. As with observational study designs, having appropriate controls and adequate replication are key.

Experimental studies can be conducted in natural settings or in the laboratory. Field experiments are powerful in their natural context but can be challenging to implement. Laboratory studies can be effective for isolating individual variables, but they have the added challenge of demonstrating that results are relevant to natural systems. Sometimes researchers look for a middle ground by establishing controlled replicates in relatively natural settings (for example, using mesocosms as shown in Figure 1.6). Overall, it is important to remember that experimental approaches are by no means limited to short-term laboratory studies. Large-scale and long-term field studies have provided some of the most important insights in Global Change Biology, and we will highlight many of these throughout the text.

Experimental approaches can be powerful tools for disentangling the effects of different global change stressors. However, not all species or systems lend themselves to experimental manipulation, such as highly endangered species. Further, while some iconic studies have focused on community-scale responses over multiple decades, most experimental studies focus on one or few species over short timescales. Therefore, the other approaches discussed next can be useful for generalizing organismal responses over larger geographic scales, longer timescales, or broader taxonomic scales.

Modeling Approaches

Mathematical and computational approaches can address an incredible variety of questions in Global Change Biology. Models can be used to understand and predict abiotic conditions (such as future climate), biological phenomena (such as predator–prey

FIGURE 1.6 Mesocosms on land and at sea. By housing replicated treatment groups in outdoor conditions, mesocosm experiments provide powerful opportunities to link experimental manipulation with biological realism.

Reflection: Mesocosms like this are typically used for small transportable organisms (such as invertebrates or fish). What kind of experimental approaches could maximize biological realism but also be applied to larger or more rooted species?

Source: (A) Scott Kissel; (B) Martin Oczipka IGB / HTW Dresden

dynamics), and biotic–abiotic interactions (such as how species will likely respond to a warming world). Modeling approaches are extremely flexible because they can focus on different levels of biological organization (from genes to ecosystems) and can be tailored to specific systems. For example, global vegetation models can be used to predict how future climate conditions will impact hydrological and biogeochemical cycles; species distribution models can be used to predict the geographic distribution of different species under future environmental scenarios; and socioecological models can be used to understand the interplay between environmental conditions and human behavior. We will take a detailed look at many different modeling approaches in upcoming chapters.

As with all approaches, modeling methods have limitations. Inferences from models are extremely sensitive to initial assumptions. Moreover, models cannot incorporate all possible factors. For example, a model predicting species response to climate change could underestimate overall extinction risk because other important stressors are not considered. Finally, the quality of the data entered into a model will affect the utility and relevance of predictions coming out. Thus, modeling studies are most robust when they are informed by robust data from observational or experimental studies.

Synthesis Approaches

There are numerous ways to synthesize scientific information. Perhaps the most common synthesis method in Global Change Biology research is data aggregation, commonly referred to as **meta-analysis**. Meta-analyses gather together data from numerous individual studies and reanalyze these data in a unified framework. By bringing together individual datasets collected from different species, geographic regions, or time points, meta-analyses can reveal larger-scale and more general patterns. As shown in Figure 1.7, since meta-analyses were first popularized in ecology, evolution, and conservation biology (Fernandez-Duque & Valeggia 1994; Arnqvist & Wooster 1995), their use has increased dramatically (Vetter et al. 2013).

Meta-analyses can reveal general patterns that cannot always be seen from individual studies. However, there are limitations for meta-analyses. For example, Global Change Biology, like all fields of science, is affected by the "file-drawer problem." Unexpected or "negative" results (where an effect was expected but not found) can be difficult to publish and thus left in a researcher's file drawer (Rosenthal 1979). This publication bias can impact meta-analyses if they draw from an incomplete or biased sample of published studies. Moreover, like for the modeling approaches discussed earlier, the quality of the input data (in this case, the individual studies) will greatly affect the outcome of the meta-analysis. In addition, choice of statistical approaches will influence results, and new approaches are always being developed to more robustly evaluate shared patterns—and also variation—among individual studies (e.g., Nakagawa et al. 2015). Thus, as Figure 1.7 shows, it is important to have a formal approach for a meta-analysis that ensures results are as unbiased as possible.

Participatory Approaches

Participatory methods are those that engage community members and/or community-based organizations in research (Fortmann 2008). At its most committed, participatory research partners with communities in all phases from project conception, to data collection, to results interpretation, to communication of findings. Participatory approaches can increase trust and transparency, build partnerships across sectors and help translate results into action (e.g., Ramirez-Andreotta et al. 2015).

Citizen science is one participatory approach that engages the general public in scientific research (e.g., Dickinson et al. 2010; Soleri et al. 2016). Citizen science projects can be orchestrated with a rigorous experimental design or can be solicited opportunistically. An example of opportunistic citizen science is eButterfly (e-butterfly.org), which provides a platform for butterfly enthusiasts to record and share their sightings. Research comparing eButterfly observations to those of professional biologists demonstrates that citizen science can provide valuable and complementary data on the responses of species to global change (e.g., changes in species distributions and migratory timing; Soroye et al. 2018). Overall, citizen science can foster public enthusiasm for science and also provide a significant boost to research efforts. For example, one study estimated that the public provided more than 1 million volunteers and up to $2.5 billion of in-kind financial support annually to biodiversity monitoring efforts alone (Theobald et al. 2015).

In Sum

Of course, these different approaches can be used in concert. For example, observational studies might generate hypotheses that are then tested with experimental approaches. Experimental data might be used to parameterize models to make predictions about future outcomes. Modeling results from numerous studies might be synthesized in meta-analyses. Synthesis studies might generate new predictions that can be tested

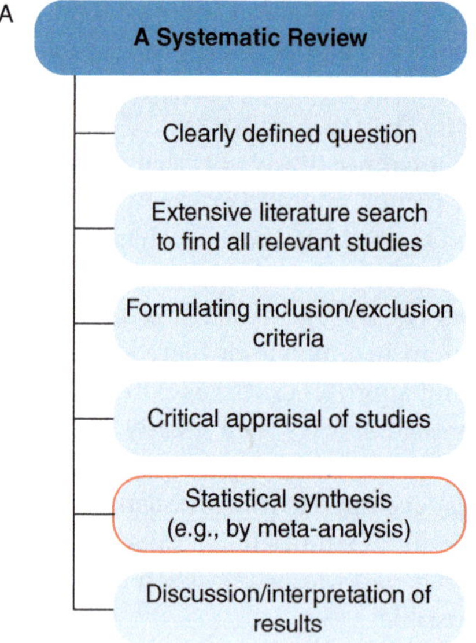

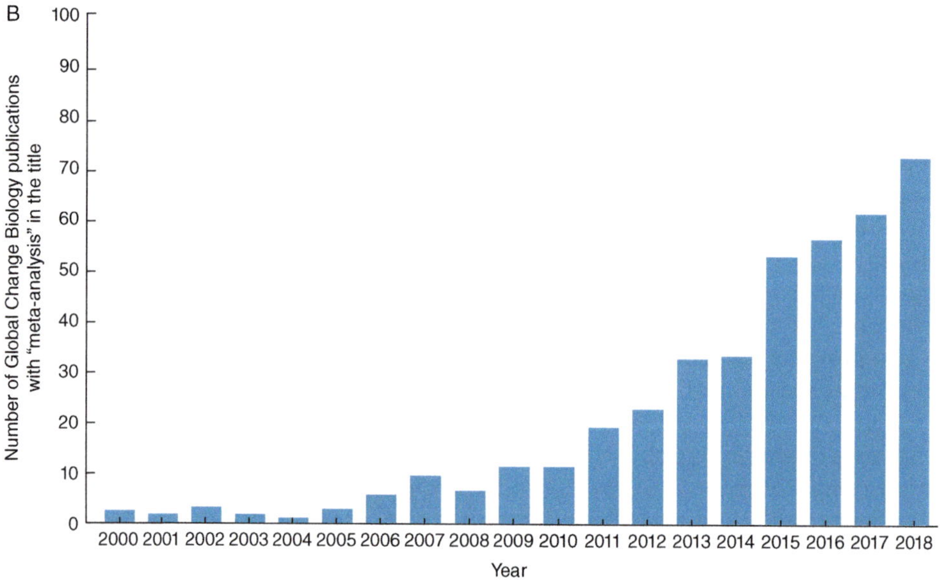

FIGURE 1.7 Meta-analyses play an important role in revealing large-scale patterns in Global Change Biology. (A) Meta-analyses need to take a systematic approach to avoid biases. (B) Meta-analyses have become increasingly important in this field, demonstrated by the increase in global change publications that have "meta-analysis" in the title.

Reflection: What is a specific way that multiple research approaches could be used in concert, for example in our heavy metal pollution case study?

Source: (A) Vetter et al. 2013 (https://esajournals.onlinelibrary.wiley.com/doi/10.1890/ES13-00062.1); (B) Citation report from Web of Science

with observational data. Participatory methods could be used in conjunction with any of the other approaches. Thus, global change biologists can apply multiple perspectives and approaches simultaneously or sequentially to generate a more comprehensive understanding of how natural systems respond to environmental change.

WHAT KEY TOOLS ARE USED IN GLOBAL CHANGE BIOLOGY?

As we discussed earlier in this chapter, humans have been observing and measuring patterns in the natural world for thousands of years. However, the research tools used for understanding patterns of abiotic and biotic change have become increasingly sophisticated, and technological innovations continue to revolutionize the study of Global Change Biology. Here we introduce a few exemplars to show how new technologies provide unprecedented opportunities to understand how environmental change impacts life on Earth. Then throughout the upcoming chapters, we will highlight leading-edge research tools and the ways they help scientists address previous intractable questions in Global Change Biology.

Environmental Monitoring Tools

From measuring the fine-scale flow of gases through the soil to measuring global atmospheric changes, tools for measuring abiotic conditions allow global change biologists to collect ever more detailed information about changing environmental conditions. Take, for example, technological innovations for measuring marine environments illustrated in Figure 1.8. To measure conditions in the deep ocean, submersible vehicles can now descend nearly 20,000 feet to the ocean floor. Equipped with sonar and cutting-edge data loggers, these vehicles can record abiotic conditions (such as temperature, salinity, and dissolved oxygen) while generating detailed maps of remote oceanic terrain. Similarly, to measure conditions on the open ocean, swarms of small robots that float like plankton use sophisticated sensors to measure abiotic factors under normal conditions or when algal blooms or oil spills threaten protected marine areas (Jaffe et al. 2017). Throughout this textbook we will highlight tools like these that allow global change biologists to record marine, terrestrial, and atmospheric conditions to document environmental change and its consequences for life on Earth.

Organismal Monitoring Tools

In addition to monitoring abiotic conditions, technological innovations allow global change biologists to measure ever-increasing facets of organismal morphology, physiology, and behavior. For example, the deep-sea vessels we just discussed can carry sophisticated acoustic and optic equipment to record whale communication or detect glowing bioluminescent organisms. In fact, some sensors can be manufactured at such small scales that even insects can carry them. For example, as shown in Figure 1.8, light-weight radio-transmitters can be attached to bees to understand patterns of foraging, dispersal, and reproduction. On the other end of the spectrum, remote sensing technology can provide information about vegetation at global scales. For example, the Landsat 8 satellite (shown in Figure 1.8) images the entire Earth in just over 2 weeks,

FIGURE 1.8 (A–C) From drifting sensors, to micro radio-transmitters, to global imaging satellites, tools for monitoring environmental conditions and organismal response to environmental change are becoming ever more sophisticated.

Reflection: If you could invent any tool or technology to use as a global change biologist, what would it be used for and what difficult question could it help answer?

Source: (A–C) Andrew McRobb/Royal Botanical Gardens, Kew; Andrey Armyagov/Shutterstock; https://www.shutterstock.com/image-photo/space-satellite-orbiting-earth-elements-this-363654452

with data freely and publicly available within 24 hours. These images, which can have a resolution of up to 1 meter, can then be used to infer functional vegetation traits (like photosynthetic capacity). These and other tools give global change biologists unprecedented opportunities to understand the impact of environmental change at the organismal level across the tree of life.

Molecular Tools

One of the most dramatic scientific advances in the last few decades has been the widespread application of genetic and genomic tools. The capacity of DNA sequencing technologies has risen exponentially while costs have declined exponentially. As shown in Figure 1.9, the equivalent of an entire human genome can now be sequenced in ~10 minutes for ~$1,000. This means that incredible amounts of genetic information

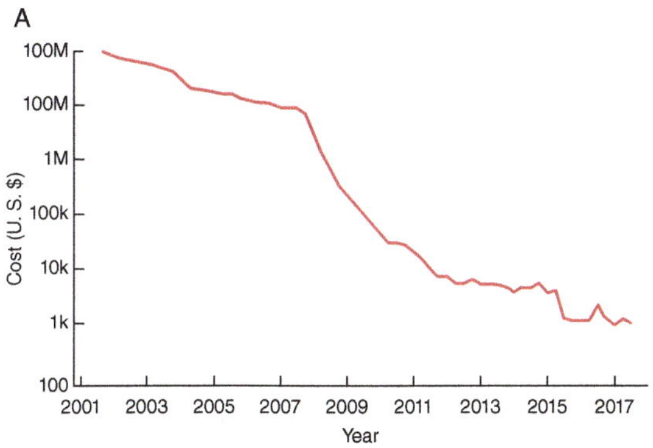

FIGURE 1.9 (A) Declining costs of DNA sequencing allow researchers to use genetic data for diverse questions in Global Change Biology. (B) Biopsy darts can be used to take small tissue samples without harming the endangered Antarctic blue whale. DNA can be isolated from these tissue samples and can be used to identify subspecies, study migration routes, and estimate population sizes.

Reflection: What are other noninvasive sampling methods can be used to obtain DNA from study species without causing any harm?

Source: (A) Ben Moore/Wikipedia; (B) Nino Pierantonio/Tethys Research Institute

can be used to address myriad questions in Global Change Biology. Moreover, the sensitivity of new DNA sequencing technologies means that vanishingly small amounts of starting material are required. Thus, global change biologists can use **noninvasive sampling** methods to collect DNA from feathers, fur, or scat without harming their study species, a topic we will return to later (e.g., Byrne et al. 2017; Natesh et al. 2019). Environmental DNA (e-DNA) can also be filtered from soil or water samples, providing incredible opportunities to understand how living organisms—from microbes to mammals—respond to global change stressors. The ability to measure—and even manipulate—the molecular and cellular building blocks of life provides a complex and powerful set of tools in Global Change Biology, which we will revisit in upcoming chapters.

Computational Tools

We are living during a revolution in computing technology, with ever-faster processing and ever-larger data storage capacities. The most powerful supercomputers in the world now contain millions of processing units and can be used to simulate millions of years in the universe's history. Bringing that power to bear on questions in Global Change Biology has opened many new research avenues. Advanced computational systems provide unprecedented opportunities for mathematical modeling and data synthesis. Organismal and environmental data can be explicitly integrated across spatial and temporal scales and used to make predictions about future environmental conditions and organismal responses. Computing infrastructure also allows researchers to access digital data across many different scales. For example, many museums are digitizing their records and even their biological collections, allowing researchers around the world electronic access to millions of biological records. With interactive three-dimensional visualizations of specimens, researchers can also now more efficiently look back in time to directly study organismal responses to changing environments. In upcoming chapters, we will continue highlighting ways that computational tools create new opportunities for integrative research in Global Change Biology.

In Sum

With so many sophisticated tools generating so much data, the information available to global change biologists is unprecedented. The development of large, publicly available databases has been essential to facilitate rapid, robust, and reproducible research. For example, in the United States, the National Oceanic and Atmospheric Administration (NOAA) provides access to data from glaciers and ice caps around the world, the National Institutes of Health (NIH) hosts hundreds of millions of DNA sequence archives, and the National Science Foundation (NSF) provides open access to an amazing array of empirical ecosystem data through the National Ecological Observatory Network. These are just a few of the repositories around the globe contributing to the information revolution in Global Change Biology.

CORE CONCEPTS

HOW ARE DATA DISPLAYED?

In each chapter you will find a *Core Concepts* feature that reviews foundational scientific concepts. Here we will briefly review the fundamentals of graphical data interpretation. Data can be displayed in numerous ways, including charts, scatter plots, bar graphs, line graphs, and box plots. Throughout this book you will be interpreting data from visual displays. When you encounter a data display, read the title and labels and analyze any general trends. If you understand the graph, explain what it is saying in your own words. If you do not understand the graph, articulate specifically what is difficult to understand.

Graphs show the relationship between variables—factors that vary and are measured to understand their relationship. Typically, graphs are displayed with two axes: a horizontal x-axis and a vertical y-axis. The x-axis represents the **independent variable**—a factor being measured whose value does not depend on the other factor(s) studied. The y-axis represents the **dependent variable**—a factor whose value depends on the independent variable. For example, in Box Figure 1.1, time is displayed on the x-axis and population size is displayed on the y-axis. Variables can be continuous (measures that have an infinite number of possible values, such as population size) or categorical (measures that have a limited number of possible values, such as presence versus absence).

The relationship between variables can take on many forms. Box Figure 1.1A shows examples of linear and exponential relationships. Of course, other relationships are possible—for example, cyclical associations. When assessing trends, it is important to pay attention to the scale presented on the axes. For example, if one variable is presented on the log scale, an exponential relationship can appear misleadingly linear (as shown in Box Figure 1.1B). When many measurements are taken, scientists often calculate the **mean** (the average value) and the **variance** (the variation around the mean). Box Figure 1.1 graphically demonstrates changes in mean and variance. The **statistical significance** of relationships between variables can be assessed to determine whether an association can be explained simply by chance. To demonstrate that a relationship cannot be explained by chance, scientists often present a **p-value** or error bars, as shown in Box Figure 1.1C. Traditionally, a p-value of <0.05 is considered statistically significant.

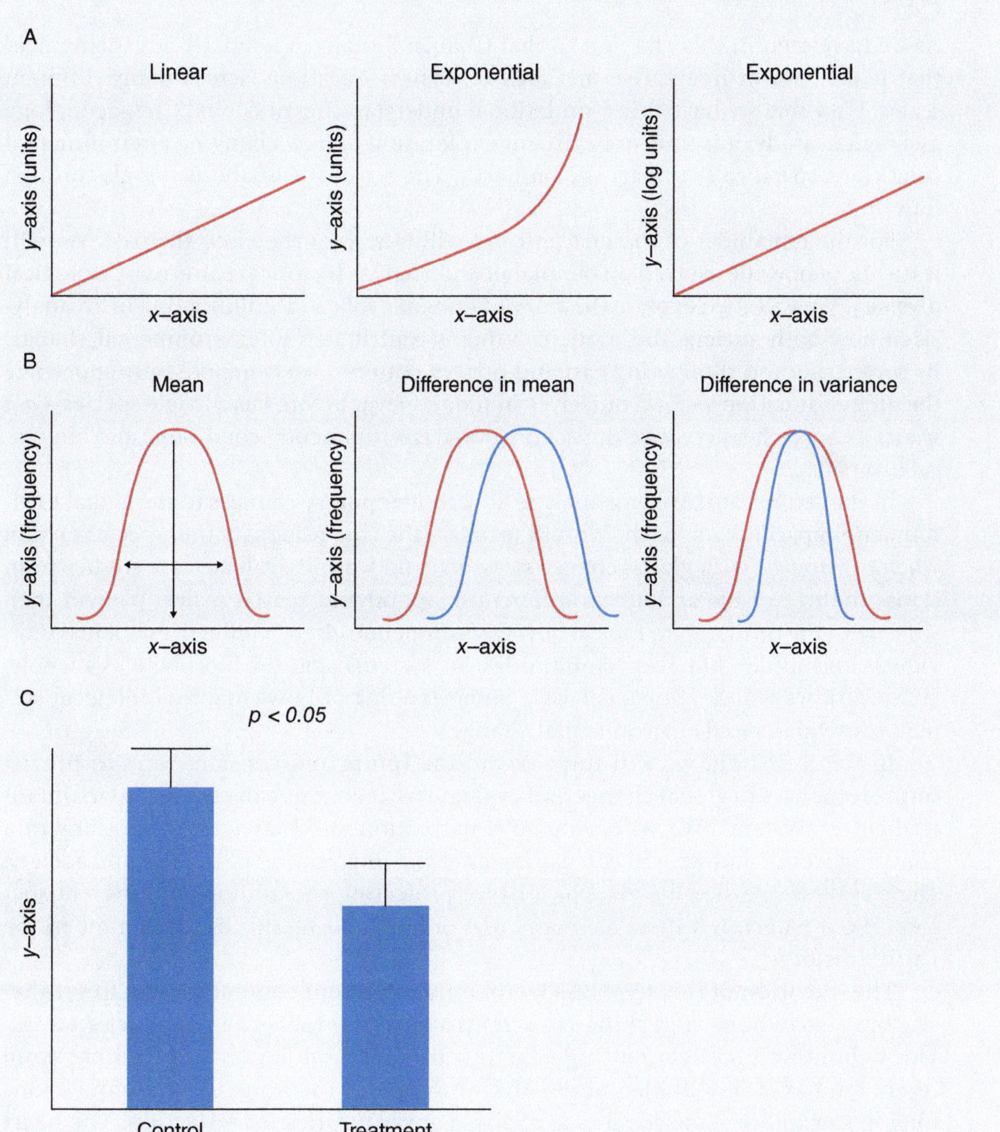

BOX FIGURE 1.1 (A) Example of several possible relationships between two variables (displayed with arbitrary units). Always orient yourself to graphs by looking at the axes. In this example, an exponential relationship can appear linear when the y-axis is presented on a log scale. (B) Graphical description of mean and variance. (C) Example of error bars and p-values used to demonstrate statistical significance.

Reflection: Draw an example graph in the style of the middle row where there is a difference in both mean and variance.

Source: Courtesy of author

CONCLUSION

As we have seen in this chapter, Global Change Biology is a rapidly advancing field that makes use of integrative methods to answer questions across many different scales. Now that we have this foundational understanding of Global Change Biology as a field of study, it is time to begin our exploration of how changing environmental conditions specifically impact life on Earth. The arc of the textbook travels through four units.

For the remainder of the first unit, we will take a journey into the past. We will learn about how life evolved on our planet and analyze historical patterns of biological diversity. We will then explore the story of human evolution, culminating in an analysis of how both ancient and modern humans contributed to environmental change. By comparing and contrasting past and present patterns, we can more fully appreciate the unique situation we find ourselves in today. Never before has a single species—*our species*—had such a pervasive impact on global environmental conditions and biological diversity.

In the second unit, we will analyze how contemporary changes to the global environment impact life on Earth. We will focus on the core responses that species exhibit when confronted with global change stressors, and we will evaluate case studies from across the tree of life and across different temporal and spatial scales. We will then scale up in the third unit to look at how environmental change influences not just individuals and species but also communities, ecosystems, and the biosphere as a whole. Ultimately, we will develop a cohesive understanding of how complex biological systems respond to rapid environmental change.

In the final unit, we will turn toward the future and consider ways to predict future responses to global change and evaluate what actions can be taken to maintain resilient ecosystems. We will evaluate conservation and management options in a changing world, and we will discuss issues at the intersection of science and society. There are few one-size-fits-all solutions for protecting the millions of different species on our planet, but there are many rays of hope during this dynamic time in the Earth's history.

The intention of this textbook is not only to present content for you to synthesize but also to hone your skills as a scientist and your tools as a planetary steward. Throughout we highlight cutting-edge research methodologies and findings from recent studies. You will also notice that throughout the book there are questions for you to explore. Take advantage of these opportunities for reflection. The heart of the scientific process is in developing and addressing questions. Every time you grapple with a question, you will consolidate your existing knowledge, develop new skills, and practice applying what you know to real-world problems. We live in a historic time, one where your unique talents and skills will be needed in science and society. Whether you pursue a career in biology or not, the goal of this book is to assist you in developing a conceptual base and analytical skills that will serve you well in whatever ways you may contribute to sustainable solutions on our planet.

MEET THE DATA THE ECONOMIC VALUE OF NATURE

In every chapter you will find a *Meet the Data* feature, which will give you experience interpreting and applying real-world data. *Meet the Data* features take a deep dive into one particular Global Change Biology study so you can directly experience how to generate predictions, collect data, and articulate the significance of research findings.

Who Are the Scientists and What Did They Set Out To Do?

Here we will begin to explore the theme of the value of nature. In particular, we will look at two related studies that set out to quantify the *economic value* of our natural environment. Box Figure 1.2 introduces the scientists. The first paper, "The Value of the World's Ecosystem Services and Natural Capital" by Costanza et al., was published in the journal *Nature* in 1997. The second paper, "Global Estimates of the Value of Ecosystems and Their Services in Monetary Units" by de Groot et al., was published in the journal *Ecosystem Services* in 2012 and sought to reevaluate the conclusions of Costanza et al. 15 years later.

Of course, ecosystems provide a myriad of essential functions that have intrinsic value. Many of these functions—from the release of oxygen, to the provisioning of clean water, to the production of soil—serve as a life support system for all species on Earth. When these **ecosystem functions** directly benefit humans, they are referred to as **ecosystem services**. The focal studies set out to quantify the economic value of the benefits humans derive from Earth's natural resources.

Both studies were **meta-analyses** that gathered previously published data into a common statistical framework. The original Costanza et al. (1997) paper generated a phenomenal amount of interest and subsequent research activity in the topic of ecosystem valuation. Since its publication, it has since been referenced by other authors over 5,000 times. Thus, by the time de Groot et al. revisited the topic, hundreds of additional studies had been published and could be included in an updated analysis.

What Are *Your* Predictions?

Before you read on, take a few minutes to apply what you have learned in this chapter about ecosystem services and the history of human land use. Start by considering the following questions:

- *What is your best guess as to the annual value in dollars of global ecosystem services?*
- *Do you expect the value of ecosystem services from particular biomes or particular parts of the world to be especially high?*
- *Do you expect the value of ecosystem services to have changed in the 15 years between the two studies? If so, in what direction?*

Now write a summary prediction statement about your expectations of the value of ecosystem services and whether that value might vary over regions of the world or time periods.

What Were the Scientists' Predictions?

The last paragraph of the Costanza et al. (1997) paper established a prediction that was then tested in de Groot et al. (2012): *"As natural capital and ecosystem services become more stressed and more 'scarce' in the future, we can only expect their value to increase."*

BOX FIGURE 1.2 Drs. Robert Costanza (A) and Rudolf de Groot (B) set out to understand the value of our natural world. Provisioning clean water (C) is one example of an ecosystem service.

Source: (A) By permission of Robert Costanza; (B) By permission of Rudolf de Groot; (C) Vaclav Uhlir/ Shutterstock

(Continued)

What Data Were Collected?

The authors conducted a literature review and collected data from prior publications. The 1997 study collated estimates of the value of ecosystem services from over 100 previous studies. Between 1997 and 2012, the number of ecosystem valuation studies increased dramatically, and the later study used data from more than 300 individual studies. The authors converted all valuation metrics to common spatial, temporal, and currency units: international dollars per hectare per year (Int$/ha/year). They then evaluated the value of ecosystem services over different land use types.

What Is *Your* Interpretation of the Data?

Before you read on, take a few minutes to interpret the results from the study shown in Box Figure 1.3. Consider the following questions as you look at the data:

- *How does the value of ecosystem services vary by biome?*
- *How might you explain differences in the mean and variation across biomes?*
- *What are the key trends across time and how might you explain them?*

Write a sentence or two describing the *key findings* of this study and their *significance*.

What Was the Scientists' Interpretation of the Data?

The total value of global ecosystem services was estimated in 1997 to be approximately 33 trillion US dollars per year and in 2012 to be approximately 125 trillion US dollars per year. The scale of these numbers is staggering. Take, for example, that the World Bank estimated the 2012 global gross domestic product (GDP)—the financial value of all goods and services produced—to be approximately 74 trillion US dollars per year (http://data.worldbank.org). Thus, ecosystem services are of incredible value, surpassing the monetary worth of all other traded goods combined.

In addition, some biomes provide an exceptional amount of ecosystem services per unit area. For example, coastal systems (including coral reefs and coastal wetlands), terrestrial forests, and terrestrial wetlands generate tremendous value. Moreover, most ecosystem services are not traded in the current market system—for example, nutrient cycling, which alone is estimated to contribute 17 trillion US dollars of value per year.

The value of ecosystem services in many biomes changed from 1997 to 2012, and there are several possible explanations for observed patterns. The overall increase in the valuation of global ecosystem services is likely due to better valuation methods. More studies have accumulated in more regions of the world, providing more comprehensive data from which to estimate value. However, land-use changes also alter the value of ecosystem services. For example, between 1997 and 2012, more than 20 trillion US dollar per year in ecosystem services were lost due to decreased area of highly productive ecosystems like coral reefs and tropical forests (Costanza et al. 2014). Also between 1997 and 2012, the human population increased more than 15% from ~5.9 billion to ~7 billion (United Nations). Thus, as ecosystems are under increasing pressure from the human population, ecosystem services will become more and more priceless.

Some researchers have argued that placing ecosystem services in an economic framework could lead to increasing commodification or privatization of natural resources. However, Constanza et al. and de Groot et al. are not arguing that ecosystem services should be valued and traded like other commodities. Rather, they are drawing attention to the incredible value of our natural world and cautioning against creating a false dichotomy between the environment and the economy. In fact, our economy, our well-being, and ultimately our survival all depend on ecosystem services.

What Are *Your* Ideas for Future Research Directions?

Given that Constanza et al. and de Groot et al. provided evidence that the value of ecosystem services is high and likely increasing, what next steps do you envision for this research program? If you had been involved in this study, how would you follow up? Start by considering the following questions.

- *Are there more detailed questions that could now be asked about particular biomes?*
- *Are there predictions about the future that could be made and later tested?*
- *Are there related economic or philosophical questions that could be addressed?*

Now write a few sentences about one possible future direction on this research theme.

(Continued)

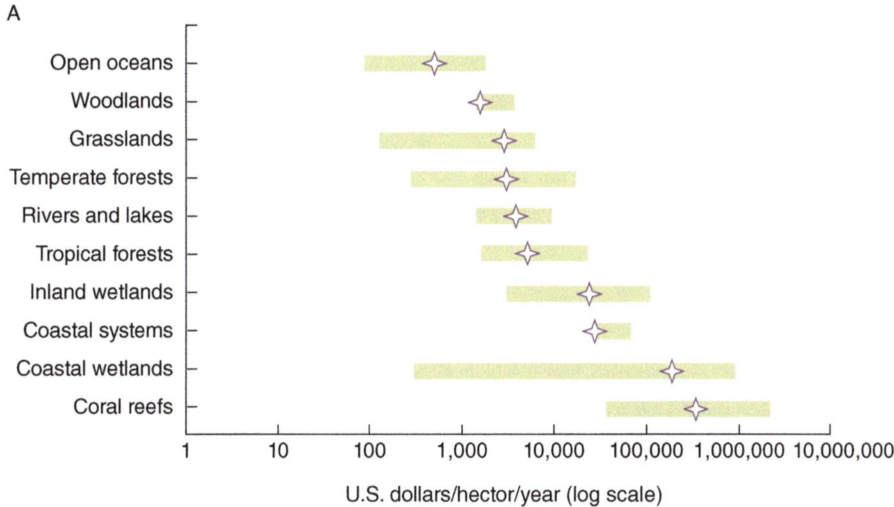

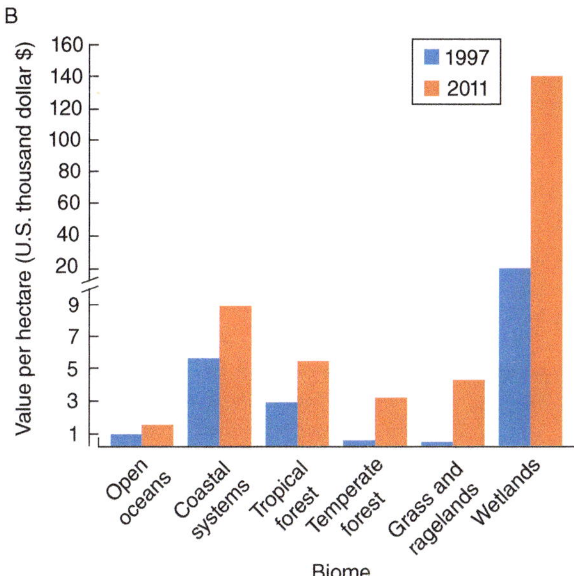

BOX FIGURE 1.3 (A) Mean value (star) and range (bar) for ecosystem services in different biomes. Values are given in dollars per hectare per year. (B) Changes in mean value in different biomes from 1997 to 2011 rounded to the nearest hundred. Value is given in dollars per year (with dollars scaled to 2007 currency to correct for inflation), and total value is given in *billions* of dollars per year.

Source: (A) "de Groot et al., Global estimates of the value of ecosystems and their services in monetary units, Ecosystem Services Volume 1, Issue 1, July 2012, Pages 50–61"; (B) "de Groot et al., Global estimates of the value of ecosystems and their services in monetary units, Ecosystem Services Volume 1, Issue 1, July 2012, Pages 50–61"

TAKING A CLOSER LOOK THE VALUE OF BIOLOGICAL DIVERSITY

In every chapter you will find a *Taking a Closer Look* feature, which offers an opportunity to evaluate integrative topics that cross themes from multiple chapters. Sometimes these features will discuss a particular global change stressor, ecosystem, or group of organisms, and other times these features will examine a topic with particular relevance to society.

As we discussed in this chapter, Global Change Biology focuses on understanding the impact of environmental change on biological diversity. But why should we care about biological diversity in the first place? It may seem self-evident that life on Earth has value, but there are many reasons it is import to understand—and protect—the biotic heritage of our planet. Here we will look briefly at several of these reasons, which are not exhaustive or mutually exclusive. As you read, see which of the reasons resonate most strongly with you personally. These themes will reemerge throughout the book and also be highlighted again in Unit IV.

Why Should We Care About Biological Diversity?

SCIENTIFIC VALUE The biological world has tremendous scientific value. Studying biological diversity allows us to learn about the processes that shape life as we know it. Scientific inquiry can shed light on diverse topics from the origins of life, to survival in extreme environments, to the factors that lead to large-scale losses of biodiversity. Only by studying life on Earth can we learn to protect the biological heritage of our planet.

VALUE TO HUMAN HEALTH The natural world is essential for human survival and well-being. We require sunlight, water, air, food, and shelter for our very survival, but we also rely on the natural world for an enormous range of materials to improve longevity and standards of living. For example, many plants and animals produce compounds that can be used—or mimicked—to treat human diseases. Moreover, human health is ever more intimately linked with that of other species on the planet. Not only do we rely on ecosystems to provide clean water and air, but with globalized travel and trade, it is increasingly easy for pathogens to circulate the globe, as we have seen in recent years. Thus, diseases can spread quickly and occasionally even cross species boundaries. This has led to the concept of "One Health," which emphasizes the interdependence of the health of people, other animals, and the environment.

ECONOMIC VALUE The natural world has unfathomable economic value. The commercial value of biodiversity is concretely evident in the goods and services that are monetized through commerce and trade. However, biodiversity also provides services that are of almost infinite value, such as the air we breathe and the water we drink. The topic of economic valuation of biodiversity is treated in more detail in this chapter's *Meet the Data* feature.

CULTURAL VALUE The natural world provides a tremendous anchor for cultural systems around the world. Since antiquity, cultures have been deeply oriented around cycles in the natural word. Symbols of culturally important plants and animals have been used in art and ceremonies since antiquity. In modern times, we also obtain cultural value from biodiversity through exploration, relaxation, and recreation in nature, whether in laying in the grass in a small urban park or looking for rare birds in a large nature preserve.

ETHICAL VALUE Many individuals and societies express a deep ethical imperative toward planetary stewardship. This ethical perspective is more than a simple pragmatic need to protect the natural systems upon which we depend. Rather, it stems from a moral sense of the meaning and value of biological diversity. The ethical impetus toward stewardship has deep roots in many cultures around the world. For some, this imperative has become even stronger over the last decades as individuals and communities begin to take responsibility for the consequences of human actions on the planet.

INTRINSIC OR SPIRITUAL VALUE Another compelling reason to care about biodiversity on our planet is because we intrinsically value life and value diversity. Many drive a great sense of solace, awe, and inspiration from the natural world. Some may feel this as an aesthetic appreciation of beauty, while others may feel it as a connection to a sense of divinity. Regardless, many encounter a sense of connection and wholeness when experiencing themselves as an indivisible part of the totality of biological diversity on Earth. Thus, without invoking any other reasons, we can simply hold the natural world in great regard because we are inextricably linked to it.

IN SUM Although it may seem straightforward to identify reasons to value biological diversity, there are numerous

(Continued)

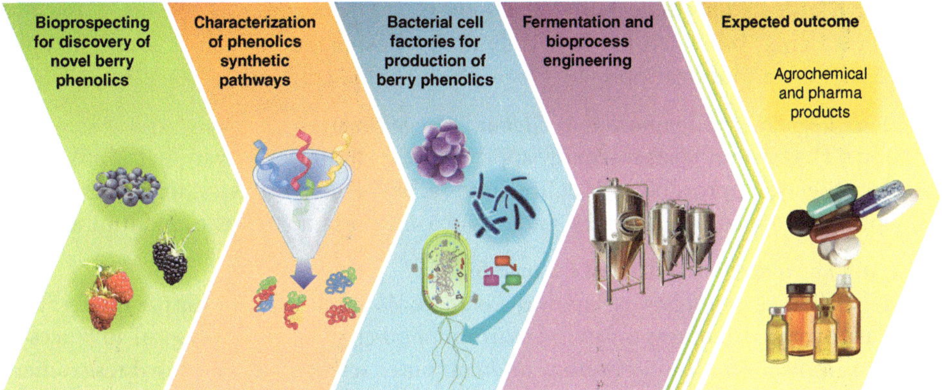

BOX FIGURE 1.4 Bioprospecting projects aim to develop new products and contribute to rural development through the commercialization of biodiversity. In this example, a consortium of international researchers funded by a European Commission works to identify new phenolic compounds from berry species that could be used in the food and pharmacology industries.

Reflection: Make a list of obvious benefits for human societies, other species, and the environment that could come from these uses of bioprospecting. Make a list of risks (unintended negative consequences) that could occur from these uses of bioprospecting. Are there ways to facilitate the benefits while reducing the risks?

Source: http://www.bachberry.eu

complexities raised by these valuation approaches. For example, the economic value of natural resources can lead to overexploitation, a topic we will return to in Chapter 4. The monetary value of natural resources can also lead to conflicts—even armed conflicts—over which individuals and communities have access to those resources. Similarly, the value of biodiversity to human health can lead to bioprospecting—the search for ways to commercialize biodiversity as illustrated in Box Figure 1.4. Economic incentives for finding new medicinal compounds can also motivate ethically questionable approaches, sometimes termed *biopiracy*—where indigenous knowledge is exploited by outside actors for profit. Even ethical perspectives on the protection of biological diversity can have a complex side, with emotions of guilt, fear, and hopelessness often accompanying moral motivations. We will revisit many of these complexities in future chapters and continue to highlight the relevance of issues in Global Change Biology for individual and collective decision making in our societies.

KEY CONCEPTS

What is Global Change Biology?

- Global Change Biology focuses on understanding environmental change and its effects on—and interactions with—the biosphere. It integrates across many different spatial and temporal scales and draws from many different fields of science.

How did the field of Global Change Biology develop?

- The field of Global Change Biology is one youngest branches of biology. It was motivated by evidence that accumulated (beginning in the late 1800s) that humans were rapidly and drastically altering the global environment.

How are Global Change Biology studies designed?

- Global change biologists use key study design best practices such as controls, replication, and randomization.

What are key research approaches in Global Change Biology?

- Global change biologists use many approaches, including observational approaches, experimental approaches, modeling approaches, and synthesis approaches. These approaches are complementary, and each has important strengths and limitations.

What are key tools used in Global Change Biology?

- Global change biologists use many different tools to monitor abiotic parameters and measure organismal responses to environmental change. Technological advances—for example, in molecular biology and computer science—allow global researchers to apply emerging technologies to questions in the field.

Core concepts: How are data displayed?

- Interpreting Global Change Biology data requires a familiarity with foundational statistical concepts.

Meet the data: The economic value of nature

- Two meta-analyses evaluated the economic value of ecosystem services and found that the benefits humans derive from natural ecosystems are worth >100 trillion US dollars per year, a figure that is increasing as ecosystems become more impacted by human activities.

Taking a closer look: The value of biological diversity

- There are many implicit and explicit motivations for valuing the biological diversity on our planet—from the economic value of biomedicines to the cultural value of recreation.

CONSOLIDATE YOUR KNOWLEDGE

Answer the following questions to consolidate your knowledge, assess your progress meeting the learning outcomes, and reveal any areas in need of further exploration:

1. In your own words, write a one- or two-sentence synthesis of the big-picture take-away point of this chapter.
2. Revisit your answer to the *Blank Page* exercise in the beginning of this chapter. Would you refine your answer now based on knowledge you integrated from this chapter?
3. What key factors influenced the emergence of Global Change Biology as a field of study?
4. In your own words, how would you define the field of Global Change Biology?
5. Compare and contrast the four key approaches in Global Change Biology. Define each approach, discuss its strengths and limitations, and give an example of how you could use that approach in the copper mining example introduced in this chapter.

6. Your colleague is planning an experiment to determine whether pesticides harm fish in the Sierra Nevada Mountains. She has received permission to add pesticide to one lake and is planning to monitor the fish after adding the pesticide. What are the two most important ways that she could improve the experimental design of this study?
7. How could a meta-analysis approach be used to complement your colleague's experimental study in the Sierra Nevada Mountains to increase generality? If you were designing the meta-analysis study, what data would you look to aggregate and what criteria would you use for inclusion?
8. What are three examples of tools or technologies used in Global Change Biology that were not available 100 years ago? Pick one of these and provide a specific example of how its use has deepened what scientists can learn about organismal response to changing environments.
9. Which of the reasons to value biodiversity explored in the *Taking a Closer Look* section resonated the most strongly with you? Are there other reasons you value biodiversity that were not mentioned?
10. What does the term *ecosystem service* mean? If you were trying to convince someone of the economic value of ecosystem services, what three main arguments would you make?
11. In your own words, define the bolded terms in this chapter.
12. What are some questions that you have about the content in this chapter? If you found some content particularly challenging or particularly interesting, identify these as areas for additional reflection or reading.

LITERATURE CITED

Alexander, Bynum, Johnson, King, Mustonen, Neofotis, Oettlé, et al. 2011. "Linking Indigenous and Scientific Knowledge of Climate Change." *BioScience* 61 (6): 477–84. doi.org/10.1525/bio.2011.61.6.10.

Arnqvist & Wooster. 1995. "Meta-Analysis: Synthesizing Research Findings in Ecology and Evolution." *Trends in Ecology & Evolution*. 10 (6): 236–40. doi.org/10.1016/S0169-5347(00)89073-4.

Arrhenius. 1897. "On the Influence of Carbonic Acid in the Air upon the Temperature of the Earth." *Publications of the Astronomical Society of the Pacific* 9 (54): 14–24. doi.org/10.1086/121158.

Berkes. 2009. "Indigenous Ways of Knowing and the Study of Environmental Change." *Journal of the Royal Society of New Zealand*. 39 (4): 151–56. doi.org/10.1080/03014220909510568.

Brulle & Pellow. 2006. "Environmental Justice: Human Health and Environmental Inequalities." *Annual Review of Public Health* 27 (1): 103–24. doi.org/10.1146/annurev.publhealth.27.021405.102124.

Byrne, Rothstein, Poorten, Erens, Settles, & Rosenblum. 2017. "Unlocking the Story in the Swab: A New Genotyping Assay for the Amphibian Chytrid Fungus *Batrachochytrium Dendrobatidis*." *Molecular Ecology Resources* 17 (6): 1283–92. doi.org/10.1111/1755-0998.12675

Callendar. 1938. "The Artificial Production of Carbon Dioxide and Its Influence on Temperature." *Quarterly Journal of the Meteorological Society* 64 (275): 223–40.

Carson. 1962. *Silent Spring*. Boston: Houghton Mifflin.

Chen. 2020. "Researchers around the World Prepare to #ShutDownSTEM and 'Strike For Black Lives'." *Science News*. doi:10.1126/science.abd2504. www.sciencemag.org/news/2020/06/researchers-around-world-prepare-shutdownstem-and-strike-black-lives

Costanza, D'Arge, De Groot, Farber, Grasso, Hannon, Limburg, et al. 1997. "The Value of the World's Ecosystem Services and Natural Capital." *Nature* 387 (6630): 253–60. doi.org/10.1038/387253a0.

Costanza, de Groot, Sutton, van der Ploeg, Anderson, Kubiszewski, Farber, & Turner. 2014. "Changes in the Global Value of Ecosystem Services." *Global Environmental Change* 26 (1): 152–58. doi.org /10.1016 /j.gloenvcha.2014.04.002.

De Groot, Brander, van der Ploeg, Costanza, Bernard, Braat, Christie, et al. 2012. "Global Estimates of the Value of Ecosystems and Their Services in Monetary Units." *Ecosystem Services* 1 (1): 50–61. doi.org/10.1016/j.ecoser.2012.07.005.

Dickinson, Zuckerberg, & Bonter. 2010. "Citizen Science as an Ecological Research Tool: Challenges and Benefits." *Annual Review of Ecology, Evolution, and Systematics* 41: 149–72. doi.org/10.1146 /annurev-ecolsys-102209-144636.

Dresher. 2019. "Urban Heating and Canopy Cover Need to be Considered as Matters of Environmental Justice." *Proceedings of the National Academy of Sciences* 116 (52): 26153–54.

Fernandez-Duque & Valeggia. 1994. "Meta-Analysis: A Valuable Tool in Conservation Research." *Conservation Biology* 8: 555–61. doi .org/10.1046/j.1523-1739.1994.08020555.x.

Fortmann. 2008. *Participatory Research in Conservation and Rural Livelihoods: Doing Science Together.* John Wiley & Sons, West Sussex, UK

Jaffe, Franks, Roberts, Mirza, Schurgers, Kastner, & Boch. 2017. "A Swarm of Autonomous Miniature Underwater Robot Drifters for Exploring Submesoscale Ocean Dynamics." *Nature Communications* 8: 1–8. doi.org/10.1038 /ncomms14189.

Kelly-Reif & Wing. 2016. "Urban-Rural Exploitation: An Underappreciated Dimension of Environmental Injustice." *Journal of Rural Studies* 47: 350–58. doi.org /10.1016/j.jrurstud.2016.03.010.

Kuras, Warren, Zinda, Aronson, Cilliers, et al. 2020. "Urban Socioeconomic Inequality and Biodiversity often Converge, but not Always: A Global Meta-Analysis." *Landscape and Urban Planning* 198: 103799. doi .org/10.1016/j.landurbplan.2020.103799.

Leong, Dunn, & Trautwein. 2018. "Biodiversity and Socioeconomics in the City: A Review of the Luxury Effect." *Biology Letters* 14 (5): 20180082 doi.org/10.1098 /rsbl.2018.0082

Makondo & Thomas. 2018. "Climate Change Adaptation: Linking Indigenous Knowledge with Western Science for Effective Adaptation." *Environmental Science and Policy* 88 (October): 83–91. doi.org /10.1016/j.envsci.2018.06.014.

Middleton, Talaugon, Young, Wong, Fluharty, Reed, Cosby, & Myers. 2019. "Bi-Directional Learning: Identifying Contaminants on the Yurok Indian Reservation." *International Journal of Environmental Research and Public Health* 16 (19): 3513. doi.org/10.3390/ijerph 16193513.

Nagajyoti, Lee, & Sreekanth. 2010. "Heavy Metals, Occurrence and Toxicity for Plants: A Review." *Environmental Chemistry Letters* 8 (3): 199–216. doi.org/10.1007/s10311 -010-0297-8.

Nakagawa, Poulin, Mengersen, Reinhold, Engqvist, Lagisz, & Senior. 2015. "Meta-Analysis of Variation: Ecological and Evolutionary Applications and Beyond." *Methods in Ecology and Evolution* 6 (2): 143–52. doi .org/10.1111/2041-210X.12309.

Natesh, Taylor, Truelove, Palumbi, Hadly, Petrov, & Ramakrishnan. 2019. "Empowering Conservation Science and Practice with Efficient and Economical Genotyping from Poor Quality Samples." *Methods in Ecology and Evolution* 10 (6): 853–9. doi.org /10.1111/2041-210X.13173.

Puritty, Strickland, Alia, Blonder, Klein, et al. 2017. "Without Inclusion, Diversity Initiatives May Not Be Enough." *Science* 357 (6356): 1101–2. DOI: 10.1126/science .aai9054.

Ramirez. 2019. *Environmental Health, Community Engagement, and Environmental Justice. In Environmental and Pollution Science*, 3rd Edition. Academic Press, London.

Ramirez-Andreotta, Brusseau, Artiola, Maier, & Gandolfi. 2015. "Building a Co-Created

Citizen Science Program with Gardeners Neighboring a Superfund Site: The Gardenroots Cast Study." *International Public Health Journal* 7 (1): 13.

Rosenthal. 1979. "The File Drawer Problem and Tolerance for Null Results." *Psychological Bulletin* 86 (3): 638–41.

Schell, Guy, Shelton, Campbell-Staton, Sealey, Lee, & Harris. 2020a. "Recreating Wakanda by Promoting Black Excellence in Ecology and Evolution." *Nature Ecology and Evolution*. doi.org/10.1038/s41559-020-1266-7.

Schell, Dyson, Fuentes, De Roches, Harris, Miller, Woelfle-Erskine, & Lambert. 2020b. "The Ecological and Evolutionary Consequences of Systemic Racism in Urban Environments." *Science*: eaay4497. doi.org/10.1126/science/aay4497.

Selin. 2008. *Encyclopaedia of the History of Science, Technology, and Medicine in Non-Western Cultures*. Springer Netherlands.

Soleri, Long, Ramirez-Andreotta, Eitemiller, & Pandya. 2016. "Finding Pathways to More Equitable and Meaningful Public-Scientist Partnerships." *Citizen Science: Theory and Practice* 1 (1): 9. doi.org/10.5334/cstp.46.

Soroye, Ahmed, & Kerr. 2018. "Opportunistic Citizen Science Data Transform Understanding of Species Distributions, Phenology, and Diversity Gradients for Global Change Research." *Global Change Biology* 24 (11): 5281–91.

Theobald, Ettingera, Burgess, DeBeya, Schmidt, et al. 2015. "Global Change and Local Solutions: Tapping the Unrealized Potential of Citizen Science for Biodiversity Research." *Biological Conservation* 181: 236–44.

Vaughan, Miegroet, Pennino, Pressler, Duball, Brevik, Berhe, & Olson. 2019. "Women in Soil Science: Growing Participation, Emerging Gaps, and the Opportunities for Advancement in the USA." *Soil Science Society of America Journal* 83 (5): 1278–89. doi.org/10.2136/sssaj2019.03.0085.

Vetter, Rucker, & Storch. 2013. "Meta-Analysis: A Need for Well-Defined Usage in Ecology and Conservation Biology." *Ecosphere* 4 (6): 1–24. doi.org/10.1890/ES13-00062.1.

Brief History of Life on Earth

Learning Outcomes

After working with this chapter, you will be able to:

- Describe how life on Earth first evolved and subsequently diversified.
- Analyze which evolutionary processes led to the accumulation—and loss—of species on our planet.
- Examine fluctuations in biodiversity over time, including historical mass extinctions.
- Apply your knowledge to real-world case studies and interpret data from recent scientific studies

THE BLANK PAGE

Before reading this chapter, take a moment to reflect on your current understanding of the history of life on Earth. On a blank page, draw a timeline starting with the formation of our planet and ending with the present day. Populate your timeline with key biological events in the history of life on our planet. Think about the time elapsed between events and label your timeline with some approximate dates.

INTRODUCTION

When drawing a representation of the history of life, students often first draw a horizontal line punctuated by three major landmarks approximately equally spaced in time: a single cell, a dinosaur, and a cartoon person labeled "me." The goal of this chapter is to provide a more nuanced understanding of the critical landmarks in the history of life on Earth. Understanding the history of biodiversity on our planet is itself an important goal of Global Change Biology. It also is essential as preparation for analyzing current changes occurring on our planet. If you have minimal background in evolutionary biology or earth science, this chapter will help build a foundation for our journey. If you have already studied these fields, this chapter will provide an opportunity to deepen and apply your prior knowledge to the study of life on a rapidly changing planet.

WHAT KEY TRANSITIONS LED TO THE EMERGENCE OF LIFE ON EARTH?

The planet we call home is one of more than 100 billion planets in our galaxy. With more than 200 billion galaxies in the universe, the number of planets that could harbor life is astounding (e.g., Petigura et al. 2013). How Earth formed and how life as we know it came to flourish on this particular planet is a long and amazing story. This chapter does not provide an exhaustive account of what modern science has revealed about the origins of life on our planet—rather, it highlights key transition points along the journey. Figure 2.1 shows a timeline with early events in the history of life on Earth.

Origin of the Universe

Our story begins nearly 14 billion years ago with the cosmological event referred to as the Big Bang. The Big Bang is a model that describes the process by which the universe rapidly expanded from an exceedingly hot and extraordinarily dense state, essentially creating all light and matter that we observe today. Only minutes after the universe's birth—when the temperature was still approximately 1 billion kelvin—protons and neutrons combined into the first elemental nuclei. It would take hundreds of thousands of years for those nuclei to combine with electrons to create atoms and ultimately the elemental building blocks of life.

As the universe continued to expand and cool, matter attracted nearby matter, creating areas of density. These areas of density slowly created structure in the young universe. One billion years after its origin, the universe already exhibited many of its current features, with stars, galaxies, galaxy clusters, and superclusters. Our own solar system was born approximately 4.5 billion years ago from the collapse of a part of a molecular cloud (a dense interstellar area of gas and dust; see Figure 2.2). Most of the matter from this collapse became centralized into our sun, but a disk of gas and dust surrounded the young star. This matter slowly began to come together into small celestial bodies, and gravitational forces led to their continued collisions and growth, eventually giving rise to planets.

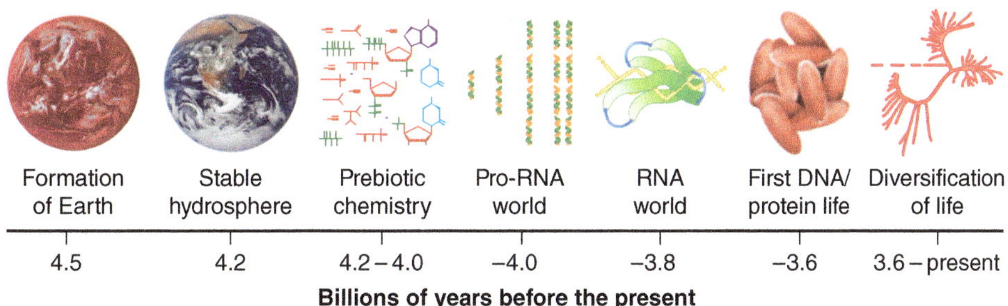

FIGURE 2.1 A timeline of early events in the history of life on Earth.

Reflection: The timeline shows key events but does not illustrate the processes responsible for these events. Pick three of the events displayed and brainstorm what processes might have contributed to their origin.

Source: Joyce. 2002. The antiquity of RNA-based evolution. Nature. vol: 418 pp: 214–221

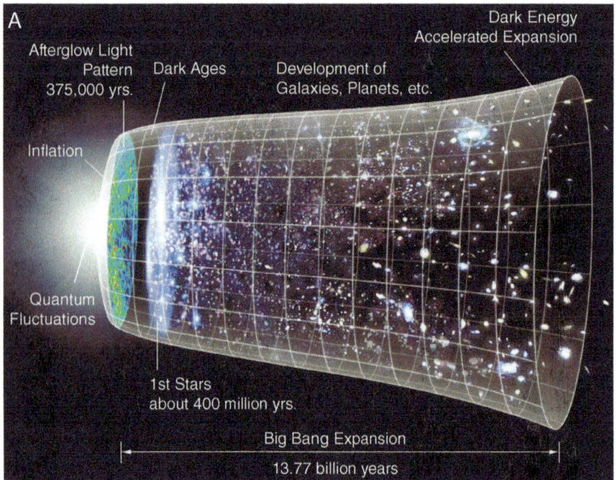

FIGURE 2.2 The birth of our world. (A) A model of the Big Bang expansion of the universe. The known universe continues to expand from the Big Bang nearly 14 billion years ago. Although the first stars appeared several hundred million years after the Big Bang, our own solar system did not originate until approximately 4.5 billion years ago. (B) The "Pillars of Creation" as imaged by the Hubble Space Telescope in 2014. This example of a molecular cloud may be similar to that which gave birth to our solar system.

Reflection: What is one big-picture question you have about our universe and its formation?

Source: (A) Wikipedia public domain (https://en.wikipedia.org/wiki/Big_Bang#/media/File:CMB_Timeline300_no_WMAP.jpg); (B) ©NASA

Formation of the Earth

The Earth took shape approximately 4.5 billion years ago, but after forming it was bombarded by interstellar debris for millions of years. The heat produced by these early impacts was likely extreme enough to sterilize the entire planet. The surface was largely molten without any liquid water. Thus, during the first half billion years of our planet's history, there is no record of life on Earth. Even if life could have evolved during this time, any early life would have been decimated by the extreme conditions.

Development of a Crust and Hydrosphere

The necessary conditions for life were in place on our planet within a billion years of its formation. After several hundred million years, much of the debris in our solar system had cleared away or contributed to the formation of other planets. Around the same time, Earth began to form an atmosphere that protected it from constant interstellar bombardment. This early atmosphere was primarily composed of gases that would spew from volcanoes formed in early tectonic plate activity, including hydrogen sulfide, methane, and carbon dioxide. As a result, Earth began to cool and form a solid crust. As the planet cooled, water could be sustained in a liquid form, establishing conditions more suitable for sustaining life.

Prebiotic Chemistry

Contemporary scientific studies suggest that the elements present in the early Earth's atmosphere were sufficient to form the building blocks of life. Prebiotic chemistry—the environmental conditions before life evolved—cannot be studied directly. But experimental approaches can be used to understand and recreate critical steps in the development of life on Earth.

One of the key demonstrations that organic compounds can be synthesized spontaneously from inorganic precursors was the Miller-Urey experiment (Miller 1953; Miller & Urey 1959). Drs. Stanley Miller and Harold Urey sealed elements and compounds that were present in the early Earth's atmosphere inside flasks (e.g., water, methane, ammonia, and hydrogen). They subjected these flasks to heat and electricity in an attempt to simulate the energy sources present early in Earth's history. After only one week, a significant proportion of the carbon was incorporated into organic compounds. Sugars, lipids, and some building blocks for nucleic acids were formed, and a small amount of the carbon formed amino acids. There is debate over whether the Miller-Urey experimental conditions adequately simulated the Earth's early atmospheric conditions and whether the yields of organic compounds would have been sufficient to facilitate other prebiotic processes. However, the Miller-Urey experiment and more than 50 years of subsequent work in this field have demonstrated the feasibility of developing the building blocks of life from inorganic molecules (Robinson 2005; McCollom 2013), a key step toward the emergence of life on Earth.

Origin of Heredity

The next landmark on the journey toward life on Earth was the development of a molecule that could replicate itself. In the words of evolutionary biologist and author Richard Dawkins (1976): *"At some point a particularly remarkable molecule was formed. . . . We will call it the Replicator. It may not have been the biggest or the most complex molecule around, but it had the extraordinary property of being able to create copies of itself."* What was the first molecule that could replicate itself? There are three macromolecules (large, complex molecules) that are key to contemporary life on earth: **ribonucleic acid** (RNA), **deoxyribonucleic acid** (DNA), and **proteins**. We will learn more about these building blocks of life in Chapter 6. Of these macromolecules, scientists have deduced that RNA was likely the first on the Earth's stage. RNA can both store and process information and is widely considered to have been a key step toward the origin of life on Earth (Gilbert 1986; Joyce 2002; Powner et al. 2009). The first RNA molecules likely appeared approximately 3.8 billion years ago.

There are many steps between the appearance of the first replicator molecular and complex cellular life. But simply put, once there was a mechanism for **heredity**, or a means by which genetic information could be passed from parent to offspring, early life could evolve via a process of replication, variation, competition, and natural selection. For example, the occasional and inevitable replication mistakes—termed *mutations*—would have generated variation in the pool of RNA molecules. Variants that replicated more quickly or were more stable would have had an advantage and thus disproportionately increased in the RNA population. Over long time periods, this process generated the major transitions toward cellular life, including the emergence of RNA's more stable cousin, DNA, and the development of cellular membranes, which could selectively filter certain materials in and out of the cell and allow for a large, diverse group of reactions to occur more efficiently.

HOW DID CELLULAR LIFE EVOLVE AND DIVERSIFY?

Less than a billion years after Earth's tumultuous formation, cellular life first emerged. Figure 2.3 shows landmarks in the evolution of life on our planet.

Evolution of Bacteria and Archaea

The first living cells appeared on Earth approximately 3.5 billion years ago. Then, for more than a billion years, single-celled organisms flourished. These first organisms contained DNA but did not have any membrane-enclosed compartments. Often, single-celled organisms without a membrane-enclosed nucleus are lumped together and referred to as **prokaryotes**. However, the term *prokaryote* does not accurately describe any unified segment of the tree of life. In fact there were two early and distinct lineages: **Archaea** and **Bacteria** (Woese et al. 1990). Archaea (sometimes referred to as Archaebacterium) were initially classified as bacteria, but they are now recognized as a unique group with a complex evolutionary history of their own. Bacteria (sometimes referred to as eubacteria or "true bacteria") are likely the oldest cellular organisms on Earth.

The first single-celled organisms played an essential role in the evolution of life on Earth. Not only were they ancestors of all subsequent life on Earth, but they also altered the abiotic and biotic conditions in ways that set the stage for evolution of additional life forms. For example, oxygen-releasing photosynthesis evolved more

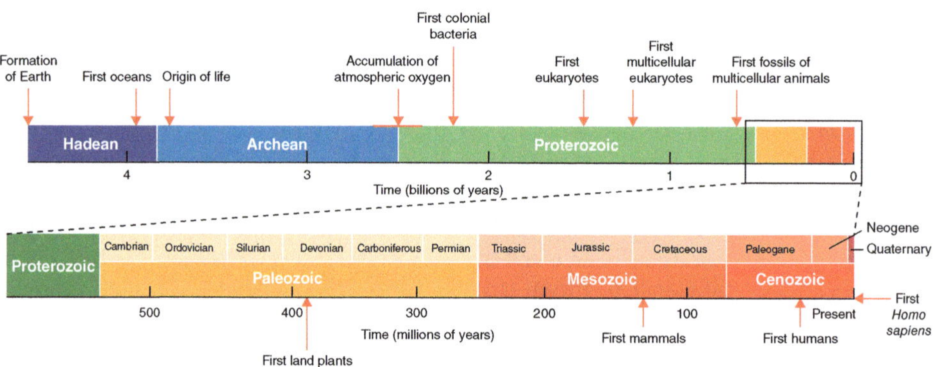

FIGURE 2.3 A timeline illustrating some of the major transitions of life on Earth. Note the arrival of humans - and our particular species - is incredibly recent in the context of the Earth's entire history.

Reflection: Throughout the history of life on Earth, biotic and abiotic processes have affected each other. Take the example of "atmospheric oxygen" shown in this figure. What biotic events could have *caused* the accumulation of oxygen in the atmosphere more than 2.5 billion years ago? What *consequences* could this then have had on the evolution of life on Earth?

Source: Modified from Figure 17.10 from Hillis, et al. Principles of Life 3e (Sinauer Associates)

than 2.5 billion years ago in cyanobacteria (Leslie 2009). The resulting slow accumulation of atmospheric oxygen would have been toxic to most anaerobic organisms living at the time, altering the evolutionary landscape and changing the course of life on Earth.

Evolution of Eukaryotes

Approximately 2 billion years ago, **eukaryotes** appeared. Like their ancestors, the first eukaryotes were single-celled organisms. But the eukaryotes had a nucleus—a membrane-enclosed compartment where the core genetic material was stored. The Eukaryotic genome likely resulted from a fusion of genetic material from two highly distinct ancestors—Bacteria and Archaea (Rivera & Lake 2004; McInerney et al. 2015). The *Meet the Data* feature at the end of this chapter explores this ancient fusion in more depth.

Most eukaryotic cells also contain other membrane-enclosed organelles in addition to the nucleus, some of which contain DNA distinct from that found in the nucleus. The acquisition of mitochondrial genomes and chloroplast genomes reflects ancient and extensive interaction among the earliest life forms on Earth, often referred to as **endosymbiosis** (Margulis 1971; Martin et al. 2015). The theory of endosymbiosis proposes that some of the organelles (cell components) that compose eukaryotes are related to ancient prokaryotes that were absorbed into a larger cell. Over time, these prokaryotes lost components of their genome and became dependent upon the larger cell to function. It's been suggested that the chloroplasts in plant cells, for example, are related to cyanobacteria. Understanding the relationships between the earliest branching lineages of life on Earth is thus complex not only because it involves looking back so deeply in time but also because of ancient sharing of genetic information among bacteria, archaea, and eukaryotes. A **lineage** is a group of organisms (both living and extinct) that share a common ancestor.

Evolution of Multicellularity

Living as a single cell is an incredibly successful evolutionary strategy, and the vast majority of life on Earth—even today—is still represented by single-celled organisms. However, multicellularity also provides evolutionary benefits, and the first multicellular life forms may have appeared almost 2 billion years ago (Albani et al. 2010). Figure 2.4 illustrates some of these early colonial organisms found in the fossil record. Multicellularity is favored under many different ecological conditions, given that clusters can provide buffering from environmental stress, protection from predation, and enhanced opportunities to use nutrients (Libby et al. 2016). Recent studies with the yeast *Saccharomyces cerevisiae* show that unicellular organisms can evolve complexity in the lab over short timescales, and that the transition to multicellularity can have a simple genetic basis (Ratcliff et al. 2012, 2015). Therefore, the evolution of multicellularity could have occurred relative quickly, as aggregations of individual cells lost their autonomy over evolutionary timescales.

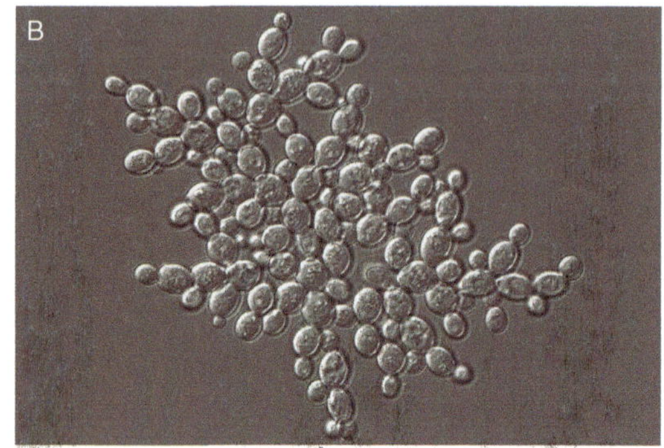

FIGURE 2.4 Evolution of multicellularity. Almost 2 billion years ago, colonies of cells were living together and creating complex coordinated structures. (A) The (unnamed) colonial organisms shown, which are reconstructions based on fossils, likely represent some of the earliest multicellular life forms on Earth. (B) Even in the laboratory, single-celled organisms like yeast can be induced to grow in colonies, suggesting that the evolutionary transition to multicellularity may have been easy under some conditions.

Reflection: What are two evolutionary benefits of multicellularity?

Source: (A) Albani, A., Bengtson, S., Canfield, D. et al. Large colonial organisms with coordinated growth in oxygenated environments 2.1 Gyr ago. Nature 466, 100–104 (2010) (B) Jennifer Pentz/Georgia Tech

Early multicellular organisms were simple in structure, and only recently in Earth's history did multicellular plants and animals evolve and take forms that are familiar to us today. Approximately 500 million years ago, there was a rapid and dramatic increase in organismal complexity, often termed the Cambrian Explosion. During this period, nearly all major groups of animals (referred to as phyla) evolved within approximately 10 million years (Valentine et al. 1999). Dramatic morphological and developmental changes accompanied this rapid diversification (Knoll & Carroll 1999; Marshall 2006). For example, many new species of hard-shelled animals emerged—including mollusks and arthropods—lineages that would become some of the most biodiverse on our planet.

Another example of rapid diversification occurred 400 million years ago when plants first colonized land (Bateman et al. 1998). These early colonizers were descendants of green algae and underwent dramatic adaptations to survive on dry terrestrial conditions. For example, some lineages developed a waxy cuticle to limit desiccation and stomata to facilitate gas exchange, while others evolved complex vascular tissue allowing water and nutrients to flow up to great heights. Thus, interaction between biotic and abiotic factors continued to drive the evolution of life on our planet.

Although most major phyla were established hundreds of millions of years ago, many groups that we intuitively consider ancient are actually quite recent additions to the tree of life. For example, as shown in Figure 2.3, the first mammals evolved only ~100 million years ago (Kemp & Kemp 2005). In addition, our own species, *Homo sapiens*, is vanishingly recent from a geological perspective, a topic we will return to in the next chapter.

WHAT EVOLUTIONARY PROCESSES SHAPE BIOLOGICAL DIVERSITY?

The diversity of life on Earth has been shaped over millennia by the gain and loss of lineages. Here we briefly review two fundamental mechanisms of evolution (**natural selection** and **genetic drift**) and two key processes shaping the tree of life (**speciation** and **extinction**). Typically, scientists consider natural selection and genetic drift as population-level processes, while speciation and extinction relate to the species level. Of course, there are other important evolutionary mechanisms (such as mutation and gene flow) and numerous ecological processes (such as nutrient cycles, species interactions, and ecological succession) that shape patterns of biological diversity, which we will evaluate in later chapters. Unit II of the book explores key processes that impact biodiversity at the individual, population, and species levels, while Unit III focuses on community-, ecosystem-, and biosphere-level processes, including key biotic–abiotic feedbacks.

Natural Selection

Natural selection is the process by which populations become adapted to their biotic and abiotic environment. More specifically, natural selection is the differential survival or reproduction of organisms with different traits based on their fit to their environment. Charles Darwin, the grandfather of evolutionary theory, described several conditions necessary for evolution by natural selection to occur (1859). First, there must be variation among individuals in traits that affect survival or reproduction. Second, there must be a mechanism for heredity of these characteristics—traits must be inherited from parent to offspring. Third, there must be a reason that not all individuals survive and reproduce equally well. This could be competition for mates, limitation of a key abiotic resource, or challenging physiological conditions. Over time, differential survival and reproduction can lead to trait changes at the population level. Natural selection can occur very rapidly over few generations or very slowly over long geological timescales. We will evaluate how natural selection leads to adaptation in more detail in Chapter 7.

Genetic Drift

Genetic drift is the process by which the genetic composition of a population changes over time due to random chance. Survival and reproduction are not only due to how well suited an individual is to its environment but also to stochastic (randomly determined) effects. Figure 2.5 illustrates genetic drift in contrast to natural selection. Genetic drift occurs in all populations, but its effects are especially pronounced in small populations. Extreme examples include when a population experiences a large reduction in size (referred to as a bottleneck) or a small number of individuals colonize a new environment (referred to as the founder effect). In these cases, the resulting population shows not only reduced genetic variation but also nonrandom sampling of variation from the original population. We will study the effects of genetic drift on small populations in more detail in Chapter 8.

Speciation

Speciation is the diversifying process leading to the formation of new lineages. As we saw earlier, natural selection and genetic drift clearly affect the fate of

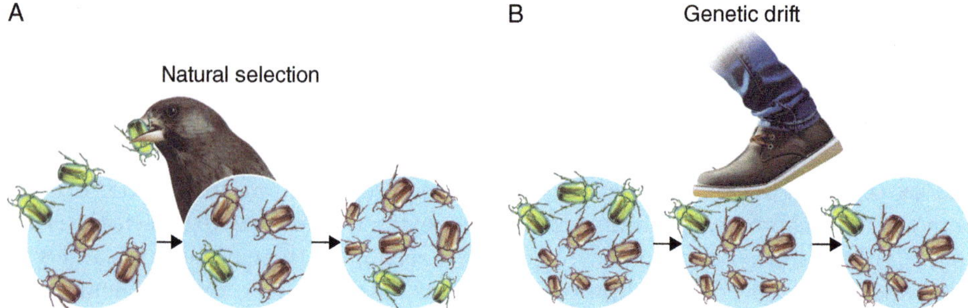

FIGURE 2.5 Contrast between (A) natural selection and genetic (B) drift. Both natural selection and genetic drift can lead to genetically based trait changes in populations over time. In this example, we see an increase in brown beetles and a decrease in green beetles over time in both panels (arrows represent beetle generations). However, natural selection leads to adaptation: organisms become better suited to their environment (for example, if brown beetles are better camouflaged than green beetles). Conversely, genetic drift is a stochastic process, which does not necessarily lead to a better match between organisms and their environment (but does lead to small changes in the genetic composition of the population across generations). Ultimately, although the frequency of green beetles is the same in the final generation depicted, the causal processes are different.

Reflection: These images show natural selection and genetic drift leading to changes *within* a population. Can you draw an example of how natural selection or genetic drift could lead to differences *between* populations of the same species?

Source: https://evolution.berkeley.edu/evolibrary/article/evo_16

populations, as their genetic composition changes over time. However, both processes can also generate differences between populations. Over longer timescales, these differences between populations can accumulate and ultimately lead to the formation of new species. Speciation can result from geographic separation and genetic drift—for example, if populations are separated by geological uplift of a new mountain range (Avise 2000). In contrast, speciation can result from natural selection even in close geographic proximity—such as when populations become adapted to different environments and intermixed offspring do not survive well in either environment (Nosil et al. 2005). The factors that promote speciation vary across lineages and depend on many things, including whether or not lineages reproduce sexually.

Extinction

Extinction is the permanent loss of a species. As we will consider more deeply in future chapters, the causes of extinctions are varied. Ultimately, abiotic and biotic conditions change over time, and eventually, the conditions that once supported a particular species are no longer present. It is estimated that more than 99% of species that ever lived have gone extinct (e.g., Raup 1991a). Although extinction rates vary dramatically over time

FIGURE 2.6 The fate of all species is eventually extinction, but some species survive for exceptionally long periods of time, for example (A) the echidna (an egg-laying mammal) and (B) the tuatara (an ancient reptile). Other examples of "living fossils" include the horseshoe crab and the ginkgo tree.

Reflection: What factors affect extinction rates? Brainstorm characteristics of a species itself and its abiotic and biotic environment that could make a species particularly extinction prone.

Source: (A) Mark Walshe/Shutterstock; (B) San Diego Zoo (https://animals.sandiegozoo.org/animals/tuatara

and across lineages, it appears that species have a typical lifespan of 1–10 million years (e.g, Raup 1991b). However, Figure 2.6 shows some species—often referred to as living fossils—that have evaded extinction for hundreds of millions of years.

WHEN HAVE SPECIATION AND EXTINCTION RATES BEEN PARTICULARLY HIGH?

Scientists study the processes shaping the diversity of life using a variety of methods. To understand patterns of past speciation and extinction, scientists have historically relied heavily on classifying and dating fossils. However, recent advances in molecular genetics also allow scientists to infer historical patterns with genetic data in addition to morphological data. Additionally, advances in computational approaches allow scientists to understand past patterns of speciation and extinction by creating sophisticated computer models that can be compared to real-world patterns. Most of these methods rely on reconstructing the relationships among species using a **phylogenetic tree**, a concept that is reviewed in the *Core Concepts* feature of this chapter.

What have scientists discovered using diverse approaches to studying historical biodiversity? One key finding is that speciation and extinction rates vary through time. There have been periods in the Earth's history with especially high speciation rates and periods with especially high extinction rates. These periods are exceptional when compared to background rates of speciation and extinction. By synthesizing many different studies and many different types of data, scientists can determine average rates of speciation and extinction for particular lineages of organisms in particular geological periods (Sepkoski 1998; McPeek & Brown 2007; Barnosky et al. 2011). Looking for deviations from background rates can then reveal elevated speciation or extinction rates in particular lineages, particular regions of the world, or particular geological periods.

Evolutionary Radiations

Sometimes speciation occurs especially quickly or frequently, leading to the rapid accumulation of diversity in a lineage. These bursts of diversification are referred to as **evolutionary radiations** (or adaptive radiations if the role of natural selection is clear). Elevated speciation rates often correspond to periods of ecological opportunity, where abiotic and/or biotic conditions provide opportunities for species to expand into new ecological roles (Yoder et al. 2010).

CORE CONCEPTS

WHAT IS A PHYLOGENETIC TREE?

Reconstructing the relationships among species is central for nearly all approaches to studying historical patterns of biodiversity. Box Figure 2.1 shows a **phylogenetic tree** depicting a subset of the major lineages in the tree of life. A phylogenetic tree is a description of the ancestor–descendant relationships among living organisms. The tips of the tree represent organisms, and the branches of the tree represent the evolutionary relationships of these organisms. As you trace branches backward on the tree, there are nodes where branches meet, representing the common ancestor for those organisms. A common ancestor and all of its descendants are referred to as a lineage. Some trees depict time clearly, with branch-lengths scaled to units of time or particular amounts of divergence. Some trees distinguish between **extant** organisms (those that are alive today) and **extinct** organisms (those that have disappeared).

Phylogenetic trees are typically displayed as bifurcating trees, where a single common ancestor gives rise to two descents at each node. However, it is important to keep in mind that this convention cannot capture a number of important processes that have shaped life on our planet. For example, genetic information can be passed not only vertically (from ancestor to descendant) but also horizontally (from one lineage to another). In fact, horizontal transfer of genetic information can occasionally occur between even distant relatives. The *Meet the Data* feature in this chapter highlights the key role of horizontal flow of genetic information for even the very early lineages of life on Earth.

One important role of phylogenetic reconstruction is to accurately understand and classify the relationships among segments of the tree of life. Most consider the Swedish botanist Carl Linnaeus (1707–1778) to be the father of modern taxonomy (the science of classifying and naming living organisms). Linnaeus introduced a hierarchical classification system (with class, order, genus, and species) and a binomial naming system (with a two-part Latin name in the form of *Genus species*) (Linnaeus 1735, 1753).

However, Linnaeus only addressed biological diversity within plants and animals. Thus, the hierarchical classification system has itself evolved over the past 300 years to better align with our current understanding of the phylogeny of life on Earth (e.g., Ruggiero et al. 2015). Box Figure 2.1 shows this nested classification system. Scientists currently recognize three domains (Bacteria, Archaea, and Eucarya), which become seven kingdoms once the diversity within Eucarya is recognized (Bacteria, Archaea, Protozoa, Chromista, Plantae, Fungi, and Animalia). Any of these hierarchical levels (from a species to a domain) can be referred to as a lineage.

Changes in nomenclature—and in the nomenclature system itself—are to be expected as our methods for inferring relationships among organisms become increasingly sophisticated. As we discussed in Chapter 1, advances in genomics and computational methods have revolutionized many areas of Global Change Biology. The science of building and interpreting phylogenetic trees is no exception. With more data and ever-finer resolution, scientists can peer back in evolutionary time more accurately than ever before. Thus, not only is the tree of life constantly changing through speciation and extinction, so, too, is our understanding of it.

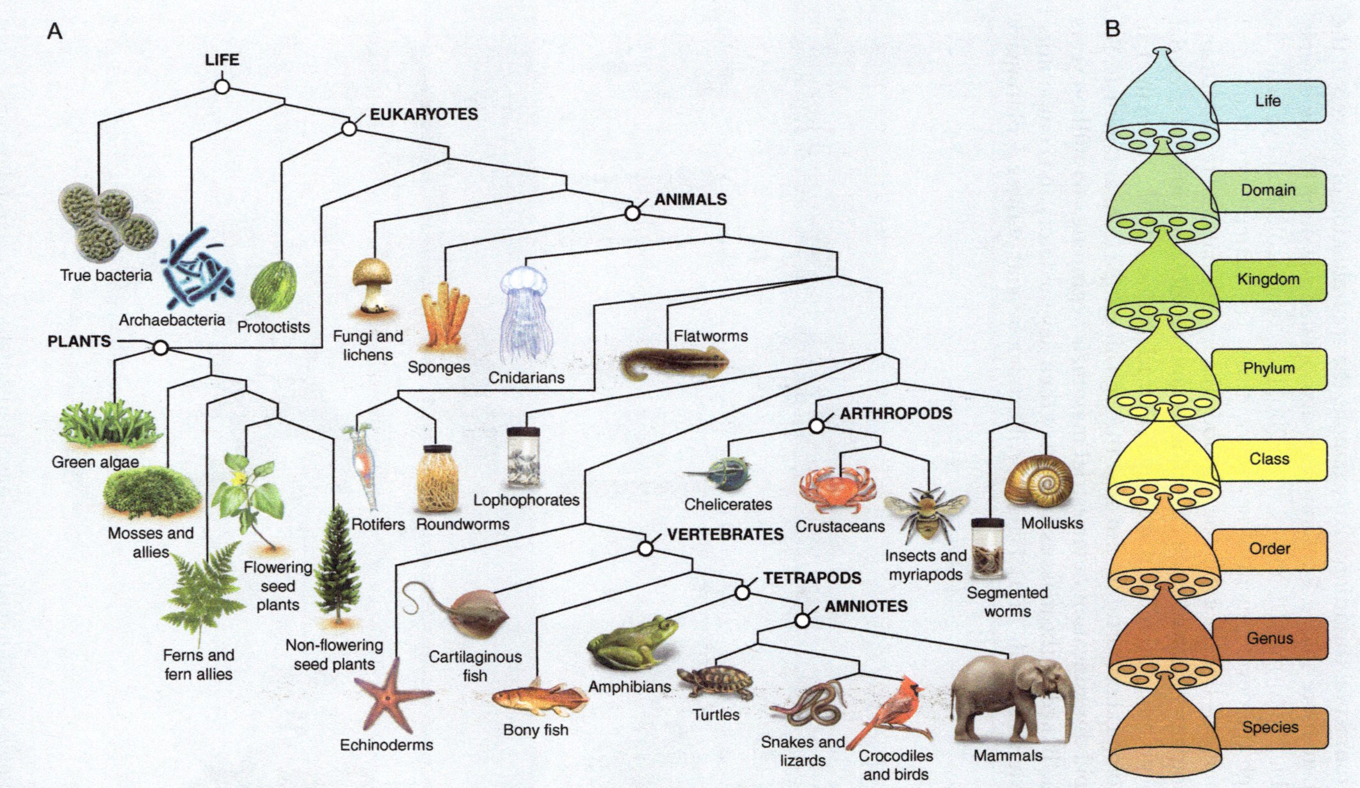

BOX FIGURE 2.1 (A) A phylogenetic tree depicting a subset of the major lineages in the tree of life. (B) A simple nested classification system for biological diversity.

Reflection: Brainstorm factors that may make it difficult to reconstruct accurate phylogenetic trees. Consider how you might address one of these factors.

Source: (A) www.biology.unm.edu/ccouncil/Biology_203/Summaries/Phylogeny.htm; (B) https://en.wikipedia.org/wiki/Taxonomy_(biology)#/media/File:Biological_classification_L_Pengo_vflip.svg (public domain)

Examples from the history of life on Earth demonstrate that ecological opportunity can arise in multiple ways. First, a new ecological opportunity could emerge due to shifts in environmental conditions. For example, the accumulation of oxygen in the Earth's atmosphere more than 2.5 billion years ago opened the door for the diversification of aerobic life forms (Dismukes et al. 2001). Second, ecological opportunity can arise through the development of key innovations. For example, structural and physiological changes—like the evolution of woody tissue, internal fluid transport, and specialized sex organs—allowed plants to colonize and rapidly diversify in terrestrial environments beginning 400 million years ago (Kenrick & Crane 1997; Bateman et al. 1998). Third, ecological opportunity can arise through an escape from antagonists. For example, the rapid diversification of terrestrial mammals beginning ~65 million years ago was likely facilitated by the mass extinction of their key predators and competitors: the dinosaurs (Meredith et al. 2011). Figure 2.7 illustrates several of these evolutionary radiations.

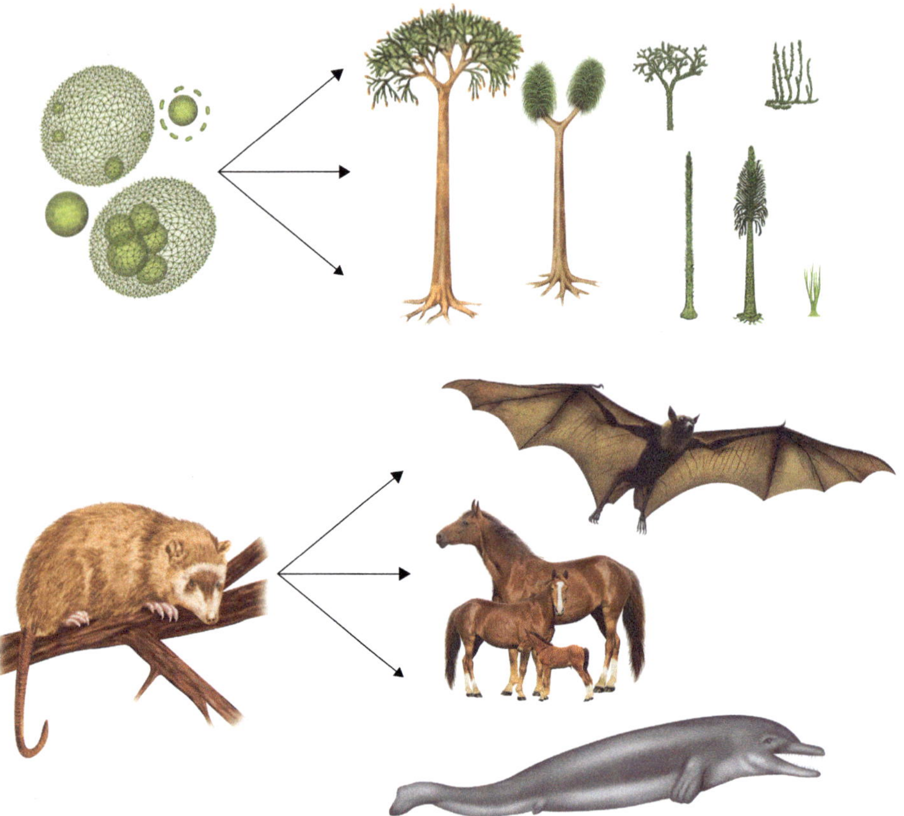

FIGURE 2.7 Examples of evolutionary radiations. (Top) Radiation of plants following colonization of land. (Bottom) Radiation of mammals following extinction of the dinosaurs.

Reflection: What biotic or abiotic changes can lead to evolutionary radiations?

Mass Extinctions

Scientists refer to periods of time when many lineages in many parts of the world show high rates of extinction as **mass extinctions.** These periods of elevated extinction often correspond to major abiotic shifts in the environment. For example, volcanic activity and extraterrestrial **bolide** impacts can alter environmental conditions quickly over large regions, rendering previously suitable habitat unsuitable for many species. Environmental changes precipitated by these abiotic events are thought to have contributed to several mass extinction events, including the end-Cretaceous extinction that wiped out the dinosaurs (Schulte et al. 2010). Biotic factors also influence the abiotic environment and thus extinction rates. For example, the photosynthetic activity of early single-celled organisms altered the concentration of oxygen in the Earth's atmosphere, likely killing many anaerobic life forms. Similarly, the diversification of photosynthetic land plants significantly altered the Earth's climate, ultimately altering what species could survive on our planet (Algeo et al. 2001). Table 2.1 summarizes the five historical mass extinctions generally recognized by scientists. Keep in mind that although mass extinctions are often described as "events," periods of elevated extinction rates do not occur instantaneously and may represent a cluster of changes that are difficult to disentangle.

Understanding historical patterns of extinction provides important context for analyzing contemporary dynamics of biodiversity loss. Today, we have entered a period of elevated extinction dramatic enough to be considered the sixth mass extinction

TABLE 2.1 The "Big Five" Historical Mass Extinction Events

Geological Epoch	Approx. End Date (mya)	Percent Species Lost	Description
Ordovician	445	86%	In the *Ordovician*, most of life was oceanic: marine organisms flourished, species like corals appeared, the first plants appeared, the first jawless fish appeared. At the end of the Ordovician, movement of tectonic plates, massive glaciation, and dramatic changes in sea levels were the likely causes of the second largest known mass extinction.
Devonian	360	75%	In the *Devonian*, terrestrial life began to flourish: land-living tetrapods like amphibians emerged, arthropods appeared, and seed plants flourished. At the end of the Devonian, diversification of land plants reduced concentration of carbon dioxide in the atmosphere, altering the climate and air and water chemistry.
Permian	250	96%	In the *Permian*, the earliest dinosaur-reptiles emerged. At the end of the Permian, volcanic activity, and possibly extraterrestrial bolide impact, caused the largest mass extinction ever recorded.
Triassic	220	80%	In the *Triassic*, tetrapods recovered from the Permian extinction: reptiles, amphibians, and the very earliest mammals dominated terrestrial landscapes. At the end of the Triassic, volcanic activity likely precipitated extinctions.
Cretaceous	65	76%	At the end of the *Cretaceous*, volcanic activity, tectonic uplift, and bolide impact likely caused the mass extinction that wiped out dinosaurs and most large land mammals.

Note: Approximate date in millions of years ago (mya), percent species lost, descriptions, and causes are shown. The extinction of anaerobic microbes following the increase in atmospheric oxygen is not included in the "Big 5."

Reflection: Compare and contrast the causes and consequences of the Big 5 historical mass extinctions. Are there predictable biotic or abiotic causes to mass extinctions?

Source: Barnosky et al. 2011

(Barnosky et al. 2011). None of the "Big 5" was precipitated by the activities of a single species, but the contemporary extinction crisis has been precipitated by the activities of a single species, *Homo sapiens*. We will analyze this modern extinction spasm in detail in Chapter 8.

CONCLUSION

It has taken billions of years of evolution to assemble the biologically diverse world we live in today. As geological and chemical conditions change over time, so, too, has life evolved, increasing in diversity and complexity over Earth's history. The evolution of life on our planet has not been a linear process; it is best represented by a deeply branching tree connecting all life forms in a shared story. That story is now being dramatically influenced by an evolutionary newcomer: *Homo sapiens*. Thus, before we turn our attention to current Global Change Biology stressors, we need to understand the history of our own species and how we came to alter the fate of so many other species on our planet.

MEET THE DATA THE RING OF LIFE

Who Are the Scientists and What Did They Set Out To Do?

Here we will evaluate an influential paper: "The Ring of Life Provides Evidence for a Genome Fusion Origin of Eukaryotes" by Drs. Maria Rivera and James Lake, which was published in the journal *Nature* in 2004. Box Figure 2.2 introduces the scientists. The authors focus on an issue at the heart of our understanding of the evolution of eukaryotic life. How did eukaryotes evolve and what are the relationships among the ancient lineages of life on Earth?

As we learned earlier in this chapter, the earliest lineages of cellular life were single-celled organisms often referred to as "prokaryotes." There are two main lineages of prokaryotes: the Archaea and the Bacteria. These single-cell microbes have cell membranes but no internal cellular compartments. Prokaryotic lineages dominated Earth for nearly 2 billion years before eukaryotes evolved, and they still compromise the vast majority of biodiversity on our planet. The transition to eukaryotic life ~2 billion years ago entailed a number of dramatic transformations. For example, genetic material was condensed into a membrane-enclosed nucleus, and additional organelles like mitochondria were acquired (Margulis 1971; Martin et al. 2015). When and how these transitions occurred have been hotly debated, and there remain a number of unresolved questions about the ancestry of the eukaryotic genome.

Questions about eukaryotic origins are implicitly thorny, given how difficult it is to look back through billions of years of evolution. However, the increasing availability of genomic data opened a new window to study ancient evolutionary transitions. Genome sequences could now be used to infer the ancestry of elements in the eukaryotic genome. Early molecular studies of eukaryotic genes gave seemingly contradictory results. Some analyses found genetic similarity between eukaryotic and archaeal genes, while other analyses suggested a closer relationship between eukaryotic and bacterial genes (Rivera et al. 1998). What could explain these seemingly contradictory results?

As we discussed earlier in this chapter, phylogenetic methods used to reconstruct the relationships among lineages typically assume a bifurcating process of lineage formation. Genetic information is assumed to be inherited vertically—from one ancestor to its descendants. But what if horizontal processes play an important role in lineage formation? We know that there are a number of mechanisms for **horizontal gene transfer** (also sometimes called lateral gene transfer)—the movement of genetic information between individuals that do not have a vertical (parent–offspring or ancestor–descendant) relationship. For example, genetic information can move between even distantly related lineages facilitated by plasmids, viruses, and transposable elements. Hybridization between distinct lineages can also result in genomes with mixed ancestry. Rivera and Lake thus set out to understand whether genetic interactions among distinct lineages could have played a role in the origin of eukaryotic life on Earth.

BOX FIGURE 2.2 (A) Lead author Dr. Maria Rivera (shown) and coauthor Dr. James Lakeset out to understand the origins of eukaryotic life. (B) Examples of modern eukaryotes.

Source: (A) Maria Rivera; (B) Wikipedia https://en.wikipedia.org/wiki/Eukaryote#/media/File:Eukaryota_diversity_2.jpg, public domain

What Are *Your* Predictions?

Before you read on, take a few minutes to think about how to apply what you have learned in this chapter about the history of life on Earth to the question of eukaryotic evolution. Start by considering the following questions:

- *What lineages of life were on Earth before eukaryotes evolved and could have contributed to eukaryotic evolution?*
- *Did the first eukaryotes necessarily have a simple origin with a single ancestor?*
- *What types of processes can create a pattern of mixed ancestry?*

Now write a summary prediction statement about what you think the relationship might have been between the early lineages of life on Earth.

What Were the Scientists' Predictions?

Rivera and Lake predicted that limitations in prior data analysis methods made it difficult to detect genome fusions. They predicted that applying new analytical methods would reveal a mixed heritage for eukaryotic genomes with both archaeal and bacterial contributions.

What Data Were Collected?

The authors selected eukaryotic, bacterial, and archaeal genomes for their analysis to represent the major lineages of life on Earth. They chose relatively simple species with similar total number of genes (~2,400–6,400 genes). They chose two Eukaryotes (both yeast species), four Bacteria, and four Archaea.

The authors first used a computation approach to identify genes that were valid to compare across species. There are several reasons why genes may exhibit similarity at the sequence level. They searched for orthologs—genes copies in different lineages that are similar because of shared ancestry. They distinguished orthologs from paralogs—gene copies that are similar because of gene duplication. Once they had a focal set of genes that met their criteria, they used a method to infer the relationships among the species that allow for genome fusions, not simply bifurcation. Based on ~2,500 genes, Rivera and Lake generated a number of ancestry reconstructions and statistically assessed the most likely historical scenario for the origin of eukaryotes.

What Is *Your* Interpretation of the Data?

Before you read on, take a few minutes to look at the summary diagram Rivera and Lake provided of their findings in Box Figure 2.3. Consider the following questions as you look at the figure:

- *Does the ancestry reconstruction look like a bifurcating tree?*
- *Do the eukaryotes appear to have a single ancestor?*
- *What appear to be the closest relatives of the eukaryotic species?*

(Continued)

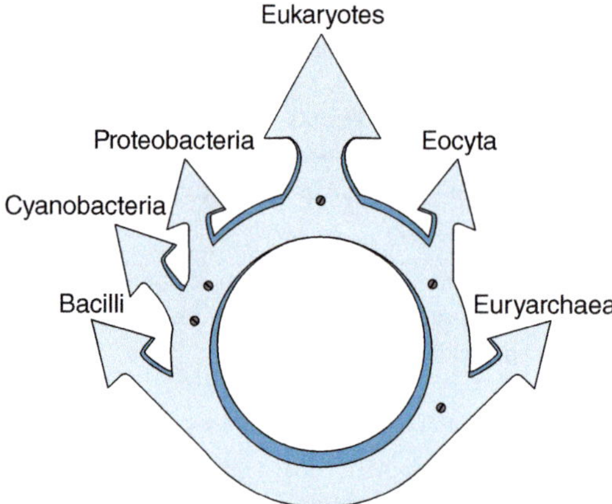

BOX FIGURE 2.3 Summary of the findings from ancestry reconstructions based on ~2,500 genes.

Source: Rivera, M., Lake, J. The ring of life provides evidence for a genome fusion origin of eukaryotes. Nature 431, 152–155 (2004) doi:10.1038/nature02848

Write a sentence or two describing the *key findings* of this study and their *significance*.

What Was the Scientists' Interpretation of the Data?

All of the ancestry reconstructions Rivera and Lake recovered showed a ring structure rather than a simple bifurcating tree structure. River and Lake state: *"From these results we infer that the eukaryotic nuclear genome was formed from the fusion of the genomes of a relative of a proteobacterium and a relative of an archaeal eocyte."* Thus, input from both ancient prokaryotic lineages was necessary for the evolution of the first eukaryotes. Moreover, the contributions of the bacterial and the archaeal ancestors can still be traced in modern eukaryotic genomes. For example, different classes of genes in the eukaryotic genome show different patterns of ancestry. Information processing genes (with key functions in transcription and translation) are typically inherited from the archaeal ancestor, whereas genes involved in cellular metabolism (such as amino acid biosynthesis) are typically inherited from the Eubacterial ancestor (Rivera et al. 1998).

Although the Rivera and Lake study focused on mixed ancestry for the core nuclear genome of eukaryotes, organelles in eukaryotic cells also resulted from ancient interactions between different lineages. **Endosymbiosis**—defined as a mutually beneficial relationship where one organism lives inside another—played a crucial role in the early evolution of the eukaryotic cell (Margulis 1971; Martin et al. 2015). For example, organelles like mitochondria and chloroplasts resulted from the uptake of previously free-living bacteria. Over long evolutionary timescales, many of these bacterial genes were transferred to the eukaryotic nucleus (Ku et al. 2015). Thus, endosymbiosis and gene transfer provide one specific mechanism to help explain the chimeric origin of the eukaryotic nuclear genome.

What Are *Your* Ideas for Future Research Directions?

Given that Rivera and Lake provided evidence for extensive genomic interaction among the three main domains of life, what next steps do you envision for this research program? If you had been involved in this study, how would you follow up? Start by considering the following questions:

- *Does this study provide a new way of thinking about evolution that could motivate a future study?*
- *Are there more detailed questions that can now be asked in this system?*
- *How could the study be broadened taxonomically to achieve more generality?*

Now write a few sentences about future directions on this research theme.

TAKING A CLOSER LOOK BIOLOGICAL LEVELS OF CHANGE

In the *Core Concepts* feature of this chapter, we looked at how our modern classification system grew out of the work of Carl Linnaeus in the 1700s. However, scientists have been studying and classifying the natural world for millennia. One of the earliest known classification systems came from Aristotle (384–322 BCE). As illustrated in Box Figure 2.4, Aristotle's *Scala Naturae* (or Ladder of Life) represented all living things in a single linear hierarchy with humans on top. The belief that species were fixed entities that did not evolve and change persisted throughout much of antiquity.

It was not until the 1800s that scientists began recognizing how much change occurs on our planet. Geologists like Charles Lyell described how shifting geological forces

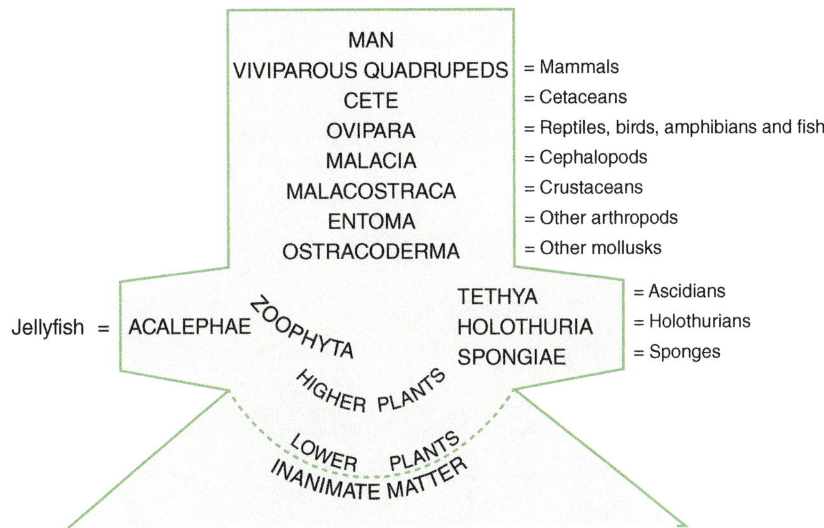

BOX FIGURE 2.4 Illustration of Aristotle's *Scala Naturae*. Note that this "Ladder of Life" classification scheme is at odds with our modern understanding of the tree of life (not only in its structure and the proposed relationship among groups but also in its depiction of species as static entities).

Reflection: What are some specific differences between Aristotle's classification of life and our modern understanding? What key discoveries do you think challenged the idea that species were fixed, unchanging entities?

Source: http://palaeos.com/systematics/greatchainofbeing/scala_naturae.html

shape the planet over time, emphasizing that the same forces acting on our planet today also were active millions of years ago (Lyell 1830). Against this shifting geological landscape, biologists realized that living things, too, would be subject to the law of change. In his book *The Origin of Species*, Darwin famously proposed the first mechanistic explanation of biological evolution (Darwin 1859). At the same time, Thoreau was discussing the succession of forest trees (Thoreau 1860), highlighting the shared theme of biological change over time.

Over the last 150 years, biologists have developed a detailed understanding of mechanisms shaping life on Earth and discovered the depth of our interconnectedness. In an evolutionary sense, species are not fixed entities in a linear hierarchy. Instead, the tree of life is shaped by the branching and pruning of interacting lineages over millennia. Similarly, from an ecological sense, species interact in complex biotic communities under shifting abiotic conditions. Even from a biochemical perspective, the energy and matter moving through our bodies have been recycled through countless other life forms over billions of years. Thus—from myriad perspectives—life is an interrelated web of being.

What Are the Biological Levels of Change?

Even in such an interdependent biosphere, it is useful to define nested levels of organization. As illustrated in Box Figure 2.5, we can study patterns and processes at different biological scales: from genes to organisms to populations to species to communities to ecosystems to the biosphere. Fundamentally, these levels are interconnected. However, focusing attention on specific levels can provide a more refined understanding of the diverse ways life on Earth responds to environmental change.

MOLECULES AND CELLS All living things are patterned by molecular building blocks. Information encoded at the genetic level and passed down across generations is largely responsible for the development of traits observed at other levels of biological organization. Molecules are organized into cells, the fundamental structural and functional unit of life on Earth. In multicellular organisms, groups of cells comprise tissues, organs, and organ systems. Many factors at the molecular and cellular level—such as rates of mutation and cellular respiration—can influence and/or be influenced by environmental stressors.

(Continued)

BOX FIGURE 2.5 Levels of biological organization (from bottom to top): molecular and cellular; individual; population; community; ecosystem; biosphere.

Reflection: What are two specific examples of how it can be difficult to clearly delineate among biological levels?

Source: From Figure 1.05, from Hillis, et al. Principles of Life 3e (Sinauer Associates)

ORGANISM An organism is an individual life form and the fundamental unit of survival and reproduction. We typically consider each organism to be a single individual, but most multicellular organisms are host to numerous symbionts and therefore ultimately collections of multiple life forms. For example, the human body is home to tens of thousands of bacterial species and contains approximately the same number of human cells as bacterial cells (Sender et al. 2016). Numerous traits at the organismal level—such as generation time, body size, feeding behavior, dispersal ability, and reproductive mode—can affect and/or be affected by environmental stressors.

POPULATION A population is a group of individuals of the same species living together in the same area. For sexually reproducing organisms, a population is further defined as a group of interbreeding individuals. Although we consider populations as distinct segments, populations can be connected to each other through migration or dispersal. Myriad factors at the population level—such as population size, sex ratio, age, class structure—can influence and/or be influenced by environmental change.

SPECIES A species can be broadly defined as an independently evolving lineage in the tree of life. This definition is referred to as the **lineage species concept** (de Queiroz 1999) and will be used throughout this book because it can be applied inclusively to organisms on our planet whether they reproduce sexually or asexually. For sexually reproducing organisms, a species can be further understood as the sum of all populations that can interbreed (referred to as the biological species concept). Although species are typically defined as reproductively isolated from each other, it is important to remember that there can be exceptions. For example, **hybridization** and **horizontal gene transfer** can occur across species boundaries, and contemporary studies continually demonstrate that even distantly related lineages can exchange genes (e.g., Belahbib et al. 2001; Pace et al. 2008). Many factors at the species level—such as range size—can affect and/or be affected by environmental change.

COMMUNITY A community is a set of interacting species in a particular area. We typically think of ecological communities at local geographic scales (such as the community of fish in a coral reef), but some communities may be nested in others (such as the community of microbes in the human gut). Many factors at the community level—such as dynamics of competition, predation, parasitism, and facilitation—can influence and/or be influenced by environmental change.

ECOSYSTEM An ecosystem is comprised of the biological community and the surrounding abiotic environment. Biotic and abiotic factors are intimately entwined. Sunlight, air, water, soil, and nutrients are required to support life. In turn, organisms affect abiotic conditions through energy flows and nutrient cycling. Many elements at the ecosystem level—such as rates of photosynthesis, primary productivity, and decomposition, can alter and/or be altered by environmental change.

BIOSPHERE The biosphere is the sum total of all ecosystems on the planet. This includes all environments containing and supporting life on Earth. Although some biotic, nutrient, and energy cycles have predominantly local consequences, many contribute to global cycles that affect all living beings. For example, photosynthetic output from tropical trees or marine algae affects oxygen and carbon dioxide concentrations in the atmosphere. Within the biosphere, scientists often distinguish among different biomes (including tundra, deciduous forests, deserts, grasslands, riparian wetlands, and the marine intertidal). These biomes represent communities that occur in multiple geographic regions but exhibit similar biotic and abiotic characteristics. Many factors at the biome and biosphere levels—such as dynamics of global gas, nutrient, water, and energy cycles—can influence and/or be influenced by environmental change.

It is important once again to remember that while we differentiate among levels for clarity and convenience, ultimately these different levels are all inextricably nested and linked. For example, an individual organism does not exist without the molecules that encode it or without the ecosystem that sustains it. Thus, all levels are connected to each other and influenced by both biotic and abiotic factors. We will be exploring these levels of change throughout our Global Change Biology journey, and you can elaborate on the framework in this special feature as you build knowledge and connections in future chapters.

KEY CONCEPTS

What key transitions led to the emergence of life on Earth?

- The universe originated ~14 billion years ago (bya), and the Earth formed ~4.5 bya. It took hundreds of millions of years for conditions to stabilize enough for the early stirrings of life to occur.

How did cellular life evolve diversify?

- The first living cells appeared ~3.5 bya, eukaryotes emerged ~2 bya, and multicellular life followed shortly after.

What evolutionary processes shape biological diversity?

- Natural selection (whereby populations become better suited to their environment) and genetic drift (whereby populations change over time due to random chance) both lead to evolution of lineages over time.
- Speciation (the formation of new lineages) and extinction (the loss of lineages) ultimately shape the tree of life.

When have speciation and extinction rates been particularly high?

- Speciation and extinction rates vary through time. Periods of elevated speciation (evolutionary radiations) or elevated extinction (mass extinctions) can be triggered by changes in biotic or abiotic conditions.
- Over the last 500 million years, there have been five major historical mass extinctions where more than 75% of species on the planet were lost.

Core concepts: What is a phylogenetic tree?

- Phylogenetic trees illustrate bifurcating ancestor–descendant relationships among lineages.
- Modern classification systems recognize seven kingdoms embedded in three domains (Bacteria, Archaea, and Eucarya).

Meet the data: The ring of life

- Scientists analyzed genomes from the three domains of life and found that eukaryotic genomes resulted from a fusion of archaeal and bacterial ancestors, demonstrating that evolutionary transitions can rely not only on vertical inheritance from ancestor to descendant but also on horizontal mixing of genomes from distinct lineages.

Taking a closer look: Biological levels of change

- The prevailing view through antiquity was that species were fixed entities that did not change. However, we now know—from both evolutionary and ecological perspectives—that life on Earth is an ever-changing and interconnected web.
- We define nested and interdependent biological levels (from molecules to organisms, populations, species, communities, ecosystems, and ultimately the biosphere) as a framework for understanding how life on Earth responds to environmental change.

CONSOLIDATE YOUR KNOWLEDGE

Answer the following questions to consolidate your knowledge, assess your progress meeting the learning outcomes, and reveal any areas in need of further exploration:

1. In your own words, write a one- or two-sentence synthesis of the big-picture takeaway point of this chapter.
2. Revisit your answer to the *Blank Page* exercise in the beginning of this chapter. Would you refine your answer now based on knowledge you integrated from this chapter?
3. Place the following key events in the correct order and add approximate dates (in billions of years before present). For each, describe its importance.
   Photosynthesis
   First cellular life
   First eukaryotic life
   First replicating macromolecule
   Formation of Earth
   First multicellular organisms
   First animals
   Big Bang
4. Compare and contrast natural selection and genetic drift.
5. Approximately what percent of species that have ever been alive on Earth are now extinct? Enumerate the difference between background extinction and mass extinction and possible causes of each.
6. What is ecological opportunity and how can it lead to diversification?
7. What is the biological species concept and what are some of its limitations?
8. What is the evidence for extensive genomic interaction among divergent lineages in the tree of life?
9. Provide a specific example of how abiotic factors can influence biotic evolution and how biotic factors can influence the abiotic environment.
10. Discuss the reasons why understanding the history of life on Earth is important for the study of contemporary biodiversity change.
11. In your own words, define the bolded terms in this chapter.
12. What are some questions that you have about the content in this chapter? If you found some content particularly challenging or particularly interesting, identify these as areas for additional reflection or reading.

LITERATURE CITED

Albani, Bengtson, Canfield, Bekker, MacChiarelli, Mazurier, Hammarlund, et al. 2010. "Large Colonial Organisms with Coordinated Growth in Oxygenated Environments 2.1 Gyr Ago." *Nature* 466 (7302): 100–104. doi.org/10.1038/nature09166.

Algeo, Scheckler, & Maynard. 2001. "Effects of the Middle to Late Devonian Spread of Vascular Land Plants on Weathering Regimes, Marine Biotas, and Global Climate." In *Plants Invade the Land: Evolutionary and Environmental Perspectives*, 213–36. New York: Columbia University Press.

Avise. 2000. *Phylogeography: The History and Formation of Species*. Cambridge, MA: Harvard University Press.

Barnosky, Matzke, Tomiya, Wogan, Swartz, Quental, Marshall, et al. 2011. "Has the Earth's Sixth Mass Extinction Already Arrived?" *Nature* 471: 51–57. doi.org/10.1038/nature09678.

Bateman, Crane, DiMichele, Kendrick, Rowe, et al. 1998. "Early Evolution of Land Plants: Phylogeny, Physiology, and Ecology of the Primary Terrestrial Radiation." *Annual Review of Ecology and Systematics* 29 (1): 263–92. www.annualreviews.org/doi/abs/10.1146/annurev.ecolsys.29.1.263

Belahbib, Pemonge, Ouassou, Sbay, Kremer, & Petit. 2001. "Frequent Cytoplasmic Exchanges Between Oak Species That Are Not Closely Related: Quercus Suber and Q. Ilex in Morocco." *Molecular Ecology* 10 (8): 2003–12. doi.org/10.1046/j.0962-1083.2001.01330.x.

Darwin. 1859. *On the Origin of Species by Means of Natural Selection, or Preservation of Favoured Races in the Struggle for Life*. London: John Murray.

Dawkins. 1976. *The Selfish Gene*. London: Oxford University Press.

De Queiroz. 1999. "The General Lineage Concept of Species and the Defining Properties of the Species Category." *Species: New Interdisciplinary Essays*, November: 49–89. http://hdl.handle.net/10088/4651

Dismukes, Klimov, Baranov, Kozlov, DasGupta, & Tyryshkin. 2001. "The Origin of Atmospheric Oxygen on Earth: The Innovation of Oxygenic Photosynthesis." *Proceedings of the National Academy of Sciences of the United States of America* 98 (5): 2170–75. doi.org/10.1073/pnas.061514798.

Gilbert. 1986. "Origin of Life—The RNA World." *Nature* 319: 618.

Joyce. 2002. "The Antiquity of RNA-Based Evolution." *Nature* 418 (6894): 214–21. doi.org/10.1038/418214a.

Kemp & Kemp. 2005. *The Origin and Evolution of Mammals*. London: Oxford University Press.

Kenrick & Crane. 1997. "The Origin and Early Evolution of Plants on Land." *Nature* 389 (6646): 33–39. doi.org/10.1038/37918.

Knoll & Carroll. 1999. "Early Animal Evolution: Emerging Views from Comparative Biology and Geology." *Science* 284 (5423): 2129–37. doi.org/10.1126/science.284.5423.2129.

Ku, Nelson-Sathi, Roettger, Sousa, Lockhart, Bryant, Hazkani-Covo, McInerney, Landan, & Martin. 2015. "Endosymbiotic Origin and Differential Loss of Eukaryotic Genes." *Nature* 524 (7566): 427–32. doi.org/10.1038/nature14963.

Leslie. 2009. "On the Origin of Photosynthesis." *Science* 323 (5919): 1286–87. science.sciencemag.org/content/323/5919/1286.short.

Libby, Conlin, Kerr, & Ratcliff. 2016. "Stabilizing Multicellularity Through Ratcheting." *Philosophical Transactions of the Royal Society B: Biological Sciences* 371 (1701). doi.org/10.1098/rstb.2015.0444.

Linnaeus. 1735. *Systema Naturae*. Nieuwkoop.

Linnaeus. 1753. *Species Plantarum*. Stockholm: Laurentius Salvius.

Lyell. 1830. *Principles of Geology*. London: John Murray.

Margulis. 1971. "The Origin of Plant and Animal Cells." *American Scientist* 59: 230–35.

Marshall. 2006. "Explaining the Cambrian 'Explosion' of Animals." *Annual Review of Earth and Planetary Sciences* 34 (1): 355–84. doi.org/10.1146/annurev.earth.33.031504.103001.

Martin, Garg, & Zimorski. 2015. "Endosymbiotic Theories for Eukaryote Origin." *Philosophical Transactions of the Royal Society B: Biological Sciences* 370 (1678). doi.org/10.1098/rstb.2014.0330.

McCollom. 2013. "Miller-Urey and Beyond: What Have We Learned About Prebiotic Organic Synthesis Reactions in the Past 60 Years?" *Annual Review of Earth and Planetary Sciences* 41 (1): 207–29. doi.org/10.1146/annurev-earth-040610-133457.

McInerney, Pisani, & O'Connell. 2015. "The Ring of Life Hypothesis for Eukaryote Origins Is Supported by Multiple Kinds of Data." *Philosophical Transactions of the Royal Society B: Biological Sciences* 370 (1678). doi.org/10.1098/rstb.2014.0322.

McPeek & Brown. 2007. "Clade Age and Not Diversification Rate Explains Species Richness Among Animal Taxa." *The American Naturalist* 169 (4): E97–106. doi.org/10.1086/512135.

Meredith, Janečka, Gatesy, Ryder, Fisher, Teeling, Goodbla, et al. 2011. "Impacts of the Cretaceous Terrestrial Revolution and KPg Extinction on Mammal Diversification." *Science (New York, N.Y.)* 334 (6055): 521–24. doi.org/10.1126/science.1211028.

Miller. 1953. "A Production of Amino Acids Under Possible Primitive Earth Conditions." *Science* 117 (3046): 528–29. doi.org/10.1126/science.117.3046.528.

Miller & Urey. 1959. "Organic Compound Synthesis on the Primitive Earth." *Science* 130: 245–51.

Nosil, Vines, & Funk. 2005. "Reproductive Isolation Caused by Natural Selection against Immigrants from Divergent Habitats." *Evolution* 59 (4): 705–19. doi.org/10.1111/j.0014-3820.2005.tb01747.x.

Pace, Gilbert, Clark, & Feschotte. 2008. "Repeated Horizontal Transfer of a DNA Transposon in Mammals and Other Tetrapods." *Proceedings of the National Academy of Sciences* 105 (44): 17023–28. doi.org/10.1073/pnas.0806548105.

Petigura, Howard, & Marcy. 2013. "Prevalence of Earth-Size Planets Orbiting Sun-like Stars." *Proceedings of the National Academy of Sciences* 110 (48): 19273–78. doi.org/10.1073/pnas.1319909110.

Powner, Gerland, & Sutherland. 2009. "Synthesis of Activated Pyrimidine Ribonucleotides in Prebiotically Plausible Conditions." *Nature* 459 (7244): 239–42. doi.org/10.1038/nature08013.

Ratcliff, Denison, Borrello, & Travisano. 2012. "Experimental Evolution of Multicellularity." *Proceedings of the National Academy of Sciences* 109 (5): 1595–600. doi.org/10.1073/pnas.1115323109.

Ratcliff, Fankhauser, Rogers, Greig, & Travisano. 2015. "Origins of Multicellular Evolvability in Snowflake Yeast." *Nature Communications* 6 (May 2014): 1–9. doi.org/10.1038/ncomms7102.

Raup. 1991a. *Extinction: Bad Genes or Bad Luck?* New York: Norton.

Raup. 1991b. "A Kill Curve for Phanerozoic Marine Species." *Paleobiology* 14: 34–48.

Rivera, Jain, Moore, & Lake. 1998. "Genomic Evidence for Two Functionally Distinct Gene Classes." *Proceedings of the National Academy of Sciences of the United States of America* 95 (11): 6239–44. doi.org/10.1073/pnas.95.11.6239.

Rivera & Lake. 2004. "The Ring of Life Provides Evidence for a Genome Fusion Origin of Eukaryotes." *Nature* 431 (7005): 152–55. doi.org/10.1038/nature02848.

Robinson. 2005. "Jump-Starting a Cellular World: Investigating the Origin of Life, from Soup to Networks." *PLoS Biology* 3 (11): 1860–63. doi.org/10.1371/journal.pbio.0030396.

Ruggiero, Gordon, Orrell, Bailly, Bourgoin, Brusca, Cavalier-Smith, Guiry, & Kirk. 2015. "A Higher Level Classification of All Living Organisms." *PLoS ONE* 10 (4): 1–60. doi.org/10.1371/journal.pone.0119248.

Schulte, Alegret, Arenillas, Arz, Barton, et al. 2010. "The Chicxulub Asteroid Impact and Mass Extinction at the Cretaceous-Paleogene Boundary." *Science* 327 (5970): 1214–1218. science.sciencemag.org/content/327/5970/1214.short.

Sender, Fuchs, & Milo. 2016. "Are We Really Vastly Outnumbered? Revisiting the Ratio of Bacterial to Host Cells in Humans." *Cell* 164 (3): 337–40. doi.org/10.1016/j.cell.2016.01.013.

Sepkoski. 1998. "Rates of Speciation in the Fossil Record." Edited by A. E. Magurran and R. M. May. *Philosophical Transactions of the Royal Society of London. Series B: Biological Sciences* 353 (1366): 315–26. doi.org/10.1098/rstb.1998.0212.

Thoreau. 1860. "*The Succession of Forest Trees, and Wild Apples.*" New York: The New York Weekly Tribune.

Valentine, Jablonski, & Erwin. 1999. "Fossils, Molecules and Embryos: New Perspectives on the Cambrian Explosion." *Development* 126 (5): 851–59. http://www.ncbi.nlm.nih.gov/pubmed/9927587.

Woese, Kandlert, & Wheelis. 1990. "Towards a Natural System of Organisms: Proposal for the Domains Archaea, Bacteria, and Eucarya (Euryarchaeota/Crenarchaeota/Kingdom/Evolution)." *Proceedings of the National Academy of Sciences USA* 87 (June): 4576–79. doi.org/10.1136/bmj.1.4540.46.

Yoder, Clancey, Des Roches, Eastman, Gentry, Godsoe, Hagey, et al. 2010. "Ecological Opportunity and the Origin of Adaptive Radiations." *Journal of Evolutionary Biology* 23 (8): 1581–96. doi.org/10.1111/j.1420-9101.2010.02029.x.

Rise of the Humans

Learning Outcomes

After working with this chapter, you will be able to:

- Summarize the evolutionary history of early hominid lineages.
- Reconstruct the origin and spread of *Homo sapiens*.
- Examine the impacts of early humans on their environment.
- Apply your knowledge to real-world case studies and interpret data from recent scientific studies.

THE BLANK PAGE

All species affect their surroundings. For example, beavers engineer their local environments by building dams, earthworms contribute to regional patterns of nutrient cycling, and trees influence global concentrations of atmospheric gases. Before reading this chapter, take a moment to reflect on the impact of our species on the Earth. Do you think that humans have a disproportionate effect on the Earth's environment relative to other species? If so, when in human history do you think our activities first began to fundamentally alter ecosystems? Draw a representation of key time periods or events in early human history that increased our impact on the environment.

INTRODUCTION

To situate the arrival of humans on the evolutionary timeline we explored in the previous chapter, consider that our species, *Homo sapiens*, has been on Earth for only 0.005% of our planet's history, an evolutionary blink of an eye. Despite our relatively recent arrival, our species has had a dramatic effect on the biosphere. The goal of this chapter is to provide an overview of the history of humans on our planet and set the stage for understanding the contemporary effects of *H. sapiens*.

WHEN AND HOW DID EARLY HOMINIDS EVOLVE?

Divergence from Great Ape Common Ancestor

Humans belong to a family of primates called the **Hominidae** (or *hominids*). Figure 3.1 illustrates the evolutionary relationships among the living members of this family, and

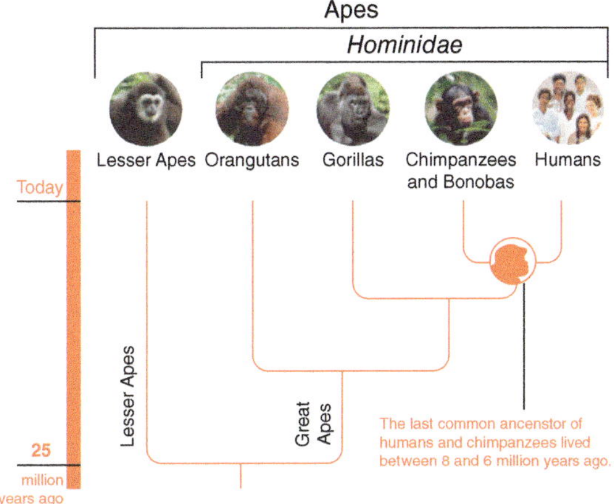

FIGURE 3.1 Family tree of humans and their living primate relatives. Note that this phylogenetic tree contains only living (extant) species and thus does not capture the full complexity of evolution in the *Hominidae*.

Reflection: Using the figure and what you learned about phylogenetic trees in the last chapter, describe why it is not true that humans evolved *from* chimpanzees.

Source: Human Origins Program, Smithsonian Institution

we review naming conventions for this group in this chapter's *Core Concepts* feature. In addition to our species, the Hominidae family contains chimpanzees (genus *Pan*), gorillas (genus *Gorilla*), and orangutans (genus *Pongo*). Of these great apes, our closest living relatives are the chimpanzees. Humans and chimpanzees shared a common ancestor approximately 6–8 million years ago (Chen & Li 2001; Langergraber et al. 2012). It is important to recognize that humans did not evolve *from* chimpanzees; rather, both humans and chimpanzees evolved from a shared common ancestor and have been diverging along separate evolutionary trajectories for the last several million years.

Although we view ourselves as quite distinct from the great apes, our evolutionary connection is relatively recent. Consider that humans are much more closely related to chimpanzees than mice are to rats, which diverged >25 million years ago (www.timetree.org). Humans and chimpanzees are also still remarkably similar at the genetic level, with fewer than 2% of the sites in the genome differing (The Chimpanzee Sequencing and Analysis Consortium 2005).

Overlapping Early Human Lineages

After the split from the human–chimpanzee common ancestor, human evolution did not proceed—as it is often depicted in cartoons—in a linear, directional progression to modern humans. Although modern *Homo sapiens* is the only human species alive today, there were many early human lineages. These early lineages gave rise to later lineages or died out. In total, scientists recognize more than a dozen different early human species (Boyd & Silk 2018). Working with an incomplete fossil record, it is difficult to fully resolve a precise number of ancient species, their relationship to each other, and their full geographic distributions. However, paleoanthropologists generally agree that ancient

CORE CONCEPTS

WHAT IS IN A NAME?

In the last chapter, we introduced two different hierarchical systems for understanding biological organization. We looked at nested levels of taxonomic classification (species, genus, family, order, class, phylum, kingdom, domains). We also evaluated nested levels of biological change across scales (cell, organism, population, species, community, ecosystem, biosphere).

Of course, our species, *Homo sapiens*, also fits into these classification systems. As shown in Box Figure 3.1, we belong to the genus *Homo*, in which *Homo sapiens* is the only surviving member. At one level up in the hierarchy, we are members of the family *Hominidae*, which includes all of the great apes (orangutan, gorilla, chimpanzee, bonobo, humans).

Beyond this, the nomenclature becomes dizzying. Scientists recognize a "superfamily" above *Hominidae* and a "subfamily" below. The superfamily *Hominoidea* includes the great apes plus the gibbons, and the subfamily *Homininae* excludes orangutans. Scientists also sometimes refer to "tribes" and "subtribes" below the subfamily—such as *Hominini* and *Hominina*—to further differentiate between our closest living relative (chimpanzee) and other extinct and extant lineages descended from our shared common ancestor (e.g., Ardipithecus, Australopithecus, Homo).

These names reflect nested segments of our extended family tree (i.e., lineages containing a single common ancestor and its descendants). However, the nested names—which all sound quite similar—can be extremely confusing. Moreover, the usage of these names has shifted over time, as conventions and our understanding of human evolution have changed.

Therefore, in this book, we follow current scientific convention and refer to all the great apes as "hominid." For simplicity we avoid the terms "hominoid" and "hominin." We use the term "human" to refer to all species in the genus *Homo* and differentiate between ancient (extinct) and modern (extant) species.

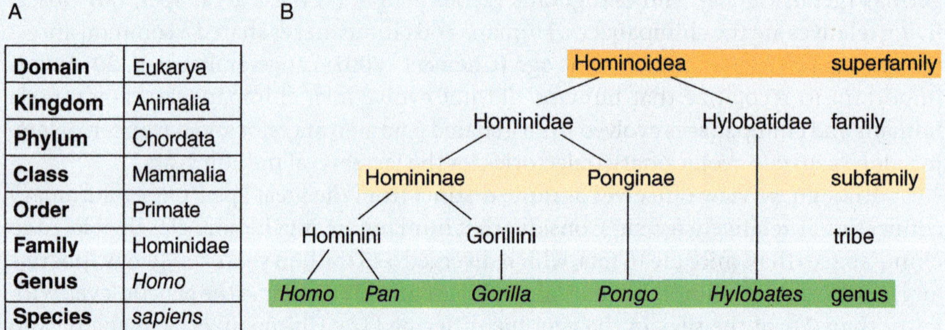

A

Domain	Eukarya
Kingdom	Animalia
Phylum	Chordata
Class	Mammalia
Order	Primate
Family	Hominidae
Genus	*Homo*
Species	*sapiens*

BOX FIGURE 3.1 Classification of our species with an illustration of additional "superfamily," "subfamily," and "tribe" terms used to describe evolutionary groups within the primates.

Reflection: What are two or three big-picture questions about human evolution that you are curious about?

Source: Modified from Wikipedia, https://en.wikipedia.org/wiki/Ape#/media/File:Hominoid_taxonomy_7.svg, https://en.wikipedia.org/wiki/Homo_sapiens#Name_and_taxonomy

human species fall into four general lineages (Ardipithecus, Australopithecus, Paranthropus, and Homo).

Figure 3.2 illustrates the timeline and relationships among these four lineages. The earliest diverging group was the Ardipithecus *group*, which was present approximately 4–6 million years ago (Boyd & Silk 2018). **Ardipithecus** species were most similar to the human–chimpanzee common ancestor, and fossil evidence suggests these earliest humans already had the ability to walk upright on two legs (Lovejoy et al. 2009). The **Australopithecus** group was present approximately 2–4 million years ago. The famous fossil Lucy was from the best-studied species in this group, *Australopithecus afarensis*, a species that walked the Earth for nearly 1 million years. The **Paranthropus** group was present approximately 1–3 million years ago. All three of

FIGURE 3.2 Early human evolution. A stylized tree depicting four early human lineages. There was temporal overlap of many different lineages. The approximate timing of major milestones in human evolution (e.g., bipedalism, tool use, fire use, brain size expansion, agriculture) shown with the timeline on the left.

Reflection: Would you consider hominids an *evolutionary radiation*? Why or why not?

Source: Human Origins Program, Smithsonian Institution

these early lineages lived only in Africa, as indicated by fossil evidence (Lewin & Foley 2004; Boyd & Silk 2018).

Evolution of the Homo Group

The **Homo** group is the only surviving lineage in the human family tree and the only lineage to have migrated out of Africa (Lewin & Foley 2004; Boyd & Silk 2018). The Homo group contains a number of named species, several of which demonstrate important landmarks in human evolution. *Homo habilis* originated in Africa more than 2 million years ago and is referred to as "handy man" because of some evidence for stone tool use in this group. *Homo erectus* appeared a few hundred thousand years later and was one of the longest persisting early human lineages, with a >1 million year history. *Homo erectus* was also the first human lineage thought to have migrated out of Africa to Asia. *Homo heidelbergensis* walked the Earth in the last million years, dying out ~200,000 years ago. This species was distributed in both Africa and Europe and demonstrated use of spears for hunting large game and construction of simple shelters. *Homo neanderthalensis* is our most contemporary relative, originating around the same time as *Homo sapiens* and dying out more than 25,000 years ago. *Homo neanderthalensis* exhibited many behaviors we associated with modern humans, including making sophisticated tools, using symbols, and living in complex social groups. Another species also living in the last hundred thousand years, *Homo floresiensis*, was only recently described and had a restricted distribution to Indonesia (Brown et al. 2004).

Early human evolution was not a simple processional from one form to the next, and some early human groups inhabited the Earth at the same time as each other. Some lineages overlapped not only temporally but also geographically, raising the possibility of interaction between groups. For example, *Homo habilis* was thought to be the direct ancestor of *Homo erectus* until new fossil evidence indicated that these two species likely coexisted in East Africa for hundreds of thousands of years (Spoor et al. 2007). For a more contemporary example, there is evidence for the interaction between *Homo neanderthalensis* and *Homo sapiens*. These two species overlapped geographically, and there is evidence for limited interbreeding between *H. sapiens* and *H. neanderthalensis* (Green et al. 2010). Our *Meet the Data* feature in this chapter takes a closer look at how these rare interbreeding events have left trace Neanderthal DNA in modern human genomes.

Like the evolution of all species on our planet, early human evolution was strongly influenced by the environment. Early human evolution took place on a backdrop of a cooling world (Feakins & de Menocal 2010). Not only were global temperatures dropping, but cyclic climate dynamics created highly impactful periods of wet and dry extremes (Maslin et al. 2014). Climate shifts altered not only temperature but also water availability and vegetation regimes (such as the formation of lakes and transitions between forests and grasslands). These changes, in turn, altered resource availability and dispersal routes for early humans. Beyond climate, numerous other abiotic factors (such as landscape topography) and biotic factors (such as predator and prey dynamics) shaped the transitions between forms. In fact, many key changes during human evolution (such as shifts in habitat use, diet, bipedalism, and brain size) were influenced by local, regional, and global environmental factors (Levin 2015). We will further explore the climate system in Chapter 4.

WHEN AND HOW DID MODERN HUMANS SPREAD AROUND THE WORLD?

Our own species, *Homo sapiens*, originated several hundred thousand years ago in Africa. Fossil, linguistic, and genetic data have been used to understand the origin and spread of our species (Chen et al. 1995; Cavalli-Sforza 1998; Gunz et al. 2009). Other creative data sources, like genetic data from parasites that travel with humans (such as *Plasmodium falciparum*, which causes malaria) have also helped reconstruct patterns of early human dispersal (Tanabe et al. 2010). The story of our dispersal around the globe is constantly being updated based on new evidence. Although there are still hotly debated questions about the timing and sequence of human dispersal, key elements of the narrative are well resolved by these multiple sources of data.

Figure 3.3 illustrates the fundamental **Out of Africa hypothesis** that is well supported by diverse datasets. *Homo sapiens* initially evolved in Africa. For many years, the origin date for modern *H. sapiens* in Africa was inferred to be approximately 200,000 years ago. However, new evidence suggests that our species may have originated even earlier. Ancient genomes from southern African and new fossil evidence from north Africa suggest that *H. sapiens* could have evolved as long as 350,000 years ago (Hublin et al. 2017; Richter et al. 2017; Schlebusch et al. 2017).

At some point—likely more than 100,000 years ago—a tremendous expansion of *Homo sapiens* around the world began. The expansion likely occurred via successive waves—where relatively small groups of humans colonized new environments. Modern *Homo sapiens* likely first spread through Africa and across the Arabian Peninsula. From there, their range expanded southward into Southeast Asia and northward into Eurasia. The arrival of *H. sapiens* into the Americas occurred much more recently across the Bering land bridge after the last major glacial period. A signature from these serial founder events is observed in modern human genomes: genetic diversity tends to decrease in populations more geographically distant from the cradle of human diversity in Africa (Ramachandran et al. 2005).

With the advent of whole genome sequencing and increasingly sophisticated computational approaches, our understanding of the nuances of early human expansion are constantly being refined. For example, recent research suggests that expansion out of Africa likely occurred earlier than previously believed and also likely occurred in more than one pulse. Specifically, there may have been two pulses of migration of modern humans out of Africa (Reyes-Centeno et al. 2015), as illustrated in Figure 3.3. First, there was likely a migration from Africa along a southern route to southern Asia much earlier than previously thought (~130,000 years ago). Subsequently, there was the well-studied migration from Africa to northern Eurasia ~50,000 years ago along a northern route. Because of subsequent replacement (or interbreeding) of migrants in the first wave by migrants in the second wave throughout Asia, only Australian and Melanesian populations retain a strong signal of the initial migration out of Africa along the southern route. Recent fossil evidence also suggests that waves of *H. sapiens* may have begun far earlier than previously thought (Harvati et al. 2019).

Again climate and geography played key roles in ancient human movement. For example, harsh climatic conditions were likely a factor precipitating dispersal events (Levin 2015). In addition, the route of early human dispersal likely followed particular

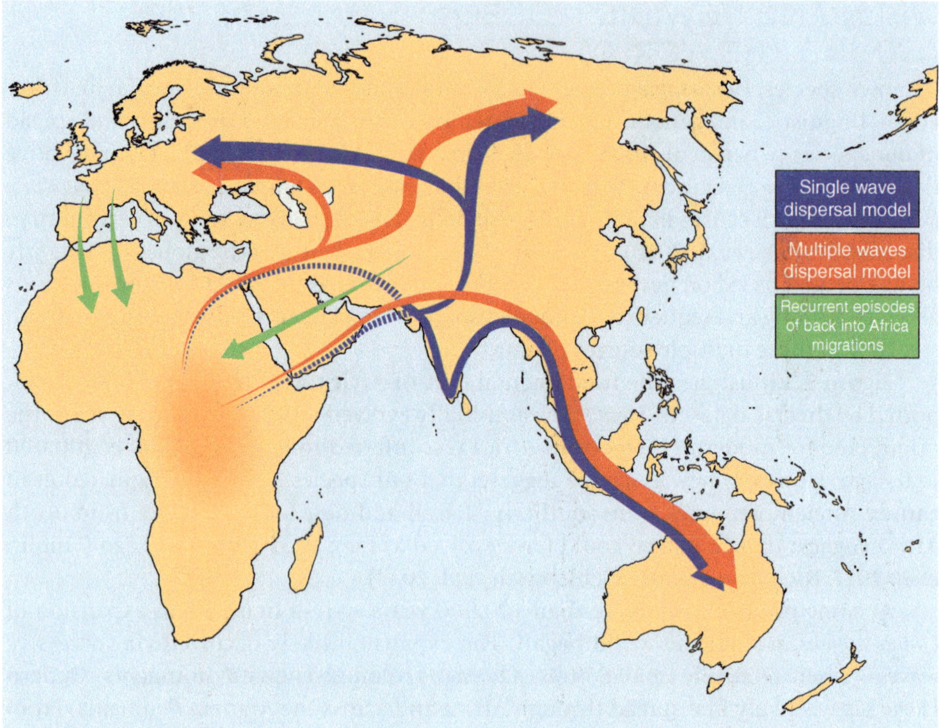

FIGURE 3.3 Human migration out of Africa. For a long time, scientists believed that there was a single wave of *Homo sapiens* dispersal out of Africa. However, new data suggest there were at least two waves of migration out of Africa and likely some recurrent migration back into Africa. In the multiple waves model illustrated in red, a southern dispersal event (~130,000 years ago) and a northern dispersal event (~50,000 years ago) are shown. But recent studies suggest these dispersal events may have been even older.

Reflection: As *Homo sapiens* expanded out of Africa, did they likely encounter other early human lineages? If so, which other lineages?

Source: López et al. 2015

landscape features, like rough, hilly landscapes where groups could hunt and shelter (King & Bailey 2006). As human populations settled in new areas of the world, they continued to adapt to local conditions, with changes in myriad traits from diet to immune function to skin color (e.g., Teaford & Ungar 2000; Jablonski & Chaplin 2017).

Ultimately, by ~15,000 years ago, modern *Homo sapiens* were present on nearly every major landmass. As *H. sapiens* spread, they came into contact with other early human lineages, a topic we will explore further in the *Meet the Data* feature of this chapter. Modern *H. sapiens* typically out-competed their ancient relatives (but also occasionally interbred with them) until our species was the only remaining human species on the planet. The factors that help account for the evolutionary success of our species are evaluated in the *Taking a Closer Look* feature in this chapter.

HOW DID EARLY HUMAN CIVILIZATIONS IMPACT THE ENVIRONMENT?

Although some early human species used tools, controlled fire, built dwellings, and hunted, dramatic human impacts on the Earth's environment did not occur until more recently in the evolutionary history of our own species, *Homo sapiens*. As *H. sapiens* spread around the world, their geographic expansion and population growth began to have large-scale influences on the environments they colonized. Before we turn to the contemporary effects of humans, we will take a brief look at several significant changes in our relationship to the land and to other species that occurred over the last tens of thousands of years.

Hunting and Megafaunal Extinctions

Human lineages have been hunting for millennia. The earliest evidence for hunting dates back nearly 2 million years—well before *Homo sapiens* evolved—when humans began using stone tools to kill and deflesh other animals (Plummer et al. 2009; Ferraro et al. 2013). But when did subsistence hunting begin to have irreversible impacts on wildlife populations?

One clear example is the dramatic megafaunal extinctions that occurred at the end of the Quaternary period. Approximately 50,000 years ago, our planet was replete with large-bodied species. For example, as shown in Figure 3.4, woolly mammoths roamed in Eurasia, giant ground sloths lived in South America, saber-tooth cats prowled in North America, and 20-foot-long monitor lizards could be found in Australia. However, between 50,000 and 10,000 years ago, dramatic die-offs of large animals occurred. With the exception of Africa, where some large-bodied species survived, 100% of terrestrial species weighing larger than 1 ton were lost (Koch & Barnosky 2006). Within mammals alone, approximately 90 *genera* of large mammals became extinct (Koch & Barnosky 2006).

The late **Quaternary megafaunal extinctions** occurred at different times on different continents. However, the first arrival of *H. sapiens* to each continental land mass corresponded roughly to these die-offs, suggesting that overkill by humans played a role in the loss of large-bodied species (Koch & Barnosky 2006; Allentoft et al. 2014; Sandom et al. 2014). Figure 3.4 shows this association between human arrival and megafaunal extinctions. Studies from smaller islands also reveal a similar pattern: massive biodiversity loss following colonization by *Homo sapiens* (e.g., Steadman 1995).

It is important to remember that a *correlation* between *H. sapiens* migration patterns and megafaunal species extinctions does not demonstrate *causation*. During this time the climate was also changing rapidly. Many megafaunal extinctions coincided either with periods of cooling before—or warming after—the last glacial maxima (Barnosky et al. 2004). Thus, a scholarly debate has raged as to the relative importance of hunting and climate change in the late Quaternary extinctions. Some recent studies suggest less chronological overlap in the New World between humans and megafaunal species than previously thought, indicating that large-bodied species may already have been in decline before human arrival (Lima-Ribeiro & Diniz-Filho 2013; Boulanger & Lyman 2014). But other studies show strong chronological overlap and demonstrate that human

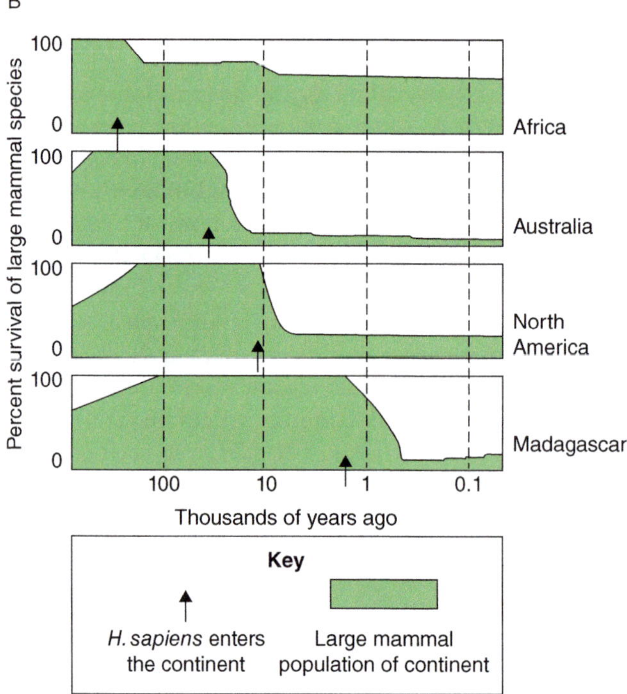

FIGURE 3.4 (A) Examples of extinct megafauna (wooly mammoth from Eurasia, giant sloth from South America, saber-tooth tiger from North America, and monitor lizards from Australia. (B) The approximate arrival of humans on major continents and the corresponding timing and extent of megafauna extinctions on that continent.

Reflection: Megafaunal extinctions often followed *Homo sapiens* arrival on new continents. Overhunting is an obvious culprit, but what other specific human activities could have impacted native large animals?

Source: (A) Reimar/Shutterstock, AuntSpray/Shutterstock, Daniel Eskridge/Shutterstock, Jaime Chirinos/Science Source; (B) Martin & Klein 1989

hunting better explains patterns of extinction than climate shifts (Allentoft et al. 2014; Sandom et al. 2014).

Overall, it is clear that human hunting played an important role in the late Quaternary extinctions, at minimum by delivering the final blow to species that were already in decline. Moreover, case studies from island systems that control for other factors clearly demonstrate the link between human arrival and biodiversity loss (e.g., Duncan et al. 2002). Thus, by 10,000 years ago, hunting by humans already had a significant impact on the survival of other species on our planet.

In addition to megafaunal extinctions, early human hunting had impacts on many other species. For example, hunting and harvesting can lead to changes in population size or structure and shifts in organismal morphology or behavior. Even invertebrate species were affected. For instance, numerous species of limpet (which are aquatic snails) decreased dramatically in shell size over the last tens of thousands of years because foragers preferentially selected larger individuals (Sullivan et al. 2017). Thus, early hunting and harvesting had diverse impacts, some subtle and some dramatic.

Agriculture and Domestication

The development of agriculture is another landmark in the increasing impact of *Homo sapiens* on the terrestrial biosphere. Early humans gathered plant foods and altered local plant communities with the use of fire. But the cultivation of plants as crops did not emerge until approximately 10,000 years ago. Agricultural practices spread from several centers of origin, with the earliest agrarian centers in the Fertile Crescent, the Yangtze and Yellow River Basins, and the New Guinea highlands. Figure 3.5 shows these major agricultural centers and the approximate timing of their origin.

There are many competing hypotheses for the emergence of agrarian societies (Weisdorf 2005; Rowthorn & Seabright 2010). Whether farming arose out of necessity or opportunity, it is clear that both environmental factors (such as changing climate patterns) and social factors (such as trends away from nomadism) played important roles. Regardless of the causes, the emergence of agriculture—often termed the **Neolithic Revolution**—altered human societies and their relationship with the land.

The domestication of plants and animals precipitated shifts in patterns of human settlement and trade, property ownership, disease transmission, cultural practices, and the arts. Many human societies moved away from nomadic lifestyles and toward more permanent settlements. Food cultivation allowed human population size and density to increase dramatically. Agriculture also led to direct modification of the landscape. For example, irrigation practices have been used for thousands of years, altering the flow of water bodies and the productivity of surrounding landscapes. Similarly, clearance of forests for agricultural practices has been practiced for thousands of years. In fact, some authors assert that even small-scale forest clearance may have had a global influence on the carbon dioxide balance in the Earth's atmosphere well before contemporary times (Ruddiman 2003). Thus, the development of agriculture—and the accompanying changes in social structure—would fundamentally alter the course of human history.

Early agriculture had dramatic impacts on the land itself and also on cultivated and domesticated species. Over the last 10,000 years, we have selected for particular plant

A

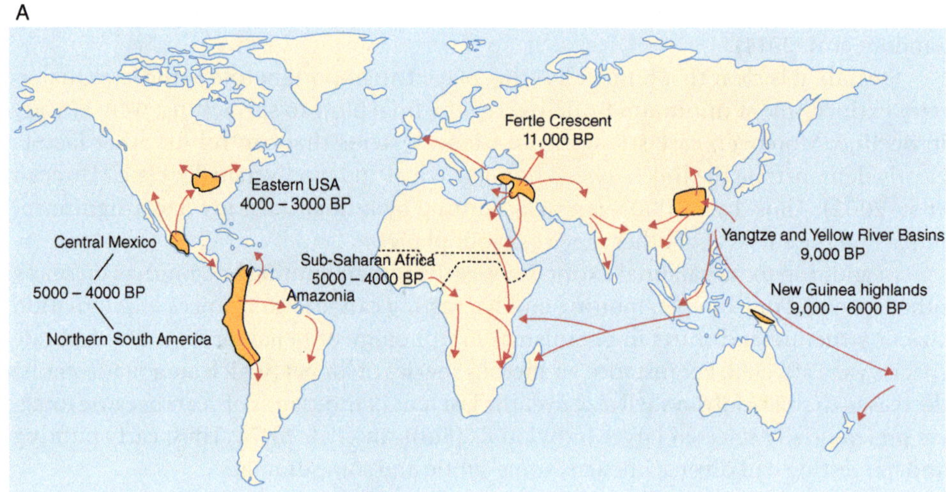

B

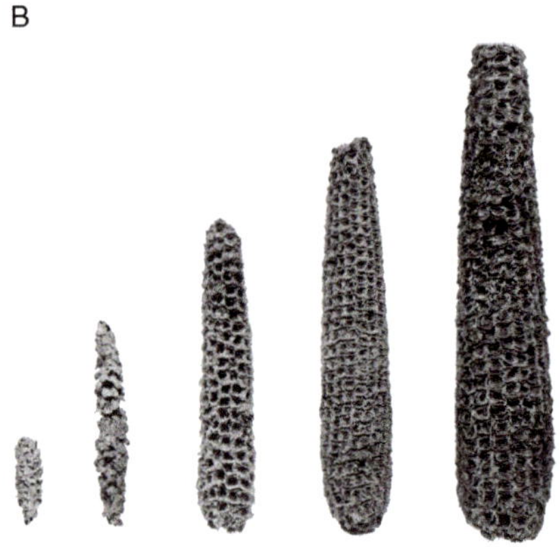

FIGURE 3.5 (A) Map showing major agricultural centers of origin and approximate dates before present (BP). (B) Maize cobs found by archaeologists in different historical time periods illustrate how artificial selection and domestication can dramatically alter species traits. The wild ancestor of maize (a grass named teosinte) was domesticated more than 8,500 years ago.

Reflection: How did the emergence of agriculture change human society?

Source: (A) Farmers and Their Languages: The First Expansions. BY JARED DIAMOND, PETER BELLWOOD. SCIENCE 25 APR 2003 : 597-603; (B) © Robert S. Peabody, Institute of Archaeology, Phillips Academy, Andover, Massachusetts. All Rights Reserved.

and animal traits and altered the morphology and behavior of numerous species. In some cases, selection for increased size or increased seed set render the crop nearly impossible to identify from wild relatives. Modern maize (*Zea mays*), for example, has a 4,000% greater number of kernels per ear than its wild predecessor, the grass teosinte. Thus, humans have been altering the evolutionary trajectory of other species on the planet in a targeted way for thousands of years.

Urbanization and Population Growth

Although urbanization and population growth are often recognized as contemporary issues, there have been urban centers with large populations for thousands of years. Following the Neolithic Revolution, the first cities sprung up in Mesopotamia, the Nile Valley, and the Indus Basin approximately 5,000 years ago. By 2,000 years ago, Rome became the first million-strong city (Oates 1934). Early cities were quite variable in their size, layout, economic institutions, and social and religious practices. However, each fit the definition of "city" as a "relatively large, dense, and permanent settlement of socially heterogeneous individuals" (Wirth 1938).

With dramatic shifts in agricultural and human settlement patterns, population growth rates began to accelerate. The human population had been increasing in size consistently for tens of thousands of years as *Homo sapiens* dispersed and established populations throughout the world. However, around the Neolithic Revolution, population growth rates increased. Figure 3.6 illustrates the shift toward exponential growth of the human population during the Neolithic Revolution. At the beginning of the Neolithic ~10,000 years ago, the world population was estimated to be between 1 and 10 million strong (United States Census Bureau 2013; United Nations Department of Economic and Social Affairs 2015). By 2,000 years ago, the population had grown to several

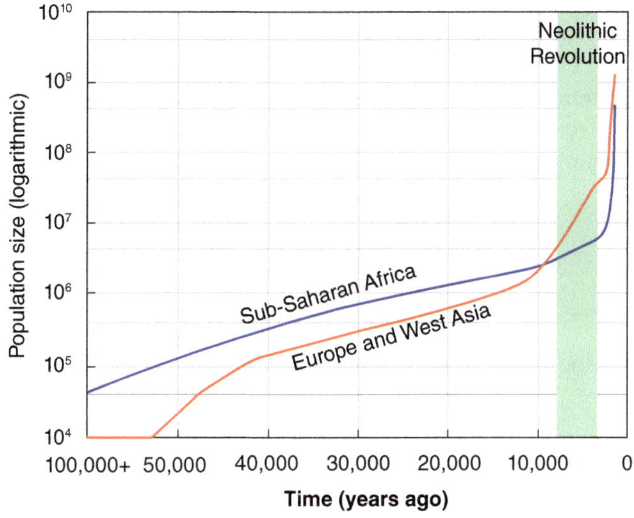

FIGURE 3.6 Ancient *Homo sapiens* population growth. Population size is depicted on a logarithmic scale, so each transition is an order of magnitude change.

Reflection: What explains the differences in population growth patterns in Africa versus Europe and West Asia? How stable were historical population sizes over the last hundred thousand years? What explains major inflection points on the graph?

Source: Recent acceleration of human adaptive evolution John Hawks, Eric T. Wang, Gregory M. Cochran, Henry C. Harpending, Robert K. Moyzis Proceedings of the National Academy of Sciences Dec 2007, 104 (52) 20753-20758. Copyright (2007) National Academy of Sciences, U.S.A.

hundred million. This means that the population had increased >20-fold in less than 5% of the time *Homo sapiens* had walked the earth.

Some authors argue that the Agricultural Revolution created the conditions for population expansion (Gignoux et al. 2011). Others contend that population growth began before the Neolithic Revolution, creating an impetus for agricultural development (Zheng et al. 2011). Still others argue that population growth occurred *despite* the shift to an agrarian society. For example, agricultural practices increased access to food, but a higher density and more sedentary lifestyle may have increased risk of infectious disease and nutritional deficiencies (Armelagos et al. 1991). Regardless, it is clear that demographic shifts such as shortened intervals between births and increased number of total children per female began to facilitate population growth (Armelagos et al. 1991; Bocquet-Appel 2011). Over thousands of years, that population growth has led to the dramatic human impacts on the environment that we will turn to in the next chapter.

CONCLUSION

In the last 200,000 years (<0.005% of Earth's history), *Homo sapiens* rose from being an evolutionary newcomer to becoming a planetary force. In waves of migration out of Africa, modern humans spread to nearly every part of the world. As *H. sapiens* arrived in new places, they displaced earlier hominid lineages and altered the landscape through hunting, agriculture, and urban development. By the end of the Neolithic Revolution, the stage was set for our species to have truly planetary impacts. Having a foundational understanding of human evolutionary history will allow us to more effectively evaluate and contextualize contemporary human impacts on the planet. We will now look more closely at impacts over the last few hundred years, the period where anthropogenic influences on the environment became so dramatic and pervasive.

MEET THE DATA ICE AGE GENETICS

Who Are the Scientists and What Did They Set Out To Do?

Here we will take a closer look at a recent paper by Dr. Qiaomei Fu et al., "The Genetic History of Ice Age Europe," published in the journal *Nature* in 2016. Box Figure 3.2 introduces the scientists. The study brought together more than 60 scientists from around the world: >40 researchers were involved in assembling archaeological material, >20 assisted with laboratory genetics work, and >10 contributed to data analysis. Thus, this study highlights the intensely collaborative nature of contemporary science in the field of human genetics.

As we learned earlier in this chapter, there were numerous different hominid lineages that walked the Earth before going extinct. The closest extinct relative to modern *Homo sapiens* was the Neanderthal. *Homo neanderthalensis* was found in Eurasia during the **Pleistocene** (40,000–200,000 years ago), the geological epoch of the most recent Ice Age. In an overlapping time period, modern humans migrated out of Africa into Eurasia, likely arriving in southern Asia ~130,000 years ago and in northern Eurasia ~50,000 years ago (Green et al. 2010; Reyes-Centeno et al. 2014).

Given that Eurasian populations of modern humans overlapped in time and space with Neanderthals, these two species likely interacted. In fact, there is evidence for occasional interbreeding, which—even if quite rare—can leave a genetic signature. Therefore, although Neanderthals did not survive to the present day, it is possible that some of their DNA survived in our **genomes**.

With the advent of modern sequencing techniques, it is now feasible to sequence entire genomes even of ancient specimens. The first draft Neanderthal genome was published

BOX FIGURE 3.2 (A) Dr. Qiaomei Fu (right in photo) was the lead scientist in the study. Dr. Fu along with Dr. David Reich (left in photo), Dr. Johannes Krause, and Dr. Svante Pääbo co-supervised an international team of researchers to understand how much Neanderthal DNA can be found in modern human genomes. (B) The researchers used sophisticated techniques to sequence ancient DNA from a 45,000 year-old Siberian thighbone.

Source: (A) Jon Chase/Harvard Staff Photographer; (B) Alexander Maklakov

in 2010 based on DNA from Neanderthal bone samples (Green et al. 2010). Since then, several studies have looked for—and found—evidence of Neanderthal genetic material in *H. sapiens* genomes. Modern human genomes (from non-African populations) typically contain ~2% Neanderthal DNA (Green et al. 2010; Prüfer et al. 2014). But what patterns would be expected if we looked back in time at the genomes of earlier *Homo sapiens*? Fu et al. set out to answer this question and determine how the proportion of Neanderthal ancestry in modern human genomes has changed over time.

What Are *Your* Predictions?

Before you read on, take a few minutes to apply what you have learned in this chapter about the history of early hominid lineages to the question of how much Neanderthal ancestry you predict to find in the genomes of early *Homo sapiens*. Start by considering the following questions:

- *Do you expect equal amounts of Neanderthal DNA in modern human genomes from different parts of the world? Why?*
- *Is there a particular part of the world where you would predict Neanderthal ancestry in modern human genomes would be particularly high? Why?*
- *Do you expect the proportion of Neanderthal ancestry to increase, decrease, or stay constant over the last 45,000 years? Why?*

Now write a summary prediction statement about how much Neanderthal DNA you would expect to find in the genomes of *Homo sapiens* from different continents and time periods.

What Were the Scientists' Predictions?

Fu et al. did not present formalized predictions, but they set out to quantify the percentage of Neanderthal ancestry in Eurasian *Homo sapiens* samples from multiple time points beginning shortly after Neanderthals and modern humans likely came into first contact (~45,000 years ago) and ending well after the extinction of Neanderthals (~7,000 years ago).

What Data Were Collected?

The authors analyzed genomes from 51 modern humans over a time series from when modern humans first arrived in Europe (~45,000 years ago) to more recent samples (~7,000 years ago). Box Figure 3.3 shows the sampling design for the study. Samples were obtained from archaeological sites throughout Eurasia. Radiocarbon dating confirmed that samples spanned the last ~45,000 years.

Working with ancient samples presents a number of challenges. DNA from old specimens is often degraded, contaminated with microbial DNA, and occasionally contaminated with DNA from present-day humans that have handled the specimens. Thus, the authors meticulously extracted DNA in dedicated "clean rooms" and used an enrichment process to select for human DNA. They sequenced the DNA samples on an Illumina platform, a sequencing machine that generates incredible amounts of data for small fragments of DNA. After sequencing, they filtered out samples that showed evidence for contamination with modern human DNA.

Using computational approaches, the authors identified **single nucleotide polymorphisms** (SNPs), which are sites in the genome that show variation across the individuals

(Continued)

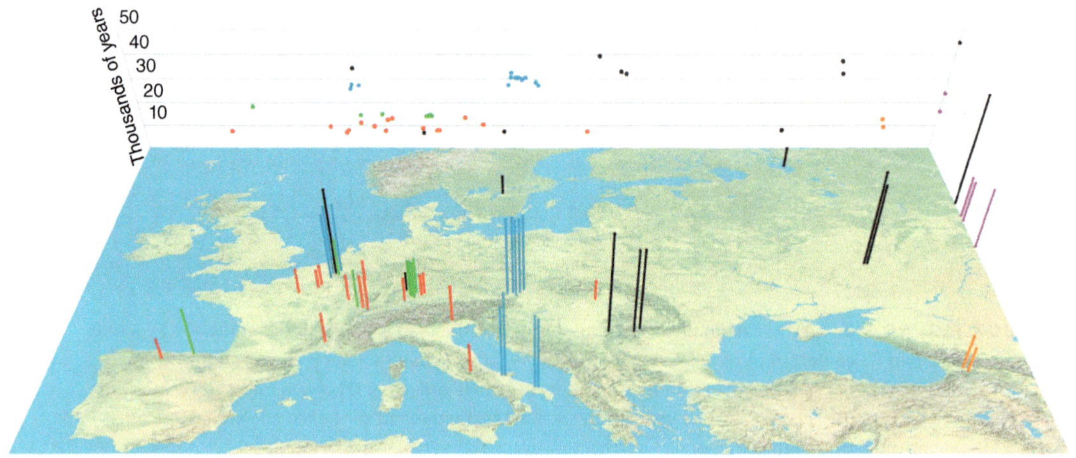

BOX FIGURE 3.3 Location and age of the samples used in this study. Each sampling location is shown in front of a timeline illustrating the age of the samples (in thousands of years).

Source: Fu et al. 2016

sampled. For many of the samples, the authors retrieved hundreds of thousands of SNPs throughout the genome. Statistical analyses of these SNPs were then used to evaluate the proportion of Neanderthal ancestry in the Eurasian genomes and whether this proportion changed through time.

What Is *Your* Interpretation of the Data?

Before you read on, take a few minutes to interpret the results from the study shown in Box Figure 3.4. Consider the following questions as you look at the figure:

- *What do the x- and y-axes of the figure represent?*
- *What do the data points represent?*
- *Is there a trend over time?*

Write a sentence or two describing the *key findings* of this study and their *significance*.

What Was the Scientists' Interpretation of the Data?

Box Figure 3.4 shows that the percentage of Neanderthal ancestry in modern human genomes has decreased over time in Eurasia. The earliest samples included in this study (from ~45,000 years ago) contained ~4% DNA of Neanderthal origin. Neanderthal ancestry has significantly decreased since then, and contemporary Eurasian genomes contain ~2% Neanderthal DNA.

What could explain the decrease in Neanderthal DNA in modern human genomes over time? Genetic changes in a population over time can be due to **genetic drift**, **natural selection**, and/or **gene flow**. In this case, the authors inferred that selection against Neanderthal DNA may have played a role. Specifically, they found that Neanderthal DNA has persisted in less constrained parts of the genome but has been lost from functionally important regions of the genome.

Some of the conclusions of the Fu et al. study have recently been re-evaluated with additional sequence data and different statistical methods (Petr et al. 2019). These analyses were made possible by a new Neanderthal genome (Prüfer et al. 2017), which was not available when Fu et al. conducted their study. The new results suggest that the proportion of Neanderthal ancestry has not changed as dramatically as previously thought, and that patterns of gene flow (rather than selection) may have been responsible for previous findings. However, the key conclusion remains: occasional interbreeding during the Pleistocene left a signature of Neanderthal ancestry in modern human genomes. Our genomes carry the legacy of our past, and our understanding of human history will continue to be refined as we excavate the story recorded in our DNA.

What Are *Your* Ideas for Future Research Directions?

Given that Fu et al. provided evidence for a detectable—but decreasing—proportion of Neanderthal DNA in modern Eurasian *H. sapiens* genomes, what next steps do you envision for this research program? If you had been involved in this study, how would you follow up? Start by considering the following questions:

- *Are there more detailed questions about the distribution of Neanderthal DNA in modern human genomes that can now be addressed?*
- *How could the study be broadened temporally or geographically to ask more general questions?*
- *Are there predictions about the future that could be made and later tested?*

Now write a few sentences about future directions on this research theme.

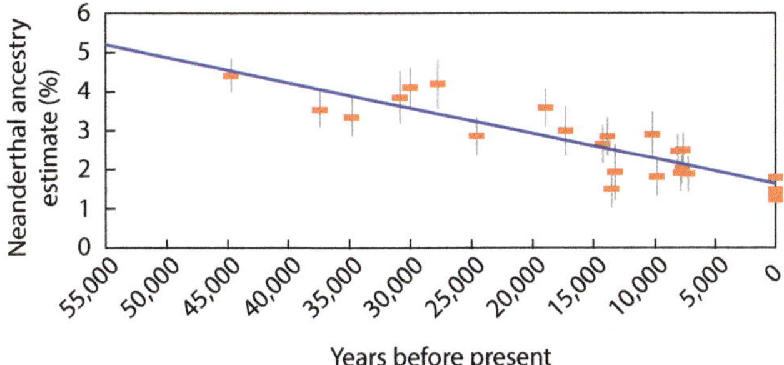

BOX FIGURE 3.4 Percentage of Neanderthal ancestry over time. Each red bar with black whiskers shows the average percentage of Neanderthal ancestry for one individual based on analysis of at least 200,000 SNPs. Dates before present are inferred from radiocarbon dating. The blue line is the fit of a line to the data and has a significantly negative slope.

Source: Fu et al. 2016

TAKING A CLOSER LOOK THE EVOLUTIONARY SUCCESS OF HUMANS

Scientists have named and classified more than a million species on planet Earth. However, the vast majority of biological diversity on our planet remains undescribed. Older studies estimated that there were likely more than 8 million total species on Earth (Mora et al. 2011), but newer research suggests that there may be as many as 1 trillion microbial species alone (Locey & Lennon 2016). Just within primates, there is incredible biological diversity, with more than 400 species of primates alive today. Scientists also estimate that more than 95% of all species that have ever lived on Earth have gone extinct. For example, in our own genus—*Homo*—there are more than a dozen named species that died out over the last several million years.

What Traits Contributed to the Success of *Homo sapiens*?

Of the millions—or trillions—of species that have evolved on our planet, why did one species—*our* species—rise to global dominance? Of course, this is a difficult and multifaceted question to answer. But scientists have identified a number of key landmarks in early human history that may have led to our incredible evolutionary success. Many of these traits had emerged in hominid lineages before our own. For example, bipedalism likely originated in our precursors more than 4 million years ago, and the use of stone tools more than 3 million years ago. However, the evolution of many morphological and behavioral traits appears to have accelerated along the *Homo sapiens* branch of the tree. Here we will look at a few exemplars.

BRAIN EVOLUTION We have many morphological traits that distinguish us from other mammals. For example, *Homo sapiens* exhibit a more upright posture, less body hair, and opposable thumbs. But perhaps our most dramatic morphological adaptation is our brain. We do not have the largest brain on the planet (that would be sperm whales), nor even the largest relative to body weight (that would be dolphins). But the human brain is still impressively large relative to body size. It also boasts a phenomenal number of neurons per unit volume and a dramatically expanded cerebrum. The cerebrum and cerebral cortex play a vital role in our exceptional information processing, problem solving, and language development abilities.

Primates in general exhibit some expansion of the brain relative to many other groups of mammals, but there are many lines of evidence demonstrating an accelerated expansion in our own lineage. As Box Figure 3.5 shows, brain size has exponentially increased over the last 2 million years of evolution in the *Homo* lineage. Moreover, scientists have identified specific genes involved in regulating brain size, and some of these show dramatic recent evolution. In one of these genes (dubbed *ASPM* for "abnormal spindle-like microcephaly-associated gene"), there is a new genetic variant that appeared less than 6,000 years ago and has increased in frequency in modern humans due to natural selection (Mekel-Bobrov et al. 2005). Thus, the human brain has undergone—and may still be undergoing—rapid evolutionary changes.

(Continued)

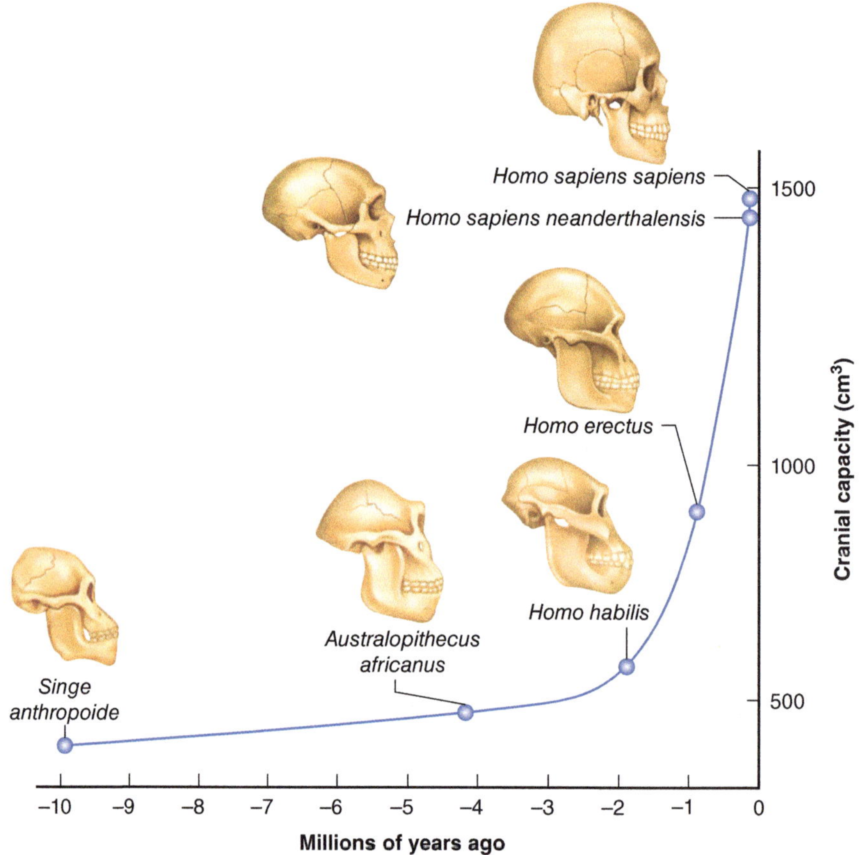

BOX FIGURE 3.5 Exponential expansion of cranium capacity in the *Homo* lineage.

Reflection: Why is a bigger brain better? How specifically might an expanded brain size have facilitated human survival and spread around the world?

Source: http://www.linternaute.com/science/biologie/dossiers/06/0608-memoire/8.shtml

USE OF FIRE One factor that may have contributed to our accelerated brain evolution was the advent of controlled fire. Early humans began to use fire approximately 1 million years ago. The use of fire likely provided a number of essential advantages. Fire provided light, warmth, and deterring of predators. It also—for the first time—allowed early humans to cook their foods. Cooked foods are easier to chew and digest, providing more calories per unit effort. Researchers believe that this facilitated brain evolution as human diets shifted to include more meat, fats, and proteins, essential to maintaining an energetic investment in cognitive processes (Wrangham 2009).

SOCIAL STRUCTURE AND CULTURE Humans have created complex social structures largely due to our incredible capacity for language and symbolic communication. Many other species live in social groups and have sophisticated means of communication. For example, social bees communicate about food resources using different "dance" elements, while social prairie dogs have distinct distress calls to alert their colonies to different predators. But humans use language in unprecedented ways (Henshilwood & d'Errico 2011). As language developed, so, too, did societies increasingly share food, responsibilities for childrearing, and the development of cultural means of learning.

Why did human societies become increasingly collaborative? Many researchers believe that at some point in human evolution natural selection began to increasingly favor sociality rather than aggression. This hypothesized "survival of the friendliest" may have precipitated a variety of other morphological and behavioral changes. In fact, some researchers suggest that humans show a number of correlated traits found in other domesticated species (Hare 2017, 2018). Box Figure 3.6 draws a parallel between the evolution of archaic to modern humans and the evolution of wolves to dogs. For example, in both cases, the juvenile developmental phase is expanded and the production of particular hormones (like serotonin and oxytocin) is increased. Thus, *Homo sapiens* may have essentially "self-domesticated" as they developed increasingly complex social structures and capacity for communication.

Are Humans Still Evolving Today?

Even after our species stepped onto the evolutionary stage, we continued to adapt and evolve. As *H. sapiens* colonized new areas, we rapidly adapted to novel and changing environmental conditions. From a common ancestor hundreds of thousands of years ago, we have diverged in countless ways, and we continue to evolve. Today, we see an amazing variety of adaptation to different altitudes, different climates, and different diets.

Humans have always evolved in response to shifts in our surroundings, but increasingly it is our own species that is responsible for precipitating dramatic environmental changes. As we change the environmental conditions on our planet, we alter the evolutionary trajectories of multitudes of species, including our own. We have long intentionally influenced the evolution of other species through **artificial selection**—the process of breeding plants and animals to accentuate desirable traits. Over the last several thousand years, we bred dogs from wolves, developed corn from wild teosinte, and evolved a variety of vegetable staples from wild plants as

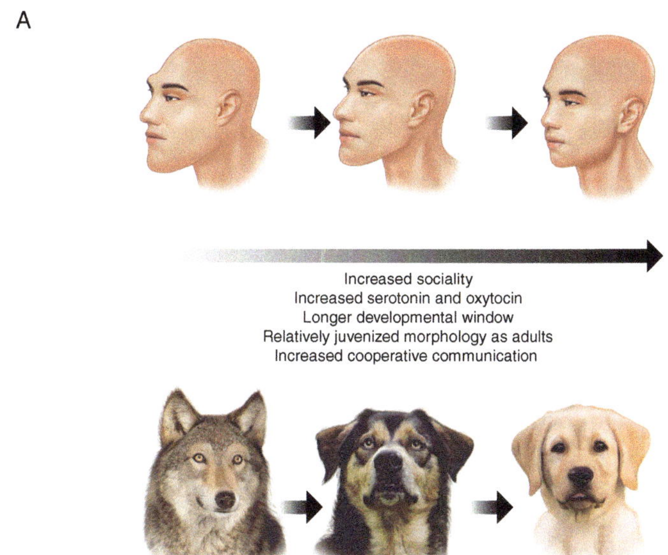

BOX FIGURE 3.6 Examples of domestication and artificial selection. (A) Parallels between the evolution of archaic to modern humans and the evolution of wolves to dogs. Modern humans show numerous trait trajectories that are commonly observed during domestication. (B) Modern humans have bred many other species for desirable traits. For example, from a single wild ancestor, humans have generated incredible variety in the cabbage family by selecting on different traits.

Reflection: The domestication of dogs and the cultivation of different cabbage varieties are examples where humans have directly—and intentionally—influenced the evolution of another species. What is a specific example where human activities have unintentionally influenced the evolution of another species?

Source: (A) "Survival of the Friendliest: Homo sapiens Evolved via Selection for Prosociality Brian Hare Annual Review of Psychology 2017 68:1, 155-186"

(Continued)

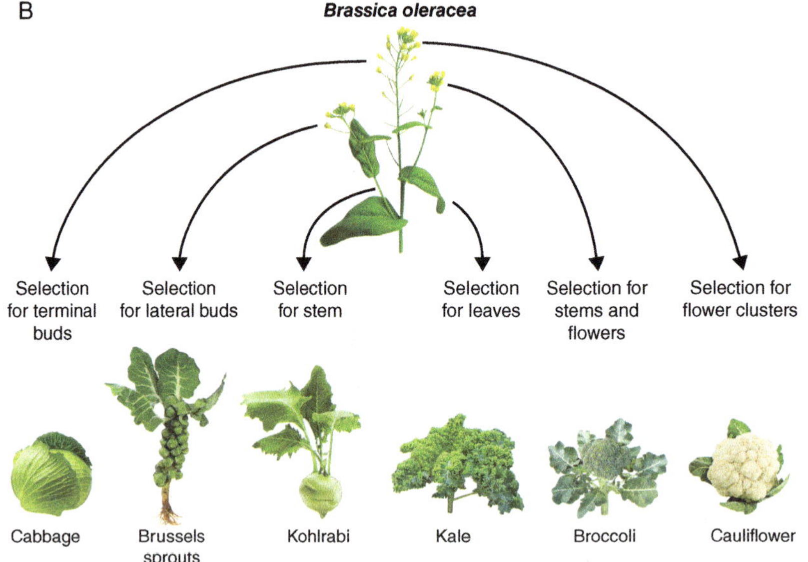

BOX FIGURE 3.6 Continued.

illustrated in Box Figure 3.6. But in today's world, *H. sapiens* is also unintentionally directing the evolution of biological diversity through changes to the environment (Palumbi 2001). For example, our use of antibiotics and chemical pesticides has selected for resistance in many microbes and invertebrates over very short timescales. Other examples abound, and we will evaluate contemporary human activities and how they alter the biosphere in the next chapter.

KEY CONCEPTS

When and how did early hominids evolve?

- Human and the chimpanzee lineages diverged from a common ancestor ~6–8 million years ago.
- Four human lineages are recognized (Ardipithecus, Australopithecus, Paranthropus, and Homo); each of these lineages contained multiple species, some of which overlapped temporally and geographically.

When and how did modern humans spread around the world?

- Our species—*Homo sapiens*—evolved in Africa more than 200,000 years ago and spread around the world likely in successive waves.

How did early human civilizations impact the environment?

- *Homo sapiens* have had large-scale impacts on the environment for thousands of years due to activities like hunting, agriculture, urbanization, and population growth.
- In the last 10,000 years, the global human population has grown from <10 million to >7 billion, an exponential increase with dramatic effects.

Core concepts: What is in a name?

- Humans belong to a family called the *Hominidae.*

Meet the data: Ice age genetics

- A global team of scientists sequenced dozens of genomes from *Homo sapiens* and found small—and decreasing—amounts of Neanderthal DNA in modern human genomes, indicating that Neanderthals and *Homo sapiens* interacted—and occasionally interbred—in Eurasia during the late Pleistocene (~50,000 years ago).

Taking a closer look: The evolutionary success of humans

- *Homo sapiens* have been remarkably successful due to a number of key innovations, including rapid evolution of brain size and complexity, use of fire, and development of symbolic communication.

CONSOLIDATE YOUR KNOWLEDGE

Answer the following questions to consolidate your knowledge, assess your progress meeting the learning outcomes, and reveal any areas in need of further exploration:

1. In your own words, write a one- or two-sentence synthesis of the big-picture take-away point of this chapter.
2. Revisit your answer to the *Blank Page* exercise in the beginning of this chapter. Would you refine your answer now based on knowledge you integrated from this chapter?
3. Organize the following key events in hominid evolution in chronological order and add approximate dates (in thousands or millions of years before present). Add three other key events of your choice and explain their contribution to the history of hominids.
   Split of human and great ape lineages
   Evolution of *Homo sapiens*
   Interaction between *H. sapiens* and *H. neanderthalensis*
   First use of fire
   First large-scale agriculture
   First human-mediated megafaunal extinctions
4. Often people erroneously think that humans evolved *from* apes. Describe the relationship between humans and chimpanzees using an evolutionary perspective.
5. Where and when did *H. sapiens* evolve? How and when did they spread around the world?
6. *H. sapiens* is just one of more than a dozen named human species. Do you think humans should be considered an evolutionary radiation?
7. What were some key ways that early human civilizations began to impact their surrounding environment even thousands of years ago?
8. Why do you think the Quaternary megafaunal extinctions are not considered in the "Big 5 Mass Extinction" category that we discussed in the last chapter?
9. What evidence can you provide that *H. sapiens* and *H. neanderthalensis* interbred? When in history and where geographically did this likely occur?
10. What are some factors that have contributed to the evolutionary success of modern humans? What are two additional factors that were not discussed in detail but that you think are traits that make humans evolutionarily unique?

11. In your own words, define the bolded terms in this chapter.
12. What are some questions that *you* have about the content in this chapter? If you found some content particularly challenging or particularly interesting, identify these as areas for additional reflection or reading.

LITERATURE CITED

Allentoft, Heller, Oskam, Lorenzen, Hale, Gilbert, Jacomb, Holdaway, & Bunce. 2014. "Extinct New Zealand Megafauna Were Not in Decline Before Human Colonization." *Proceedings of the National Academy of Sciences* 111 (13): 4922–27. doi.org/10.1073/pnas.1314972111.

Armelagos, Goodman, & Jacobs. 1991. "The Origins of Agriculture: Population Growth During a Period of Declining Health." *Population and Environment* 13 (1): 9–22. doi.org/10.1007/BF01256568.

Barnosky, Koch, Feranec, Wing, & Shabel. 2004. "Assessing the Causes of Late Pleistocene Extinctions on the Continents." *Science* 306 (5693): 70–75. doi.org/10.1126/science.1101476.

Bocquet-Appel. 2011. "When the World's Population Took Off: The Springboard of the Neolithic Demographic Transition." *Science* 333 (6042): 560–61. doi.org/10.1126/science.1208880.

Boulanger & Lyman. 2014. "Northeastern North American Pleistocene Megafauna Chronologically Overlapped Minimally with Paleoindians." *Quaternary Science Reviews* 85: 35–46. doi.org/10.1016/j.quascirev.2013.11.024.

Boyd & Silk. 2018. *How Humans Evolved*. New York: W. W. Norton & Company.

Brown, Sutikna, Morwood, Soejono, Jatmiko, Saptomo, & Due. 2004. "A New Small-Bodied Hominin from the Late Pleistocene of Flores, Indonesia." *Nature* 431 (7012): 1055. doi.org/10.1038/nature02999.

Cavalli-Sforza. 1998. "The DNA Revolution in Population Genetics." *Trends in Genetics* 14 (2): 60–65. doi.org/10.1016/S0168-9525(97)01327-9.

Chen & Li. 2001. "Genomic Divergences Between Humans and Other Hominoids and the Effective Population Size of the Common Ancestor of Humans and Chimpanzees." *The American Journal of Human Genetics* 68 (2): 444–56. doi.org/10.1086/318206.

Chen, Torroni, Excoffier, Santachiara-Benerecetti, & Wallace. 1995. "Analysis of MtDNA Variation in African Populations Reveals the Most Ancient of All Human Continent-Specific Haplogroups." *American Journal of Human Genetics* 57 (1): 133–49.

Diamond & Bellwood. 2003. "Farmers and Their Languages: The First Expansions." *Science* 300 (5619): 597–603. https://doi.org/10.1126/science.1078208.

Duncan, Blackburn, & Worthy. 2002. "Prehistoric Bird Extinctions and Human Hunting." *Proceedings of the Royal Society B: Biological Sciences* 269 (1490): 517–21. doi.org/10.1098/rspb.2001.1918.

Feakins & de Menocal. 2010. "Global and African Regional Climate During the Cenozoic." In *Cenozoic Mammals of Africa*, edited by Lars Werdelin, 45–55. Berkeley: University of California Press.

Ferraro, Plummer, Pobiner, Oliver, Bishop, Braun, Ditchfield, et al. 2013. "Earliest Archaeological Evidence of Persistent Hominin Carnivory." *PLoS ONE* 8 (4): e62174. doi.org/10.1371/journal.pone.0062174.

Fu, Posth, Hajdinjak, Petr, Mallick, Fernandes, Furtwängler, et al. 2016. "The Genetic History of Ice Age Europe." *Nature* 534 (7606): 200–205. doi.org/10.1038/nature17993.

Gignoux, Henn, & Mountain. 2011. "Rapid, Global Demographic Expansions After the Origins of Agriculture." *Proceedings of the National Academy of Sciences* 108 (15): 6044–49. doi.org/10.1073/pnas.0914274108.

Green, Krause, Briggs, Maricic, Stenzel, Kircher, Patterson, et al. 2010. "A Draft Sequence of the Neandertal Genome." *Science*

328 (5979): 710–22. doi.org/10.1126/science.1188021.

Gunz, Bookstein, Mitteroecker, Stadlmayr, Seidler, & Weber. 2009. "Early Modern Human Diversity Suggests Subdivided Population Structure and a Complex Out-of-Africa Scenario." *Proceedings of the National Academy of Sciences* 106 (15): 6094–98. doi.org/10.1073/pnas.0808160106.

Hare. 2017. "Survival of the Friendliest: *Homo Sapiens* Evolved via Selection for Prosociality." *Annual Review of Psychology* 68: 155–86. doi.org/10.1146/annurev-psych-010416-044201.

Hare. 2018. "Domestication Experiments Reveal Developmental Link Between Friendliness and Cognition." *Journal of Bioeconomics* 20 (1): 159–63. doi.org/10.1007/s10818-017-9264-9.

Harvati, Röding, Bosman, Karakostis, Grün, Stringer, Karkanas, et al. 2019. "Apidima Cave Fossils Provide Earliest Evidence of Homo Sapiens in Eurasia." *Nature* 571: 500–504. doi.org/10.1038/s41586-019-1376-z.

Henshilwood & d'Errico. 2011. *Homo Symbolicus: The Dawn of Language, Imagination and Spirituality*. Philadelphia, PA: John Benjamins.

Hublin, Ben-Ncer, Bailey, Freidline, Neubauer, Skinner, Bergmann, et al. 2017. "New Fossils from Jebel Irhoud, Morocco and the Pan-African Origin of Homo Sapiens." *Nature* 546 (7657): 289–92. doi.org/10.1038/nature22336.

Jablonski & Chaplin. 2017. "The Colours of Humanity: The Evolution of Pigmentation in the Human Lineage." *Philosophical Transactions of the Royal Society B: Biological Sciences* 372 (1724). doi.org/10.1098/rstb.2016.0349.

King & Bailey. 2006. "Tectonics and Human Evolution." *Antiquity* 80 (308): 265–86. doi.org/10.1017/S0003598X00093613.

Koch & Barnosky. 2006. "Late Quaternary Extinctions: State of the Debate." *Annual Review of Ecology, Evolution, and Systematics* 37 (1): 215–50. doi.org/10.1146/annurev.ecolsys.34.011802.132415.

Langergraber, Prufer, Rowney, Boesch, Crockford, Fawcett, Inoue, et al. 2012. "Generation Times in Wild Chimpanzees and Gorillas Suggest Earlier Divergence Times in Great Ape and Human Evolution." *Proceedings of the National Academy of Sciences* 109 (39): 15716–21. doi.org/10.1073/pnas.1211740109.

Levin. 2015. "Environment and Climate of Early Human Evolution." *Annual Review of Earth and Planetary Sciences* 43: 405–29. doi.org/10.1146/annurev-earth-060614-105310.

Lewin & Foley. 2004. *Principles of Human Evolution*. 2nd ed. Malden, MA: Blackwell.

Lima-Ribeiro & Diniz-Filho. 2013. "American Megafaunal Extinctions and Human Arrival: Improved Evaluation Using a Meta-Analytical Approach." *Quaternary International* 299: 38–52. doi.org/10.1016/j.quaint.2013.03.007.

Locey & Lennon. 2016. "Scaling Laws Predict Global Microbial Diversity." *Proceedings of the National Academy of Sciences* 113 (21): 5970–75. doi.org/10.1073/pnas.1521291113.

López, van Dorp, & Hellenthal. 2015. "Human Dispersal out of Africa: A Lasting Debate." *Evolutionary Bioinformatics* 11: EBO-S33489. doi.org/10.4137/EBo.s33489.

Lovejoy, Suwa, Spurlock, Asfaw, & White. 2009. "The Pelvis and Femur of *Ardipithecus ramidus:* The Emergence of Upright Walking." *Science* 326 (5949): 71–71e6. doi.org/10.1126/science.1175831.

Martin & Klein. 1989. *Quaternary Extinctions: A Prehistoric Revolution*. Tuscon, AZ: University of Arizona Press.

Maslin, Brierley, Milner, Shultz, Trauth, & Wilson. 2014. "East African Climate Pulses and Early Human Evolution." *Quaternary Science Reviews* 101: 1–17. doi.org/10.1016/j.quascirev.2014.06.012.

Mekel-Bobrov, Gilbert, Evans, Vallender, Anderson, Hudson, Tishkoff, & Lahn. 2005. "Ongoing Adaptive Evolution of ASPM, a Brain Size Determinant in Homo Sapiens." *Science* 309 (5741): 1720–22. doi.org/10.1126/science.1116815.

Mora, Tittensor, Adl, Simpson, & Worm. 2011. "How Many Species Are There on Earth and in the Ocean?" *PLoS Biology* 9 (8): e1001127. doi.org/10.1371/journal.pbio.1001127.

Oates. 1934. "The Population of Rome." *Classical Philology* 29 (2): 101–16.

Palumbi. 2001. "Humans as the World's Greatest Evolutionary Force." *Science* 293 (5536): 1786–90. doi.org/10.1126/science.293.5536.1786.

Petr, Pääbo, Kelso, & Vernot. 2019. "Limits of Long-Term Selection Against Neandertal Introgression." *Proceedings of the National Academy of Sciences* 116 (5): 1639–44. doi.org/10.1073/pnas.1814338116.

Plummer, Ditchfield, Bishop, Kingston, Ferraro, Braun, Hertel, & Potts. 2009. "Oldest Evidence of Toolmaking Hominins in a Grassland-Dominated Ecosystem." *PLoS ONE* 4 (9): e7199. doi.org/10.1371/journal.pone.0007199.

Prüfer, De Filippo, Grote, Mafessoni, Korlević, Hajdinjak, Vernot, et al. 2017. "A High-Coverage Neandertal Genome from Vindija Cave in Croatia." *Science* 358 (6363): 655–58. doi.org/10.1126/science.aao1887.

Prüfer, Racimo, Patterson, Jay, Sankararaman, Sawyer, Heinze, et al. 2014. "The Complete Genome Sequence of a Neanderthal from the Altai Mountains." *Nature* 505 (7481): 43. doi.org/10.1038/nature12886.

Ramachandran, Deshpande, Roseman, Rosenberg, Feldman, & Cavalli-Sforza. 2005. "Support from the Relationship of Genetic and Geographic Distance in Human Populations for a Serial Founder Effect Originating in Africa." *Proceedings of the National Academy of Sciences* 102 (44): 15942–47. doi.org/10.1073/pnas.0507611102.

Reyes-Centeno, Ghirotto, Detroit, Grimaud-Herve, Barbujani, & Harvati. 2014. "Genomic and Cranial Phenotype Data Support Multiple Modern Human Dispersals from Africa and a Southern Route into Asia." *Proceedings of the National Academy of Sciences* 111 (20): 7248–53. doi.org/10.1073/pnas.1323666111.

Reyes-Centeno, Hubbe, Hanihara, Stringer, & Harvati. 2015. "Testing Modern Human Out-of-Africa Dispersal Models and Implications for Modern Human Origins." *Journal of Human Evolution* 87: 95–106. doi.org/10.1016/j.jhevol.2015.06.008.

Richter, Grün, Joannes-Boyau, Steele, Amani, Rué, Fernandes, et al. 2017. "The Age of the Hominin Fossils from Jebel Irhoud, Morocco, and the Origins of the Middle Stone Age." *Nature* 546: 293–96. doi.org/10.1038/nature22335.

Rowthorn & Seabright. 2010. "Property Rights, Warfare and the Neolithic Transition." TSE Working Papers 10-207, Toulouse School of Economics (TSE).

Ruddiman. 2003. "The Anthropogenic Greenhouse Era Began Thousands of Years Ago." *Climatic Change* 61 (3): 261–93. doi.org/10.1023/B:CLIM.0000004577.17928.fa.

Sandom, Faurby, Sandel, & Svenning. 2014. "Global Late Quaternary Megafauna Extinctions Linked to Humans, Not Climate Change." *Proceedings of the Royal Society B: Biological Sciences* 281 (1787). doi.org/10.1098/rspb.2013.3254.

Schlebusch, Malmström, Günther, Sjödin, Coutinho, Edlund, Munters, et al. 2017. "Southern African Ancient Genomes Estimate Modern Human Divergence to 350,000 to 260,000 Years Ago." *Science* 358 (6363): 652–55. doi.org/10.1126/science.aao6266.

Spoor, Leakey, Gathogo, Brown, Antón, McDougall, Kiarie, Manthi, & Leakey. 2007. "Implications of New Early Homo Fossils from Ileret, East of Lake Turkana, Kenya." *Nature* 448 (7154): 688–91. doi.org/10.1038/nature05986.

Steadman. 1995. "Prehistoric Extinctions of Pacific Island Birds: Biodiversity Meets Zooarchaeology." *Science* 267 (5201): 1123–31. doi.org/10.1126/science.267.5201.1123.

Sullivan, Bird, & Perry. 2017. "Human Behaviour as a Long-Term Ecological Driver of Non-Human Evolution." *Nature Ecology and Evolution* 1 (3): 0065. doi.org/10.1038/s41559-016-0065.

Tanabe, Mita, Jombart, Eriksson, Horibe, Palacpac, Ranford-Cartwright, et al. 2010. "Plasmodium Falciparum Accompanied the Human Expansion out of Africa." *Current Biology* 20 (14): 1283–89. doi.org/10.1016/j.cub.2010.05.053.

Teaford & Ungar. 2000. "Diet and the Evolution of the Earliest Human Ancestors." *Proceedings of the National Academy of Sciences*

97 (25): 13506–11. doi.org/10.1073/pnas.260368897.

The Chimpanzee Sequencing and Analysis Consortium. 2005. Initial sequence of the chimpanzee genome and comparison with the human genome. *Nature* 437: 69–87. doi.,org/10.1038/nature04072

United States Census Bureau. 2013. "Historical Estimates ofWorld Population." 2013. https://www.census.gov/population/international/data/worldpop/table_history.php.

United Nations Department of Economic and Social Affairs. 2015. World Population Prospects. https://population.un.org/wpp/Publications/Files/Key_Findings_WPP_2015.pdf

Weisdorf. 2005. "From Foraging to Farming: Explaining the Neolithic Revolution." *Journal of Economic Surveys* 19 (4): 561–86. doi.org/10.1111/j.0950-0804.2005.00259.x.

Wirth. 1938. "Urbanism as a Way of Life (L'urbanesimo Come Modo Di Vita)." *American Journal of Sociology* 44 (1): 1–24. doi.org/10.1086/217913.

Wrangham. 2009. *Catching Fire: How Cooking Made Us Human*. New York: Basic Books.

Zheng, Yan, Qin, Wang, Tan, Li, & Jin. 2011. "Major Population Expansion of East Asians Began Before Neolithic Time: Evidence of MtDNA Genomes." *PLoS ONE* 6 (10): e25835. doi.org/10.1371/journal.pone.0025835.

The Anthropocene

Learning Outcomes

After working with this chapter, you will be able to:

- Describe how the Anthropocene is unique from past eras.
- Analyze the impacts of contemporary human societies on biodiversity and the environment.
- Compare and contrast causes of historical and contemporary climate change.
- Apply your knowledge to real-world case studies and interpret data from recent scientific studies.

THE BLANK PAGE

Over the last several hundred years, human influence on the planet has accelerated dramatically. What changed? Before reading this chapter, brainstorm what events or innovations brought humans from the end of the Neolithic Revolution (where we left off in the last chapter) to today. Then make a list of contemporary human activities that have *global* effects. When you are making this list, consider how local or regional activities can have global impacts.

INTRODUCTION

Human populations have always influenced their ecosystems. Simply by eating, breathing, and excreting nutrients into the soil, humans—like all animals—alter their *local* surroundings. But over time, human activities began having a disproportionate influence on the Earth's environment. As we saw in the previous chapter, widespread hunting, farming, urbanization, and population growth in the last 10,000 years began having *regional* effects. However, over the last several hundred years, human impacts on the planet magnified to have systemic effects at a *global* scale. The goal of this chapter is to evaluate the anthropogenic stressors of the last 200 years as *Homo sapiens* have become a planetary force. These impacts have urgent implications not only for human society but for all biological diversity on Earth.

WHAT IS THE ANTHROPOCENE AND WHEN DID IT BEGIN?

The **Anthropocene** is the interval of time where human activities dominate key processes on Earth. The scientific community has recognized for well over 100 years that humans have large-scale impacts on the global environment (Marsh 1864). In the last several decades, the concept of the Anthropocene has been used to formalize our understanding of human effects on the biosphere (e.g., Crutzen 2002; Zalasiewicz et al. 2011). Considering the Anthropocene as a formal geological era is a recognition that our species has altered the planet in dramatic ways—at a magnitude comparable to the changes seen between other geological epochs. In fact, a proposal to declare the Anthropocene as a formal geological epoch was made in 2016 to the International Geological Congress.

Figure 4.1 shows the increase in anthropogenic activities on our planet and the transitions between geological epochs. The last transition between geological epochs

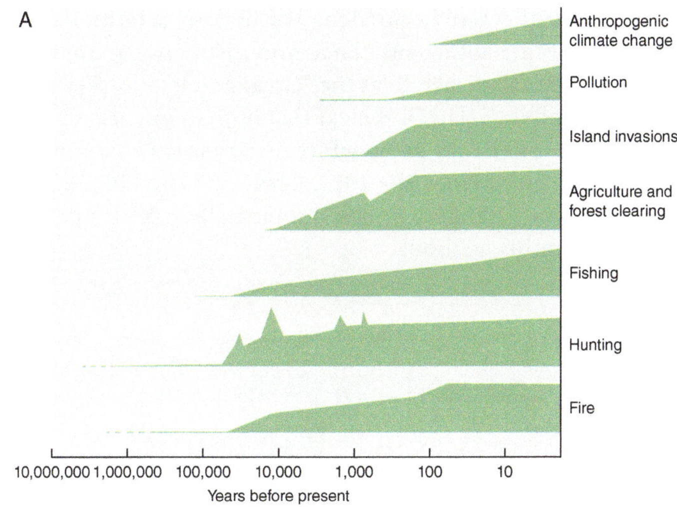

B

ERA	PERIOD	EPOCH	BEGAN (Years ago)
Cenozoic	Quaternary	Anthropocene	??
		Holocene	11,700
		Pleistocene	2.5 M
	Tertiary	Pliocene	5.3 M
		Miocene	23 M
		Oligocene	34 M
		Eocene	56 M
		Paleocene	65.5 M
Mesozoic	Cretaceous		146 M
	Jurassic		200 M
	Triassic		251 M
Paleozoic			542 M
Proterozoic			4.5 B

FIGURE 4.1 (A) The development of key anthropogenic activities over time. (B) Geological epochs, including the proposed Anthropocene (M = millions and B = billiions of years ago).

Reflection: Based on the increase in key anthropogenic modifications to the global environment, when do you think the Anthropocene began?

Source: (A) "Global Biodiversity Change: The Bad, the Good, and the Unknown Henrique Miguel Pereira, Laetitia Marie Navarro, and Inês Santos Martins Annual Review of Environment and Resources 2012 37:1, 25–50"; (B) Climate & Capitalism (https://climateandcapitalism.com)

was from the Pleistocene to the Holocene approximately 12,000 years ago. The Pleistocene was an **Ice Age** with repeated glacial cycles. At the largest glacial extent, up to 30% of the Earth's surface was covered in ice, with ice sheets in some places multiple kilometers thick (Clark & Mix 2002). Scientists now recognize that human activities are indeed changing our planet in ways comparable to other major geological transition points. Humans are leaving a mark on the geological record, and as Dr. Anthony Barnosky says, *"We have become a geological force greater than what ends an Ice Age."*

When did the Anthropocene begin? There is a rich scientific debate about dating the Anthropocene. Earlier studies focused on the Industrial Revolution (~1750 Common Era) as a logical start date for the Anthropocene (Steffen et al. 2011). More recent studies have suggested that the geological signal of human activities on the biosphere and climate system became sufficiently large and distinctive more recently—around 1950—ushering in a period referred to as the "Great Acceleration" (Steffen et al. 2015). Regardless of the specific start date, it is clear that human impacts on the Earth's environment likely met the criteria for an epoch-scale transition sometime in the last 400 years (Zalasiewicz et al. 2011; Barnosky 2013; Lewis & Maslin 2015). The *Core Concepts* feature provides an introduction to climate, a topic we will return to at the end of this chapter and throughout this textbook.

CORE CONCEPTS

WHAT IS CLIMATE AND HOW IS IT MEASURED?

Climate can be defined as "the mean and variability of temperature, precipitation and wind over a period of time, ranging from months to millions of years" (Intergovernmental Panel on Climate Change 2014). This definition is more accurate than the simplistic definition of climate as "average weather" because it captures the importance of defining which abiotic elements are under consideration, what timescale is being evaluated, and what is measured about these elements (for example, mean versus variability). However, even this definition does not capture the complexity of the climate system. Ultimately, as Box Figure 4.1 shows, the climate system is complex and interactive and is influenced by the atmosphere, the terrestrial surface of the Earth, the aquatic bodies of water, the cryosphere (snow and ice), and also by living things.

Instrumentation to conduct detailed measurements of climate has only been available for the last ~150 years. Instruments such as thermometers, data loggers, weather stations, and satellites can provide accurate measurements of our contemporary climate, but how do scientists study past climate? Reconstruction of past climate requires proxies as shown in Box Figure 4.2 records that can be directly observed and then used to infer past conditions. Some examples of climate proxies include the following.

GLACIAL EVIDENCE Glacial evidence such as moraines (rock and soil debris) can be used to infer the presence of past ice sheets. For example, terminal moraines are often found at the end of a glacier, thereby providing a historical record of the timing of glacial advance and retreat.

ICE AND SEDIMENT CORES Vertical samples called cores can be used as stratigraphic records to reconstruct past climates. Ice and sediment cores contain evidence such as pollen, plankton, air bubbles, or ice crystals that can all be used to infer past environmental conditions.

BANDING PATTERNS A historical record of climate is preserved in the physical structure of some living things. For example, temperate tree rings record information about moisture and temperature. Similarly, some tropical corals exhibit banding patterns that are influenced by water temperature and nutrient availability.

FOSSILS The distribution of climate-sensitive species can be inferred from pollen and fossil records. For example, fossil evidence of aquatic turtles and alligators in far northern latitudes suggests that the Arctic Circle was dramatically warmer 50 million years ago.

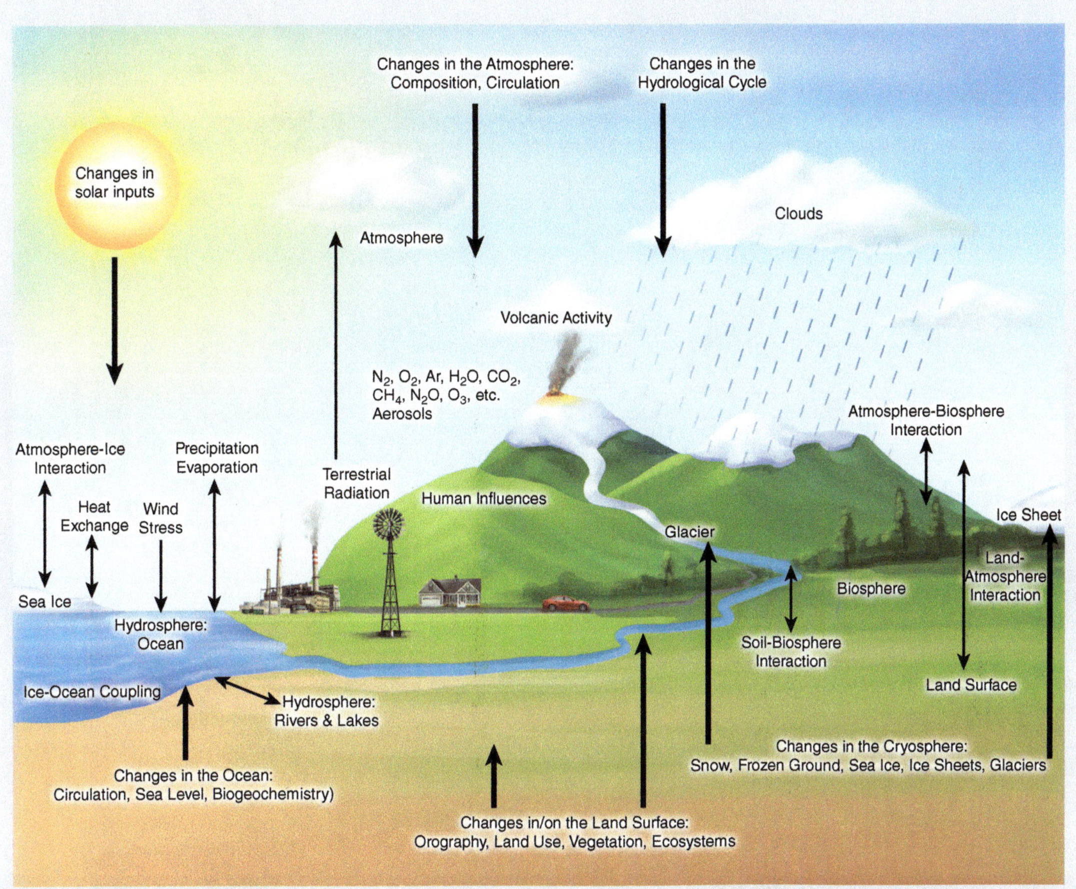

BOX FIGURE 4.1 The climate system is a complex interactive system that is impacted by changes in numerous systems and cycles.

Reflection: Give three examples of changes in the climate system that occurred before humans arrived.

Source: Figure 1.2 from Le Treut, H., R. Somerville, U. Cubash, Y. Ding, C. Mauritzen, A. Mokssit, T. Peterson, and M. Prather, 2007: Historical Overview of Climate Change: The Physical Science Basis. Contribution of Working Group I to the Fourth Assessment Report of the Intergovernmental Panel on Climate Change [Solomon, S., D. Qin, M. Manning, Z. Chen, M. Marquis, K.B. Averyt, M. Tignor and H.L. Miller (eds.)]. Cambridge University Press, Cambridge, United Kingdom and New York, NY, USA.

No one method can provide detailed past climate data across all time frames and all geographic locations. However, methods for climate reconstruction are reliable enough to understand the history of our planet's climate over hundreds of millions of years, an adequate timescale to study patterns and trends in the Earth's climate before humans came on the scene.

In addition, modeling approaches for studying climate have improved dramatically in their complexity and resolution over the last several decades, as illustrated in Box Figure 4.2. Scientists can now understand with impressive nuance how changes in biotic and abiotic factors (e.g., patterns of vegetation, cloud cover, ocean circulation) impact climate metrics (e.g., temperature, precipitation, snow melt). These advances allow superior reconstruction of past climate and more accurate projection of future climate. We discuss historical and contemporary patterns of climate change further in this chapter's *Taking a Closer Look*.

(Continued)

BOX FIGURE 4.2 How scientists measure and model the climate system. (A) Examples of evidence scientists use to reconstruct past climate, including glacial moraines, tree banding patterns, and sedimentation or ice cores. (B) Climate models have increased in resolution and accuracy over the last few decades so that climate scientists can model climate processes with a high degree of complexity and realism. Models increasingly use finer scale grids to understand local climate processes. Models also increasingly incorporate interactions among different factors that influence the climate system.

Reflection: Describe how scientists can use evidence from both abiotic and biotic components of the ecosystem to reconstruct past climate.

Source: (A) NPS Photo/James W. Frank; (B) https://www.visualisingdata.com/2015/02/dendrochronology-visualisation-literacy/

WHAT ARE PATTERNS OF CONTEMPORARY POPULATION GROWTH?

As we saw in the last chapter, *Homo sapiens* began to have a significant impact on the environment tens of thousands of years ago as activities like hunting and agriculture altered landscapes and patterns of biodiversity. Human impacts accelerated during the Neolithic Revolution as agricultural innovation led to dense human settlements and ultimately tremendous population growth. By 2,000 years ago, ancient cities like Rome were thriving, and the global human population had grown to several hundred million. However, the shift from subsistence living and hand production techniques (like pottery and carpentry) to manufacturing and large-scale production would not begin in earnest for thousands of years.

The **Industrial Revolution**, which began around 1750, would alter human society and, ultimately, the planet. During this period, a number of innovations first appeared, including steam engines, textile manufacturing, more efficient methods for mining and resource extraction, gas lighting, cement, roads, and the first railways. The global human population reached the 1 billion mark in the early 1800s (United States Census Bureau 2013; United Nations Department of Economic and Social Affairs 2015). From the days of the Roman Empire to the end of the Industrial Revolution (<1% of human history), the population increased more than fivefold.

One hundred years later, the Second Industrial Revolution (sometimes referred to as the Technological Revolution) brought large-scale factory production, electricity, automobiles, and the first telecommunication systems. Steam turbines and diesel engines facilitated an increasingly globalized transportation network across both land and sea. By the mid-20th century, the human population had again doubled.

Now, another 100 years later, human activities have expanded to nearly all regions of the planet. Table 4.1 shows that over the last 200 years, the interval of time to add 1 billion people has been shrinking. There are now more than 300,000 human births per day and the global population is now well over 7 billion (United States Census Bureau 2013; United Nations Department of Economic and Social Affairs 2015). This most recent fivefold population increase occurred in only 0.1% of *H. sapiens* history.

TABLE 4.1 Global Human Population Growth Since the Industrial Revolution with Growth Rate Depicted as Time to Add an Additional 1 Billion People

World Population Milestone	Approximate Year	Time to Add 1 Billion
1 billion	1800	>100,000 years
2 billion	1930	123 years
3 billion	1960	33 years
4 billion	1975	14 years
5 billion	1990	13 years
6 billion	2000	12 years
7 billion	2011	11 years

Reflection: Calculate the time it took for the human population to double from 1 to 2 billion and from 2 to 4 billion. What is your prediction—and rationale—for when the population will have doubled again to 8 billion?

Source: United States Census Bureau 2013

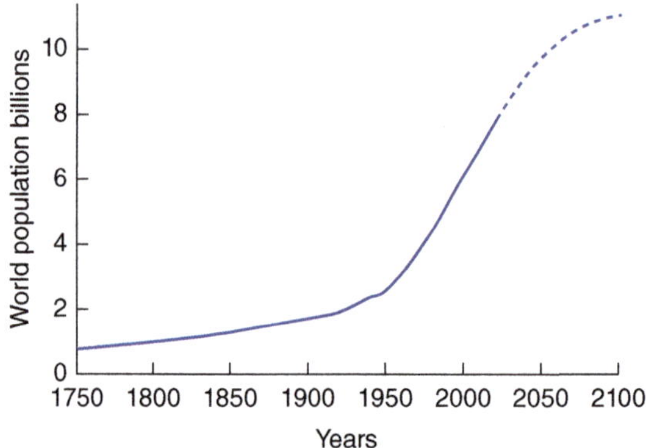

FIGURE 4.2 World population growth over the last several hundred years (solid line) and projected future population growth (dotted line).

Reflection: If the human population increases by nearly 50% in the next hundred years, which specific natural resources do you think will come under the most strain?

Source: https://ourworldindata.org/world-population-growth

The human population is projected to continue growing into the next century, as illustrated in Figure 4.2. Current models predict the global population will exceed 11 billion by 2100. However, most forecast that population growth will occur at a slower rate. Although life expectancy has increased rapidly over the last century, fertility rates have declined, slowing overall growth. The global average for number of children per woman is currently less than 2.5, less than half the fertility rate 50 years ago. Of course, mortality and fertility rates vary dramatically across regions, and future population trajectories depend dramatically on many biological, socioeconomic, and cultural factors. Most models show only a small probability that the world population will begin to stabilize or fall by 2100. Thus, at least in the short term, the human population will continue to grow on a planet with finite space and natural resources.

HOW ARE CONTEMPORARY HUMAN CIVILIZATIONS IMPACTING THE ENVIRONMENT?

The exponential growth of the human population has led to increasing pressure on natural systems. In the last chapter, we looked at examples of how *early* human civilizations impacted their environment. We evaluated several key activities (e.g., hunting, agriculture, urbanization) that began altering the natural environment at a *regional scale* over the last 50,000 years. To complement that analysis, we will now briefly look at examples of how *contemporary* human civilizations are modifying the planetary environment. Here we will review three exemplar activities that have implications at a *global scale* in contemporary times.

Land-Use Change

In the last chapter, we looked at how agriculture and urbanization began to have regional effects thousands of years ago. Habitat modification is not new, as humans have long altered landscapes by using fire, clearing land for farming, and building settlements. However, in the last several hundred years, the scale and pace of land-use change have dramatically accelerated.

Figure 4.3 shows that the period of first significant land-use change for most regions of the planet occurred within the last few hundred years. In fact, by the year 2000, nearly 75% of the terrestrial surface of the planet was altered by human activities and more than 55% experienced heavy use as settlements, croplands, or rangelands

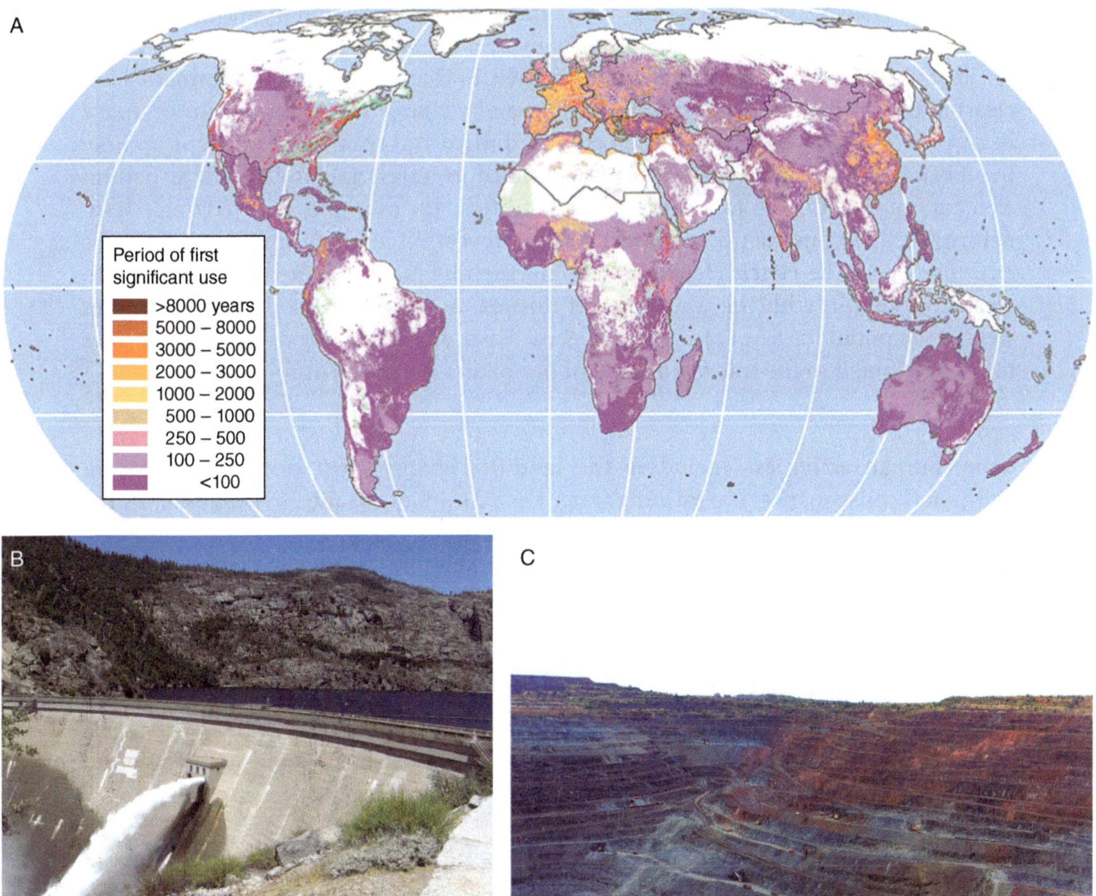

FIGURE 4.3 Modern humans have dramatically altered the terrestrial landscape. (A) Anthropogenic changes to the terrestrial landscape over the last 8,000 years, highlighting period of first significant use. (B–C) Land-use change can take many forms, including creating dams and reservoirs (such as the flooding of Hetch Hetchy Valley in Yosemite National Park) and iron and gold mining (as illustrated by a mine in Brazil).

Reflection: In the top panel, compare and contrast the pattern of anthropogenic development over time in different regions of the world. Which regions have experienced more or less habitat modification over time, and what might explain regional differences in period of first significant land-use change? Have you personally observed any land use change within your own lifetime?

Source: (A) "Used planet: A global history Erle C. Ellis, Jed O. Kaplan, Dorian Q. Fuller, Steve Vavrus, Kees Klein Goldewijk, Peter H. Verburg Proceedings of the National Academy of Sciences May 2013, 110 (20) 7978–7985."; (B) Inklein/Wikipedia; (C) TR STOK/Shutterstock

(Ellis 2011). Some biomes have seen disproportionate impacts; for example, <10% of native grasslands remain, most having been altered to range or croplands (Ellis 2011).

Terrestrial land-use change can take many forms. Many activities, like deforestation, mining, and urban development, alter terrestrial habitat structure in obvious and often irreversible ways. One dramatic example is land-use change associated with resource extraction. As the human population has grown, so, too, have industrialized systems for food production and extraction of natural resources. Figure 4.3 shows an example of land-use change associated with an iron and gold mining region in Brazil, where the footprint of mining has expanded dramatically over the last 20 years (Sonter et al. 2014).

It is also important to note that even activities that do not involve dramatic land clearing can have large-scale effects. For example, changes in irrigation or drainage in croplands can alter patterns of erosion, sediment transport, and nutrient cycling. In addition, habitat modification is not restricted to terrestrial landscapes. Structural elements of aquatic habitats can also be modified, for example, by dredging, channelization, and construction of reservoirs and levees. Figure 4.3 provides a dramatic example where the Hetch Hetchy Valley in Yosemite National Park was dammed to create a reservoir, with the water moved through an aqueduct more than 150 miles west to the populous San Francisco Bay Area. Thus, a wide variety of human activities have contributed to the transformation of our planet in the Anthropocene.

Pollution

Many human activities—including the land-use changes just discussed—have an additional consequence: the introduction of contaminants to the environment. In the extreme, synthetically altered substances can be accidentally released, causing large-scale devastation. One dramatic example is illustrated in Figure 4.4. During the 1991 Gulf War, sustained gushes from oil wells in Kuwait and Iraq released millions of gallons of crude oil, creating hundreds of standing oil lakes. Another example occurred more recently in your lifetime: the devastating BP oil spill in the Gulf of Mexico in 2010. This oil spill released more than 200 million gallons of oil (approximately a half million *tons*) and is likely the largest marine oil spill in history. Similar large-scale contamination events—from train derailments to nuclear waste leaks—have affected many different ecosystems around the world. These catastrophic events can have devastating consequences for wildlife, human health, and local economies.

Chemical contamination can occur from a **point source**—as in an oil spill—but can also come from diffuse inputs. For example, coastal waters are highly impacted by run-off from adjacent terrestrial landscapes. Literally thousands of different synthetic chemicals are found in aquatic environments, including insecticides, flame retardants, gasoline additives, pharmaceuticals, and illicit drugs (Dachs & Mejanelle 2010). Some of these contaminants, like pesticides and fertilizers, were intentionally added to residential and agricultural landscapes. But others simply wash unintentionally through our streets into our storm drains (or through our bodies into our septic system). These chemical contaminants, which leach into the soil and water (including freshwater, marine, and groundwater systems), can have both short- and long-term effects on human and ecosystem health.

Pollution can also be an intensification of an otherwise naturally occurring element, whether a chemical compound or an energetic input (such as light, heat, or noise). Perhaps the most dramatic example is the anthropogenic release of large

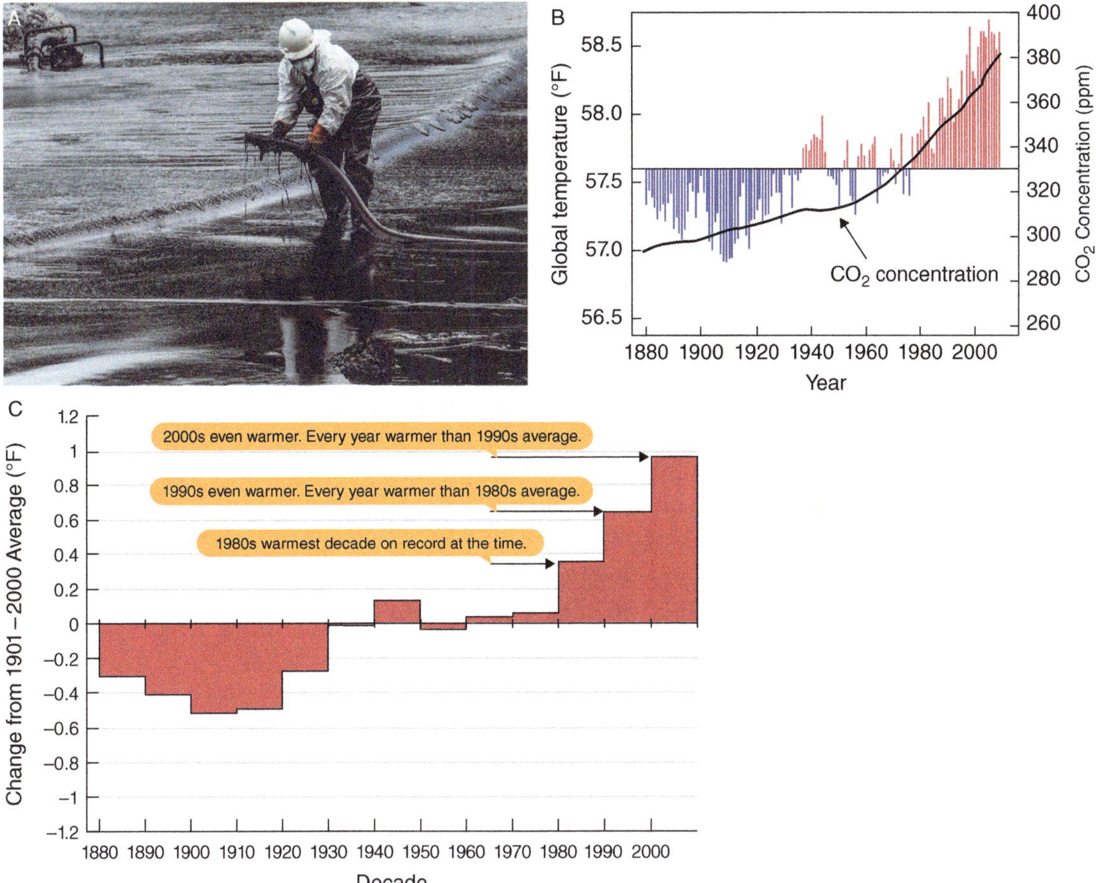

FIGURE 4.4 Pollutants have impacts on terrestrial, marine, and atmospheric systems. (A) Point source pollution can occur through regular release of contaminants into the environment or from large perturbations like oil spills. (B) Anthropogenic emissions of carbon dioxide has led to increasing carbon dioxide concentration in the atmosphere and increased global temperatures. One graph shows increasing carbon concentrations (black line) and average global temperature (colored bars). The other shows how each of the last several decades has been the warmest on record.

Reflection: Roughly how much temperature change has occurred over the last 100 years? What factors might determine how much temperature change will occur in the next 100 years?

Source: (A) Shutterstock.com; (B) Wu et al. (2017), https://onlinelibrary.wiley.com/doi/full/10.1002/advs.201700194; (C) https://cellcode.us/quotes/temperature-world-average-change.html

quantities of carbon dioxide into our atmosphere. Carbon dioxide is a naturally occurring greenhouse gas, but the concentration of carbon dioxide in the atmosphere has increased steadily over the last 150 years (Intergovernmental Panel on Climate Change 2014). The increase is due primarily to burning fossil fuels such as oil, coal, and natural gas in the transportation and industry sectors. However, deforestation also contributes by disrupting photosynthesis and releasing stored carbon through biomass burning.

Increasing carbon dioxide has led to an unambiguous perturbation of the climate system: global warming. Figure 4.4 illustrates the link between increasing carbon dioxide concentration and global temperature. Global mean surface temperatures have risen approximately 1 degree Celsius over the last century, and every decade since the 1980s has been the warmest decade on record. Anthropogenic changes to the climate system are an imminent and global threat. The *Taking a Closer Look* feature at the end of this chapter provides a more thorough introduction to anthropogenic climate change, a pressing issue we will return to throughout the book.

Globalization

As we explored in Chapter 3, it took generations of early *Homo sapiens* tens of thousands of years to disperse around the Earth. Today, a single individual can circumnavigate the globe in days. **Globalization** is the interaction and increasing interdependence of people, organizations, and governments worldwide.

At its heart, globalization has resulted from the increasing movement of people, goods, services, technologies, and knowledge around the world. Figure 4.5 illustrates the extent of our contemporary terrestrial road, marine shipping, and air travel networks. The economic, cultural, and political impacts of this globalized transportation system are enormous (Rodrigue et al. 2016), but most relevant to our exploration are effects of globalization on the natural world.

Some impacts of globalization on natural systems relate to the previous topics considered: globalized infrastructure and trade clearly lead to both land-use change and pollution. However, globalization also presents unique stressors to biotic systems. Not only do humans intentionally move plants and animals across international borders (e.g., for agriculture and the pet trade), but we also unintentionally move an incredible number of species (including pests and pathogens) around the world.

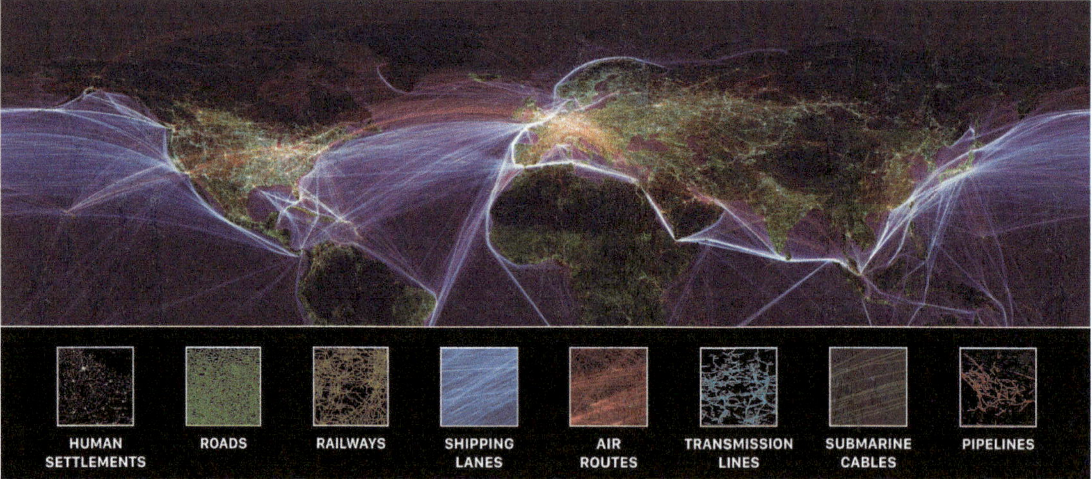

FIGURE 4.5 Globalized transportation systems. Terrestrial roads, marine shipping routes, and air travel networks cover the globe and have led to an increasingly interconnected world.

Reflection: List three intentional and three unintentional consequences of our globalized transportation system.

Source: globaia.org

Thus, globalization has increased the number of species introduced to new regions and also facilitated the spread of numerous diseases of humans, wildlife, and agriculturally important crops. One compelling example is *Toxoplasma gondii*, a parasite that can infect the majority of mammal and bird species worldwide and can be fatal to humans. Although there are different lineages of *T. gondii* found around the world, one of these lineages has recently radiated globally (Lehmann et al. 2006). Transatlantic slave trading ships in the 16th century are a likely culprit as their dense populations of rodents and cats likely provided this parasite with incredible opportunities for expansion to new places. We will return to the topics of invasive species and novel disease spread in later chapters.

Additional Stressors

It is important to remember that there are many additional anthropogenic activities that have global impacts on living systems. Moreover, there are nuances even within the broad categories we have discussed (for example, land-use change for agriculture will have different effects than land-use change for urbanization). We will address many other specific global change stressors—and the potential for complex relationships among them—throughout the remainder of the book. In particular, we will analyze the *effects* of these stressors on living systems at different scales in Units II and III.

HOW DO ANTHROPOGENIC STRESSORS INTERACT WITH EACH OTHER?

Many anthropogenic stressors are interrelated, and different stressors can act in concert. Imagine an ecosystem experiencing three stressors—habitat loss, hunting, and climate change—at the same time. Figure 4.6 illustrates three different possible relationships between these stressors and their combined impact. This example uses biodiversity loss as an impact metric, but many different outcomes could be measured.

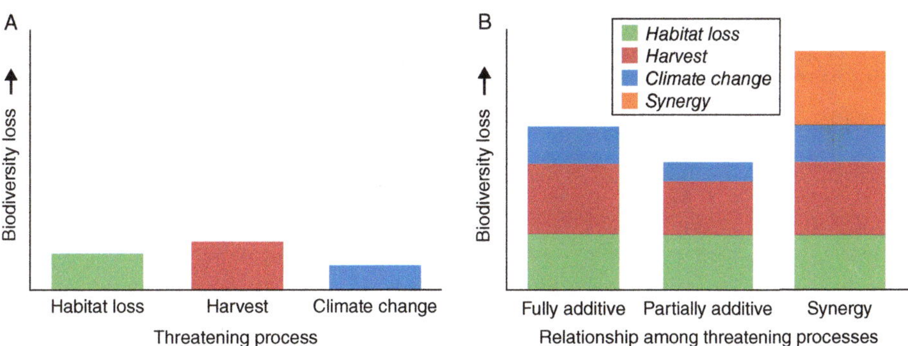

FIGURE 4.6 Anthropogenic stressors can act individually (A) or in combination (B). When multiple stressors impact the same system, they can exhibit different effects, illustrated here with hypothetical data. Fully additive is where the total effect is exactly the sum of the individual effects. Partially additive is where the total effect is less than the sum of the individual effects. Synergy is when the total effect is more than the sum of individual effects.

Reflection: What is a realistic example of a situation where a species might be resilient to individual threats but not to their combined effects?

Source: Brook, Sodhi & Bradshaw. Synergies among extinction drivers under global change. Trends in Ecology and Evolution 23 (8): 453–460.

The combined effect of multiple stressors can be equal to, less than, or greater than the sum of their individual effects. A **fully additive effect** is when the total impact of the stressors is exactly the sum of the individual effects. A **partially additive effect** is where the total effect is less than the sum of the individual effects. Partially additive effects can occur because a species cannot be lost twice (e.g., once to overharvest and again to climate change). A **synergistic effect** is when the combined effect of multiple factors is greater than the sum of their individual impacts. Synergies among stressors can be small or large and can be caused by any number of interactions. For example, perhaps habitat loss was due to new road construction and the roads provided improved access for hunters or harvesters. Or perhaps some species were stressed by warming temperatures, leaving them more vulnerable to capture by hunters. Ultimately, there are many ways in which the impacts of one stressor can be amplified by another.

Synergistic effects are common in the Anthropocene and can also result from complex reverberations through biotic systems. We will revisit these ecosystem-level interactions (such as lag effects, feedback effects, and threshold effects) in more detail in Unit III.

WHAT INFLUENCES OVERALL VULNERABILITY TO GLOBAL CHANGE PRESSURES?

Ultimately, predicting whether species or ecosystems will be vulnerable to particular stressors—and whether they might experience synergistic effects—is difficult. **Vulnerability** is the degree to which a species or system is susceptible to environmental change. Vulnerability is determined by numerous related factors, and many partially overlapping terms are used in Global Change Biology to describe the components of vulnerability. Figure 4.7 provides one framework, but let us look at several overarching factors that influence whether species or ecosystems will be particularly vulnerable to environmental change.

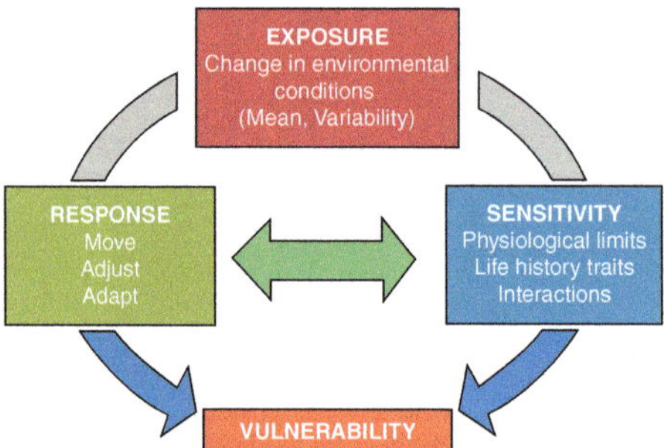

FIGURE 4.7 Many factors influence vulnerability to global change stressors. Here vulnerability to anthropogenic impacts is mediated through exposure (e.g., degree of environmental change), sensitivity (e.g., physiological tolerances), and capacity to respond (e.g., ability to adapt).

Reflection: Is vulnerability to global change stressors a fixed trait? What is a specific example of how vulnerability itself could change over space or time?

Source: Modified from Moritz, C., & Agudo, R. (2013). The future of species under climate change: resilience or decline? *Science* 341 (6145), 504–508. https://science.sciencemag.org/content/341/6145/504

Exposure

Exposure is the likelihood that a particular species or ecosystem experiences a particular global change stressor. Exposure is typically defined based on the magnitude and velocity of anthropogenic change. However, species or ecosystem characteristics can also affect exposure. For example, exposure to global warming is greater for organisms inhabiting the shallow seas relative to the deep ocean.

Sensitivity

Sensitivity is the degree to which a species or system is affected by a particular global change stressor. More sensitive species will experience a greater reduction in performance, reproduction, or survival than less sensitive species. For example, a diet or habitat specialist is likely to be more sensitive than a generalist. Similarly, a species with narrow physiological tolerance is likely to be more sensitive than one that can thrive in a broader array of environmental conditions.

Capacity to Respond

Capacity to respond is the ability of a species or system to react to environmental change in a way that ameliorates the stressor. Species or systems with higher capacity to respond will be less vulnerable than those with limited capacity to respond. For example, a species that can easily move, acclimate, or adapt to new conditions will have a greater capacity to respond. We will evaluate these core organismal responses in detail in our next unit.

These components of vulnerability are not fixed characteristics of species or ecosystems. For example, an initial response (like moving to a new area) can then reduce exposure and/or sensitivity to current or future stressors. Alternatively, an initial response could exacerbate vulnerability. Environmental change can occur slowly or quickly over short or long timescales. Similarly, species and ecosystem responses can be gradual or abrupt. Thus, it is important to consider the dynamic nature of vulnerability through space and time. In Unit II, we will consider the core responses that contribute to vulnerability in more detail.

CONCLUSION

In the Anthropocene, a single species—*H. sapiens*—has risen to dominate and modify the planetary system. As the human population has grown, so, too, has pressure on natural systems from agriculture, industry, urbanization, and other anthropogenic enterprises. Many of these activities have provided incredible benefits to human societies as longevity and standards of living increase. However, they also have complex effects on the millions—or billions—of nonhuman life forms on our planet.

We are now completing Unit I, which provided an introduction to Global Change Biology as a field, a glimpse into the history of life on our planet, and an overview of the historical and contemporary impacts of our species. We will now begin to explore how these impacts reverberate throughout the biosphere. In the next unit, we will address the question: When confronted with a changing world, how do individuals, populations, and species respond?

Each chapter in Unit II focuses on one core response to environmental change. Simply put, as conditions change, organisms can move, adjust, adapt, or die. "Move" responses encompass all changes in the distribution of individuals, populations, and species. "Adjust" responses consist of all types of acclimatization that can alter the characteristics of individuals, populations, and species. "Adapt" responses include all trait evolution due to heritable genetic changes. "Die" responses comprise all losses of individuals, populations, or species. Of course, the core responses are not mutually exclusive and can occur over different temporal and spatial scales. In future units, we will scale up further to evaluate how integrated earth systems respond to environmental change and how we can best address global change stressors.

MEET THE DATA POLLINATORS AND PESTICIDES

Who Are the Scientists and What Did They Set Out To Do?

As we come to the close of Unit I, let us begin to synthesize and apply what we have learned about global change study design and anthropogenic stressors to a real-world biological system. Here we focus on a study entitled "Seed Coating with a Neonicotinoid Insecticide Negatively Affects Wild Bees," published by Rundlöf and collaborators in 2015 in the journal *Nature*. The study takes an experimental approach to understand current threats to insect pollinators. It also leveraged the efforts of many different stakeholders working together for a common goal (including scientists, commercial seed growers, beekeepers, and collaborative farmers). Lead author Maj Rundlöf is shown in Box Figure 4.3.

Insects are a phenomenally diverse lineage, accounting for more than half of all described Eukaryotic species. Insects play essential roles in natural ecosystems and also provide hundreds of billions of dollars per year in ecosystem services like pollination (Gallai et al. 2009). Over the last 50 years, dramatic declines have been observed in both managed and wild populations of insect pollinators. Bee declines in particular have garnered attention as the threat of a "pollination crisis" looms: where agricultural production could become limited by loss of pollinators (Holden 2006).

In the Anthropocene, bee populations have been threatened by a number of interacting factors, including habitat loss, increased exposure to parasites and pathogens, climate change, competition with introduced species, and pesticide exposure (Goulson et al. 2015). As Box Figure 4.4 shows, these stressors can have interacting effects. For example, declines in native vegetation force bees to forage in agricultural settings where their exposure to pesticides is greater. In turn, exposure to pesticides can weaken immune response when confronted with pathogens. Pesticide exposure has become an important focal point in bee decline research as more than 150 different pesticides have been found in honeybee colonies alone (Mullin et al. 2010), and several of these have been directly

BOX FIGURE 4.3 Lead author Dr. Maj Rundlöf conducting field research.

Source: Miranda Rundlöf/Christian Krinte.

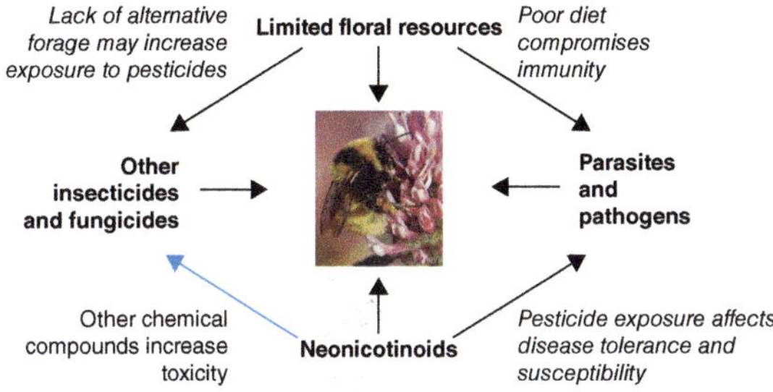

BOX FIGURE 4.4 Interacting stressors contributing to declines of bee pollinators.

Source: "Bee declines driven by combined stress from parasites, pesticides, and lack of flowers. by dave goulson, elizabeth nicholls, cristina botías, ellen l. rotheray. science 27 mar 2015"

implicated in declines in both native and managed bee populations (Goulson et al. 2015).

Here we take a closer look at an experimental study that set out to assess the role of a particular class of pesticides in bee declines. Neonicotinoids (neonics) are chemical pesticides that target the nervous system of insects. These insecticides are used commonly to coat seeds in agricultural settings. While neonics may help protect seeds from agricultural pests, they also can have devastating effects on bees. Previous studies had focused primarily on artificial exposure to neonics in managed honeybee populations, but Rundlöf et al. wanted to understand effects of these pesticides in native bee populations foraging in real landscapes.

What Are *Your* Predictions?

Before you read on, take a few minutes to apply what you have learned in this unit about methods in Global Change Biology to the bee system. Start by considering the following questions:

- *If you were going to take an experimental approach to understanding neonic effects on wild bees, what would be the key elements of your study design?*
- *What treatment(s) might you apply and what response variables might you measure?*
- *How would ensure you had adequate controls, replication, and randomization?*

Now write a summary statement about how you might best test for the effects of neonicotinoids on wild bees in real working landscapes.

What Were the Scientists' Predictions?

The authors did not include a formal prediction but were explicit about the key question guiding their research. They stated: "*The key question is how wild bees, which may differ from honeybees in response to insecticides, are affected by neonicotinoids when foraging in real agricultural landscapes.*"

What Data Were Collected?

Rundlöf et al. used a powerful paired study design. They chose eight replicated pairs of agriculture landscapes as shown in Box Figure 4.5. They randomly assigned one field in each pair as the control and one field as the treatment and exposed seeds in the treatment field to standard doses of neonics. They then measured a number of response variables of wild bees, focusing particularly on two species of native bee (the solitary bee *Osmia bicornis*, the colony-forming bumblebee *Bombus terrestris*) and the European honeybee (*Apis mellifera*), which is managed and introduced regularly to agricultural landscapes around the world. First, they measured the density of wild bees in control versus treatment plots. Next, they compared the reproductive success of wild bees. For the solitary *O. bicornis*, they measured whether females built brood cells. For the colony-forming *B. terrestris*, they measured colony growth and production of both queens and workers. For the managed honeybee *A. mellifera*, they assessed the number of bees in control versus treatment plots.

What Is *Your* Interpretation of the Data?

Before you read on, take a few minutes to interpret the results from the study shown in Box Figure 4.5. Consider the following questions as you look at the figures:

- *How does neonic seed coating affect overall wild bee density?*
- *How does the neonic treatment affect reproductive output and why are results displayed differently for the solitary species versus the colony-forming species?*
- *Is the effect similar or different for wild bees versus managed honeybees?*

Write a sentence or two describing the *key findings* of this study and their *significance*.

What Was the Scientists' Interpretation of the Data?

The first key conclusion from the research is that neonic insecticides have negative effects on wild bees. Populations exposed to the neonic treatment exhibited significantly reduced total density of wild bees. Wild bees also showed decreased reproductive output in the treatment group. In the solitary species (*O. bicornis*), females showed reduced nesting and failure to build brood cells when exposed to

(*Continued*)

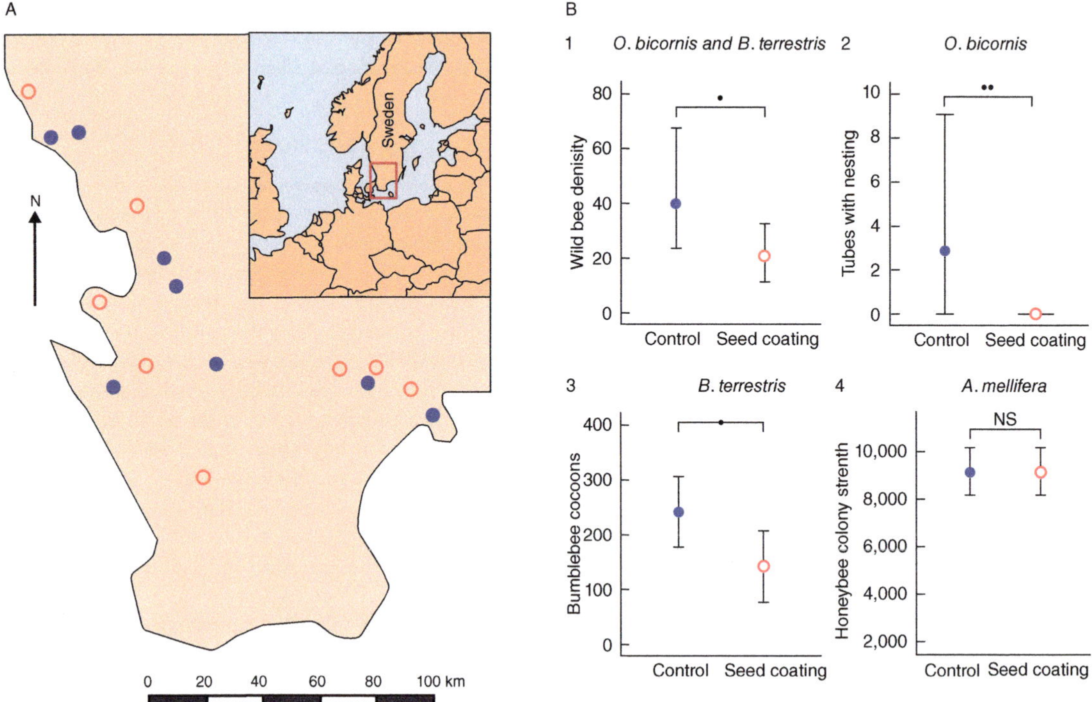

BOX FIGURE 4.5 (A) Research sites in agricultural fields of southern Sweden, showing a paired, replicated study design. Control and treatment fields are shown with open red and closed blue circles, respectively. (B) Results from the study. Plots show comparisons between control versus neonic-exposed sites for (1) bee density for the two wild species (*O. bicornis* and *B. terrestris*); (2) reproductive output (number of tubes with successful female nesting) in the solitary species (*O. bicornis*); (3) reproductive output (number of male workers) in the colony-forming species (*B. terrestris*); and (4) number of adults in colonies of the honeybee (*A. mellifera*). In all plots, the central circle represents the mean value across replicates, and bars represent confidence intervals or data range.

Source: Rundlöf, M., Andersson, G. K., Bommarco, R., Fries, I., Hederström, V., Herbertsson, L., . . . & Smith, H. G. (2015). Seed coating with a neonicotinoid insecticide negatively affects wild bees. Nature, 521(7550), 77.

neonics. In the colony-forming species (*B. terrestris*), fewer female queens and fewer male workers were produced in fields with neonic treatment.

The second key conclusion from the study is that the effect of neonic treatment differs between the two wild bee species studied and the managed European honeybee. Honeybees are important as crop pollinators and honey producers. They are also used commonly as a **model organism** because they are easy to maintain and breed in captivity and because the *A. mellifera* genome has been sequenced (Weinstock et al. 2006). However, there are more than 15,000 bee species in the world (Danforth et al. 2006), and these species exhibit tremendous diversity in ecology, physiology, behavior, and breeding systems. Extrapolating results from one model species may therefore be problematic. As Rundlöf et al. say: "*The use of honeybees as model organisms in environmental risk assessments of neonicotinoids may not allow generalizations to other bee species.*" Thus, this study underscores the risks of assuming that the response of one model species to a global change stressor will accurately represent the response of its relatives in natural systems.

Finally, this study emphasizes the importance of developing an appropriate study design for a particular biological question. The authors discuss the value of studying not only short-term but also long-term effects of global change stressors. They also highlight the value of conducting experimental studies in natural working landscapes. As Rundlöf et al. conclude: "*We question whether prevailing risk assessment standards, where predominantly short-term and lethal effects are assessed on model species under laboratory conditions, can be used to predict real-world consequences of pesticide use for populations, communities and ecosystems."*

What Are *Your* Ideas for Future Research Directions?

Given the findings of Rundlöf et al. about the detrimental effects of neonics on wild bees, what next steps do you envision for this research program? If you had been involved in this study, how would you follow up? Start by considering the following questions:

- *Are there more detailed questions that could now be asked about the mechanism by which neonics affect bees?*
- *Are there ways to increase the generality of the findings by studying additional species, localities, time frames, or stressors?*
- *Are there complementary approaches (observational, experimental, mathematical, synthesis) that could be brought to bear?*

Now write a few sentences about future directions on this research theme.

TAKING A CLOSER LOOK HISTORICAL AND CONTEMPORARY CLIMATE CHANGE

Climate change is one of many contemporary anthropogenic stressors on the biosphere. However, it is a defining environmental issue of the 21st century and an inescapable backdrop against which all other changes are occurring. The goal of this special feature is to build a foundational understanding of our planet's past climate history and the dynamics of contemporary anthropogenic climate change. This primer is not intended as a comprehensive guide to atmospheric sciences, and there is a wealth of resources for those interested in taking a deeper dive into understanding climate change. We will also take a more comprehensive look at how human activities modify large-scale earth systems in Chapter 10.

What Do We Know about Earth's Climate History?

Earth's climate exhibits nested levels of climate variability, from seasonal dynamics to millennial cycles to long-term changes over hundreds of millions of years. Box Figure 4.6 shows temperature as a key marker of changes in the climate system over geological time. Over the last 500 million years, cool **Ice House** and warm **Hot House** periods have alternated on our planet. Temperature has fluctuated as much as 15 degrees Celsius between these warm and cool periods.

Consider how dramatically a Hot House planet would differ from our relatively cool contemporary world. For example, in the early Tertiary period approximately 50 million years ago, the Arctic was ice-free with summers reaching a balmy 20 degrees Celsius (70 degrees Fahrenheit). The landscape was more similar to the lush forests and swamps of the present-day southeastern United States (Eberle & Greenwood 2012), a far cry from contemporary Arctic conditions.

Within epochs, there also have been temperature fluctuations. For example, the most recent cool period was the Pleistocene, which began approximately 2 million years ago. Box Figure 4.6 shows cyclical temperature fluctuations during the Pleistocene that occurred on a roughly 100,000-year cycle. These fluctuations correspond to glacial and interglacial periods, with the last glacial period ending roughly 20,000 years ago.

What Caused Historical Climate Variability?

Climate conditions result from a delicate balance of incoming and outgoing radiation. In general, the amount of **incoming solar radiation** is roughly balanced by the amount of **reflected solar radiation** and **outgoing longwave radiation**. Box Figure 4.7 presents a schematic of this radiation balance. Solar radiation enters our atmosphere as shortwave radiation. Some incoming solar radiation is reflected back to space as shortwave radiation by clouds, atmospheric gases, or the surface of the Earth. Some solar radiation is absorbed in the atmosphere or by the surface of the Earth and is subsequently emitted as longwave radiation (primarily as thermal energy). Notice that our atmosphere plays a key role in maintaining radiation balance. Atmospheric gases not only reflect solar energy but also trap energy close to the planet's surface. This **greenhouse effect** warms the planet and helps support life as we know it.

Anything that alters the radiation balance of the climate system is referred to as a **forcing**. By changing the radiation balance, forcings ultimately impact the temperature of our planet. Here we will look at forcings on the three key components of the global radiation balance that contributed to historical shifts in climate.

(Continued)

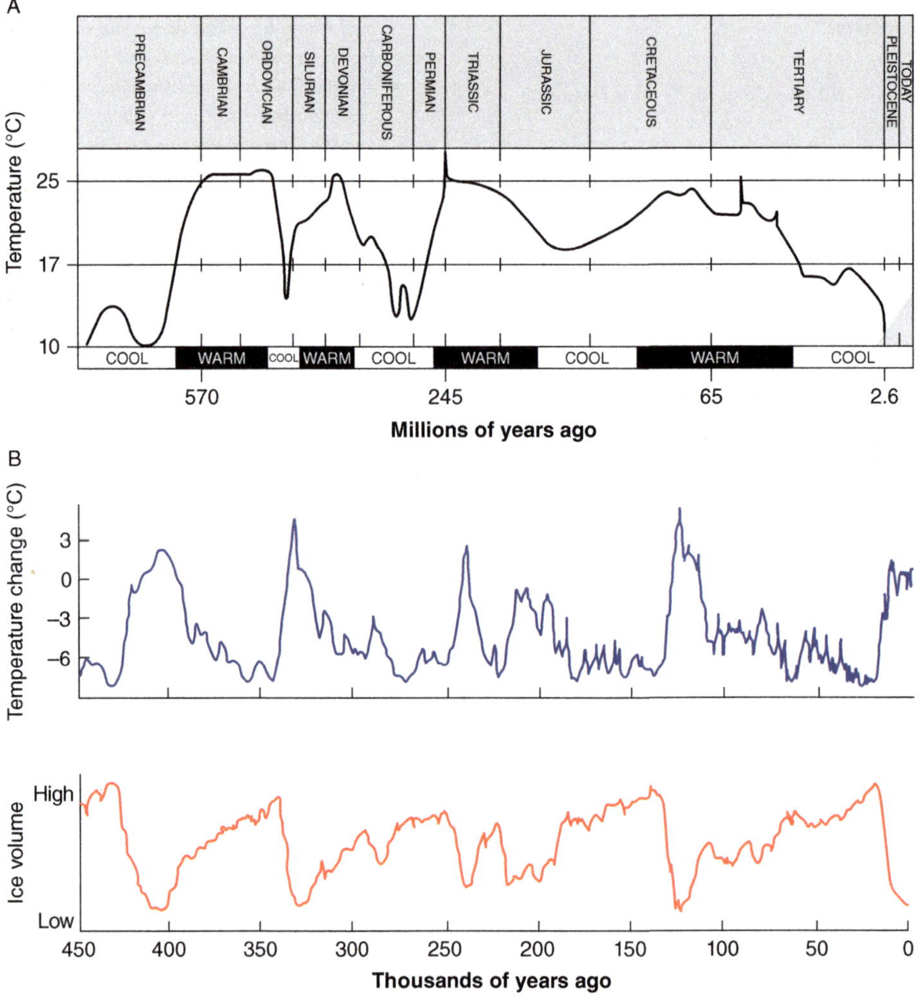

BOX FIGURE 4.6 Temperature through time. (A) Inferred temperature over the Earth's known climate history. Warm and cool periods have alternated over the major geological epochs. (B) Even within geological periods, temperatures have fluctuated. Temperature and ice volume were inversely proportional during the glacial and interglacial periods of the Pleistocene.

Reflection: Compare the top and bottom panels—how much temperature change was there *between* major Ice House and Hot House periods in the last 500 million years, and how much temperature change was there *within* the cool Pleistocene period?

Source: (A) PALEOMAP Project ; (B) Image created by R. A. Rohde (2005a) from publicly available data, https://www2.palomar.edu/anthro/homo/homo_3.htm

INCOMING SOLAR RADIATION The amount of incoming solar radiation can be altered by several factors. The simplest example is that long-term dynamics of our climate system are influenced by changes in solar output. Stars have life cycles, and our sun is a middle-aged star. At the time of our planet's formation, the sun emitted only ~70% as much solar radiation as it does today, and it is projected to die in ~5 billion years. In addition to the overall birth-death process of stars, shorter term changes in solar output can be recorded by the number and intensity of sunspots (Hathaway et al. 2002).

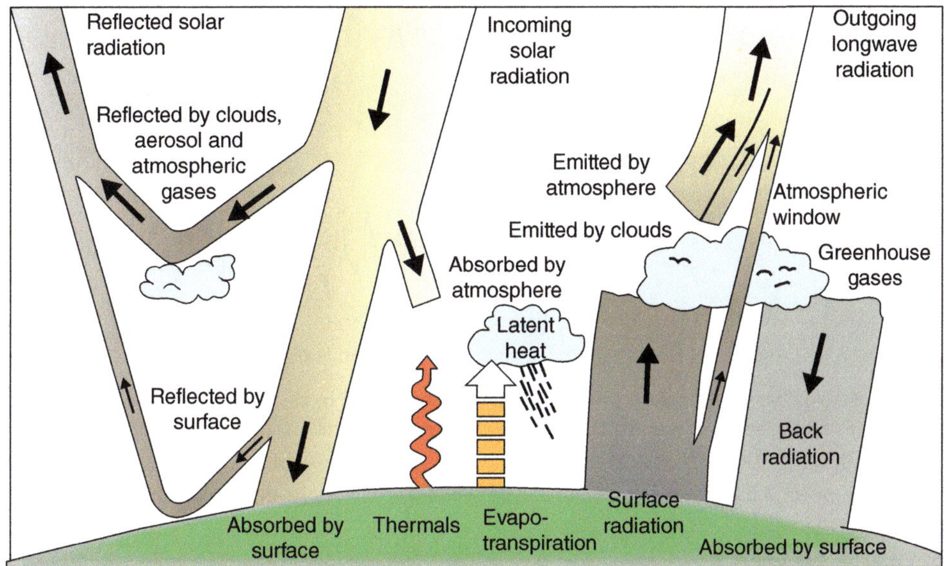

BOX FIGURE 4.7 Schematic of the radiation balance on Earth. Incoming solar radiation can be reflected by clouds and atmospheric gases or can be absorbed and re-emitted by the surface of the planet. For a stable climate, incoming solar radiation must be roughly balanced by reflected solar radiation plus outgoing longwave radiation.

Reflection: What is the difference between the two types of *outgoing solar radiation* (reflected solar radiation and outgoing longwave radiation)?

Source: Skeptical Science/IPCC

The amount of incoming solar radiation is also influenced by the Earth's orbit, which exhibits cyclical changes. These changes, called Milankovitch cycles, describe variation in three aspects of Earth's orbit. First, the shape of the Earth's orbit changes from more circular to more elliptical (eccentricity). Second, the tilt of the Earth on its rotational axis shifts by a small but appreciable degree (tilt). Third, the orientation of Earth's rotational axis shifts (precession). Box Figure 4.8 shows how the Milankovich cycles can be aligned to known changes in Earth's climate. For example, Milankovitch cycles affect the length and variability of the seasons, which in turn affect glacial cycles. These orbital variations help explain many of the glacial and interglacial cycles in Earth's recent history (Lisiecki 2010; Huybers 2011). The details of the Milankovitch cycles are not critical for our purposes, but they are one key mechanism driving Earth's climate patterns.

REFLECTED SOLAR RADIATION The amount of reflected solar radiation can be altered by a number of factors, including changes in cloud cover or atmospheric composition. For example, impacts of **bolides**—astronomical objects like asteroids or comets—on our planet can cause dust clouds, which dramatically increase particulate matter in the atmosphere. Increased particulate matter in turn increases the proportion of incoming solar radiation that is reflected. An important example of global cooling after a bolide impact occurred at the end of the Cretaceous period and likely contributed to the mass extinction of the dinosaurs (Vellekoop et al. 2014).

OUTGOING LONGWAVE RADIATION The amount of outgoing longwave radiation that leaves the Earth's atmosphere can be altered by a number of factors. For example, volcanic activity can affect concentrations of particles and gases in the atmosphere, which in turn can trap heat close to Earth's surface. Moreover, changes in surface reflectance alter how much solar energy is absorbed versus reflected. The proportion of light and heat reflected by a surface is termed **albedo**. Fresh snow (high albedo) will reflect more than dark soil (low albedo). As glacial cycles

(Continued)

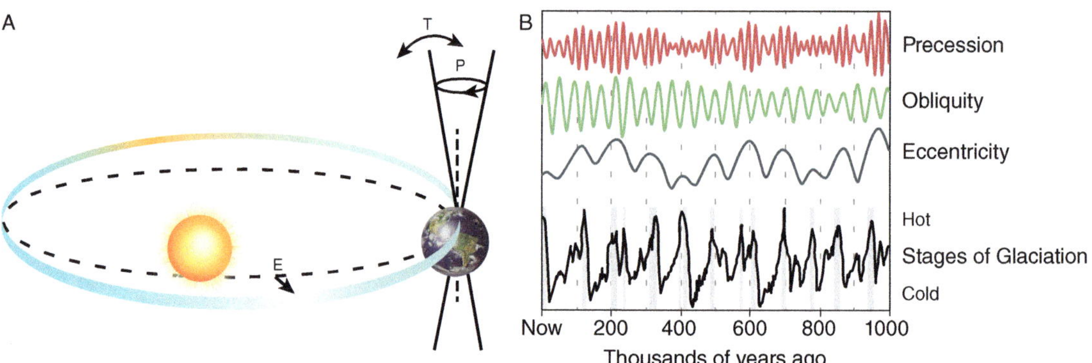

BOX FIGURE 4.8 (A) Changes to incoming solar radiation can be caused by aspects of the Earth's orbit that shift over time: eccentricity (E), tilt (T), and precession (P). (B) Cyclical changes in aspects of Earth's orbit are termed "Milankovitch cycles" and can be associated with climate cycles over the planet's climate history.

Reflection: Which aspects of Earth's orbit exhibit the fastest and slowest changes, and which correspond most obviously to glacial cycles?

Source: (A) Based on: Rahmstorf, S., & Schellnhuber, H. J. (2006). Der Klimawandel. Diagnose, Prognose. Therapie, 5.; (B) "figure created by Robert A Rohde (2005) from publicly available data, https://www.climate.gov/taxonomy/term/3451 Rohde's twitter: https://twitter.com/RARohde?ref_src=twsrc%5Egoogle%7Ctwcamp%5Eserp%7Ctwgr%5Eauthor"

come and go, the overall albedo of the planet changes, further impacting how much of the sun's radiation is absorbed.

The examples thus far are all abiotic, but living things have also impacted the climate for hundreds of millions of years. For example, ~2.5 billion years ago the evolution of cyanobacteria altered the composition of the Earth's atmospheric gases. The radical increase in the concentration of oxygen had dramatic effects on the Earth's climate (Kopp et al. 2005; Sessions et al. 2009).

What Is Different about Contemporary Climate Change?

The Earth has experienced massive climate fluctuations over its 4 billion year history. As discussed earlier, there are many abiotic and biotic forcings that can affect the radiation balance in the climate system. However, current climate shifts cannot be explained by these natural forcings alone, as illustrated in Box Figure 4.9. Contemporary climate change is different because of its anthropogenic causes, its velocity, and its projected impacts. Although living things have always modulated Earth's climate, never before has a single species precipitated such a rapid and significant climate perturbation.

As we saw earlier in this chapter, contemporary human activities have dramatically altered the composition of the Earth's atmosphere. The most pronounced effect of anthropogenic emissions on the climate system is global warming. Global warming is a consistent directional change observed across scores of scientific studies and has been unambiguously triggered by human activities (Oreskes 2004; Intergovernmental Panel on Climate Change 2014).

Global warming is largely due to the positive radiative forcing of increased carbon dioxide in the atmosphere. Anthropogenically mediated increases in other greenhouse gases such as methane and nitrous oxide can also contribute to global warming (Lashof & Ahuja 1990). Some human activities can lead to negative radiative forcing. For example, an increase in atmospheric concentrations of aerosols can increase the amount of solar radiation reflected from our atmosphere back into space. However, positive anthropogenic radiative forcings far outweigh negative radiative forcings, as shown in Box Figure 4.9.

While global warming is the most consistent directional signature of anthropogenic climate change, widespread changes have also been observed in the mean and variability of temperature and precipitation around the world. As we experience increased weather severity and climate extremes, the threat of anthropogenic climate change reverberates around the world. Anthropogenic climate change threatens an incredible number of sectors (from public health to food production to water resources to wildlife to national security), and evidence of these threats is only increasing (Duffy et al. 2019). We will further study the climate system—and, in particular, predictions for how much anthropogenic climate change is yet to come—in Chapter 10.

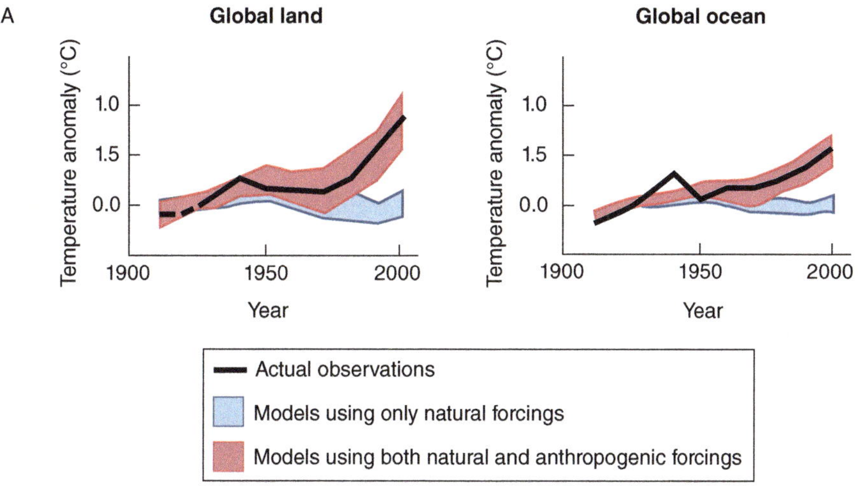

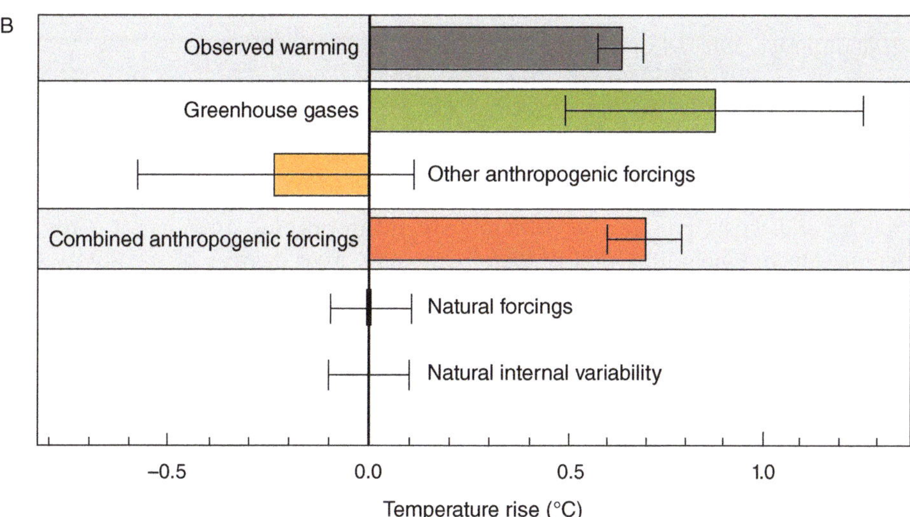

BOX FIGURE 4.9 Anthropogenic forcings have led to temperature rise over the last century. (A) Models including anthropogenic forcings are the only ones that explain observed temperature increases over the last hundred years, both at land and sea. (B) Contributions to observed surface temperature change during the period 1950–2010. Anthropogenic greenhouse gas emissions are driving observed patterns of global warming, even despite countervailing negative forcings.

Reflection: Based on the figure, write a summary in your own words about patterns of global warming over the last century.

Source: (A) Skeptical Science/IPCC, (B) Figure 1.9 from Climate Change 2014: Synthesis Report. Contribution of Working Groups I, II and III to the Fifth Assessment Report of the Intergovernmental Panel on Climate Change [Core Writing Team, Pachauri, R.K. and Meyer, L. (eds.)]. IPCC, Geneva, Switzerland.

KEY CONCEPTS

What is the Anthropocene and when did it begin?

- The Anthropocene is the current geological epoch where human activities have a dominant and global influence on the Earth's environment.

What are patterns of contemporary population growth?

- By 2,000 years ago, the first cities were thriving, and the global population was in the hundreds of millions. By the end of the Industrial Revolution, human population size reached 1 billion.
- The global population is now well over 7 billion with continued population growth projected into the next century.

How are contemporary human civilizations impacting the environment?

- Modern human societies have had dramatic and global impacts on the environment due to activities such as land-use change, pollution, and globalized travel and trade. These activities have not only changed the terrestrial surface of Earth but have altered aquatic habitats and perturbed the climate system.

How do anthropogenic stressors interact with each other?

- Synergies (when the combined effect of multiple stressors is greater than the sum of their individual effects) are common in the Anthropocene and can lead to unexpected and magnified ecosystem-level consequences.

What influences overall vulnerability to global change pressures?

- Species and ecosystems can be vulnerable to anthropogenic stressors for many different reasons, including high levels of exposure, high degrees of sensitivity, and low capacity to respond.

Core concepts: What is climate and how is it measured?

- The climate system is a complex and interactive system. Past climate can be reconstructed using proxies, and future climate can be projected using modeling approaches.

Meet the data: Pollinators and pesticides

- Scientists conducted field experiments to understand how chemicals used to kill agricultural pests might also impact native bees. They found that pesticides have negative effects on wild bee populations, and use of synthetic chemicals may jeopardize the essential pollination services provided by these insects.

Taking a closer look: Historical and contemporary climate change

- The climate system exhibits variation over many different timescales. Over the last 500 million years, temperature has fluctuated up to 15 degrees Celsius.
- Historical climate changes were precipitated by many factors that altered incoming radiation, reflected solar radiation, and/or outgoing longwave radiation.
- Contemporary climate change cannot be explained by natural forcings alone. Anthropogenic activities are unambiguously perturbing the climate system. Global warming is a key signature of anthropogenic climate change, but there are also other changes occurring in the climate system.

CONSOLIDATE YOUR KNOWLEDGE

Answer the following questions to consolidate your knowledge, assess your progress in meeting the learning outcomes, and reveal any areas in need of further exploration:

1. In your own words, write a one- or two-sentence synthesis of the big-picture take-away point of this chapter.
2. Revisit your answer to the *Blank Page* exercise in the beginning of this chapter. Would you refine your answer now based on knowledge you integrated from this chapter?
3. Define the Anthropocene and advance a specific argument for when you think it began.
4. Imagine that a civilization 1,000 years from now was looking back at the geological record from the 21st century. What epoch-scale changes do you think they might observe in our century?
5. Has human population size been increasing linearly over the last several thousand years? How about the last several hundred? Describe trends over time in human population growth and the factors that have likely shaped changes in population growth rates.
6. Provide a specific example of the effects of pollution on aquatic, terrestrial, and atmospheric systems.
7. What is a synergy? Why is it an important concept in Global Change Biology?
8. What are three different components of vulnerability and an example of each?
9. Compare and contrast the causes of historical and contemporary climate change. Provide specific examples of natural and anthropogenic forcings that have influenced Earth's climate over different timescales.
10. At first, it seems contradictory to say that we are living on an "Ice House" Earth during a period of global warming. How can both of these things be true at once?
11. In your own words, define the bolded and italicized terms in this chapter.
12. What are some questions that *you* have about the content in this chapter? If you found some content particularly challenging or particularly interesting, identify these as areas for additional reflection or reading.

LITERATURE CITED

Barnosky. 2013. "Palaeontological Evidence for Defining the Anthropocene." *Geological Society, London, Special Publications* 395 (1): 149. doi.org/10.1144/SP395.6.

Brook, Sodhi & Bradshaw. 2008. Synergies among extinction drivers under global change. Trends in Ecology and Evolution 23 (8): 453–460.

Clark & Mix. 2002. "Ice Sheets and Sea Level of the Last Glacial Maximum." *Quaternary Science Reviews* 21 (1): 1–7. doi.org /10.1016/S0277-3791(01)00118-4.

Crutzen. 2002. "Geology of Mankind." *Nature* 415 (23): 4153.

Dachs & Méjanelle. 2010. "Organic Pollutants in Coastal Waters, Sediments, and Biota: A Relevant Driver for Ecosystems during the Anthropocene?" *Estuaries and Coasts* 33: 1–14. doi.org/10.1007/s12237-009-9255-8.

Danforth, Sipes, & Fang. 2006. "The History of Early Bee Diversification Based on Five Genes plus Morphology." *Proceedings of the National Academy of Sciences* 103 (41): 15118–23. doi.org/10.1073/pnas.0604033103.

Duffy, Field, Diffenbaugh, Doney, Dutton, Goodman, Heinzerling, et al. 2019. "Strengthened Scientific Support for the Endangerment Finding for Atmospheric Greenhouse Gases." *Science* 262 (6427): eaat5982. doi.org/10.1126/science.aat5982.

Eberle & Greenwood. 2012. "Life at the Top of the Greenhouse Eocene World—A Review of the Eocene Flora and Vertebrate Fauna from Canada's High Arctic." *Bulletin of the Geological Society of America* 124 (1–2): 3–23. doi.org/10.1130/B30571.1.

Ellis. 2011. "Anthropogenic Transformation of the Terrestrial Biosphere." *Philosophical Transactions of the Royal Society A: Mathematical, Physical and Engineering Sciences* 369: 1010–35. doi.org/10.1098/rsta.2010.0331.

Ellis, Kaplan, Fuller, Vavrus, Klein Goldewijk, & Verburg. 2013. "Used Planet: A Global History." *Proceedings of the National Academy of Sciences*. 110 (20): 7978–7985 doi.org/10.1073/pnas.1217241110.

Gallai, Salles, Settele, & Vaissière. 2009. "Economic Valuation of the Vulnerability of World Agriculture Confronted with Pollinator Decline." *Ecological Economics* 68 (3): 810–821. doi.org/10.1016/j.ecolecon.2008.06.014.

Goulson, Nicholls, Botías, & Rotheray. 2015. "Bee Declines Driven by Combined Stress from Parasites, Pesticides, and Lack of Flowers." *Science* 347 (6229): 1255957. doi.org/10.1126/science.1255957.

Hathaway, Wilson, & Reichmann. 2002. "Group Sunspot Numbers: Sunspot Cycle Characteristics." *Solar Physics* 211: 357–70. doi.org/10.1023/A:1022425402664.

Holden. 2006. "Ecology. Report Warns of Looming Pollination Crisis in North America." *Science* 314 (5798): 397. doi.org/10.1126/science.314.5798.397.

Huybers. 2011. "Combined Obliquity and Precession Pacing of Late Pleistocene Deglaciations." *Nature* 480 (7376): 229–32. doi.org/10.1038/nature10626.

Intergovernmental Panel on Climate Change. 2014. *Climate Change 2014: Synthesis Report*. Geneva, Switzerland.

Kopp, Kirschvink, Hilburn, & Nash. 2005. "The Paleoproterozoic Snowball Earth: A Climate Disaster Triggered by the Evolution of Oxygenic Photosynthesis." *Proceedings of the National Academy of Sciences* 102 (32): 11131–36. doi.org/10.1073/pnas.0504878102.

Lashof & Ahuja. 1990. "Relative Contributions of Greenhouse Gas Emissions to Global Warming." *Nature* 344: 529–31. doi.org/10.1038/344529a0.

Lehmann, Marcet, Graham, Dahl, & Dubey. 2006. "Globalization and the Population Structure of Toxoplasma Gondii." *Proceedings of the National Academy of Sciences* 103 (30): 11423–28. doi.org/10.1073/pnas.0601438103.

Lewis & Maslin. 2015. "Defining the Anthropocene." *Nature* 519 (7542): 171–180. doi.org/10.1038/nature14258.

Lisiecki. 2010. "Links between Eccentricity Forcing and the 100,000-Year Glacial Cycle." *Nature Geoscience* 3 (5): 349–52. doi.org/10.1038/ngeo828.

Marsh. 1864. *Man and Nature: Or, Physical Geography as Modified by Human Action. Scribner.*

Moritz & Agudo. 2013. "The Future of Species Under Climate." *Science* 341 (6145): 504–8. doi.org/10.1126/science.1237190.

Mullin, Frazier, Frazier, Ashcraft, Simonds, vanEngelsdorp, & Pettis. 2010. "High Levels of Miticides and Agrochemicals in North American Apiaries: Implications for Honey Bee Health." *PLoS ONE* 5 (3): e9754. doi.org/10.1371/journal.pone.0009754.

Oreskes. 2004. "The Scientific Consensus on Climate Change." *Science* 306 (5702): 1686. doi.org/10.1126/science.1103618.

Peñalosa, Muñiz, & Valle. 2006. *Tópicos de Genética*. Mexico City: UAEMEX.

Pereira, Navarro, & Martins. 2012. "Global Biodiversity Change: The Bad, the Good, and the Unknown." *Annual Review of Environment and Resources* 37: 25–50. doi.org/10.1146/annurev-environ-042911-093511.

Rahmstorf & Schellnhuber. 2012. *Der Klimawandel: Diagnose, Prognose, Therapie*. doi.org/10.17104/9783406726736.

Rodrigue, Comtois, & Slack. 2016. *The Geography of Transport Systems*. New York: Routledge.

Rundlöf, Andersson, Bommarco, Fries, Hederström, Herbertsson, Jonsson, et al. 2015. "Seed Coating with a Neonicotinoid

Insecticide Negatively Affects Wild Bees." Nature 521: 77–80. doi.org/10.1038/nature14420.

Sessions, Doughty, Welander, Summons, & Newman. 2009. "The Continuing Puzzle of the Great Oxidation Event." *Current Biology* 19 (14): R567–74. doi.org/10.1016/J.CUB.2009.05.054.

Sonter, Moran, Barrett, & Soares-Filho. 2014. "Processes of Land Use Change in Mining Regions." *Journal of Cleaner Production* 84: 494–501. doi.org/10.1016/j.jclepro.2014.03.084.

Steffen, Broadgate, Deutsch, Gaffney, & Ludwig. 2015. "The Trajectory of the Anthropocene: The Great Acceleration." *The Anthropocene Review* 2 (1): 81–98. doi.org/10.1177/2053019614564785.

Steffen, Persson, Deutsch, Zalasiewicz, Williams, Richardson, Crumley, et al. 2011. "The Anthropocene: From Global Change to Planetary Stewardship." *Ambio* 40 (7): 739–61. doi.org/10.1007/s13280-011-0185-x.

United Nations Department of Economic and Social Affairs. 2015. "The World Population Prospects: 2015 Revision." www.un.org/en/development/desa/publications/world-population-prospects-2015-revision.html.

United States Census Bureau. 2013. "Historical Estimates of World Population." www.census.gov/population/international/data/worldpop/table_history.php.

Vellekoop, Sluijs, Smit, Schouten, Weijers, Sinninghe Damste, & Brinkhuis. 2014. "Rapid Short-Term Cooling Following the Chicxulub Impact at the Cretaceous-Paleogene Boundary." *Proceedings of the National Academy of Sciences* 111 (21): 7537–41. doi.org/10.1073/pnas.1319253111.

Weinstock, Robinson, Gibbs, Worley, Evans, Maleszka, Robertson, et al. 2006. "Insights into Social Insects from the Genome of the Honeybee Apis Mellifera." *Nature* 443 (7114): 31–49. doi.org/10.1038/nature05260.

Zalasiewicz, Williams, Haywood, & Ellis. 2011. "The Anthropocene: A New Epoch of Geological Time?" *Philosophical Transactions of the Royal Society A: Mathematical, Physical and Engineering Sciences* 369: 835–41. doi.org/10.1098/rsta.2010.0339.

Core Responses: Move

Learning Outcomes

After working with this chapter, you will be able to:

- Classify factors that have determined historical species distributions.
- Analyze how contemporary human activities are altering species distributions.
- Describe how scientists predict distributional changes.
- Synthesize factors influencing the "move" response and evaluate the relationship between the "move" response and the other core responses.
- Apply your knowledge to real-world case studies and interpret data from recent scientific studies.

THE BLANK PAGE

The timber rattlesnake (*Crotalus horridus*) is a venomous pit viper with a restricted distribution—it is found only in the deciduous forests of the northeastern United States. What factors likely shaped the distribution of this species? Brainstorm a list of biotic and abiotic factors that may have played a role. Now imagine that you want to predict the future distribution of the timber rattlesnake. What factors will determine where this species will occur 100 years from now? Create another list and then compare your lists of factors that have influenced—and will influence—the distribution of the timber rattlesnake. Are the lists similar or different, and why?

INTRODUCTION

Many biotic and abiotic factors influence where species are found on the planet. However, human activities are increasingly having direct effects on species distributions. For example, land-use change can reduce the area of suitable habitat for a species. Or conversely, globalized transportation can move species to new places. Human enterprises can also have indirect effects on species distributions. For example, global warming can cause geographical shifts as temperature-sensitive species track a changing climate. The goal of this chapter is to analyze the "move" response and understand why—and how—organisms move in response to global change stressors.

HOW AND WHY DO ORGANISMS MOVE?

Movement is a basic characteristic of life on Earth. Some organisms exhibit *active* locomotion such as running, swimming, or flying. Others show *passive* locomotion such as sailing or rolling with the wind. Even organisms that are **sessile**—or immobile—in one life stage typically have another mobile life stage. For example, individual trees are rooted and cannot move, but seeds, pollen, and fruit often can move long distances, assisted by the wind, winged pollinators, and herbivores (e.g., Petit & Hampe 2006).

Individuals move for many reasons and over many spatial and temporal scales, and it is useful to consider three basic types of movement. First, individuals make daily movements as they search for food, track suitable environmental conditions, and take refuge from predators over short temporal and spatial scales. Second, individuals exhibit dispersal. Dispersal events generally involve individuals moving away from their natal territory early in the life cycle before reproduction, but dispersal can also occur during the reproductive phase (e.g., when seeds or gametes move). For example, many larval marine invertebrates disperse on oceanic currents, many plant seeds disperse on the wind, and many vertebrates disperse actively by walking, flying, or swimming. Dispersal is typically a one-way trip, which can take place over short or long distances. Third, individuals can take part in coordinated long-distance movements called migration. Migrations are typically round-trip and often correspond to seasonal dynamics. Migration can be *complete* when an entire population moves together or can be *partial* whereby some individuals migrate while others do not. Moreover, not all migrations are completed in a single generation. For example, 10 generations are required for the monarch butterfly *Danaus plexippus* to complete the ~9,500 kilometer round-trip migration across North America (Flockhart et al. 2013; Reppert & de Roode 2018).

A number of factors influence individual movement patterns (Holyoak et al. 2008). Environmental factors like temperature, precipitation, photoperiod, nutrient availability, and predator presence can prompt daily movement, one-way dispersal, and seasonal migration. The physical environment is extremely dynamic (e.g., moment to moment, season to season, year to year), and movement patterns often closely track changing conditions. In addition, organismal traits like capacity for movement and navigation influence distance and direction traveled. As Figure 5.1 shows, many organisms have adaptations for sensing and moving through their environment, whether simple chemotaxis or sophisticated neural compasses. The internal physiological state of the organism also matters. For example, hunger, stress, body condition, and reproductive status can trigger short- or long-distance movement. Of course, these factors all interact, given that external factors (like an approaching predator or an approaching seasonal change) influence internal processes like hormone production.

Although it is individuals—or sometimes reproductive cells themselves—that move, individual movements have profound consequences at the population and species levels. For example, when individuals move and reproduce, they contribute to **gene flow**: the movement of genetic information across populations. Moreover, over long timescales, changes in dispersal or migration patterns can influence where populations and species are distributed geographically, a topic we will turn to next.

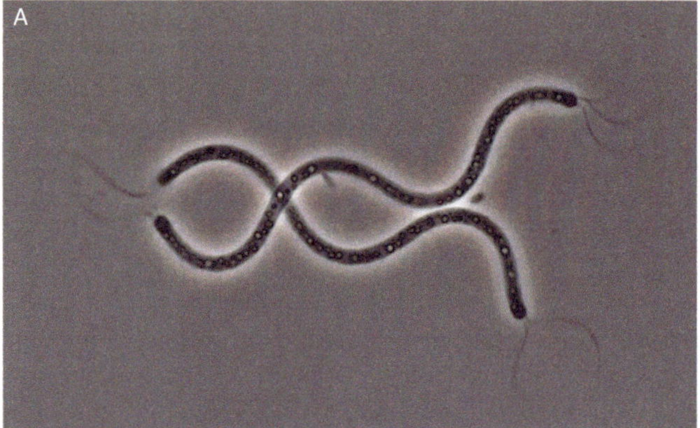

FIGURE 5.1 Sensing and moving through the environment. (A) Even single-celled organisms move toward or away from environmental stimuli, as illustrated by this bacterium propelled by a flagellum. (B) Although plants are rooted to the ground, many have reproductive cells that are dispersed by wind, water, or animals, as illustrated by this honeybee with pollen baskets on its hind legs. Similarly, many marine invertebrates that have a fixed location as adults have gametes or larva that move with ocean currents. (C) Birds and butterflies are most famous for navigating over long distances, as illustrated by these homing pigeons, but many diverse species can navigate by landmarks, by solar or stellar cues, or by Earth's magnetic fields. (D) Of course, not all organisms move long distances. For example, slender salamanders in the genus *Batrachoseps* have tiny home range sizes (often <10 meters) and lack a highly mobile life stage, making them one of the most sedentary vertebrates on the planet.

Reflection: Pick one of the examples in the figure and clearly articulate how external environmental factors, organismal traits, and internal physiological state might *interact* to determine direction and distance moved.

Source: (A) Maple Ferryman/Shutterstock; (B) Ivar Leidus/Wikipedia; (C) nitramtrebla/Wikipedia; (D) Matt Jeppson/Shutterstock

http://textbookofbacteriology.net/structure_2.html; https://en.wikipedia.org/wiki/Bee_pollen#/media/File:Apis_mellifera_-_Melilotus_albus_-_Keila.jpg; https://en.wikipedia.org/wiki/Magnetoreception#/media/File:Brieftauben_im_Anflug.jpg; https://www.horniman.ac.uk/media/w650h650/acropora-millepora-colonies.jpg (Jamie Craggs, Horniman Museum and Garden); https://www.flickr.com/photos/riffle_nature_photos/28232422679 (Spencer Riffle)

WHAT IS A GEOGRAPHIC RANGE?

Two related terms are typically used to describe the spatial arrangement of species on our planet. The **geographic range** of a species is the total spatial area in which it is found. The range is typically illustrated by a range map, which depicts the total geographic extent of a species' geographic distribution. However, it is important to remember that individuals are rarely uniformly distributed across space. Some regions within a range might have lower population densities or no populations at all. For example, as shown in Figure 5.2, the American pika (*Ochotona princeps*) has a broad range across western North America. However, it is restricted only to high-elevation sites (Beever & Smith 2014), so the range map conveys its broad geographic extent but not its fine-scale distribution across habitats.

Therefore, we refer to the **geographic distribution** as all points where individuals in that species occur. The spatial arrangement of individuals and populations within a species range is of great consequence. For example, as illustrated in Figure 5.2, the blue whale (*Balaenoptera musculus*) is found in most of the world's oceans and has a total range of >300,000,000 square kilometers (NOAA 2014). Although blue whales have a global range, their distribution is extremely patchy. Current estimates suggest that there are fewer than 25,000 individuals distributed across the vast oceanic range. Thus, a large range size does not always indicate large or dense populations.

FIGURE 5.2 Example of variation in species' range sizes. (A) American pikas occur only at high elevations in the western United States. (B) The Devil's Hole pupfish is found in only one tiny pool in a limestone cavern in southwest Nevada. Scientists are shown monitoring the population. (C) Blue whales inhabit all of the world's oceans.

Reflection: In what ways can range maps be misleading?

Source: (A) Frédéric Dulude-de Broin/Wikipedia; (B) Brett Seymour, NPS; (C1) Craig Lambert Photography/ Shutterstock; (C2) The Emirr/Wikipedia

https://en.wikipedia.org/wiki/Blue_whale#/media/File:Cypron-Range_Balaenoptera_musculus.svg; https:// www.sciencesource.com/archive/Blue-whale—Balaenoptera-musclus—SS21668315.html

We just considered a species with a tremendous range size, but range *sizes* and *shapes* vary dramatically across the tree of life. Some species are highly localized and endemic to a narrow geographic region. One example, the Devil's Hole pupfish (*Cyprinodon diabolis*), is restricted to a single freshwater spring (United States Fish and Wildlife Service 2013). One of the world's rarest vertebrates, the pupfish's survival depends on <10 square meters of shallow habitat where they can feed and reproduce. Although the population size dipped to as low as 38 individuals in 2007, a census count in 2018 revealed 187 fish. Thus, a species can be locally abundant even with a small range size.

Ultimately, the concepts of geographic distribution and geographic range are intimately connected: the geographic range encompasses the entirety of the geographic distribution. Thus, to assess vulnerability to anthropogenic stressors, it is ideal for scientists to understand not only a species total range but also the distribution of populations and individuals within that range. However, it is often difficult to obtain detailed and spatially localized knowledge of the distribution of individuals and species. Thus, many Global Change Biology studies focus on the size, shape, and boundaries of the geographic range as key ways to measure and visualize spatial changes over time.

WHAT FACTORS DETERMINE A SPECIES' GEOGRAPHIC RANGE?

Before we evaluate the anthropogenic factors that alter the distribution of species, we must first understand what determines species distributions in the absence of human activities. Let us look at three broad categories of factors that interact to influence where species are found on our planet. Ultimately, any factor that alters the distribution of individuals and populations can have an effect on the shape and size of a species' range.

Evolutionary History

All species have a place of origin, a geographic area that served as the cradle for their evolution. The geographic distribution of species can remain tied to the region of origin for millions of years. For example, the charismatic poison frogs in the genus *Mantella* illustrated in Figure 5.3 are found only on Madagascar, reflecting their evolutionary origins. Not all species maintain a narrow geographic range over long time periods, and some species achieve truly global distributions. Whether a species' range expands greatly beyond the cradle of origin depends in part on time and geography. Species that evolved relatively recently have had less time to expand. Species that originated in habitats enclosed by significant geographic barriers (e.g., mountain passes or oceanic barriers) may be more likely to remain isolated with relatively small ranges.

Species Characteristics

A number of species' attributes shape geographic distributions. One trait that is particularly important is **dispersal capability**. Dispersal capability influences species distributions, because stronger dispersers tend to have larger geographic ranges than weaker dispersers. As we already discussed, even relatively sedentary organisms can have a life stage with long-range dispersal, such as wind-dispersed seeds or water-dispersed larva. Of course, other organismal traits also play an important role. For

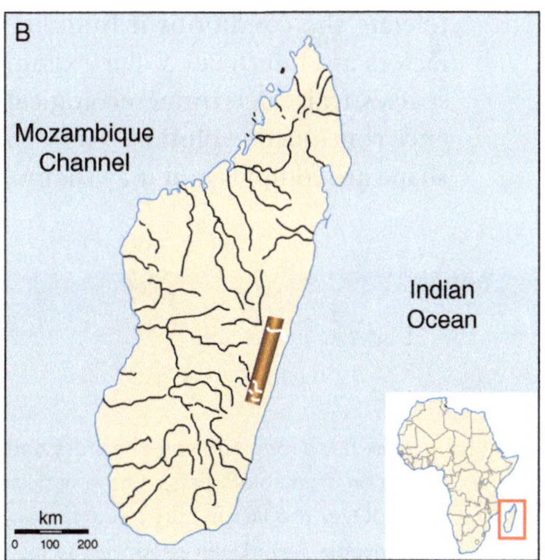

FIGURE 5.3 *Mantella baroni*, a narrow endemic, is distributed only in the east-central region of Madagascar, where the species originated.

Reflection: What is a species with an opposing pattern—where its current distribution is not limited to its region of evolutionary origin?

Source: Dudarev Mikhail/Shutterstock

example, reproductive output and generation time influence population growth rates, while physiological tolerance and diet influence specialization to particularly environmental conditions.

Ecological Requirements

All species have a range of abiotic (e.g., temperature, humidity, nutrient availability) and biotic (e.g., the presence or absence of key predators, competitors, mutualists) conditions needed for survival and reproduction. These ecological requirements are often described as a species' **niche**. Ecological requirements for survival vary dramatically from species to species. For example, species differ in physiological tolerance to temperature and pH such that some species thrive in environments that would be toxic to others (e.g., microbes in hot acidic geothermal springs). Species also differ in their tolerance range. Some species are more specialized and can only survive under a restricted set of conditions (e.g., monarch butterflies require the milkweed plant for larval development), and others are quite generalized and can survive under a broad range of conditions (e.g., omnivores such as raccoons, which opportunistically use a variety of food resources). Thus, the ecological requirements of a species largely determine what geographic areas are suitable both at regional and local scales. The *Core Concepts* box in this chapter reviews the concept of the ecological niche in more detail.

In Sum

Ultimately, the size and shape of a species' range depends on where it originates, whether it is a strong disperser, what geographic barriers it faces to dispersal, and whether it can tolerate the conditions it finds in new regions. Of course, these broad categories of factors are interrelated. For example, evolutionary history gives rise to species traits, species traits determine ecological requirements, and ecological requirements influence continued evolution. Thus, the factors that determine geographic range size and shape are complex and may themselves shift over time.

CORE CONCEPTS

WHAT IS A NICHE?

The term **niche** has a long and varied history and has received attention from biologists, mathematicians, and philosophers. Over the last hundred years, many different niche concepts have been proposed (e.g., Grinnell 1917; Elton 1927; Hutchinson 1957). Here we use a more recent framework that synthesizes earlier definitions to articulate the niche broadly as the relationship between a species and its environment (Leibold 1995; Chase & Leibold 2003). Every species has *requirements from* its environment and in turn has *effects on* its environment. Therefore, a species' niche is comprised of two fundamental components: (1) the environmental conditions that allow it to persist in a given environment and (2) the impacts it has on these environmental conditions. Let us look at these two aspects of the niche.

Requirements

All species have conditions and resources that are required for persistence. These conditions include all of the biotic and abiotic needs of a species. For example, giant sequoia trees (*Sequoiadendron giganteum*) need particular temperature, moisture, and nutrient conditions for survival, but they also require periodic fires for seed release and germination. They also have important biotic needs as beetle and squirrel activity help release seeds from cones.

Impacts

All species have an impact on the conditions and resources in their environment. Species influence both their biotic and abiotic surroundings, for example, by absorbing nutrients or by eating other species. For example, the Arctic fox (*Vulpes lagopus*) plays an important role as a predator and carrion feeder in its community. Its trophic role affects not only population densities of other species but also contributes to nutrient cycling.

Mathematical models consistently demonstrate that two species with identical niches cannot coexist in the same environment. This underscores the importance of competition as a factor that structures ecological communities. Individuals can compete for limited resources within a species (**intraspecific** competition) or between species (**interspecific** competition). Interspecific competition often explains why species are excluded from habitats that appear suitable. For example, a particular area might contain all the necessary conditions and resources for the survival of a species. However, it might also contain a competitor that dominates nutrients, physical space, sunlight, or another key resource.

Therefore, ecologists distinguish between the **fundamental niche** (the conditions and resources that a population could use under ideal conditions) and the **realized niche** (the conditions and resources that a population actually uses, given constraints such as those imposed by predators and competitors). Box Figure 5.1 illustrates a classic example from the marine intertidal zone of how one species can have a diminished realized niche because it is prevented from using resources by another species.

Finally, we can consider the ecological concept of niche over evolutionary timescales. The diversity in the tree of life is an unequivocal example of the potential for dramatic niche shifts over long timescales. However, there is also evidence that niche evolution may be relatively slow. Closely related species are often more similar in ecological role and requirements than expected by phylogenetic relatedness (Wiens et al. 2010). This tendency for species to occupy a similar niche as their ancestors is referred to as **niche conservatism**. The potential for niche evolution becomes particularly important during rapid environmental change. Under niche conservatism, species do not easily adapt to new conditions and must move to track suitable habitats, while under niche adaptation, species can more easily maintain a constant geographic distribution even while conditions change.

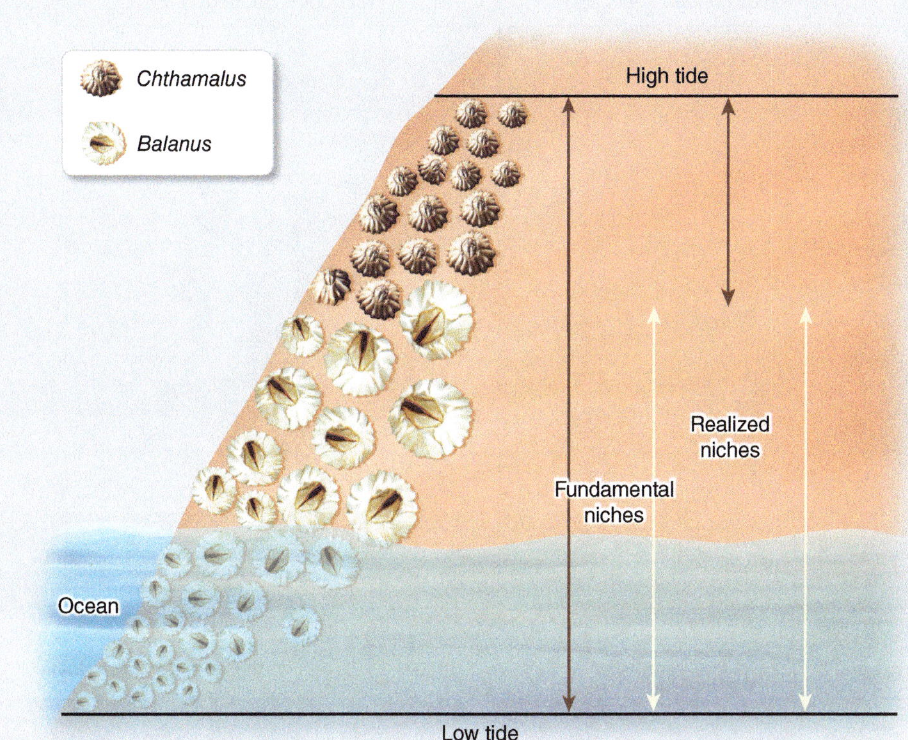

BOX FIGURE 5.1 The fundamental and realized niche. A schematic representation of Joseph Connell's famous intertidal barnacle experiment (Connell 1961). The *Chthamulus* species is restricted by desiccation at the high tide line but by competition with another barnacle species (*Balanus*) in the mid-intertidal. Therefore, the realized niche of the *Chthamulus* species is smaller than its fundamental niche due to interspecific interactions.

Reflection: Based on the figure, how would you define the term *competitive exclusion?*

Source: CAMPBELL, NEIL A.; REECE, JANE B.; TAYLOR, MARTHA R.; SIMON, ERIC J.; DICKEY, JEAN L., BIOLOGY: CONCEPTS AND CONNECTIONS, 6th Ed., ©2009. Reprinted by permission of Pearson Education, Inc., New York, New York.

DO RANGE CHANGES OCCUR EVEN WITHOUT ANTHROPOGENIC INFLUENCE?

Range changes are alterations in the geographical area where a species is found. Range changes have occurred throughout the history of our planet, even without anthropogenic influence. For example, a common pattern observed for species that have persisted through alternating warming and cooling periods is that ranges contract during glacial periods and then expand during interglacial periods. Figure 5.4 demonstrates this pattern for the Wet Tropics in Australia. At the last glacial maxima, rainforests were broken up into smaller, isolated patches—termed **refugia**. As the climate warmed and glaciers retreated, the rainforests expanded, increasing connectivity for rainforest species (Hilbert et al. 2007).

A Wet tropics (18,000 years ago) | Wet tropics (modern day)

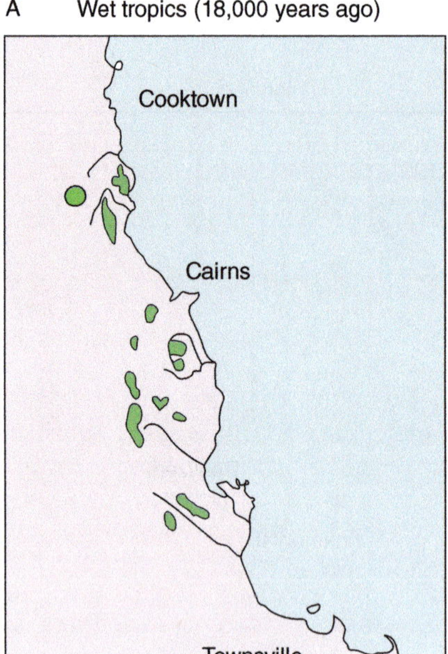

B Bering Land Bridge (21,000 years ago) | Bering Land Bridge (modern day)

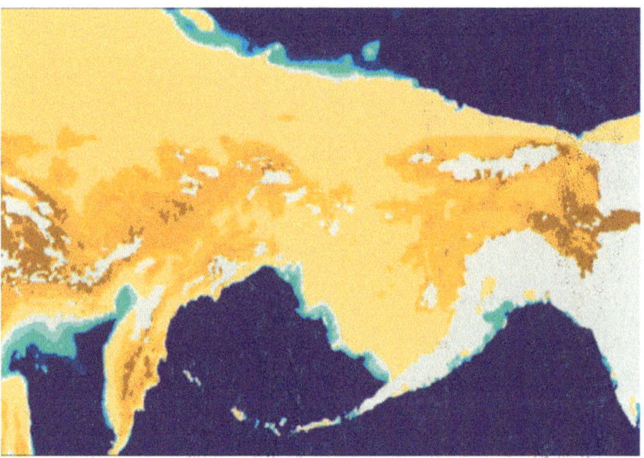

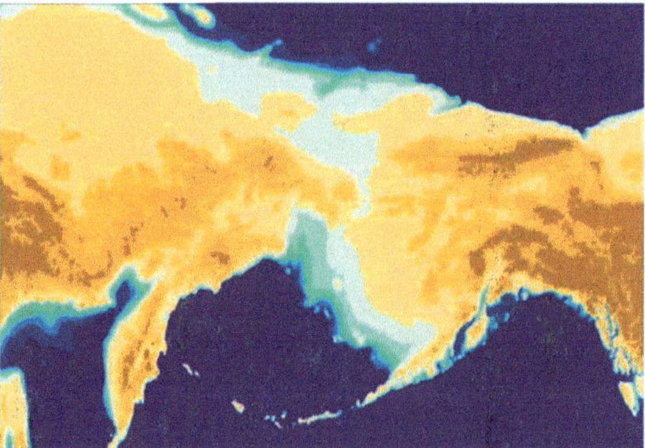

FIGURE 5.4 Geological and climatic events can trigger range changes. (A) Contraction (left) and subsequent expansion (right) of the Wet Tropics ecosystem in northeastern Australia over the last 18,000 years. The geographic distribution of many rainforest specialists can fluctuate with changing temperatures. (B) The Bering Land Bridge ~21,000 years ago (left) compared to the current coastlines of Asia and North America (right). Changes in glaciation and the tectonic movement of continents can alter dispersal corridors—and the geographic distribution—of species.

Reflection: Consider the Wet Tropics example where rainforest habitat contracted and expanded with glacial cycles. At the species level, this caused changes in range size for many rainforest specialist species. But what effect might these habitat changes have on other biological levels like population size and genetic diversity?

Source: (A) Craig Moritz Research Group; (B) NOAA/Wikipedia

The opposite pattern, where ranges expand instead of contract, can also be facilitated by glaciation. Figure 5.4 shows how the Bering Land Bridge connected North America and Asia during the last glacial maximum. Glacial expansion lowered sea levels more than 50 meters below current levels, opening a land bridge and allowing some species to significantly broaden their ranges (Hopkins 1959; Elias et al. 1997). Thus, changes over geological timescales can also create new corridors for long-range dispersal.

We just looked at an example of ancient range changes due to glaciation, but range changes can occur for many reasons. Abiotic and biotic conditions are dynamic over multiple spatial and temporal scales. For example, mountains rise and fall, rivers change course, volcanos erupt, local climates change, key nutrient or food resources shift, or new antagonists arrive. Any of these environmental changes can alter organismal survival by making formerly suitable areas inhospitable (or vice versa). Environmental changes can also alter movement patterns by opening or closing corridors. Altered patterns of survival or dispersal can ultimately lead to large-scale range changes.

Range changes can also occur due to evolution, for example, when species evolve the ability to exploit new environmental niches (see *Core Concepts* feature). It is important to remember that range changes often result from the interplay between several factors. For example, changes in the magnitude or direction of dispersal could be coincident with increased or decreased survival of dispersers in some directions, ultimately impacting the geographic distribution of species.

WHAT TYPES OF RANGE CHANGES OCCUR IN RESPONSE TO ANTHROPOGENIC PRESSURES?

Although shifts in the geographical distribution of species have occurred throughout the history of life on Earth, anthropogenic impacts have accelerated the rate and extent of range changes. As we have seen, human activities are changing global, regional, and local environmental conditions at a startling rate. Human activities also directly alter patterns of movement for many species. For example, our transportation networks literally provide vehicles for dispersal, dramatically increasing the dispersal range for many species. On the other hand, land-use change can reduce the total area of—and the connectivity between—suitable habitat patches, ultimately reducing range sizes and the potential for dispersal.

Numerous frameworks have been proposed for categorizing types of range changes (e.g., Maggini et al. 2011; Lenoir & Svenning 2015). Figure 5.5 displays the three most basic patterns. First, **range contractions** occur when a species' geographic distribution shrinks and the species occupies a subset of its historical range. Second, **range expansions** occur when a species' distribution increases and encompasses a larger geographic region than it did previously. Third, **range "marches"** occur when a species experiences a contraction in one part of its range and an expansion in another. Let us look at examples of each of these patterns.

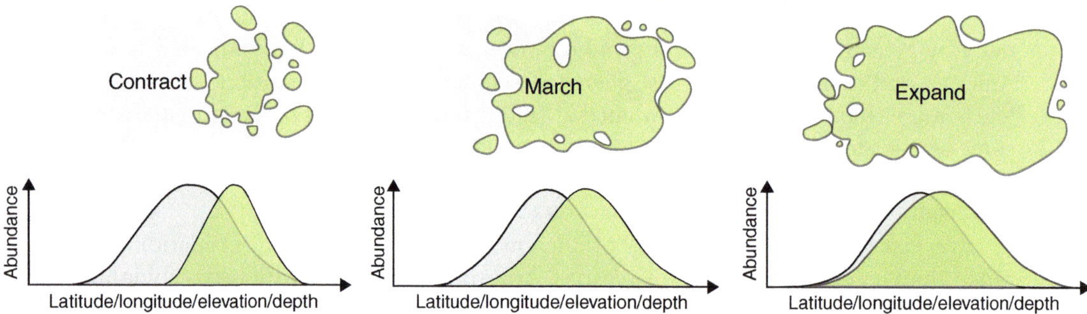

FIGURE 5.5 Three basic types of range change. Ranges are depicted as abundance along a geographic gradient. Gray indicates historical range, and green indicates shifted range. **Range contractions** can be due to overall retraction and/or fragmentation. **Range marches** occur when contraction and expansion occur simultaneously at different margins of the range. **Range expansions** reflect expansion to new geographic localities. This framework can be applied to different geographic gradients (e.g., latitude, longitude, elevation, depth) for terrestrial or marine species.

Reflection: What is an environmental change that could lead to a "march" pattern?

Source: Lenoir, J. and Svenning, J. (2015), Climate-related range shifts – a global multidimensional synthesis and new research directions. Ecography, 38: 15–28

Range Contractions

Range contractions are often a *direct* result of human activities like habitat modification or overharvesting. For example, the expansion of African oil palm (*Elaeis guineensis*) to other continents has caused range contractions for numerous endemic plant and animal species, as forests are converted into agriculturally intensive plantations (Koh & Wilcove 2008). In another example, the American bison once roamed the widespread grasslands of North America, but due to overharvesting, bison now are restricted to approximately 1% of their historic range (Sanderson et al. 2008; United States Fish and Wildlife Service 2014). Similar patterns of exploitation-driven range contractions are found in the marine realm (e.g., Worm & Tittensor 2011).

Range contractions can also be caused *indirectly* by human activities. For example, Figure 5.6 shows how the loss of polar sea ice due to anthropogenically induced climate change is projected to dramatically reduce the distribution of emperor penguins (*Aptenodytes forsteri*) (Jenouvrier et al. 2014). Range contractions can obviously have very serious implications for species survival. In the extreme, range contractions can lead to extinction, an outcome we will discuss in Chapter 8.

Range Expansions

Anthropogenically mediated range expansions are increasingly common as our transport and trade systems become globalized. Land, water, and air vessels intentionally or unintentionally move organisms to new localities. A subset of species that are moved then establish successfully in the introduced range. For example, ice plant (*Carpobrotus edulis*) is native to South Africa but was introduced to coastal California to help stabilize soils and reduce erosion (Conser & Connor 2009). Ice plant has since

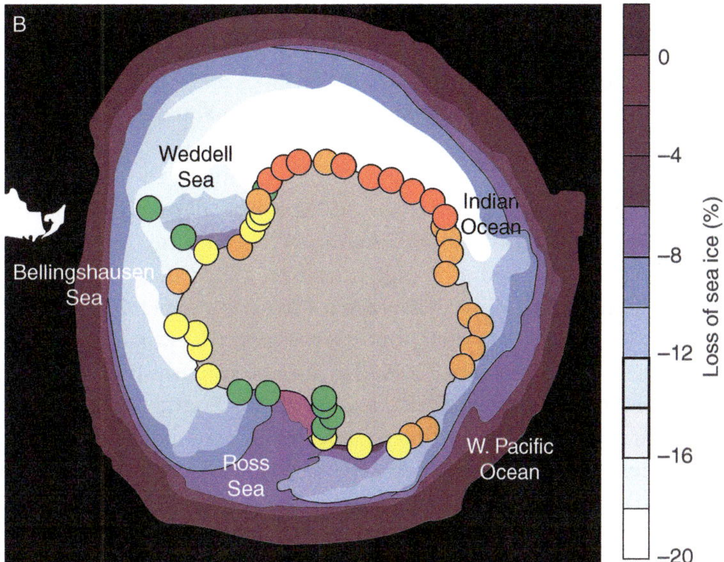

FIGURE 5.6 (A) Emperor penguins. (B) Projected and current range of emperor penguins. Purple scale bar shows mean percentage loss of sea ice over the last 100 years. Circles represent penguin colonies and are colored by conservation status (red populations are likely to be lost by 2100; orange and yellow populations are highly vulnerable; only green populations are considered not threatened).

Reflection: What would be required to reverse the range contraction of the emperor penguin?

Source: (A) Ian Duffy/Wikipedia; (B) "Projected continent-wide declines of the emperor penguin under climate change Stéphanie Jenouvrier, Marika Holland, Julienne Stroeve, Mark Serreze, Christophe Barbraud, Henri Weimerskirch & Hal Caswell, Nature Climate Change volume4, pages715–718 (2014)"

expanded its introduced range and now inhabits many Mediterranean-like ecosystems, including coastal California, Australia, New Zealand, and Southern Europe (Weber 2017).

In the marine realm, lionfish (genus *Pterois*) are native to tropical Indo-Pacific waters. However, they invaded Atlantic and Caribbean waters in force, likely released from the aquarium trade. Not only has the lionfish range expanded dramatically, but their abundance in their new habitat is staggering (Ballew et al. 2016). Species like ice plant and lionfish that not only establish in a new area but also have negative economic or ecological effects in their expanded range are considered **invasive species**. The impacts of invasive species are explored further in the *Taking a Closer Look* feature at the end of this chapter.

Range Marches

Contractions and expansions are not necessarily mutually exclusive. For example, contraction can occur at one range margin while expansion occurs at another. Thus, the overall size and shape of the range could remain relatively stable while the geographical area encompassed by the range shifts. These range "marches" are commonly observed in response to global warming. As the Earth's climate warms, species with narrow thermal tolerances may no longer be able to survive in portions of their current range. At the same time, new areas may become suitable for survival. The most consistent patterns observed with global warming are distributional shifts toward the poles, higher elevations, and deeper waters.

The observation that range changes show a consistent signature with global warming was famously formalized by meta-analysis studies published in 2003 (Parmesan and Yohe 2003; Root et al. 2003). In one of these studies, Parmesan and Yohe (2003) looked at data from over 1,000 species (spanning plants, invertebrates, and vertebrates in terrestrial and marine systems). They found that more than 80% of the species evaluated exhibited a change in the predicted direction at their upper or poleward range boundaries, with an average 6.1 kilometers per decade shift.

More recent meta-analyses suggest that range shifts are happening even faster than previously reported and also more explicitly link range marches to species tracking changing temperatures. For example, Figure 5.7 shows data that observed latitudinal changes across diverse species closely match expected latitudinal shifts based on temperature data (Chen et al. 2011). Moreover, patterns may not be equivalent across different biomes. For example, Sorte et al. (2010) studied patterns of range shifts for >120 marine species and found that range changes were in the direction predicted by

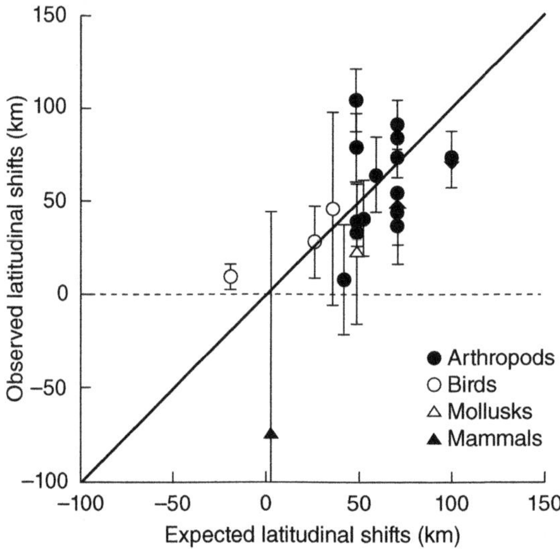

FIGURE 5.7 Observed and expected latitudinal shifts with global warming. Points represent mean responses of a particular taxonomic group in a particular region (e.g., European birds). Positive values represent shifts poleward. The diagonal represents complete concordance between observed and expected responses. Apart from mollusks (open triangle), most groups evaluated closely fit expectations of moving poleward with climate change.

Reflection: What is one hypothesis for why some groups of organisms like mollusks do not exhibit range shifts that track temperature?

Source: Rapid Range Shifts of Species Associated with High Levels of Climate Warming, BY I-CHING CHEN, JANE K. HILL, RALF OHLEMÜLLER, DAVID B. ROY, CHRIS D. THOMAS, SCIENCE 19 AUG 2011: 1024–1026

climate change and that, on average, marine species exhibited range shifts of 19 kilometers per *year*. This rate of shift is an order of magnitude faster than that observed in terrestrial systems, likely because dispersal occurs more easily in marine systems. The signature of climate-mediated range changes in marine systems is quite widespread, with examples from nearly every group of organisms from zooplankton to invertebrates to vertebrates (Poloczanska et al. 2016).

In addition to evidence from meta-analyses, there are numerous examples of range shifts for individual species associated with global warming. One classic example is the Edith's checkerspot butterfly (*Euphydryas editha*), which is distributed across much of the western United States and functions as a metapopulation. A **metapopulation** is a group of populations that are geographically separated but still interact through (common or rare) movement of individuals. In a metapopulation, extinction and recolonization are common: although individual populations may not survive year to year, the species as a whole has a stable range because populations are recolonized. In the last 100 years, Edith's checkerspot butterfly populations have experienced decreased survival at the southern range boundary and increased survival at the northern range boundary (Parmesan 1996). The net effect is that the mean range has shifted ~100 kilometers north (Parmesan 1996). This example underscores how range changes can occur through differential extinction as populations are lost at one edge of the distribution.

Although changes in latitude and longitude garner much attention, shifts in altitude (in terrestrial systems) and depth (in marine systems) are also commonly caused by climate change. One example from the Sierra Nevada Mountains of California is shown in Figure 5.8 and further highlighted in the *Meet the Data* feature at the end of this chapter. Over the last hundred years, many mid-elevation species like the piñon mouse (*Peromyscus truei*) have shifted to higher altitudes as they track cooler temperatures in a warming environment (Moritz et al. 2008). However, high-elevation species like the pika (*Ochotona princeps*) literally have nowhere to go and experience only lower elevation contraction. Thus, organisms that live in high-alpine habitat are particularly vulnerable as their populations become small, isolated, and pushed to the brink of available habitat.

In Sum

It is important to note that anthropogenically mediated changes in species ranges can have a variety of impacts. The gain or loss of one species can have repercussions for other species—and processes—in the ecosystem, a topic we will return to in Unit III. Moreover, changes in species ranges can have detrimental impacts on human societies. For example, invasive species generate billions of dollars of damage each year, an issue we will evaluate in the *Taking a Closer Look* feature in this chapter. Another clear example comes from fisheries. Many local communities and national economies depend on marine resources. As many marine species shift to higher latitudes and deeper waters, both food and economic security can be jeopardized (Allison et al. 2009; Cheung et al. 2013). Thus, it is imperative to consider the interdependence of biodiversity conservation and human well-being, a theme we will explore further in Unit IV.

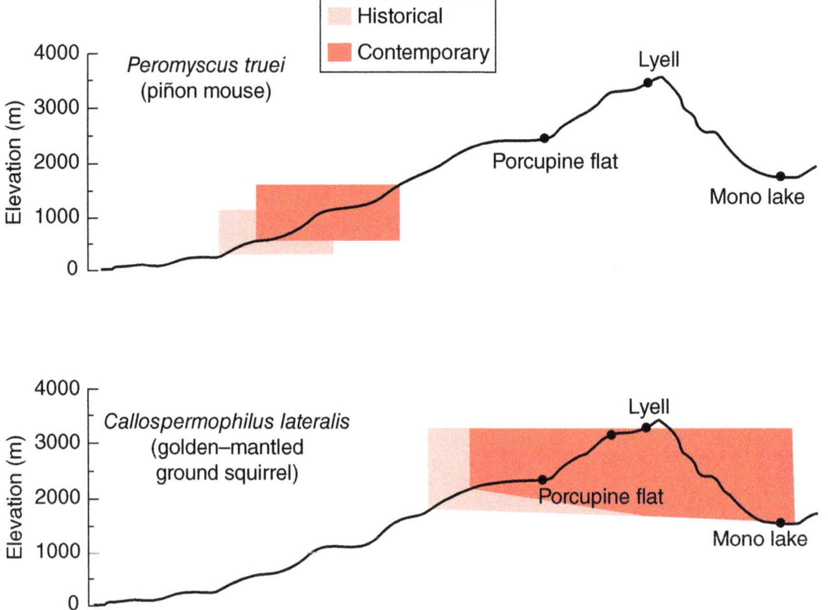

FIGURE 5.8 Elevational shifts for terrestrial species and depth shifts for marine species are a common response to global warming. For example, in the Sierra Nevada Mountains of California, many mid-elevation species, like the piñon mouse, have shifted to higher altitudes over the last 100 years. However, many higher-elevation specialists, like the ground squirrel, show only lower range contractions because mountains have tops and there is nowhere higher for temperature-sensitive species with high-elevation distributions to go.

Reflection: What do you predict will happen to the elevational distribution of these two species in another 100 years? How do you think changes in elevational distribution will translate to changes in entire species ranges?

Source: James L. Patton, UC Berkeley

HOW DO SCIENTISTS PREDICT RANGE CHANGES?

How do scientists predict whether species will exhibit distributional changes in response to global change stressors? Many different empirical, mathematical, and computational approaches are used to study spatial movement patterns of organisms (e.g., Schick et al. 2008; Yalcin & Leroux 2017). Observational and synthesis approaches are particularly effective for studying range changes that have already occurred, while modeling approaches are powerful for predicting future range changes.

Although there are numerous methods available, here we will focus on **species distribution modeling** (SDM), a widely used tool for predicting future range changes. SDMs are mathematical models that relate information about where a species is currently found to information about the environmental characteristics of those locations (Elith & Leathwick 2009). Then species occurrence under future environmental conditions can be predicted. Figure 5.9 provides a conceptual overview of the process.

To initiate an SDM, researchers typically answer the following questions to generate input data: Where does the focal species currently occur? What are the environmental characteristics of these localities? Then most SDMs produce a **habitat suitability map** as their output, which can be applied across space and time. Thus, using a species' current habitat preferences, SDMs can address the question: Where is the focal species likely to occur in the future as the environment changes?

SDM approaches are used widely to predict the effect of environmental change (like climate change) on species distributions. Specifically, scientists can forecast range changes by comparing the location of current and future suitable patches of habitat. For example, the Western Pine beetle (*Dendroctonus brevicomis*) is a deadly pathogen of pine trees in western North America. The northern extent of this species' range has likely been limited historically by cold temperatures, which decreased survival (DeMars & Roettgering 1982). However, SDMs have shown that the range of this species is expected to expand northward with global warming, which will have important consequences for forest health (Evangelista et al. 2011).

SDM approaches are extremely flexible and can be used to make predictions about species distributions over different timescales and under different environmental scenarios. SDMs can also be constructed for individual species or for communities, allowing comparison across multiple species in a region. Therefore, these modeling approaches are also often used in conservation planning, for example, to predict the extent of future spread of an invasive species (Figure 5.9) or to predict the future species composition in a protected area.

The SDMs we have discussed so far fall broadly under the category of *correlative* SDMs, models that link species distributions and environmental parameters. Despite their broad utility, correlative SDMs have limitations. For example, SDMs generally use broad-scale environmental data, which do not reflect variation at a scale relevant for many organisms. Thus, the availability of suitable microhabitat cannot readily be inferred. In addition, correlative SDMs historically focused on climate variables when determining habitat suitability. Of course, other global change drivers besides climate—like land-use change, chemical contamination, and invasive species—may cause some locations to be uninhabitable even if climate is suitable. These additional factors are increasingly being incorporated into SDM analyses (Sohl 2014). In addition, correlative SDMs focus on the indirect link between species distributions and environmental parameters. They typically do not include detailed knowledge about the functional links between organisms and their environments.

Therefore, in the last decade, there has been increasing interest in *mechanistic* SDMs, which explicitly link environmental parameters with organismal performance (e.g., Kearney & Porter 2009; Buckley et al. 2010). Mechanistic SDMs are more data intensive because they require a detailed understanding of organismal morphology, physiology, and behavior under a range of environmental conditions. For example, a recent mechanistic model looking at the endangered fish species allis shad (*Alosa alosa*)

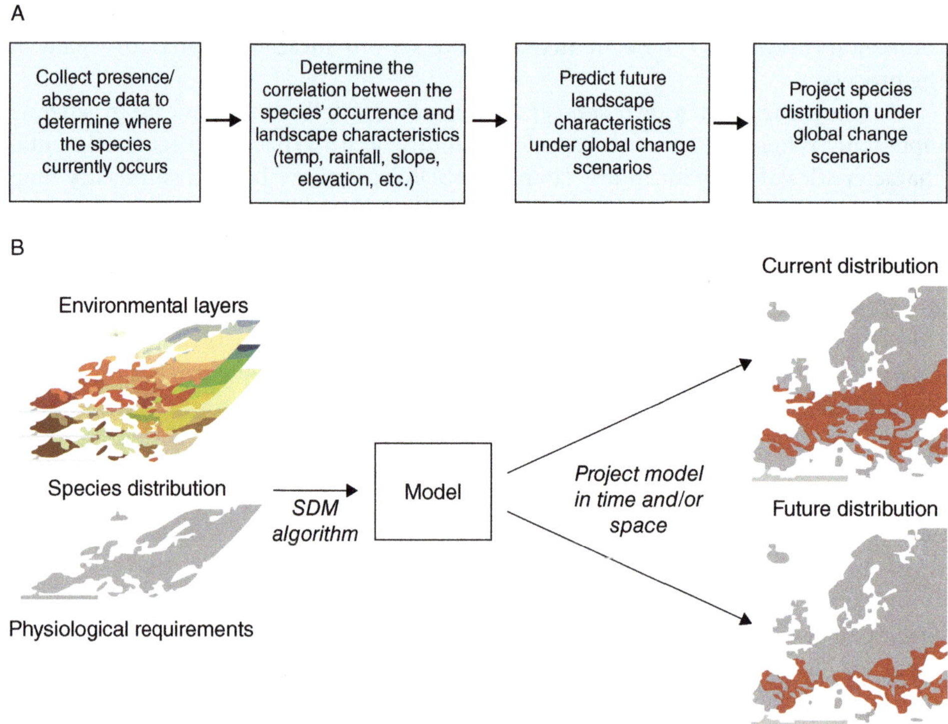

FIGURE 5.9 Species distribution modeling in theory and practice. (A) A conceptual flow chart of the species distribution modeling (SDM) approach. (B) An example of the input and output layers of SDMs. (C) A real-life example predicting the spread of an invasive species, the barred owl, into the habitat of the native spotted owl from a 2003 study. Since that time, the barred owl has invaded a significant portion of native spotted owl habitat and has had a dramatic effect on spotted owl populations due to competitive exclusion (Gutierrez et al. 2007). The interaction between these two owl species is a highly contentious issue and some management agencies are considering lethal removal of barred owls from some spotted owl habitat.

Reflection: For each step of the modeling process, brainstorm at least one challenge associated with gathering the necessary data or drawing the future inference.

Source: (A) Courtesy of the author; (B) "Applications of species distribution modeling to paleobiology, Jens-Christian Svenning, CamillaFløjgaarda, Katharine A.Marskeb, DavidNógues-Bravo, Signe Normand, Quaternary Science Reviews Volume 30, Issues 21–22, October 2011, Pages 2930–2947"; (C) "Using Ecological-Niche Modeling to Predict Barred Owl Invasions with Implications for Spotted Owl Conservation Uso de Modelos de Nicho-Ecológico para Predecir Invasiones de Strix varia y Sus Implicaciones para la Conservación de Strix occidentalis, A. Townsend Peterson C. Richard Robins, Conservation Biology, Volume 17, Issue 4, August 2003, Pages 1161–1165"

C

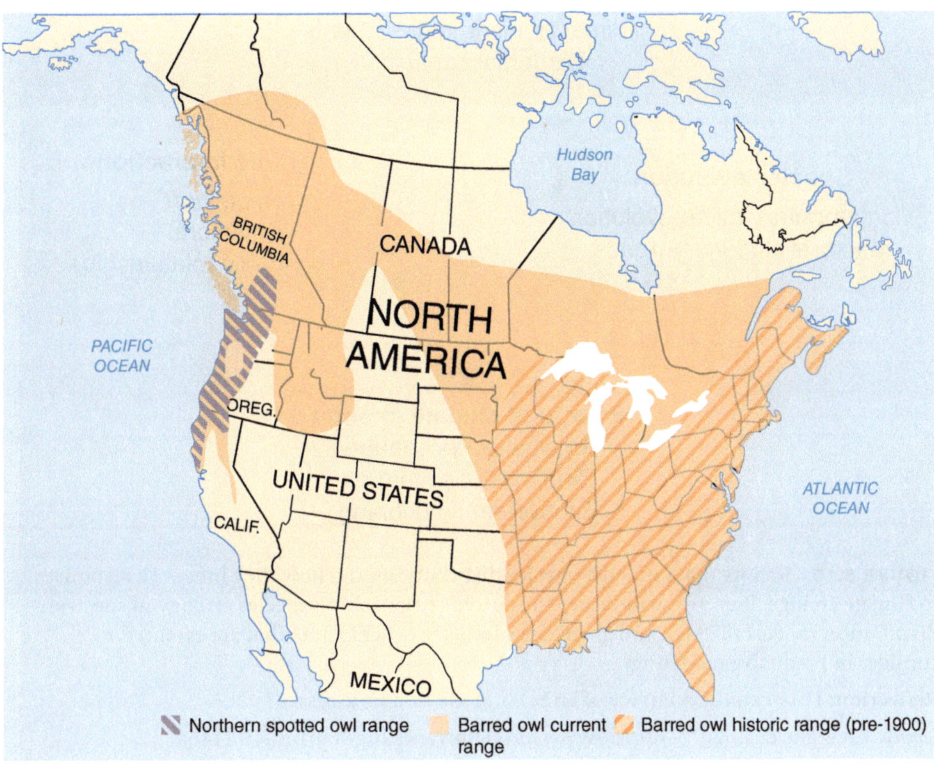

FIGURE 5.9 Continued.

incorporated not only dispersal capability but also patterns of growth, reproduction, migration, and survival across a complex life cycle (Rougier et al. 2015). By including data on fish physiology and changing water temperatures, the researchers were able to infer why the species disappeared from certain river basins. Thus, mechanistic models improve biological realism.

Whether using correlative or mechanistic SDMs, the leading edge of species distribution modeling is to incorporate additional information on ecological and evolutionary processes. As we discussed earlier, functional data can be included about the processes that influence species' survival and reproduction. In addition, SDMs can more explicitly account for biotic interactions. Predators, competitors, or mutualists can be important drivers of species distributions—for example, if a key plant pollinator is absent from a region. Finally, SDMs can also more explicitly address evolutionary processes. SDMs often assume that species' niches are unchanging over time, but organisms can adapt or adjust to new conditions. As illustrated in Figure 5.10, incorporating biotic, abiotic, and evolutionary factors that influence the potential for a move response will allow more sophisticated predictive modeling efforts (Huntley et al. 2010; Lavergne et al. 2010).

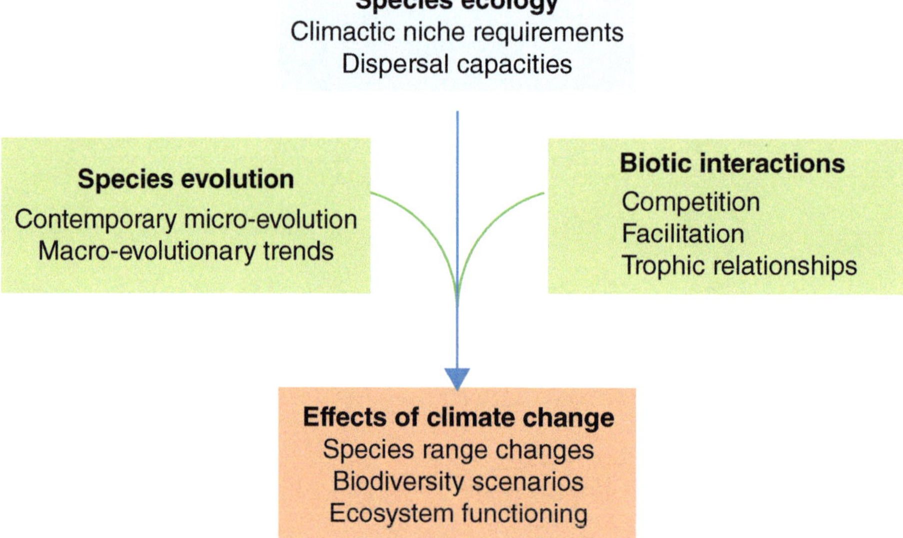

FIGURE 5.10 Factors that influence species distributions and how they influence response to climate change. Blue arrow indicates the current emphasis on species ecology in species distribution models (SDMs), and green lines indicate other factors that are essential to consider in predictive modeling.

Reflection: How can the accuracy of an SDM prediction be assessed?

Source: Lavergne et al. 2010, https://www.annualreviews.org/doi/full/10.1146/annurev-ecolsys-102209-144628

CONCLUSION

The move response can be critical for species survival in a rapidly changing world. Anthropogenic impacts have already altered the geographic distribution of many species on our planet. Moreover, predictive models suggest that there may be little geographic overlap between current species distributions and the regions that will be suitable for them in the future (Thomas et al. 2004). Thus, innate dispersal capability will continue to play an important role in the fate of many species. However, anthropogenic activities are also directly affecting the potential for organismal movement. Habitat loss and habitat fragmentation reduce dispersal for many species, while globalized transportation networks increase dispersal for others.

Ultimately, whether species will exhibit a successful "move" response to anthropogenic stressors is determined by the interaction of multiple factors. Organismal traits—like dispersal ability—influence whether species can move far enough and fast enough to track changing conditions. Other organismal traits—like reproductive mode and competitive ability—influence how likely species are to establish and proliferate in a novel environment (Estrada et al. 2015). Characteristics of the environment—like whether there are clear migration corridors—influence whether species can successfully move to a new habitat patch. The dynamics of change—like pace and magnitude—influence whether species are able to track changing conditions. Thus, the move response can be limited by intrinsic organismal traits (such as lack of a mobile life stage), landscape features (such as highly fragmented habitats), or dynamics of global change (such as rapid global warming).

Of course, the move response is not mutually exclusive from the other core responses. When organisms move, they often encounter new biotic or abiotic conditions such as novel climates or unfamiliar competitors. Newly established populations may then adjust or adapt to these conditions. In the next chapter, we turn to the adjust response, which influences whether populations can habituate to novel conditions and ultimately survive and thrive in new geographic regions.

MEET THE DATA A CENTURY OF CHANGE IN YOSEMITE

Who Are the Scientists and What Did They Set Out To Do?

Here we follow the story of two scientists, living 100 years apart but linked through time by their research (Box Figure 5.2). Joseph Grinnell was the first director of the Museum of Vertebrate Zoology (MVZ) at UC Berkeley, beginning his appointment in 1908. One of Grinnell's lasting scientific contributions was detailed faunal surveys across California, which provided important baseline data about the geographic distributions of vertebrates prior to a century of escalating anthropogenic impacts.

Grinnell's prescience is captured in one of his articles from 1910: *"At this point I wish to emphasize what I believe will ultimately prove to be the greatest value of our museum. This value will not, however, be realized until the lapse of many years, possibly a century, assuming that our material is safely preserved. And this is that the student of the future will have access to the original record of faunal conditions in California and the west, wherever we now work."* Nearly 100 years later, that "student of the future" arrived. Dr. Craig Moritz served as the Director of the MVZ from 2000 to 2012.

One of Grinnell's most sustained efforts was to collect data across an elevational transect in Yosemite National Park. Grinnell and his team collected more than 4,000 specimens, wrote more than 3,000 pages of field notes, and took more than 500 photographs while sampling across an approximately 3,000 meter rise in elevation. Grinnell's efforts were presented in a 1924 publication entitled *Animal Life in the Yosemite* (Grinnell & Storer 1924). Nearly 100 years later, Moritz initiated the "Grinnell Resurvey Project" to revisit Grinnell's transect in Yosemite and evaluate how species

BOX FIGURE 5.2 One hundred years apart, Dr. Joseph Grinnell (A) and Dr. Craig Moritz (B) shared a passion for field research and understanding how ecosystems change over time. Both scientists are shown in the field with Dr. Grinnell preparing museum specimens and Dr. Moritz searching for reptiles and amphibians.

Source: (A) The Bancroft Library/University of California; (B) Photo courtesy of Rayna Bell

(Continued)

distributions along this elevational transect changed over a century of global warming (Box Figure 5.3). Data from regional weather stations indicate a 3.7 degrees Celsius rise in average monthly minimum temperature in this area. Because Yosemite National Park has been protected since the late 1800s, Moritz and his team could focus on the effects of climate change without the confounding influence of large-scale habitat alteration.

What Are *Your* Predictions?

Before you read on, take a few minutes to think about how to apply what you have learned in this chapter about species distributions and global warming to the Yosemite case study. Start by answering the following questions:

- *Do you predict that species distributions will have changed in Yosemite over the last century?*
- *If so, what general direction do you expect to see distributional shifts?*
- *Do you expect responses to be similar across all species? For example, do you predict a similar response for low-elevation and high-elevation species?*

Now write a summary prediction statement about whether—and how—species distributions are likely to have changed in Yosemite over the last century.

What Were the Scientists' Predictions?

Here is how the authors formalized their predictions: *"Given marked regional warming over the past century, we predicted that species ranges should have shifted upward. This should manifest as upward contraction of the lower range limit for mid- to high-elevation species, upward shift of the entire range or expansion of the upper limit for low- to mid-elevation species, and altered community composition within elevational bands."*

What Data Were Collected?

On the surface, the data for this study seem simple: records of specific species at specific localities at two time periods. However, there are always many important decisions about data collection, data handling, and data analyses. For example, Moritz et al. selected a taxonomic focus of small mammals (e.g., mice, rats, shrews, chipmunks, squirrels) given the relative ease of trapping and preponderance of these species in the earlier Grinnell dataset. Then they decided on a scale for data aggregation (records were pooled into sampling sites if they were within 2 kilometers and at a 100 meter elevation).

The scientists also worked to ensure that historical and contemporary data were maximally comparable—for example, ensuring that researchers were consistent across sites and through time in their survey effort. Survey effort affects species detectability: If a species is not found at a locality during a sampling period, it may not live there or it may have been undetected. In other words, if a species is absent from the dataset, it does not necessarily mean it is absent from the sampling locality. Moritz and his team used sophisticated statistical analyses to estimate—and correct for—any difference in detectability between sampling periods. These statistical analyses allowed them to determine if range shifts over time were statistically significant.

What Is *Your* Interpretation of the Data?

Before you read on, take a few minutes to interpret data from the study shown in Box Figure 5.4. Consider the following questions as you look at the figure:

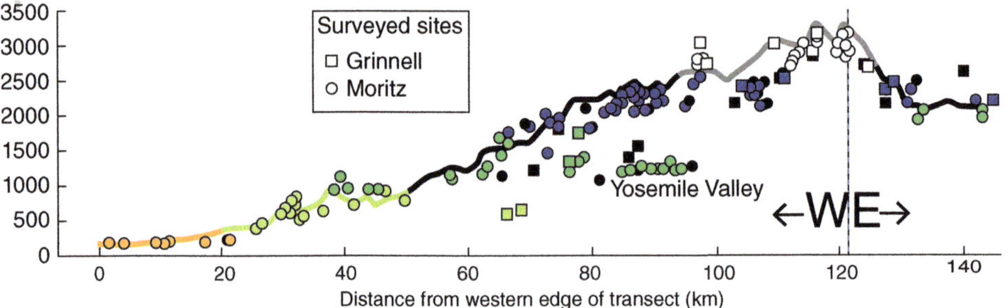

BOX FIGURE 5.3 Elevational profile of sampled sites across a transect in Yosemite National Park. Historic sites were those visited by Grinnell and his team in the early 1900s, and current sites were those visited by Moritz and his team in the early 2000s.

Source: Impact of a Century of Climate Change on Small-Mammal Communities in Yosemite National Park, USA, BY CRAIG MORITZ, JAMES L. PATTON, CHRIS J. CONROY, JUAN L. PARRA, GARY C. WHITE, STEVEN R. BEISSINGER, SCIENCE 10 OCT 2008: 261–264

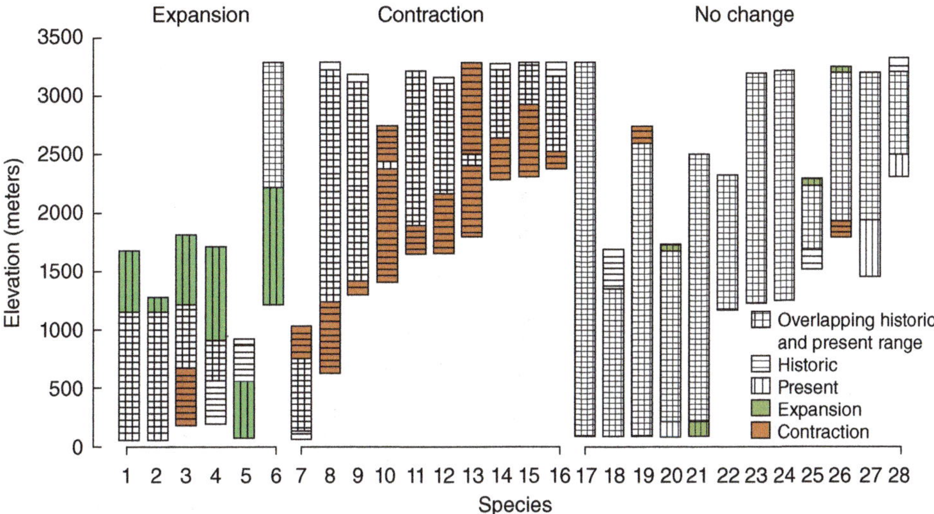

BOX FIGURE 5.4 Elevational range changes for 28 species over a century in Yosemite National Park. Statistically significant range shifts are color coded (green for expansion and brown for contraction). Species in the "no change" category did not show statistically significant range changes *or* showed range changes that were considered "biologically trivial" (<10% of previous range).

Source: Impact of a Century of Climate Change on Small-Mammal Communities in Yosemite National Park, USA, BY CRAIG MORITZ, JAMES L. PATTON, CHRIS J. CONROY, JUAN L. PARRA, GARY C. WHITE, STEVEN R. BEISSINGER, SCIENCE 10 OCT 2008 : 261-264

- *What do the axes and the color coded bars represent?*
- *How consistent are the patterns observed across species?*
- *Which species exhibit a shift in the predicted direction and which do not?*

Write a sentence or two describing the *key findings* of this study and their *significance.*

What Was the Scientists' Interpretation of the Data?

Box Figure 5.4 illustrates elevational range changes for 28 focal species. Another way to look at the same data is to pool data across species as shown in Box Figure 5.5. Moritz and his team found that species distributions had shifted for more than half of the focal species, and the general trend was range shifts upward in elevation. Affected species showed an average increase in elevation of 500 meters over the centennial timescale. This key finding is consistent with initial predictions and trends observed for species around the world. However, not all species achieved this upper range shift in the same way. In general, low-elevation species were more likely to show range expansions (expanding into a higher elevational distribution), and high-elevation species were more likely to show range contractions (losing the lower part of their elevational distribution). These findings support initial predictions and the reality that high-elevation species have nowhere higher to go. The significance of this study is in showing elevational range changes coincide with a century of climate warming across an entire community in a relatively pristine habitat.

What Are *Your* Ideas for Future Research Directions?

Given what the Grinnell Resurvey Project accomplished, what next steps do you envision for this research program? If you had been involved in the Moritz et al. study, how would you follow up? Start by considering the following questions.

- *Does this study provide baseline data that could be used for a future study?*
- *How could the study be broadened taxonomically or geographically to achieve more generality?*
- *Are there predictions about the future that could be made and later tested?*

Now write a few sentences about future directions on this research theme.

(Continued)

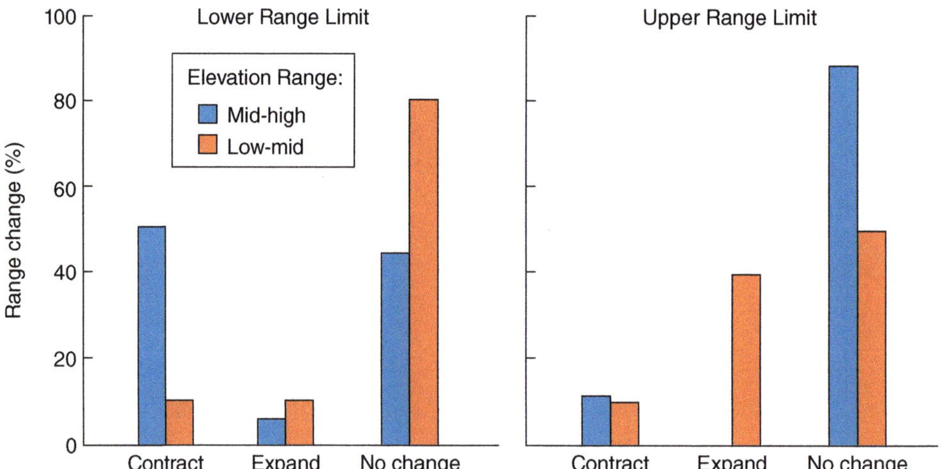

BOX FIGURE 5.5 Elevational range changes for 18 species that were historically found at mid–high elevations (blue bars) and 10 species that were historically found at low–mid elevations (orange bars).

Source: Impact of a Century of Climate Change on Small-Mammal Communities in Yosemite National Park, USA, BY CRAIG MORITZ, JAMES L. PATTON, CHRIS J. CONROY, JUAN L. PARRA, GARY C. WHITE, STEVEN R. BEISSINGER, SCIENCE 10 OCT 2008 : 261-264

TAKING A CLOSER LOOK GLOBALIZATION AND INVASIVE SPECIES

Globalized transport and trade have altered the distribution of countless species on our planet. The **native range** of a species is its geographic distribution prior to (direct or indirect) human influence. Many species could survive outside their native range but have historically been dispersal limited. However, human activities have increased the dispersal range of many species, providing opportunities for species to arrive—and potentially thrive—outside their native range. Box Figure 5.6 illustrates the increase of non-native species in Europe over the last several hundred years.

The issue of invasive species is clearly linked to the move response because the root cause of many biological invasions is anthropogenic movement of species outside of their native ranges. However, the dynamics of invasive species also relates to other core responses. For example, the ability to adapt or adjust may facilitate invasion, and invasive species can trigger losses of native species.

What Is an Invasive Species?

The USDA defines invasive species as species that are "1) non-native to the ecosystem under consideration and 2) whose introduction causes or is likely to cause economic or environmental harm or harm to human health."

What Are the Impacts of Invasive Species?

Invasive species can cause harm in different ways. From an ecological perspective, invasive species can be damaging as they enter naive ecosystems as predators, competitors, or disease vectors. In fact, many analyses suggest that next to habitat loss and climate change, invasive species are the greatest contemporary threat to biodiversity. Some estimates suggest that nearly half of all threatened or endangered species are imperiled by invasive species (e.g., Wilcove et al. 1998).

From an economic perspective, invasive species cause more than 120 billion US dollars in damages *per year* (Pimentel et al. 2005). For example, invasive weeds cost the agricultural sector billions of dollars annually in crop losses and herbicide treatments. Invasive zebra mussels inflict millions of dollars per year in treatment costs for blocked pipelines and clogged municipal water intakes. Invasive microbial pathogens impact numerous species, with millions per year spent just to fight a single fungal pathogen

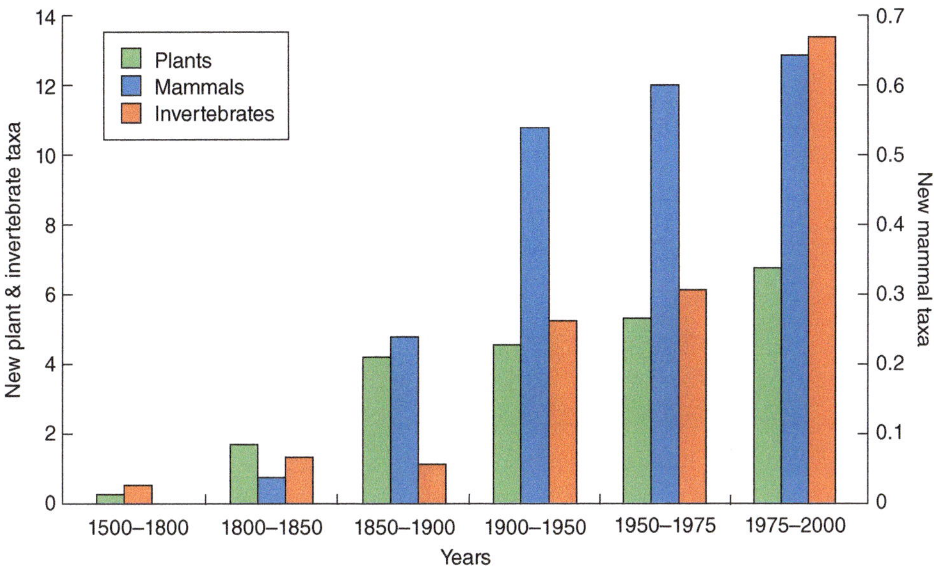

BOX FIGURE 5.6 Increase in establishment of non-native plants, invertebrates, and mammals in Europe since 1500.

Reflection: What is the biological reason there are two different scale bars in this figure?

Source: Trade, transport and trouble: managing invasive species pathways in an era of globalization, Philip E. Hulme, Journal of Applied Ecology, Volume 46, Issue 1, February 2009, Pages 10–18

affecting elm trees. Thus, dealing with the effects of even a single invasive species can cost millions or even billions of dollars.

There are hundreds of compelling examples of ecological impacts of invasive species, but let us will take a closer look at one exemplar. As shown in Box Figure 5.7, the Argentine ant (*Linepithema humile*) is native to South America. Over the last century, *L. humile* has been introduced to all continents (except Antarctica), likely through shipping of food goods. In its native range, the Argentine ant is not a dominant species and coexists with a number of other ant species. However, in parts of the introduced range, such as California, Argentine ants have rapidly displaced native ants to dominate ant communities.

One explanation for the success of *L. humile* in its introduced range is a shift in colony structure (Tsutsui et al. 2000). In many parts of the introduced range, Argentine ants are genetically similar and function as a "super colony." Instead of competing *intra*specifically with other *L. humile* colonies (as they do in their native range), they compete *inter*specifically with native ants in their introduced range. In other words, the introduced ants are highly aggressive to other species but do not keep neighboring populations of their own species in check. The invasion of Argentine ants has not only impacted native ant species but also has cascading effects on other species, including vertebrates. For example, the lizards in the genus *Phrynosoma* specialize as ant predators, but rapidly lose weight when fed a diet of Argentine ants (Suarez & Case 2002). Therefore, the invasion of a single species can have dramatic community-wide effects.

How Do Species Become Invasive?

Not all species that are transported outside their native range become invasive. To make this explicit, it is useful to consider distinct stages of invasion, as shown in Box Figure 5.8 (Lockwood et al. 2013). Of the species that are transported outside their native range, most are transported

(Continued)

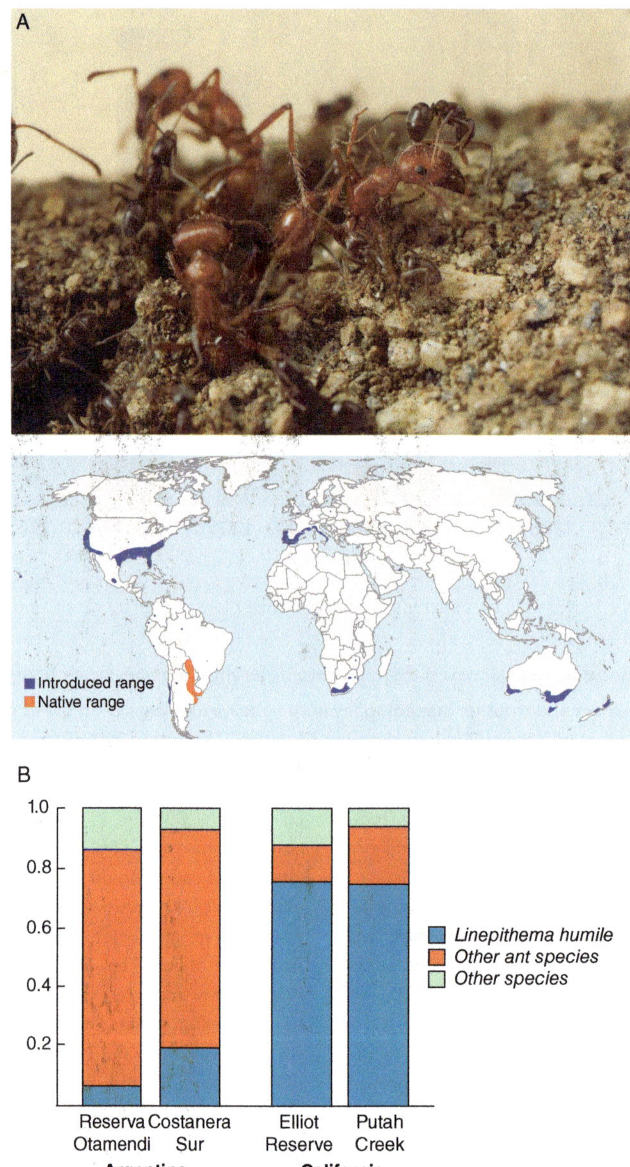

BOX FIGURE 5.7 The impact of invasive species. (A) The Argentine ant (*Linepithema humile*) is the small ant in the foreground (shown with a California native ant species) and has been introduced from South America to its current global distribution. (B) The proportion of baits dominated by different invertebrates at four different sites. The Argentine ant (shown in the blue bar) dominates the ant community in parts of its introduced range (like California) but not in its native Argentina.

Reflection: How does such a small ant have such an outsized effect in its invaded range?

Source: (A) Dong-Hwan Choe; (B) Suarez, A.V., Tsutsui, N.D., Holway, D.A. et al. Biological Invasions (1999) 1: 43

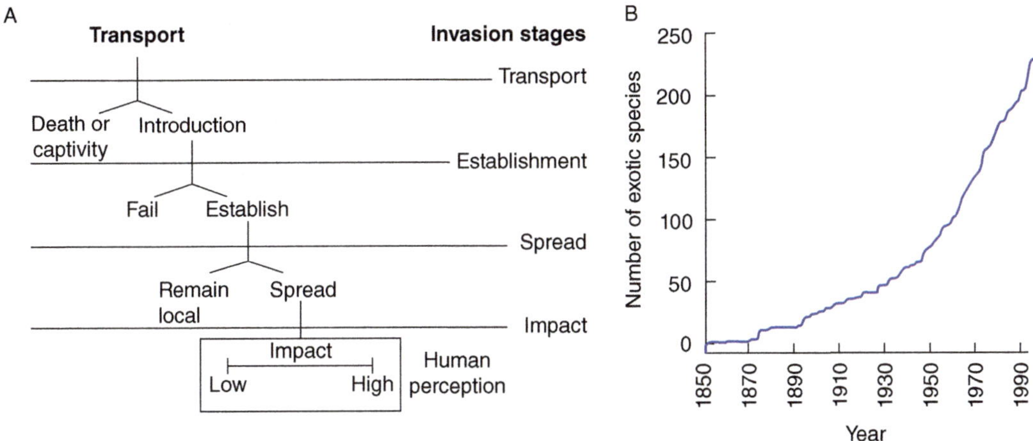

BOX FIGURE 5.8 (A) Stages of invasion (with alternative outcomes). (B) Increase of exotic species recorded in the San Francisco Bay over time.

Reflection: What is the difference between a non-native species and an invasive species? What might it mean for an introduced species to become naturalized rather than invasive?

Source: Accelerating Invasion Rate in a Highly Invaded Estuary, BY ANDREW N. COHEN, JAMES T. CARLTON, SCIENCE 23 JAN 1998: 555–558

unintentionally by our globalized trade system. For example, thousands of merchant vessels sailing the world's oceans each carry thousands of cubic meters of ballast water. This water is loaded at one port and then discharged at another, moving a tremendous amount of ocean water and any associated organisms. For example, in one sample of ballast water, more than 350 aquatic species were found (Cariton & Geller 1993). Of course, not all species that are transported will successfully establish. But even if a small proportion of transported species establish, they can have a dramatic effect. Take the example in Box Figure 5.8 of an exponential increase of non-native species in the San Francisco Bay over the last 150 years.

Of course, many species that are transported do not establish. And many that establish do not spread or cause harm. But for the small proportion of species that do become invasive, what factors explain their tremendous success in a new habitat? After all, they have left the context in which they evolved, so it would be natural to think they would be less suited to a new environment. However, changing ecological and evolutionary context can also allow species to escape from their natural enemies. In other words, invasive species can leave behind predators, competitors, and pathogens that kept their populations sizes in check in their native habitats. In fact, as Box Figure 5.9 shows, introduced species typically have fewer parasites (and fewer types of parasites) than native species (Torchin & Mitchell 2004), which can increase their chances of success in a new environment.

What Characteristics Define Invasive Species and Invaded Ecosystems?

It is difficult to generalize what makes a good invader because different traits may facilitate invasion in different species (Kolar & Lodge 2001). However, there are several categories of species-specific traits that can be important predictors of invasive potential. First, dispersal ability defines the likelihood that a species will be transported to new locations and also that a species will spread widely once introduced. Both intrinsic dispersal ability and anthropogenically mediated dispersal capability contribute. Second, capacity to thrive under varied conditions (for example, physiological tolerance) and use varied resources (for example, diet breadth) increases the likelihood of establishment. The ability to survive in diverse environments can be due to phenotypic plasticity and/or adaptive potential, topics we will explore in later chapters. Third, traits associated with reproduction can increase the likelihood of invasion, for example when species have high reproductive output, fast life cycles, and rapid growth. Finally, the likelihood of invasion increases if more individuals are released

(Continued)

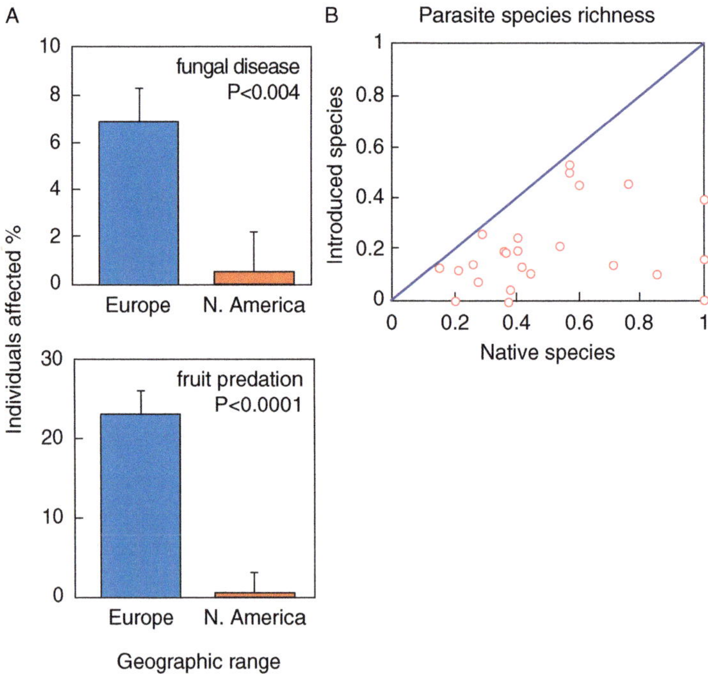

BOX FIGURE 5.9 Escape from antagonists. (A) The plant white campion (*Silene latifolia*) is subject to more predation, parasitism, and disease in its native range (Europe) relative to its introduced range (North America). (B) Meta-analysis (with many different taxa including mollusks, crustaceans, fishes, amphibians reptiles, birds, and mammals) showing that parasitism is higher in native species compared to introduced species. Parasite species richness is the number of parasite species found on a given host species. The same pattern holds for parasite prevalence, which is the percentage of hosts infected.

Reflection: What ecological or evolutionary hypothesis can you generate for why introduced species are less likely to acquire parasites than native species?

Source: (A) Wolfe 2002, https://www.journals.uchicago.edu/doi/abs/10.1086/343872; (B) Torchin et al. 2003, https://www.nature.com/articles/nature01346

and/or if there are more introduction events (Kolar & Lodge 2001). The **propagule size** matters because invasion of more individuals increase the likelihood that a founding population will persist.

Characteristics of ecosystems also determine how likely they are to be invaded. For example, more depauperate (less species-rich or biologically diverse) ecosystems are more likely to be invaded, as shown in Box Figure 5.10. This can amplify the threat of invasive species: as invasive species impact native species and reduce biological diversity, ecosystems can become even more susceptible to future invasion.

Why Do Invasive Species Lead to Biotic Homogenization?

Another feedback created by invasive species has been formalized in the **anthropogenically induced adaptation to invade (AIAI) hypothesis** (Hufbauer et al. 2012). Populations adapted to human habitats in their native range are more likely to be transported and survive in other human-altered environments outside their native range. This is intuitive when we consider that human commensals (like fruit flies and house mice) are well placed to be transported from one high-density human settlement to another and are well-adapted to survive in these human-modified environments upon arrival. Ultimately, the positive feedbacks associated with invasive species are one driver of **biotic homogenization**, the increased similarity (genetically, taxonomically, and/or functionally) of living things across locations (McKinney & Lockwood 1999). Biotic homogenization has been documented in both terrestrial and aquatic ecosystems (e.g., Rahel 2002; Olden 2006) and is often caused by the movement of invasive and domesticated species. Mitigating the effects of invasive species is a central goal of conservation efforts in many geographic regions, a topic we will return to in Chapter 11.

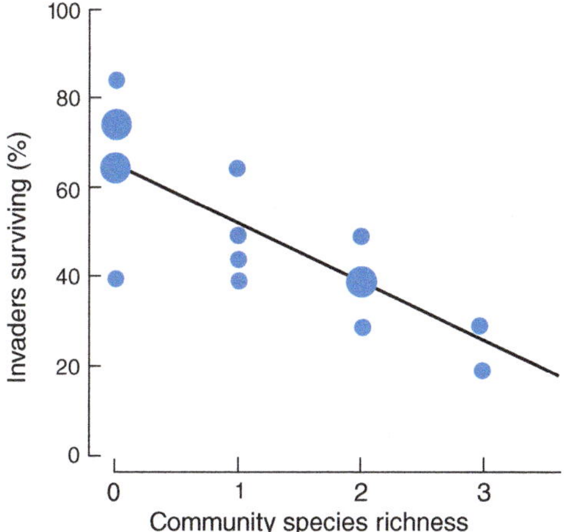

BOX FIGURE 5.10 Association between community species richness (number of species in an ecological community) and percent of invaders surviving. More depauperate communities tend to be more likely to be invaded.

Reflection: Why are more species-rich communities harder to invade?

Source: "Species Diversity and Invasion Resistance in a Marine Ecosystem BY JOHN J. STACHOWICZ, ROBERT B. WHITLATCH, RICHARD W. OSMAN, SCIENCE 19 NOV 1999: 1577–1579"

KEY CONCEPTS

What is the move response?

- Patterns of individual movement and distribution of populations and species are changing in response to global change stressors.

How and why do organisms move?

- Individuals move over many spatial and temporal scales triggered by different cues, and these individual movements affect the distribution of populations and species.

What is a geographic range?

- A geographic range is the total spatial extent of a species' distribution.

What factors determine a species' geographic range?

- Numerous factors—from evolutionary history to species traits to ecological requirements—interact to determine range sizes and shapes.

Do range changes occur even without anthropogenic influence?

- Geographic ranges are not stable through time, and range changes have occurred throughout history, even without human influence.

What types of range changes occur in response to anthropogenic pressures?

- Anthropogenic activities have accelerated the rate and extent of range changes. Range contractions are often caused by habitat loss and overharvesting, range expansions are often due to globalized transportation, and range "marches" are often triggered by global warming.

How do scientists predict range changes?

- Species distributional modeling (SDM) integrates species information (such as where a species is found) with environmental data (such as temperature and precipitation in the regions they are found) to predict range changes.

Core concepts: What is a niche?

- A niche describes the relationship between a species and its environment and consists of the species requirements from—and its impacts on—its biotic and abiotic environment.

Meet the data: A century of change in Yosemite

- Scientists used museum records and resurveys of Yosemite National Park to uncover the effects of a century of climate change on endemic small mammals. They found that most species showed distributional shifts to higher elevations, consistent with the pattern predicted by response to global warming.

Taking a closer look: Globalization and invasive species

- Invasive species are non-native species whose introduction causes economic or ecological harm.
- Invasive species cause billions of dollars in damage every year and their impact has escalated as our globalized transportation system makes it easier for species to spread around the world.
- Invasive species contribute to biotic homogenization across the globe.

CONSOLIDATE YOUR KNOWLEDGE

Answer the following questions to consolidate your knowledge, assess your progress meeting the learning outcomes, and reveal any areas in need of further exploration:

1. In your own words, write a one- or two-sentence synthesis of the big-picture takeaway point of this chapter.
2. Revisit your answer to the *Blank Page* exercise in the beginning of this chapter. Would you refine your answer now based on knowledge you integrated from this chapter?
3. Define *dispersal*. How is dispersal distinct from daily movements and from migration? Why is dispersal capability a key trait influencing species' geographic distributions?

4. Compare and contrast historical and contemporary dynamics of range changes. In what ways were range shifts prior to human influence on the biosphere similar and different from those occurring today?
5. What types of distributional changes are commonly observed in response to global change stressors? What is a specific example of each type of distributional change?
6. What is a specific example of species distribution modeling (SDM) applied to a question about an individual species? What is a specific example of SDM applied to a question about a community assemblage?
7. What factors facilitate or hinder the "move" response? What specific organismal traits, human activities, and landscape features make it more or less likely that a species will undergo an anthropogenically mediated distributional change?
8. What is a niche? What is the difference between niche conservatism and niche evolution?
9. What traits of species and ecosystems facilitate invasion?
10. What are the dominant patterns observed (in latitude and elevation) for species distribution changes under global warming?
11. In your own words, define the bolded and italicized terms in this chapter.
12. What are some questions that *you* have about the content in this chapter? If you found some content particularly challenging or particularly interesting, identify these as areas for additional reflection or reading.

LITERATURE CITED

Allison, Perry, Badjeck, Neil Adger, Brown, Conway, Halls, et al. 2009. "Vulnerability of National Economies to the Impacts of Climate Change on Fisheries." *Fish and Fisheries* 10 (2): 173–96. doi.org/10.1111/j.1467-2979.2008.00310.x.

Ballew, Bacheler, Kellison, & Schueller. 2016. "Invasive Lionfish Reduce Native Fish Abundance on a Regional Scale." *Scientific Reports* 6: 32169. doi.org/10.1038/srep32169.

Beever & Smith. 2014. "Ochotona Princeps." *The IUCN Red List of Threatened Species.* http://www.iucnredlist.org

Buckley, Urban, Angilletta, Crozier, Rissler, & Sears. 2010. "Can Mechanism Inform Species' Distribution Models?" *Ecology Letters* 13 (8): 1041–51. doi.org/10.1111/j.1461-0248.2010.01479.x.

Cariton & Geller. 1993. *Ecological Roulette: The Global Transport of Nonindigenous Marine Organisms. Science* 261 (5117): 78–82. doi.org/10.1126/science.261.5117.78

Chase & Leibold. 2003. *Ecological Niches: Linking Classical and Contemporary Approaches.* Chicago: University of Chicago Press.

Chen, Hill, Ohlemüller, Roy, & Thomas. 2011. "Rapid Range Shifts of Species Associated with High Levels of Climate Warming." *Science* 333 (*6045*): 1024–26. doi.org/10.1126/science.1206432.

Cheung, Watson, & Pauly. 2013. "Signature of Ocean Warming in Global Fisheries Catch." *Nature* 497: 365–68. doi.org/10.1038/nature12156.

Cohen & Carlton. 1998. "Accelerating Invasion Rate in a Highly Invaded Estuary." *Science* 279 (5350): 555–58. doi.org/10.1126/science.279.5350.555.

Conser & Connor. 2009. "Assessing the Residual Effects of Carpobrotus Edulis Invasion, Implications for Restoration." *Biological Invasions* 11 (2): 349–58. doi.org/10.1007/s10530-008-9252-z.

DeMars & Roettgering. 1982. "Western Pine Beetle." *US Department of Agriculture Forest Service: Forest Insect & Disease Leaflet* 1. http://www.na.fs.fed.us/spfo/pubs/fidls/we_pine_beetle/wpb.htm

Elias, Short, & Birks. 1997. "Late Wisconsin Environments of the Bering Land Bridge."

Palaeogeography, Palaeoclimatology, Palaeoecology 136 (1–4): 293–308. doi.org/10.1016/S0031-0182(97)00038-2.

Elith & Leathwick. 2009. "Species Distribution Models: Ecological Explanation and Prediction Across Space and Time." *Annual Review of Ecology, Evolution, and Systematics* 40: 677–97. doi.org/10.1146/annurev.ecolsys.110308.120159.

Elton. 1927. *Animal Ecology*. Chicago: University of Chicago Press.

Estrada, Meireles, Morales-Castilla, Poschlod, Vieites, Araújo, & Early. 2015. "Species' Intrinsic Traits Inform Their Range Limitations and Vulnerability under Environmental Change." *Global Ecology and Biogeography* 24 (7): 849–58. doi.org/10.1111/geb.12306.

Evangelista, Kumar, Stohlgren, & Young. 2011. "Assessing Forest Vulnerability and the Potential Distribution of Pine Beetles Under Current and Future Climate Scenarios in the Interior West of the US." *Forest Ecology and Management* 262 (3): 307–16. doi.org/10.1016/j.foreco.2011.03.036.

Flockhart, Wassenaar, Martin, Hobson, Wunder, & Norris. 2013. "Tracking Multi-Generational Colonization of the Breeding Grounds by Monarch Butterflies in Eastern North America." *Proceedings of the Royal Society B: Biological Sciences* 280 (1768). doi.org/10.1098/rspb.2013.1087.

Grinnell. 1917. "The Niche-Relationships of the California Thrasher." *Auk* 34: 427–33.

Grinnell & Storer. 1924. *Animal Life in the Yosemite*. Berkeley: University of California Press.

Hilbert, Graham, & Hopkins. 2007. "Glacial and Interglacial Refugia Within a Long-Term Rainforest Refugium: The Wet Tropics Bioregion of NE Queensland, Australia." *Palaeogeography, Palaeoclimatology, Palaeoecology* 251 (1): 104–18. doi.org/10.1016/j.palaeo.2007.02.020.

Holyoak, Casagrandi, Nathan, Revilla, & Spiegel. 2008. "Trends and Missing Parts in the Study of Movement Ecology." *Proceedings of the National Academy of Sciences* 105 (49): 1090–65. doi.org/10.1073/pnas.0800483105.

Hopkins. 1959. "Cenozoic History of the Bering Land Bridge." *Science* 129 (3362): 1519–28. www.jstor.org/stable/1757656.

Hufbauer, Facon, Ravigné, Turgeon, Foucaud, Lee, Rey, & Estoup. 2012. "Anthropogenically Induced Adaptation to Invade (AIAI): Contemporary Adaptation to Human-Altered Habitats within the Native Range Can Promote Invasions." *Evolutionary Applications* 5 (1): 89–101. doi.org/10.1111/j.1752-4571.2011.00211.x.

Hulme. 2009. "Trade, Transport and Trouble: Managing Invasive Species Pathways in an Era of Globalization." *Journal of Applied Ecology* 46 (1): 10–18. doi.org/10.1111/j.1365-2664.2008.01600.x.

Huntley, Barnard, Altwegg, Chambers, Coetzee, Gibson, Hockey, et al. 2010. "Beyond Bioclimatic Envelopes: Dynamic Species' Range and Abundance Modelling in the Context of Climatic Change." *Ecography* 33 (3): 621–26. doi.org/10.1111/j.1600-0587.2009.06023.x.

Hutchinson. 1957. "Concluding Remarks" *Cold Spring Harbor Symposia on Quantitative Biology* 22: 415–27. http://dx.doi.org/10.1101/SQB.1957.022.01.039

Jenouvrier, Holland, Stroeve, Serreze, Barbraud, Weimerskirch, & Caswell. 2014. "Projected Continent-Wide Declines of the Emperor Penguin under Climate Change." *Nature Climate Change* 4: 715–18. doi.org/10.1038/NCLIMATE2280.

Kearney & Porter. 2009. "Mechanistic Niche Modelling: Combining Physiological and Spatial Data to Predict Species' Ranges." *Ecology Letters* 12 (4): 334–50. doi.org/10.1111/j.1461-0248.2008.01277.x.

Koh & Wilcove. 2008. "Is Oil Palm Agriculture Really Destroying Tropical Biodiversity?" *Conservation Letters* 1 (2): 60–64. doi.org/10.1111/j.1755-263X.2008.00011.x.

Kolar & Lodge. 2001. "Progress in Invasion Biology: Predicting Invaders." *Trends in Ecology and Evolution* 16 (4): 199–204. doi.org/10.1016/S0169-5347(01)02101-2.

Lavergne, Mouquet, Thuiller, & Ronce. 2010. "Biodiversity and Climate Change: Integrating Evolutionary and Ecological Responses of Species and Communities." *Annual Review of Ecology, Evolution, and*

Systematics 41: 321–50. doi.org/10.1146/annurev-ecolsys-102209-144628.

Leibold. 1995. "The Niche Concept Revisited: Mechanistic Models and Community Context." *Ecology* 76 (5): 1370–82. doi.org/10.2307/1938141.

Lenoir & Svenning. 2015. "Climate-Related Range Shifts—A Global Multidimensional Synthesis and New Research Directions." *Ecography* 38 (1): 15–28. doi.org/10.1111/ecog.00967.

Lockwood, Hoopes, & Marchetti. 2013. *Invasion Ecology*. 2nd ed. Malden, MA: John Wiley & Sons.

Maggini, Lehmann, Kéry, Schmid, Beniston, Jenni, & Zbinden. 2011. "Are Swiss Birds Tracking Climate Change? Detecting Elevational Shifts Using Response Curve Shapes." *Ecological Modelling* 222 (1): 21–32. doi.org/10.1016/j.ecolmodel.2010.09.010.

McKinney & Lockwood. 1999. "Biotic Homogenization: A Few Winners Replacing Many Losers in the next Mass Extinction." *Trends in Ecology and Evolution* 14 (11): 450–53. doi.org/10.1016/S0169-5347(99)01679-1.

Moritz, Patton, Conroy, Parra, White, & Beissinger. 2008. "Impact of a Century of Climate Change on Small-Mammal Communities in Yosemite National Park, USA." *Science* 322 (5899): 261–64. doi.org/10.1126/science.1163428.

National Oceanic and Atmospheric Administration (NOAA). 2014. "Blue Whale (*Balaenoptera Musculus*)." http://www.nmfs.noaa.gov/pr/species/mammals/cetaceans/bluewhale.htm

Olden. 2006. "Biotic Homogenization: A New Research Agenda for Conservation Biogeography." *Journal of Biogeography* 33 (12): 2027–39. doi.org/10.1111/j.1365-2699.2006.01572.x.

Parmesan. 1996. "Climate and Species' Range." *Nature* 382: 765–66.

Parmesan & Yohe. 2003. "A Globally Coherent Fingerprint of Climate Change Impacts Across Natural Systems." *Nature* 421: 37–42. doi.org/10.1038/nature01286.

Peterson & Robins. 2003. "Using Ecological-Niche Modeling to Predict Barred Owl Invasions with Implications for Spotted Owl Conservation." *Conservation Biology* 17 (4): 1161–65. doi.org/10.1046/j.1523-1739.2003.02206.x.

Petit & Hampe. 2006. "Some Evolutionary Consequences of Being a Tree." *Annual Review of Ecology, Evolution, and Systematics* 37: 187–214. doi.org/10.1146/annurev.ecolsys.37.091305.110215.

Pimentel, Zuniga, & Morrison. 2005. "Update on the Environmental and Economic Costs Associated with Alien-Invasive Species in the United States." *Ecological Economics* 52 (3): 273–88. doi.org/10.1016/j.ecolecon.2004.10.002.

Poloczanska, Burrows, Brown, García Molinos, Halpern, Hoegh-Guldberg, Kappel, et al. 2016. "Responses of Marine Organisms to Climate Change Across Oceans." *Frontiers in Marine Science* 3: 62. doi.org/10.3389/fmars.2016.00062.

Rahel. 2002. "Homogenization of Freshwater Faunas." *Annual Review of Ecology and Systematics* 33 (1): 291–315. doi.org/10.1146/annurev.ecolsys.33.010802.150429.

Reppert & de Roode. 2018. "Demystifying Monarch Butterfly Migration." *Current Biology* 28 (17): R1009–22. doi.org/10.1016/j.cub.2018.02.067.

Root, Price, Hall, Schneider, Rosenzweig, & Pounds. 2003. "Fingerprints of Global Warming on Wild Animals and Plants." *Nature* 421: 57–60. doi.org/10.1038/nature01333.

Rougier, Lassalle, Drouineau, Dumoulin, Faure, Deffuant, Rochard, & Lambert. 2015. "The Combined Use of Correlative and Mechanistic Species Distribution Models Benefits Low Conservation Status Species." *PLoS ONE* 10 (10): e0139194. doi.org/10.1371/journal.pone.0139194.

Sanderson, Redford, Weber, Aune, Baldes, Berger, Carter, et al. 2008. "The Ecological Future of the North American Bison: Conceiving Long-Term, Large-Scale Conservation of Wildlife." *Conservation Biology* 22 (2): 252–66. doi.org/10.1111/j.1523-1739.2008.00899.x.

Schick, Loarie, Colchero, Best, Boustany, Conde, Halpin, Joppa, McClellan, &

Clark. 2008. "Understanding Movement Data and Movement Processes: Current and Emerging Directions." *Ecology Letters* 11 (12): 1338–50. doi.org/10.1111/j.1461-0248.2008.01249.x.

Sohl. 2014. "The Relative Impacts of Climate and Land-Use Change on Conterminous United States Bird Species from 2001 to 2075." *PLoS ONE* 9 (11): e112251. doi.org/10.1371/journal.pone.0112251.

Sorte, Williams, & Carlton. 2010. "Marine Range Shifts and Species Introductions: Comparative Spread Rates and Community Impacts." *Global Ecology and Biogeography* 19 (3): 303–16. doi.org/10.1111/j.1466-8238.2009.00519.x.

Stachowicz, Whitlatch, & Osman. 1999. "Species Diversity and Invasion Resistance in a Marine Ecosystem." *Science* 286 (5444): 1577–79. doi.org/10.1126/science.286.5444.1577.

Suarez & Case. 2002. "Bottom-up Effects on Persistence of a Specialist Predator: Ant Invasions and Horned Lizards." *Ecological Applications* 12 (1): 291–98. doi.org/10.1890/1051-0761(2002)012[0291:BUEOPO]2.0.CO;2.

Suarez, Tsutsui, Holway, & Case. 1999. "Behavioral and Genetic Differentiation Between Native and Introduced Populations of the Argentine Ant." *Biological Invasions* 1 (1): 43–53. doi.org/10.1023/A:1010038413690.

Svenning, Fløjgaard, Marske, Nógues-Bravo, & Normand. 2011. "Applications of Species Distribution Modeling to Paleobiology." *Quaternary Science Reviews* 30 (21–22): 2930–47. doi.org/10.1016/j.quascirev.2011.06.012.

Thomas, Cameron, Green, Bakkenes, Beaumont, Collingham, Erasmus, et al. 2004. "Extinction Risk from Climate Change." *Nature* 427: 145–48. doi.org/10.1038/nature02121.

Torchin, Lafferty, Dobson, McKenzie, & Kuris. 2003. "Introduced Species and Their Missing Parasites." *Nature* 421: 628–30. doi.org/10.1038/nature01346.

Torchin & Mitchell. 2004. "Parasites, Pathogens, and Invasions by Plants and Animals." *Frontiers in Ecology and the Environment* 2 (4): 183–90. doi.org/10.1890/1540-9295(2004)002[0183:PPAIBP]2.0.CO;2.

Tsutsui, Suarez, Holway, & Case. 2000. "Reduced Genetic Variation and the Success of an Invasive Species." *Proceedings of the National Academy of Sciences* 97 (11): 5948–53. doi.org/10.1073/pnas.100110397.

United States Fish and Wildlife Service. 2013. "Devils Hole Pupfish." http://www.fws.gov/nevada/protected_species/fish/species/dhp/dhp.html.

United States Fish and Wildlife Service. 2014. "Timeline of the American Bison." http://www.fws.gov/bisonrange/timeline.htm.

Weber. 2017. *Invasive Plant Species of the World: A Reference Guide to Environmental Weeds*. 2nd ed. Boston, MA: CABI.

Wiens, Ackerly, Allen, Anacker, Buckley, Cornell, Damschen, et al. 2010. "Niche Conservatism as an Emerging Principle in Ecology and Conservation Biology." *Ecology Letters* 13 (10): 1310–24.

Wilcove, Rothstein, Dubow, Phillips, & Losos. 1998. "Quantifying Threats to Imperiled Species in the United States." *BioScience* 48 (8): 607–15. doi.org/10.2307/1313420.

Wolfe. 2002. "Why Alien Invaders Succeed: Support for the Escape-from-Enemy Hypothesis." *The American Naturalist* 160 (6): 705–11. doi.org/10.1086/343872.

Worm & Tittensor. 2011. "Range Contraction in Large Pelagic Predators." *Proceedings of the National Academy of Sciences* 108 (29): 11942–97. doi.org/10.1073/pnas.1102353108.

Yalcin & Leroux. 2017. "Diversity and Suitability of Existing Methods and Metrics for Quantifying Species Range Shifts." *Global Ecology and Biogeography* 26 (6): 609–24. doi.org/10.1111/geb.12579.

Core Responses: Adjust

Learning Outcomes

After working with this chapter, you will be able to:

- Classify the factors influencing organismal phenotypes.
- Analyze how contemporary human activities elicit an adjust response.
- Describe how scientists measure and predict the adjust response.
- Synthesize factors influencing the adjust response and evaluate the relationship between this response and the other core responses.
- Apply your knowledge to real-world case studies and interpret data from recent scientific studies.

THE BLANK PAGE

Have you ever wondered what you would be like if you were raised in a different time, a different place, or a different culture? Make a list of your attributes that you predict would be unchanged regardless of rearing environment. Then make a list of attributes that you think have a strong environmental contribution. In what ways did the environment you were raised in influence not only your personality and psychology, but even your morphology and physiology? Can you imagine other—even dramatically different—environmental conditions that you could have adjusted to?

INTRODUCTION

All organisms are affected by their environment. In the next two chapters we will distinguish between the two fundamental ways organisms change in response to global change stressors. The goal of this chapter is to analyze the "adjust" response, which encompasses changes that occur during an individual's lifetime and are typically not passed down to offspring. In contrast, the next chapter will focus on the "adapt" response, which occurs at the population level and involves heritable genetic changes over multiple generations. Adjust responses—which are sometimes referred to as *acclimation, acclimatization*, or *plastic* responses—are a broad category because trait shifts at the individual level can occur for a diversity of traits using a diversity of mechanisms. Importantly, the adjust response can occur over short timescales without waiting for new genetic changes or many generations of natural selection. Thus, the adjust response is often critical for survival during periods of rapid environmental change.

WHAT IS PHENOTYPIC PLASTICITY?

Before we analyze the ways organisms adjust in response to global change stressors, it is essential to build a foundational understanding of phenotypic plasticity. Our attributes are influenced both by our genes and by our environment. The *Core Concepts* feature in this chapter provides a more detailed overview of the mechanisms of heredity and reviews terminology that will be used throughout the chapter. Simply put, segments of DNA, called **genes**, are inherited from parent to offspring and code for specific traits, called phenotypes. The term **phenotype** is defined broadly as any observable organismal trait and can include any measurable morphological, physiological, developmental, or behavioral attribute. The term **genotype** refers to the inherited genetic information that influences a trait.

CORE CONCEPTS

WHAT ARE THE MECHANISMS OF HEREDITY?

Ever since antiquity, scientists and philosophers have speculated about how traits were passed from parent to offspring (reviewed in Cobb 2006). It wasn't until the late 19th century that a more modern concept of heredity was developed and the mid-20th century that genetic elements of heredity were formally described. Here we will briefly review the molecular building blocks that allow for information to be passed from parent to offspring. This foundational knowledge is essential background for understanding the adjust and adapt responses.

Deoxyribonucleic acid (DNA)—illustrated in Box Figure 6.1—is the primary hereditary material for most forms of life on Earth. **Genes** are segments of DNA that code for proteins and function as units of heredity. A **locus** is a position of a gene on a chromosome (plural: loci), and an **allele** is a DNA variant at a particular locus. Genes are typically organized on **chromosomes**. Prokaryotes generally have a single circular chromosome, and eukaryotes have multiple linear chromosomes enclosed in the nucleus of the cell.

In eukaryotes, the genetic material in the nucleus of the cell is referred to as the **nuclear genome**. However, many organisms also have extrachromosomal elements, genetic material found outside the nucleus in the cytoplasm of the cell. For example, many prokaryotes have plasmids, which are autonomously replicating extrachromosomal molecules. Eukaryotes have a circular mitochondrial genome that is generally maternally inherited. Land plants (and some algae and protists) have a chloroplast genome. These extrachromosomal elements often serve specialized functions and can have dramatically different inheritance patterns than the nuclear genome.

The sum total of an organism's genetic material is referred to as its **genome**. Genomes can be extremely simple. For example, the virus *phi-X 174* has a genome that is only 5 kilobases (kb) in length and contains only 10 genes (Sanger et al. 1978). Genomes can also be extremely complex. For example, the human genome is approximately 3 *trillion* kb long and contains >20,000 genes (International Human Genome Sequencing Consortium 2004). It is also important to remember that the genome is not only composed of genes. In fact, protein coding genes comprise only ~1% of the human genome. The remaining portion of the genome was once thought of as "junk DNA," but the majority of these noncoding regions have important biochemical roles, particularly in gene regulation (Dunham et al. 2012).

Scientists are also still finding new molecular factors that influence the inheritance of traits, demonstrating that DNA nucleotide changes are not the only molecular language of inheritance. For example, at the

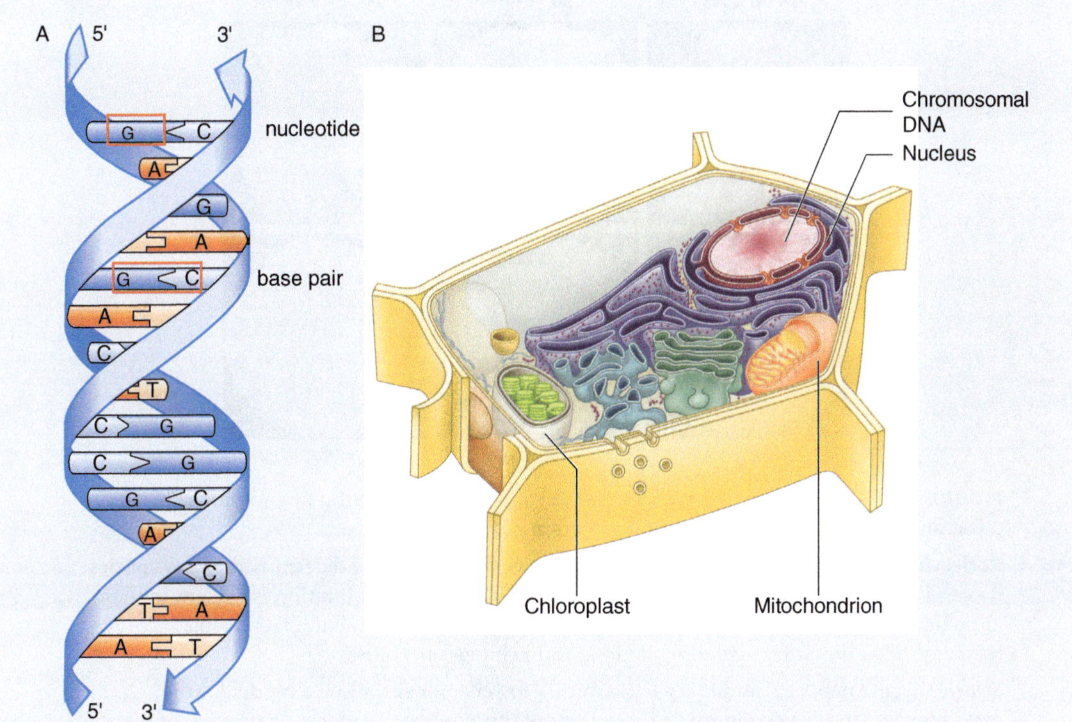

BOX FIGURE 6.1 (A) DNA is a double-stranded molecule composed of repeating units called base pairs. The repeating units are referred to as nucleotides when only one strand is considered. (B) Schematic of a plant cell showing chromosomal DNA contained in the nucleus and extrachromosomal elements in the mitochondria and chloroplast.

Reflection: Considering what we learned about early cellular evolution in previous chapters, what processes likely contributed to the acquisition of extrachromosomal elements?

Source: (A) Life, 12e Figure 4.5, Oxford University Press; (B) Life, 12e Figure 5.9, Oxford University Press

cellular level, there are a number of additional elements that play a role in heredity, including chromatin marking (such as DNA methylation) and RNA activity (such as RNA-mediated gene silencing). These non-DNA factors contribute to what is referred to as **epigenetic inheritance** (reviewed in Jablonka & Raz 2009) and represent a leading edge of research into the molecular mechanisms of heredity.

For some traits, phenotype is determined completely by genotype. For example, eye color in humans—or petal color in flowers—is shaped almost exclusively by genes. It doesn't matter if you grew up in a warm or cold environment; your eye color is predetermined by a few genetic **loci** (Sturm et al. 2008; Lui et al. 2010). Even when phenotype is determined exclusively by genotype, multiple interacting genes can be involved. Figure 6.1 shows an example from the morning glory (*Ipomoea purpurea*) where flower color is determined by the interaction of **alleles** at four loci (Epperson & Clegg 1988; Zufall & Rausher 2003).

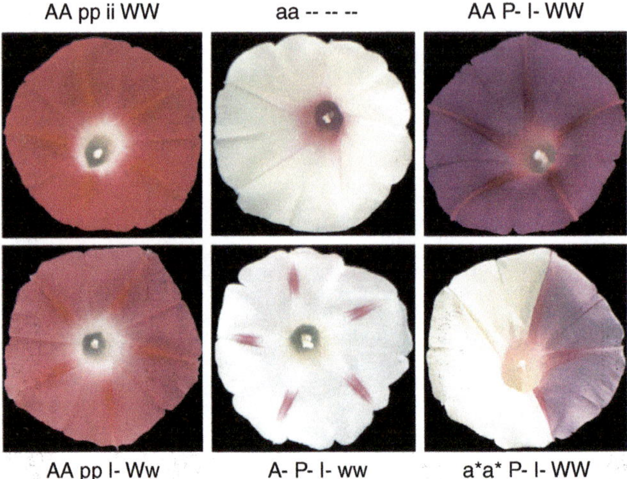

FIGURE 6.1 Morning glory (*Ipomoea purpurea*) flowers do not exhibit phenotypic plasticity in color, and color is determined by the interaction of four genes.

Reflection: Although color in morning glory flowers is genetically determined, many species do exhibit phenotypic plasticity in color. Phenotypically plastic coloration is evident in many animals from crabs to fish to lizards to birds. How could environmental factors like temperature or diet influence short- or long-term changes in color?

Source: Clegg, Michael T. and Mary L. Durbin. "Flower color variation: a model for the experimental study of evolution." Proceedings of the National Academy of Sciences of the United States of America 97 13 (2000): 7016–23

In contrast, many traits are strongly influenced by the environment. We refer to changes at the individual level in response to environmental conditions as **phenotypic plasticity**. In more precise words, *phenotypic plasticity is the capacity of a single genotype to exhibit different phenotypes in different environments.*

Two compelling examples of phenotypic plasticity are shown in Figure 6.2. First, the aquatic invertebrate *Daphnia* develops with more body armor (like "helmets") when grown in a predator-rich environment (Green 1967; Dodson 1989). The environmental stimulus appears to be chemical, and *Daphnia* will develop horns even when chemical pheromones are present without live predatory fish (Laforsch et al. 2006).

Second, *Nemoria arizonaria* caterpillars exhibit dramatic phenotypic differences depending on which season they emerge. Spring broods feed on oak catkins while summer broods feed on oak leaves, and the seasonal broods are well camouflaged in their respective environments (Greene 1989). Environmental factors—such as the levels of tannins in food resources—are thought to lead to differences in caterpillar morphology across seasons (Greene 1989). Illustrations of phenotypic plasticity from nature abound—in developmental, physiological, morphological, and behavioral traits—and we will explore other examples throughout this chapter.

There are many competing terms in use for describing aspects of phenotypic plasticity (Fusco & Minelli 2010; Forsman 2015). For example, the term ***acclimation*** is often used to describe the process of an individual adjusting to changes in its environment. Moreover, some subfields make a distinction between acclimation and acclimatization,

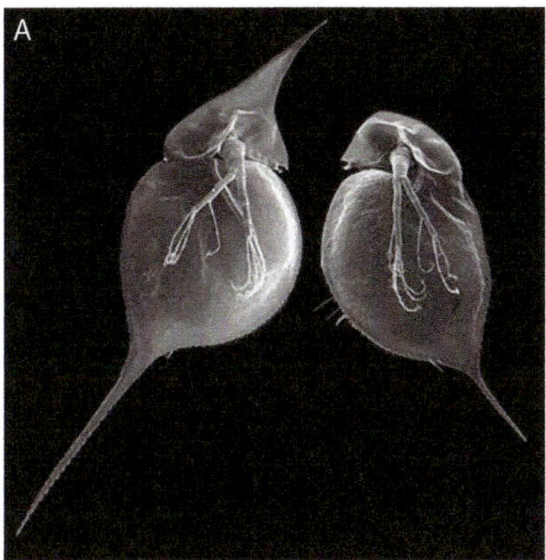

FIGURE 6.2 Examples of phenotypic plasticity. (A) The aquatic invertebrate *Daphnia* develops body armor when predators are present. Even exposure to chemical cues from a fish predator can lead to the development of body armor like horns. (B) *Nemoria* caterpillars employ different camouflage morphologies depending on which season they emerge. Summer broods feed on oak leaves and resemble twigs. Spring broods feed on oak catkins (spent pollen-filled flowers) and resemble these catkins.

Reflection: What are some hypotheses for *how* phenotypic plasticity can occur? How can organisms with the same genetic make-up develop dramatically different phenotypes?

Source: (A) Phenotypic Plasticity in the Interactions and Evolution of Species. BY ANURAG A. AGRAWAL. SCIENCE 12 OCT 2001 : 321–326; (B) Greene, J. C., Caracelli, V. J., & Graham, W. F. (1989). Toward a Conceptual Framework for Mixed-Method Evaluation Designs. Educational Evaluation and Policy Analysis, 11(3), 255–274

with **acclimation** referring to changes induced in a controlled laboratory environment and **acclimatization** referring to changes observed in natural systems. However, for clarity we will use *phenotypic plasticity* as an inclusive umbrella term throughout this chapter.

IS THE CAPACITY FOR PLASTICITY CONSISTENT ACROSS TRAITS AND SPECIES?

There is a spectrum composed of traits that are determined almost exclusively by genetics and those that are influenced almost entirely by environmental conditions. However, most traits are influenced by both genetics and the environment. It is also important to note that different traits in the same individual can show different degrees of phenotypic plasticity. For example, your eye color is not influenced by your environment, but your height is. Therefore, phenotypic plasticity is defined on a trait-by-trait basis. Similarly, traits may be influenced by certain environments but not others. For example, your weight is more likely to be directly influenced by changes to your diet than by changes in precipitation.

In addition, different genotypes might exhibit different levels of plasticity. Let us return to our copper mining example from Chapter 1. Imagine that you have been tasked with predicting which types of plants will be able to grow in soils with heavy metal contamination. You decide that your first step is to evaluate whether phenotypic plasticity might contribute to variation in plant growth rates.

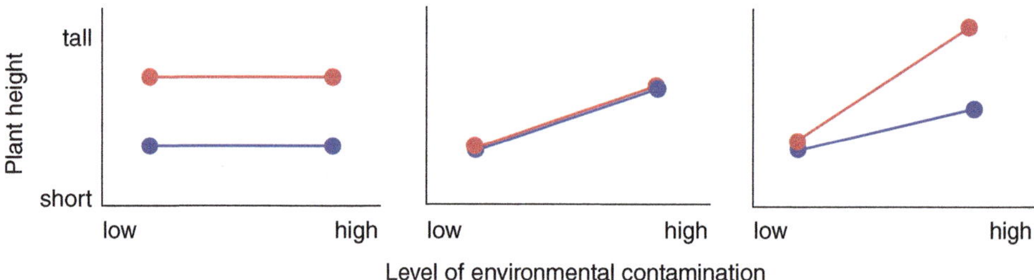

FIGURE 6.3 A hypothetical example where two plant genotypes (red and blue) are exposed to two environments (low and high contamination). In the left figure, there is an effect of genotype but not environment. In the middle, there is an effect of environment but not genotype. In the right figure, there is an effect of both genotype and environment. This genotype by environment interaction occurs when different genotypes exhibit different degrees of phenotypic plasticity.

Reflection: Circle which graphs illustrate the concept of phenotypic plasticity. Then explain the concept of *genotype by environment interaction* in your own words.

Source: Plant phenotypic plasticity in a changing climate, A.B.Nicotra, O.K.Atkin, S.P.Bonser, A.M.Davidson, E.J.Finnegan, U.Mathesius, P.Poot, M.D.Purugganan, C.L.Richards, F.Valladares, M.van Kleunen, Trends in Plant Science, Volume 15, Issue 12, December 2010, Pages 684–692

You decide to measure a trait (such as plant height) in two different environments (such as low and high contamination) for two different plant genotypes. Figure 6.3 shows three possible patterns that you could observe. First, you could find only an effect of genotype (and no phenotypic plasticity). Here one genotype always produces short plants and the other always produces tall plants, regardless of environment. Second, you could find only an effect of the environment. Both genotypes could respond in the same way by producing shorter plants when exposed to more contamination. Third, you could find that the genotypes respond differently to the two environments, what is called a **genotype by environment interaction**. The environment affects the two genotypes in different ways, so you would need to know both the genotype and the environment to predict plant height.

Again, remember that an experiment measuring plant height in two different light conditions cannot tell us whether plant height might be plastic under other conditions (for example, low and high moisture) or whether other traits might be plastic in these same environmental conditions (for example, leaf branching structure). *Thus, phenotypic plasticity must be defined with respect to a specific genotype, a specific trait, and a specific environment.*

WHAT TYPES OF PLASTICITY OCCUR IN RESPONSE TO GLOBAL CHANGE PRESSURES?

Many organisms exhibit some degree of phenotypic plasticity when confronted with rapid environmental change, and there is a diversity of plastic responses to contemporary stressors. Some plastic responses are nearly instantaneous; others occur more slowly over the course of development. Some responses are reversible; others are permanent. Let us look at specific examples for different types of organismal traits.

Shifts in Development

Anthropogenic changes to the environment can lead to dramatic shifts in how organisms grow and develop. Altered developmental rates have been observed broadly across the tree of life. Insects, in particular, provide many examples. Insects are ectotherms, and their life histories are closely tied to temperature. One key effect of global warming on insects is to speed up developmental rates. Faster developmental rates lead to faster maturity. For example, in the United Kingdom >70% of butterfly species studied over a two-decade time period exhibited earlier first appearance, and authors predict that first flight will occur 2–10 days earlier for every 1 degree Celsius increase in temperature (Roy & Sparks 2000).

Faster developmental rates due to warmer temperatures also allow faster generation times and more generations per year for many species. For example, following decades of warming in the US Rocky Mountains, the mountain pine beetle (*Dendroctonus ponderosae*) flight season has advanced by more than 1 month. Moreover, as illustrated in Figure 6.4, some populations now have two generations per year instead of one (Mitton & Ferrenberg 2012). This increases the impact of the mountain pine beetle and the fungus they transmit in pine forests. Millions and millions of acres of pine forest throughout North America have been lost as a result. The effects of global warming on the timing of life history events (e.g., spring emergence of insects and flowering of plants) is such an important topic that we address it in detail in this chapter's *Meet the Data* feature.

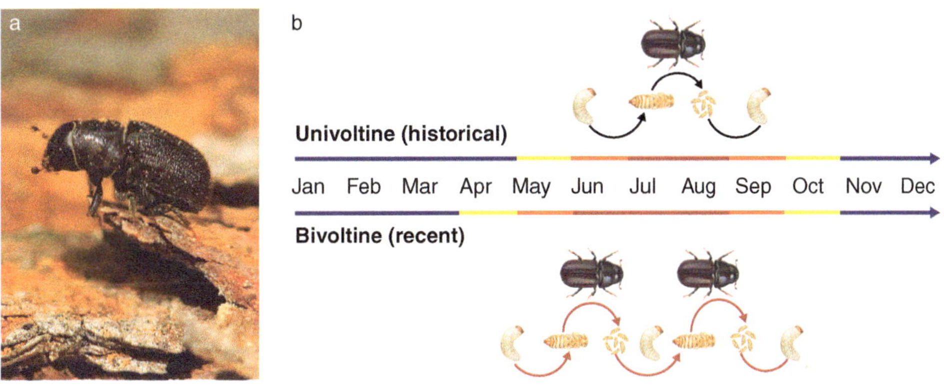

FIGURE 6.4 Historically, the mountain pine beetle in the Colorado Front Range was "univoltine," with only one reproductive cycle each year. However, some broods now are "bivoltine," with two generations per year. More generations per year means more devastation of the pine forests that are impacted by infestations of this beetle.

Reflection: What are some reasons that a pathogen or pest with a faster life cycle will have a more significant impact on its host?

Source: Jeff Mitton

Shifts in Physiology

The study of physiology encompasses a broad array of physical and chemical processes, including metabolism, respiration, and osmotic regulation, and shifts in organismal physiology are common in response to global change stressors. One example is the ability of some plants to switch modes of photosynthesis to save water when in drought-stress conditions. C_3 photosynthetic plants lose >95% of water taken up to transpiration, but CAM (crassulacean acid metabolism) photosynthetic plants keep leaf stomata closed during the day and collect carbon dioxide (CO_2) at night to use during daytime photosynthesis. A number of terrestrial plant species can actually shift modes of photosynthesis (from C_3 to CAM) when under stress (Winter & Holtum 2014). This plastic response can buffer some plant populations from anthropogenic climate stressors, including global warming and the increasing frequency and severity of drought in some biomes.

Another example comes from marine ecosystems which are experiencing not only ocean warming but also acidification (increased dissolved CO_2) and hypoxic zones (decreased dissolved oxygen). Changes in temperature and water chemistry affect myriad physiological processes for marine organisms, including cardiorespiratory function, metabolic rate, locomotor performance, and osmoregulation (Heuer & Grosell 2014; Esbaugh 2018). Many of these effects are simply detrimental, but laboratory studies suggest that some species may exhibit plastic responses that can compensate for altered conditions. For example, Antarctic emerald rockcod (*Trematomus bernacchii*) that were exposed to elevated temperatures in the lab initially showed increased cardiac, ventilation, and metabolic rates. However, after several weeks of exposure to altered conditions, these metrics decreased again, suggesting that this species could potentially adjust to moderate ocean warming (Davis et al. 2018). However, when rockcod were exposed simultaneously to increased temperature and increased CO_2 in the lab, their capacity to compensate was greatly reduced. Thus, it can be difficult for organisms to adjust to multiple global change stressors at the same time.

Shifts in Behavior

Animal behavior is highly influenced by environmental conditions, and behavioral responses to global change stressors are observed in diverse organisms. One iconic example is the effect of climate change on the foraging behavior of polar bears (*Ursus maritimus*). Climate change has impacted polar regions dramatically. One effect has been earlier ice breakup, resulting in a reduction of the ice season by almost 2 months over the past 30 years (Sahanatien & Derocher 2012). A shorter ice season affects animals—like polar bears—that depend on polar ice for hunting their preferred prey items (e.g., seals). As shown in Figure 6.5, studies comparing current polar bear diets to historical data found that polar bears have shifted their diets and are now including novel food sources like bird eggs (Gormezano & Rockwell 2013; Iverson et al. 2014). The dietary shift is the result of phenotypic plasticity in foraging behavior and can have important consequences for arctic bird populations.

Another important example comes from anthropogenically modified sensory environments. Human activities have dramatically altered the light and soundscapes of both aquatic and terrestrial environments (Swaddle et al. 2015). Noise intensities have increased around the globe largely due to transportation (e.g., traffic in cities, shipping

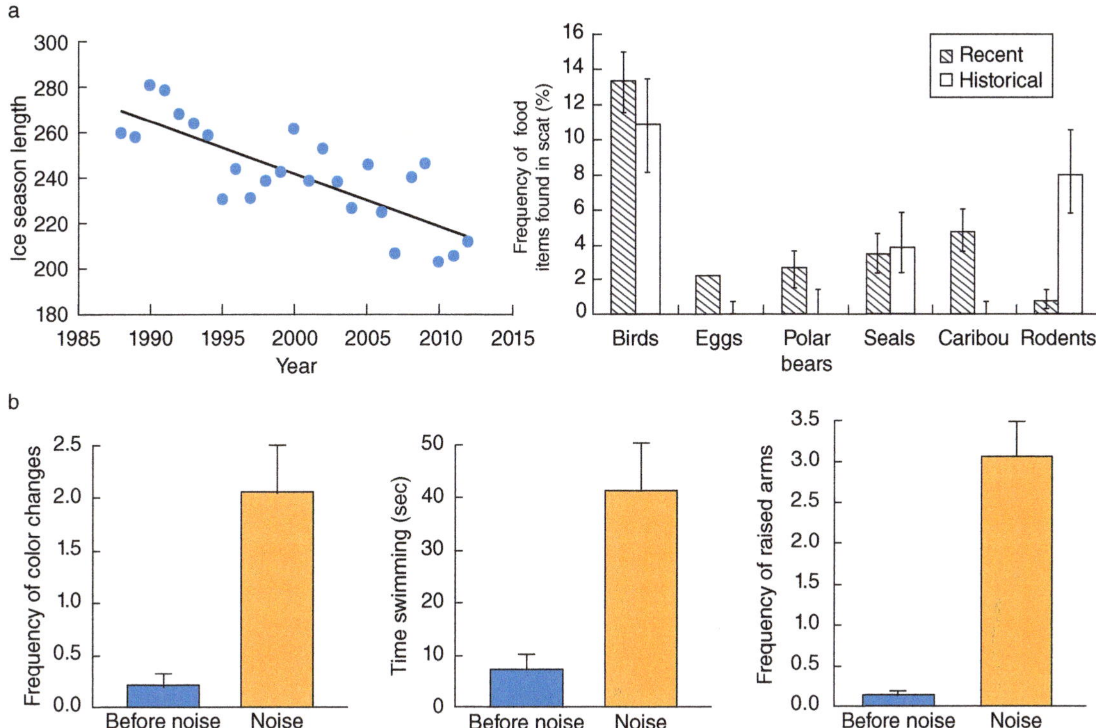

FIGURE 6.5 Phenotypically plastic shifts in behavior in response to anthropogenic stressors. (A) Global warming has led to a decline in ice season length in northern Hudson Bay (Canada). As ice disappears, polar bears need to shift their diet to more terrestrial food sources. Historical data (from 1968 to 1969) and recent data (from 2006 to 2008) on polar bear diet in Hudson Bay illustrate increasing reliance on bird eggs and caribou for nutrition. (B) Commercial shipping has led to increased noise in marine habitats. In laboratory experiments, cuttlefish change color more frequently, spend more time swimming, and raise their arms more often when exposed to audio recordings of underwater ship engine noise.

Reflection: Approximately how many days shorter per year is the ice season in Hudson Bay now compared to 30 years ago? What further changes in polar bear diet or survival might you expect 30 years from now if sea ice continues to retreat?

Source: (A) Iverson Samuel A., Gilchrist H. Grant, Smith Paul A., Gaston Anthony J. and Forbes Mark R. Longer ice-free seasons increase the risk of nest depredation by polar bears for colonial breeding birds in the Canadian Arctic 281 Proc. R. Soc. B; (B) Heather McMullen, Rouven Schmidt, Hansjoerg P. Kunc, Anthropogenic noise affects vocal interactions, Behavioural Processes Volume 103, March 2014, Pages 125–128

vessels in ocean). Many species that rely on auditory communication exhibit compensatory shifts in behavior when exposed to novel sensory environments. For example, crickets, frogs, birds, and whales all have been shown to alter their song rate, length, and/or pitch in the presence of anthropogenic noise (e.g., Slabbekoorn & Peet 2003; Morley et al. 2014; Orci et al. 2016). Moreover, input into one sensory channel can even induce plasticity in another. Figure 6.5 shows how the cuttlefish *Sepia officinalis* alters its color and behavior within minutes when exposed to anthropogenic noise (Kunc et al. 2014). Thus, an individual's behavior can adjust almost instantaneously in response to novel stimuli.

Shifts in Morphology

Environmental change can also lead to phenotypically plastic changes in morphology, or an organism's physical structure and form. In animals, perhaps the most widely discussed morphological trait affected by global change stressors is body size (e.g., Gardner et al. 2011). Many global change stressors—like climate change, habitat loss, fragmentation, or contamination—can affect the availability or quality of nutrient resources. The overall expectation is that deteriorating habitat quality would lead to decreases in body size. In one study of tree swallows (*Tachycineta bicolor*), researchers found a significant decrease in body mass (up to 8% loss in females) over a 7-year period (Paquette et al. 2014). However, the trend for body size to decrease with global change stressors is far from universal. For example, a recent analysis of 24 Australian passerine bird species (sampling museum specimens over a >40-year period) found that 38% of species showed declines in body size while 21% showed increases in body size (Gardner et al. 2014). Therefore, it is premature to make broad-scale generalizations about the effects of global change on body size.

However, some anthropogenic impacts on the environment have direct and unambiguous effects on morphology. Take, for example, the impact of contaminants on sex-specific traits. Most animal species have genotypic sex determination, but environmental cues like temperature, photoperiod, and nutrient availability influence sex-specific traits in some species (Janzen & Phillips 2006). Moreover, anthropogenic contaminants like pesticides and synthetic hormones can alter sex traits even in species that typically have genotypic sex determination. Numerous studies have demonstrated that chemicals used in agricultural landscapes, discharged from wastewater treatment plants, and even found in suburban landscapes can contribute to "sex reversal" in fish and frogs (e.g., Hayes et al. 2010; Lambert et al. 2015; Iwanowicz et al. 2016). These pollutants serve as endocrine disruptors and mimic the effects of sex hormones. Not only can genetic males develop oocytes in their testes, but in some species they actually produce viable eggs. Thus, even traits that are typically considered fixed can shift in response to global change impacts.

In Sum

It is important to remember that ultimately the categories mentioned earlier are inextricably related. For instance, shifts in morphology often result from altered development trajectories, and shifts in physiology often contribute to changes in behavior. In the example of frog sex reversal, environmental contaminants impact the endocrine system, and resulting hormonal changes alter organismal development, physiology, morphology, and behavior. Organisms are integrated systems, and phenotypically plastic responses to global change stressors can occur simultaneously in numerous organismal systems.

WHAT MECHANISMS UNDERLIE PHENOTYPIC PLASTICITY?

Often, the mechanisms underlying phenotypic plasticity seem straightforward. For example, if a plant is deprived of sunlight, moisture, or nutrients, it will not grow as well as a genetically identical plant in more optimal conditions. But how, mechanistically, can global change stressors trigger coordinated—and sometimes complex—phenotypic responses? Let us consider several examples.

Instantaneous Responses

Some responses to environmental change are essentially instantaneous. For example, many organisms experience immediate changes in behavior or physiology when temperature conditions change (e.g., plants opening or closing stomata to prevent water loss, animals shivering or seeking shade to regulate temperature). One example specific to global change stressors is the effect of anthropogenic noise on birdsong, as illustrated in Figure 6.6 for male reed buntings (*Emberiza schoeniclus*). Studying birds in their natural habitat, researchers found that males at a noisy location sing higher frequency songs than those at a quiet location, presumably so their songs can be heard above low-frequency urban background noise (Gross et al. 2010). Moreover, the same individuals adjusted their songs on noisy days or even within minutes of exposure to traffic noise. Thus, individuals can experience nearly instantaneous adjustments to changing conditions.

Gene Regulation Responses

Environmental stimuli typically affect organismal phenotypes via changes in developmental, physiological, hormonal, immunological, and/or neurochemical pathways. Ultimately, many of these effects are mediated by changes in gene regulation (Kelly et al. 2012). **Gene regulation** is the process by which cells increase or decrease the production of RNA and/or proteins from particular genes. Although the DNA sequence remains unchanged, the transcription of genes into RNA—and subsequent translation into proteins—can be modified. Thus, **gene expression** of particular genes can vary among tissues, across developmental time points, and in response to different environmental stimuli.

A classic example of the link between environmental conditions, gene expression, and organismal phenotype is shown in Figure 6.6. Dramatically different phenotypes can develop from genetically identical larvae in honeybees (*Apis mellifera*). Specifically, sterile workers and reproductive queens differ in morphology, physiology, reproductive capability, and behavior, but both develop from genetically identical larvae. The environmental trigger for this dramatic phenotypic plasticity is diet. Larvae destined to be queens are fed "royal jelly" (a substance secreted by nurse bees), whereas other larvae are fed pollen and honey. Access to royal jelly—and restriction from the phytochemicals in pollen and honey—alters the gene expression of a number of key genes, ultimately generating a queen bee (Kucharski et al. 2008; Mao et al. 2015).

A global change-specific example is plant response to drought. One recent study used greenhouse experiments to evaluate the response of a common grass (*Andropogon gerardii*) to water scarcity (Avolio et al. 2018). Researchers found changes in gene expression of specific genes known to be involved in response to water stress. These gene expression changes led to changes in physiology at the leaf level (e.g., closed stomata to prevent dehydration) and ultimately changes in overall plant growth (e.g., root vs. shoot biomass allocation). Thus, changes in gene expression can often be linked to localized and systemic responses to environmental change.

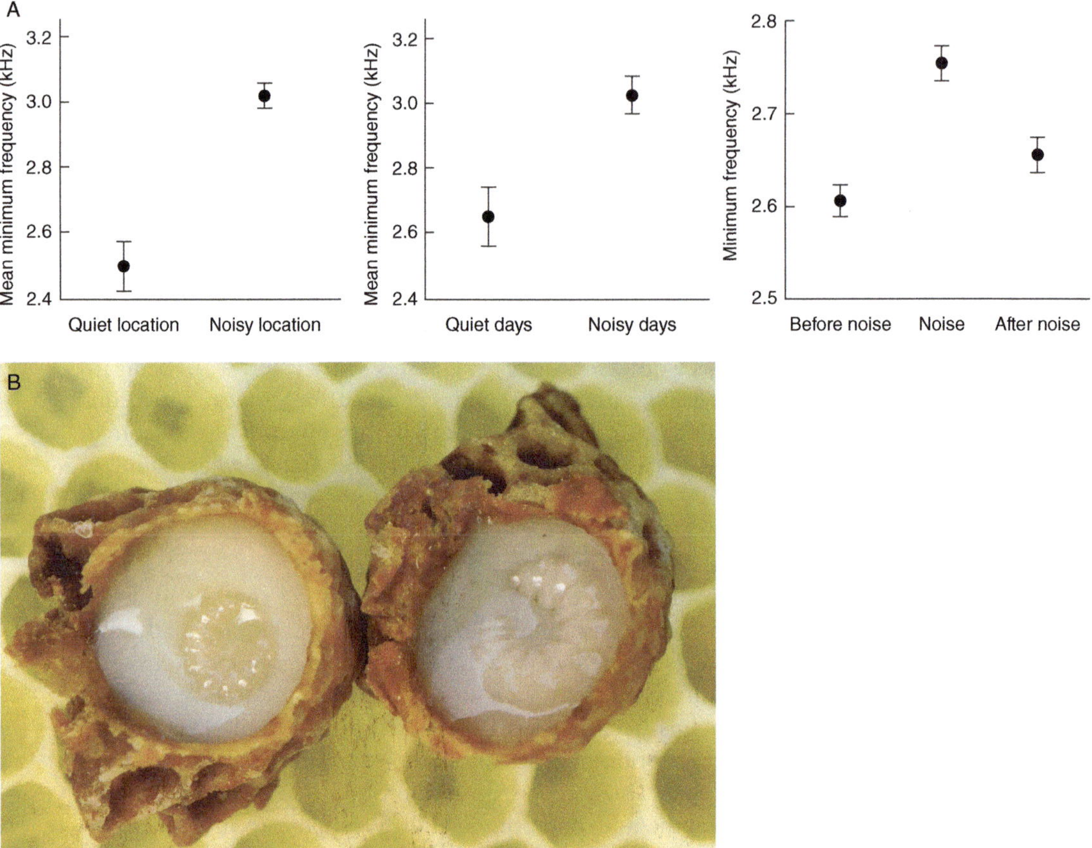

FIGURE 6.6 Plastic responses on instantaneous versus developmental timescales. (A) Rapid changes in male reed bunting (*Emberiza schoeniclus*) song in noisy versus quiet conditions. The frequency of the song increases in noisy locations and on noisy days. (B) Honeybee larvae develop into reproductive queens if reared in chambers of royal jelly. The differences between queens and workers in reproductive morphology (for example, in ovary size) result from dietary effects of royal jelly on gene expression during development.

Reflection: For each example in the figure, describe the following for the trait(s) that are phenotypically plastic: response time (does the response happen quickly or slowly?), response window (can the response happen at any stage of development?), and response permanence (can the response be reversed?).

Source: (A) Behavioral Plasticity Allows Short-Term Adjustment to a Novel Environment. Karin Gross, Gilberto Pasinelli, and Hansjoerg P. Kunc. The American Naturalist 2010 176:4, 456–464; (B) https://en.wikipedia.org/wiki/Royal_jelly#/media/File:Weiselzellen_68a.jpg and Kurcharski et al. 2008, https://science.sciencemag.org/content/319/5871/1827

Epigenetic Responses

Although changes in gene regulation are not typically inherited, there are some mechanisms by which environmentally induced shifts can affect future generations (Bossdorf et al. 2008; Verhoeven et al. 2016). Epigenetics is the study of inherited gene expression changes that do not involve changes at the DNA sequence level.

A number of different processes can influence whether genes are expressed, for example by altering whether DNA is accessible for transcription. One specific example is DNA methylation, which does not change the underlying DNA sequence but inhibits transcription. These physical inhibitions on transcription can influence the phenotype of an individual within its lifetime, but some can actually be passed down to future generations.

For example, when rodents are exposed to a common fungicide that alters patterns of DNA methylation, the effects of this toxin can still be observed on the stress response of their great grandchildren (Crews et al. 2012). Similarly, when dandelions are exposed to environmental stressors (like low nutrients and salt stress), researchers found higher rates of methylation changes, many of which were passed to offspring (Verhoeven et al. 2010). These types of methylation changes can play an important role in response to global change stressors. For example, when coral reef fish are exposed to temperatures simulating future ocean warming, they exhibit a dramatic transgenerational methylation response that may help them survive in altered conditions (Ryu et al. 2018). Thus, epigenetic inheritance blurs the line between phenotypic plasticity within an individual's lifetime and environmental effects that can influence future generations even without changes at the DNA level.

Induction and Reversal

In summary, it is important to remember that plastic responses to environmental conditions can occur in many different ways. First, *response time* can vary by species, trait, and environment. For example, some environmental cues trigger instantaneous effects, whereas others lead to a delayed response. Second, *response window* can vary. For example, some traits are inducible only during certain developmental windows, whereas others retain plasticity throughout the life span. Third, *response magnitude* can vary. Organismal phenotypes can change subtly or dramatically with global change stressors, and traits can shift in a continuous, incremental fashion or exhibit threshold shifts to a new discrete state. Fourth, *response permanence* can vary. Some plastic responses are reversible and others irreversible.

For example, the queen bee phenotype introduced earlier can only be induced in larvae, develops over several weeks, is a dramatic state change, and is irreversible. Once the developmental switch has occurred, the queen can no longer revert to a worker. However, other traits in other species can occur more quickly and can be induced and retracted multiple times. For example, many organisms can respond to unusually high temperatures (or other stressors) by inducing heat shock proteins, which play diverse cellular roles as molecular chaperones. These heat shock responses can be induced many times in the life cycle of an organism (Richter et al. 2010). Of course, many examples will fall along a spectrum of speed, magnitude, and reversibility. For example, in the classic *Daphnia* anti-predator defense we introduced in the beginning of the chapter, it takes several molts for full expression of the helmeted phenotype, and this defense typically can be reduced but not eliminated after predator cues have been removed (Tollrian & Dodson 1999). Ultimately, the plastic responses of organisms to global change stressors vary by species, trait, developmental time point, and environmental inputs (Stamps 2016; Gabriel et al. 2017).

HOW DO SCIENTISTS ASSESS AND PREDICT PHENOTYPIC PLASTICITY?

In nature, particular phenotypes are often associated only with particular environmental conditions. How can scientists then determine whether genotypic or environmental differences explain observed phenotypes? And how can scientists predict whether species will be able to adjust to global change stressors?

Experimental Approaches

Experimental approaches are perhaps the most direct initial test for phenotypic plasticity. Experiments can be used to assess the relative contribution of genetics and the environment to phenotypic patterns and to infer how genotypes might perform in future environmental conditions. Morphological, physiological, developmental, or behavioral traits can be measured following exposure to any environmental treatment whether in the lab or in the field.

Two of the most common experimental designs used to assess phenotypic plasticity are common garden and reciprocal transplant experiments, illustrated in Figure 6.7. In a **common garden experiment**, scientists rear organisms with different traits in a shared environment. Common garden experiments can have one or multiple environmental treatments and can have the "home" (native) environments of the organisms represented or not. If the trait differences persist, there are genetic differences. Alternatively, if trait differences diminish or disappear, traits are influenced at least in part by the environment.

A **reciprocal transplant experiment** is similar, but scientists explicitly rear organisms from different environments reciprocally in both "home" and "away" conditions. Again, scientists measure whether trait differences persist. Common garden experiments are often performed in lab or mesocosm settings, whereas reciprocal transplant experiments are often conducted in natural habitats. However, in practice, the design and interpretation of common garden and reciprocal transplant experiments can be quite similar.

For example, Geng et al. (2007) used common garden experiments in a greenhouse to assess the importance of phenotypic plasticity in explaining the success of the invasive alligator weed (*Alternanthera philloxeoides*). As we discussed in the previous chapter, invasive species are an important issue in global change biology, causing billions of dollars of year in damages. As shown in Figure 6.7, phenotypic differences between aquatic and terrestrial habitats are explained almost completely by phenotypic plasticity rather than genetic differences.

In the context of anthropogenic environmental change, both common garden and reciprocal transplant experiments can be used to predict whether populations might respond with phenotypic plasticity to changing environments. For example, with common garden experiments in the lab, researchers can simulate conditions projected for the future. Many studies evaluate the potential for marine species to adjust to changing oceanic temperature, pH, and nutrient availability using this approach (e.g., Habary et al. 2017; Ezzat et al. 2019). One key result is that survival and performance of marine species are especially compromised when multiple stressors act simultaneously.

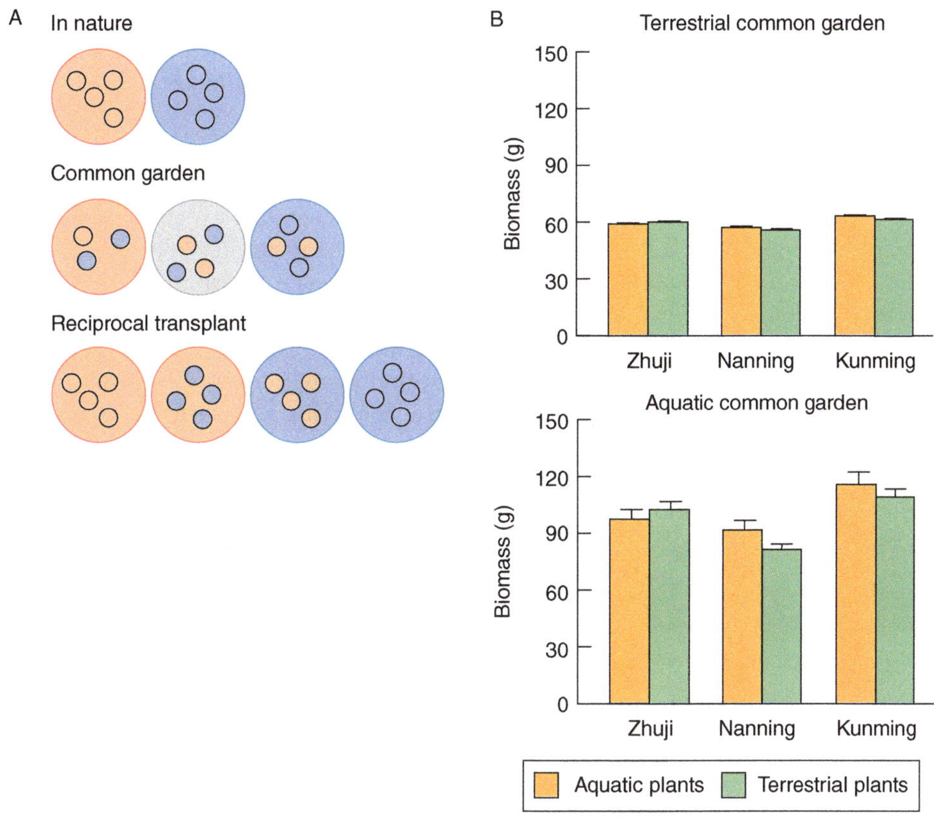

FIGURE 6.7 (A) Conceptual diagram of common garden and reciprocal transplant experimental designs. Typically, phenotypes (represented by colored circles) are found associated with particular environmental conditions, and experiments are needed to assess whether differences result from phenotypic plasticity. In a common garden experiment, individuals from different environments are reared together in one or more shared environments. In a reciprocal transplant experiment, individuals are reared in both home and away environments. (B) Results from a common garden experiment with alligator weed (*Alternanthera philloxeoides*) showing biomass for three replicates of plants from aquatic versus terrestrial habitats. When plants from either locality are grown in terrestrial conditions, they have lower biomass than when they are grown in aquatic conditions.

Reflection: In the alligator weed figure, how would the data look if genotype rather than environmental conditions determined plant biomass? How would it look if there was a genotype–environment interaction? Using the same format as the figures shown, draw an example of each scenario.

Source: (A) Courtesy of the author; (B) Geng, YP., Pan, XY., Xu, CY. et al. Biol Invasions (2007) 9: 245

With transplant experiments in the field, researchers can move individuals to habitats that provide an analog for future conditions. For example, Ishizuka and Goto (2012) planted seeds from populations of Sakhalin fir (*Abies sachalinensis*) across a habitat gradient (using a downslope transplant as proxy for climate change). They found that populations that were moved more than 200 meters downslope showed reduced growth, suggesting that global warming will likely be detrimental to this tree species (Ishizuka & Goto 2012). Thus, experimental approaches in the lab and field can be important for predicting whether plastic responses may offset impacts from global change stressors.

Molecular Approaches

Once patterns of phenotypic plasticity have been documented, a variety of molecular approaches can be used to understand underlying mechanisms (de Villemereuil et al. 2016). Specifically, researchers can look for changes in gene expression or methylation that may explain plastic responses. For example, in the killifish (*Rundulus heteroclitus*), populations exhibit plastic responses to temperature, salinity, and pollution stressors. Notably, these fish can thrive in highly different osmotic environments (from freshwater to seawater) and even in extreme pollution (such as contaminated industrial waste sites).

Whitehead et al. (2011) exposed populations of killifish to osmotic stress in controlled lab conditions by moving coastal fish to freshwater conditions and measured the resulting blood chemistry and gene expression response. Figure 6.8 illustrates the key finding of dramatic physiological plasticity: within several days the fish were able

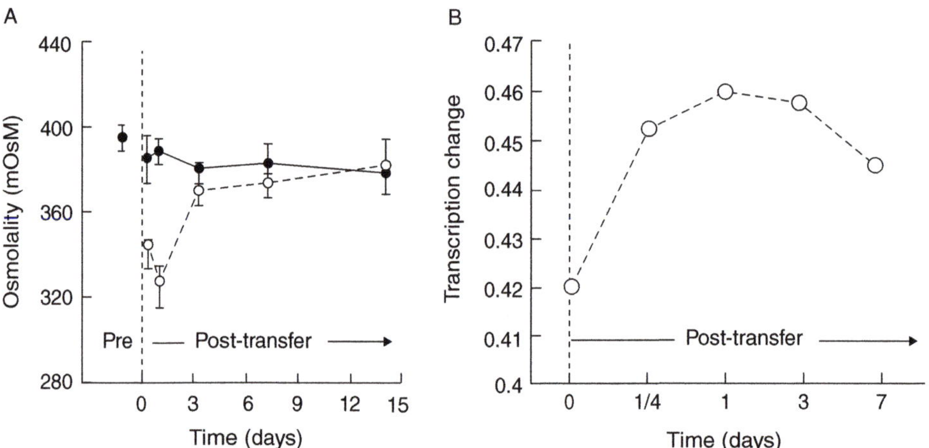

FIGURE 6.8 Altered gene expression underlies physiological plasticity in killifish. (A) Changes in blood osmolality after coastal killifish were challenged with a freshwater environment. Open circles represent fish transferred to freshwater conditions, and filled circles represent control fish kept in marine conditions. (B) The most common pattern of change for nearly 500 genes with significant variation in gene expression after osmotic shock treatment. The y-axis is a scaling from the statistical analysis rather than a particular measure of gene expression.

Reflection: How long did it take for fish to return to roughly normal blood osmolality after transfer to freshwater conditions? During that time, what was occurring at the gene expression level? Would you consider this an instantaneous or delayed response?

Source: Functional Genomics of Physiological Plasticity and Local Adaptation in Killifish, Andrew Whitehead, Fernando Galvez, Shujun Zhang, Larissa M. Williams, Marjorie F. Oleksiak, Journal of Heredity, Volume 102, Issue 5, September-October 2011, Pages 499–511

to re-establish osmotic homeostasis. Moreover, the authors identified specific regulatory pathways responsible for mounting an early stress response and a longer tissue remodeling response. Thus, molecular approaches can be combined with experimental approaches to provide an important perspective on how plastic responses are mounted. We will explore addition molecular methods for linking genotype to phenotype in the next chapter.

CAN PLASTICITY FACILITATE LONG-TERM PERSISTENCE?

Phenotypic plasticity is an important buffer for organisms during environmental fluctuations. Recent research also suggests that plasticity can also facilitate initial persistence and "buy time" for organisms to adapt to long-term environmental trends (Schlichting & Wund 2014). Thus, plasticity can be an initial response to global change stressors before populations undergo adaptive evolution.

The possibility that phenotypic plasticity might not only precede—but even promote—adaptive genetic change is referred to as the **plasticity-first hypothesis** (Schwander & Leimar 2011). In this scenario, an environmental perturbation could induce a phenotypically plastic variant with enhanced survival in new conditions. Later—via a processed termed **genetic assimilation**—this trait could evolve decreased plasticity and become genetically fixed.

The precise mechanisms of genetic assimilation are complex and actively under investigation (e.g., Diamond & Martin 2016; Ehrenreich & Pfennig 2016). However, empirical evidence shows that plastic responses to global change stressors can led to long-term evolutionary change in many systems. Evidence from butterflies (changes in wing color-pattern in response to stress; Hiyama et al. 2012), fish (changes in body size over salinity gradients; Robinson 2013), and alga (changes in tolerance to salt concentration; Lachapelle et al. 2015) all suggest that genetic assimilation may be an important process in the response to global change stressors.

It is also important to highlight that phenotypic plasticity itself can be adaptive. For example, in the predator-induced helmet in *Daphnia* example, the extra investment in horns only occurs when predators are present. Therefore, the *ability to adjust* can itself respond to natural selection, so that organisms can potentially evolve increased or decreased plasticity over time (Figure 6.9).

There are numerous examples of increased plasticity in response to anthropogenic impacts. For example, earlier in this chapter we saw how some birds can modify their calls in response to urban noise. One recent study showed that male white-crowned sparrows (*Zonotrichia leucophrys nuttalli*) from urban locations immediately adjusted their songs when exposed to city noise, whereas rural birds did not (Gentry et al. 2017). Similarly, acorn ants (*Temnothorax curvispinosus*) from urban environments exhibit more plasticity in heat tolerance traits than rural populations (Diamond et al. 2018). These examples—where the degree of plasticity varies between populations in different environments—demonstrate that plasticity itself can evolve in response to global change stressors.

In sum, plasticity can increase, decrease, or stay the same over evolutionary timescales. Sometimes plasticity can buy time for populations to adapt to new environmental conditions. In other cases, plasticity may limit future adaptive responses (Fox et al. 2019). And of course, plasticity is a critical response in and of itself, not only as a bridge to adaptation.

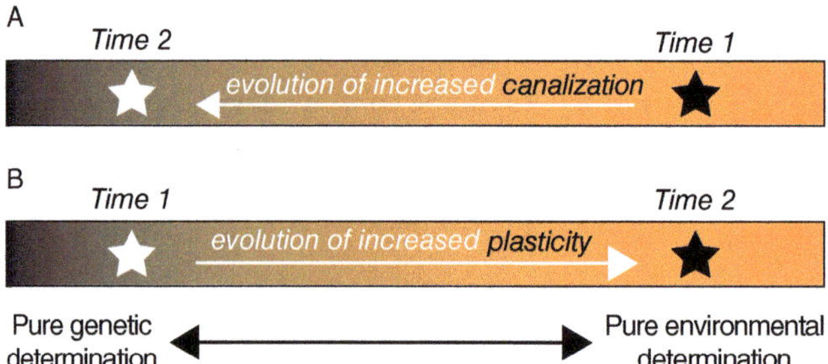

FIGURE 6.9 Phenotypic traits lie on a continuum between the extremes of pure genetic determination and pure environmental determination. The contribution of phenotypic plasticity can also change over time, with traits becoming increasingly fixed genetically (A) or increasingly plastic (B).

Reflection: Under what conditions would you expect the evolution of increased and decreased plasticity, respectively?

Source: David W. Pfennig, Matthew A. Wund, Emilie C. Snell-Rood, Tami Cruickshank, Carl D. Schlichting, Armin P. Moczek, Phenotypic plasticity's impacts on diversification and speciation, Trends in Ecology & Evolution, VOLUME 25, ISSUE 8, P459–467, AUGUST 01, 2010

CONCLUSION

Phenotypic plasticity is an important response to global change stressors for many species. As we have seen throughout this chapter, plasticity can express as instantaneous adjustments to novel stressors or slower changes over developmental timescales. Plasticity is particularly crucial for persistence for organisms that are highly sensitive to environmental conditions, less mobile, and have lower adaptive potential. The adjust response is often rapid, does not require any genetic changes, and can occur for a large variety of traits within an individual's lifetime. Whether species will adjust in response to global change stressors is determined by the degree of phenotypic plasticity in relevant traits and the magnitude of the environmental change (for example, whether traits can shift to match new environmental conditions).

Of course, the adjust response is not mutually exclusive from the other core responses. When organisms move, they often adjust to novel environmental conditions encountered. The adjust response can therefore facilitate colonization of a new geographic area or environmental envelope, and many invasive species exhibit high degrees of phenotypic plasticity (Davidson et al. 2011). As we discussed, the adjust response is also allied with the adapt response, because plasticity itself can be adaptive and because plasticity can buy time for organisms to adapt to long-term environmental trends. While adjust responses will be crucial for survival of many species, there are limits to plasticity. Not all traits can adjust to track changing conditions, and plasticity may not always adequately buffer populations from novel stressors (e.g., Binks et al. 2016). Thus, the adjust response, like all the core responses, must be evaluated from an integrated and species-specific perspective.

MEET THE DATA PHENOLOGY AND GLOBAL WARMING

Who Are the Scientists and What Did They Set Out To Do?

Here we take a closer look at two related classic papers in the field of Global Change Biology. These two meta-analyses were published in the same issue of the journal *Nature* in January 2003. The first paper, "A Globally Coherent Fingerprint of Climate Change Impacts Across Natural System," was led by Dr. Camille Parmesan, and the second paper, "Fingerprints of Global Warming on Wild Animals and Plants," was led by Dr. Terry Root (both shown in Box Figure 6.2). It is unusual for a journal to have two papers on the same topic in the same issue, but this gave power and drew attention to the findings. Combined, these papers have been cited more than 7,500 times. Moreover, they were reported on the front page of the *New York Times*, a rare honor for a scientific discovery.

The authors focused their studies on three potential biological impacts of climate change—shifts in phenology, range boundaries, and species abundances. Here we will focus on their phenology findings, as they are a clear example of the adjust response. **Phenology** is the study of cyclic or seasonal natural events, and phenological changes are shifts in the timing of such events. Abiotic factors such as day length, temperature, and humidity affect the timing of seasonal biotic events such as flowering, breeding, and migration. Therefore, many phenological traits exhibit strong phenotypic plasticity.

Understanding phenology requires documenting the environmental cues that trigger periodic life cycle events. Humans have been charting phenological events and associated environmental cues for hundreds of years. Some of the oldest examples are records of grape harvest from the Burgundy region of France and cherry tree blossoming from Japan (e.g., Chuine et al. 2004). However, farmers and nature enthusiasts have long been recording the onset of biological phenomena associated with seasonal transitions such as first crocus bloom, first butterfly flight, first frog breeding calls, and first migratory bird arrival.

Given the importance of climate for phenology, global climate change could have dramatic impacts on the timing of phenological events. In 2003, Root et al. said, "*Many studies have examined biological changes in relation to climatic change, but generally they are concentrated in particular regions or examine a limited set of taxa.*" To develop a more global perspective, both sets of authors set out to quantify the impact of global warming on biological systems using meta-analysis approaches.

What Are *Your* Predictions?

Before you read on, take a few minutes to apply what you have learned about phenotypic plasticity and global warming to phenology. Start by answering the following questions:

- *Do you predict that phenological changes will have been associated with global warming over the last century?*
- *If so, in what general direction do you expect to see phenological shifts?*

A

B

C

BOX FIGURE 6.2 Lead authors Dr. Terry Root (A) and Dr. Camille Parmesan (B). (C) Dr. Parmesan is shown conducting field research to understand how butterfly ranges are changing with global warming.

Source: (A) Photo by Stephan H. Schneider; (B–C) Michael Singer

(Continued)

- *Do you expect responses to be similar across all species? For example, do you predict similar responses for plants and animals?*

Now write a summary prediction statement about whether—and how—periodic biological phenomena are likely to have changed with global warming over the last century.

What Were the Scientists' Predictions?

Parmesan and Yohe present the following prediction: "*Expected phenological shifts for regions experiencing warming trends are for earlier spring events (for example, migrant arrival times, peak flight date, budburst, nesting, egg-laying, and flowering) and for later autumn events (for example, leaf fall, migrant departure times, and hibernation).*"

What Data Were Collected?

Both groups of researchers compiled data from empirical studies on diverse species, including numerous plant and animal taxa. They established requirements for which datasets to include in their analyses. Both sets of authors focused on studies that provided quantitative estimates of phenological change for specific species over specific time periods. Both sets of authors also looked for datasets with long temporal spans (>10 or 20 years in Root et al. 2003 and Parmesan & Yohe 2003, respectively).

Next, both sets of authors used relatively intuitive meta-analyses to assess whether overall trends were in the expected direction based on predicted phenological responses to a warming planet. The authors asked whether species showing phenological shifts consistently exhibited shifts toward earlier spring events. In other words, what proportion of species that exhibited change showed change in the predicted direction? Both sets of authors compared their results to the **null hypothesis** of no significant directional effect of global warming on phenology (e.g., an equal probability of phenological changes in either direction). The two papers used different statistical frameworks to assess the significance of their results (e.g., binomial tests, vote counting, regression-slope analyses).

What Is *Your* Interpretation of the Data?

Before you read on, take a few minutes to interpret the data in Box Figure 6.3. Consider the following questions as you look at the figure:

- *What are the overall trends observed?*
- *How consistent are these trends across groups?*
- *What are possible explanations for similarities and differences across taxonomic groups?*

Write a sentence or two describing the *key findings* of this study and their *significance*.

What Was the Scientists' Interpretation of the Data?

The key effect of climate change on phenology is the advancement of periodic biological phenomena. As global temperatures increase, many temperature-dependent processes occur earlier in the year. The meta-analyses found that phenological events were advancing several days per decade. Root et al. (2003) found a significant average advancement of 5.1 days per decade, and Parmesan and Yohe (2003) found a significant average advancement of 2.3 days per decade. These results were remarkably consistent in direction across different biological groups, including vertebrates, invertebrates, and plants. Of course, the magnitude of change varies depending on how plastic and how temperature-dependent seasonal activities are in different groups. However, 87% of species exhibited shifts in the predicted direction (Parmesan & Yohe 2003).

These results were highly influential because they demonstrated that climate change has a discernible impact on biodiversity on a global scale. Moreover, they sounded an alarm about the consequences of phenological shifts. For example, changes in phenology can disrupt ecological interactions. **Phenological mismatch**—when an organism's phenology is no longer aligned with key food or habitat resources—is an outcome that threatens the persistence of many populations. We will return to phenological mismatches and the potential for global changes stressors to disrupt species interactions in Chapter 9.

Research in the last decades has repeatedly confirmed the global trends discovered by these seminal papers. In fact, many recent studies suggest that impacts of anthropogenic climate change on phenological events are even more extreme than originally reported (e.g., Cook et al. 2012). Thus, the causes and consequences of phenological change remain an active area of research.

What Are *Your* Ideas for Future Research Directions?

Given what the meta-analysis contributed to the field of Global Change Biology, what next steps do you envision for this research program? If you had been involved in one of these studies, how would you follow up? Start by considering the following questions:

- *Are there more nuanced questions that can now be addressed in particular species or ecosystems?*
- *Are there predictions about the future that could be made and later tested?*
- *What other global signatures of anthropogenic stressors could be evaluated in a similar way?*

Now write a few sentences about future directions on this research theme.

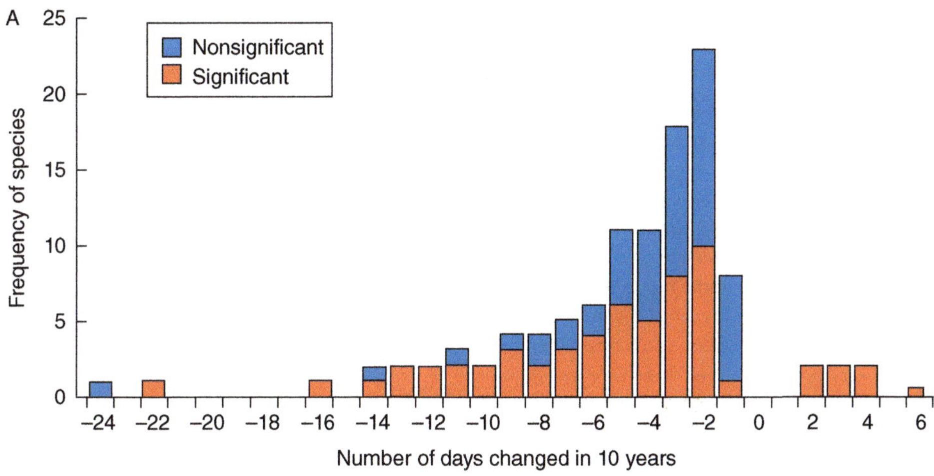

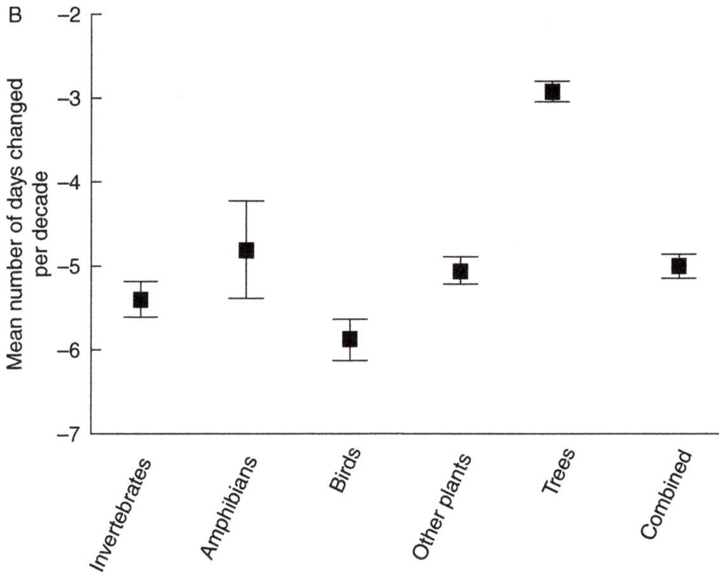

BOX FIGURE 6.3 (A) Frequency of change for phenological traits by number of days. Negative numbers indicate that phenological events occurred earlier in the year. Note that data were not analyzed for species that showed zero change over 10 years. (B) Mean number of days phenological events advanced (earlier in the spring) for different groups of species. Boxes show the mean, and whiskers show the standard error of the mean.

Source: Root, T., Price, J., Hall, K. et al. Fingerprints of global warming on wild animals and plants. Nature 421, 57–60 (2003)

TAKING A CLOSER LOOK URBANIZATION

The adjust response occurs for numerous species in myriad environments. Here we will take a closer look at how organisms adjust to one of the most extreme anthropogenically modified habitats on the planet: the city. Cities are large, dense, and long-lasting human settlements (Wirth 1938). As we explored in Chapter 3, the earliest cities were built in Mesopotamia, the Nile Valley, and the Indus Basin approximately 5,000 years ago. Rome became the first million-strong city approximately 2,000 years ago (Oates 1934).

In contemporary times, there has been an exponential increase in the land area occupied by cities and the size and density of urban populations (illustrated in Box Figure 6.4). The shift in human population demography during the 20th century was particularly dramatic. In 1900 only 10% of humans lived in cities. Now well over 50% of the global population—and more than 80% of the US population—is urban dwelling (United Nations 2014). Globally, 95% of human population growth now occurs in cities, and there are dozens of megacities with more than 10 million inhabitants (United Nations 2014).

What Effects Does Urbanization Have on the Abiotic Environment?

Urban centers have local and global effects on the environment. Following are just a few examples of how cities impact the abiotic environment (reviewed in Grimm et al. 2008 and Kowarik 2011). Of course, these abiotic shifts are not homogenous across urban environments, but they create widespread and novel pressures on numerous species inhabiting cities.

LAND USE The most obvious impact of urban development is modification of physical habitat. As buildings and roads replace forests and grasslands, the physical footprint of cities has direct and indirect effects. Of course, land-use change often extends beyond the city limits when we consider the flow of goods and services (for example, commercial materials coming in and waste products going out).

BIOGEOCHEMICAL CYCLES Urban development can have dramatic effects on water and nutrient cycling. For example,

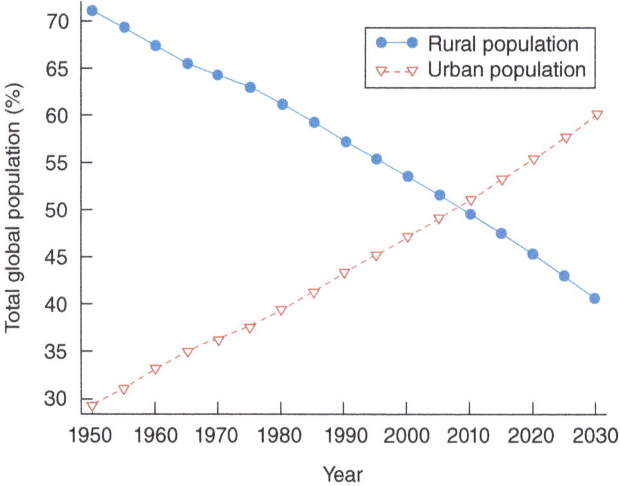

BOX FIGURE 6.4 Summary of United Nations data on urban population growth in the 20th century (note that points past the present day are future projections).

Reflection: How might recent events - like the COVID-19 pandemic - alter the future trajectory of urbanization? Has this period of time changed your personal views on where you might live in the future?

Source: "Global Change and the Ecology of Cities. BY NANCY B. GRIMM, STANLEY H. FAETH, NANCY E. GOLUBIEWSKI, CHARLES L. REDMAN, JIANGUO WU, XUEMEI BAI, JOHN M. BRIGGS. SCIENCE 08 FEB 2008 : 756–760"

urbanization can alter hydrosystems directly as dams or channels restrict the natural flow of water bodies. Urbanization can also have systemic effects on the water cycle. One key shift is the increased **impervious surface cover** in cities, which decreases water infiltration to the soil and increases water run-off, as shown in Box Figure 6.5. Increased run-off can also affect nutrient cycling, especially given that cities are generally point sources for nitrogen and phosphorus.

CLIMATE Urban environments also have a disproportionate effect on climate. Cities cover less than 5% of the terrestrial surface of Earth but produce >75% of global carbon emissions. We explored the consequences of elevated emissions on the global climate in Chapter 4. However, urban development also directly affects local climate. The most dramatic example is the **urban heat island effect**, illustrated in Box Figure 6.5. Air temperatures in urban environments can be up to 20 degrees Fahrenheit hotter than nearby nonurban environments. The heat island effect is caused by a number of factors, including absorption of solar energy by dark surfaces, reduced evaporative cooling due to low vegetation cover, and heat trapping of high buildings.

What Effects Does Urbanization Have on Biodiversity?

The abiotic changes associated with urban development have dramatic effects on biodiversity. To survive in an urban environment, species must be able to adjust to conditions different from those they evolved in. Many of the following examples are clear illustrations of phenotypic plasticity. However, genetic changes (which we will study in the next chapter) also contribute to adaptation to urban environments.

HABITAT USE Concrete and skyscrapers are dramatically different from grassland and trees. Some species are unlikely to persist in an urban environment because their habitat requirements are fundamentally mismatched. For example, ground-dwelling birds fare poorly in an urban environment due to lack of nesting habitat, reduced vegetative cover, and the exposure to predators (Marzluff et al. 2001). However, urban environments can provide analogous habitat for some species, even those threatened in their natural habitat, as illustrated in Box Figure 6.6. For example, cliff-nesting birds—such as the peregrine falcon (*Falco peregrinus*)—thrive in some cities, nesting on tall buildings and exploiting abundant food sources like pigeons (DeCandido & Allen 2006). Similarly, cavity-nesting birds—such as sparrows—can make use of urban structures for key phases of their life cycle.

COMMUNICATION Animals use many different modes of communication for finding mates, defending territories, navigating to food sources, and warning of threats. Communication signals need to stand out from the background, and signals that evolved in the context of a forest or grassland may not function well in an urban environment with artificial lighting and noise. Therefore, species that live in cities often exhibit changes to increase detectability of signals in a new sensory environment. For example, as we discussed in this chapter, many bird species adjust their songs in urban environments (e.g., Slabbekoorn & Peet 2003; Gentry et al. 2017), presumably to let their songs be heard above urban background noise.

ENVIRONMENTAL TOLERANCE The rising temperatures associated with global climate change are exacerbated in urban environments due to the urban heat island effect. Temperature-sensitive species are therefore under added thermal stress in cities. Some species are able to adapt or adjust to higher temperatures. For example, leaf-cutter ants (*Atta sexdens*) in Sao Paulo Brazil, South America's largest city, have increased heat tolerance compared to nearby ants outside of the city, as shown in Box Figure 6.6 (Angilletta et al. 2007). Although plasticity clearly contributes to increased thermal tolerance in city-dwelling organisms, several recent studies have also demonstrated an adaptive genetic component (e.g., Martin et al. 2019; Campbell-Staton et al. 2020), a topic we will study further in Chapter 7.

What Characteristics Define Species That Are Successful in Urban Environments?

There is no single trait that allows successful colonization of—and persistence in—urban environments. However, there are certain characteristics that are often shared among animals that are highly successful in cities (such as rats, pigeons, crows, squirrels, raccoons). For example, species that thrive in urban environments tend to be habitat and diet generalists and aggressive competitors. They also tend to be social or colonial such that they can coexist in large numbers (Marzluff et al. 2001). Although it is difficult to generalize across the tree of life, many species—whether dry rot fungus, dandelions, or house flies—are well suited for city living.

Species that thrive in cities also typically exhibit one of the core responses we are studying. Many species that succeed in urban environments are introduced species, a trend that we explored in the last chapter. Species that are successful in one urban environment are also more likely to be transported (intentionally or unintentionally) to—and thrive in—other urban centers. Persisting in urban environments

(Continued)

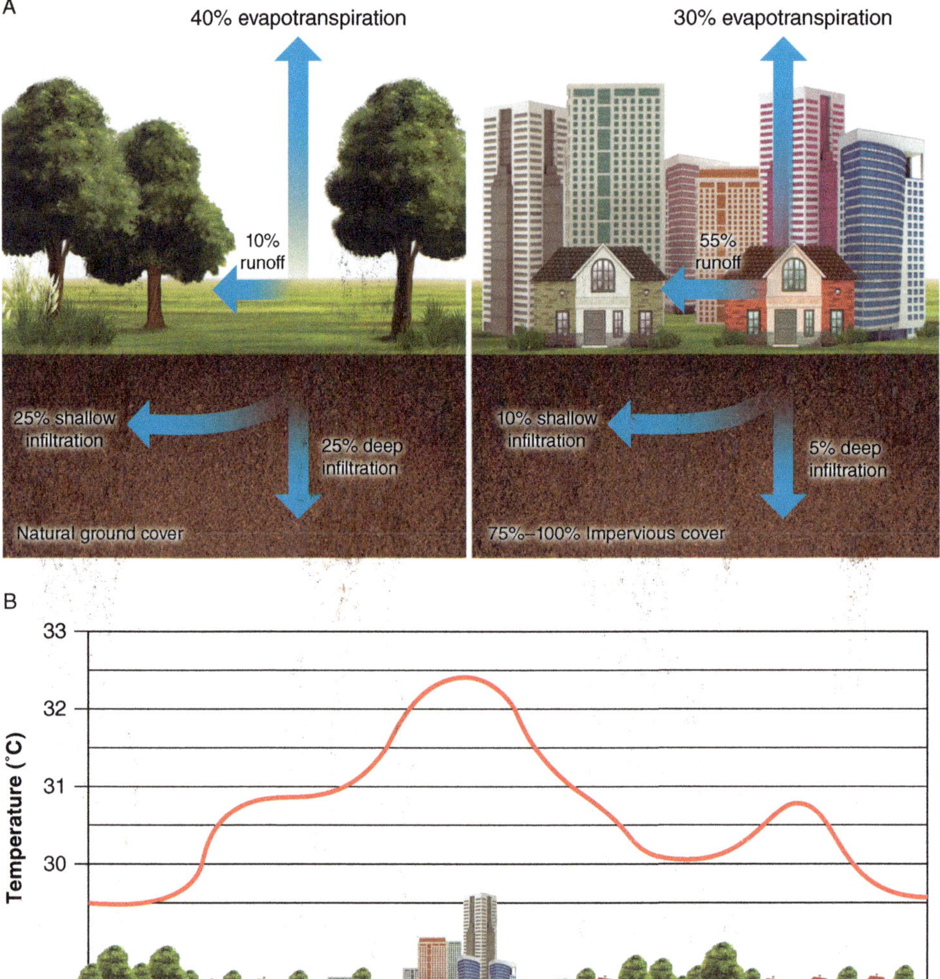

BOX FIGURE 6.5 (A) Impervious surfaces in urban environments affect the water cycle by increasing surface run-off and decreasing soil infiltration and evapotranspiration. (B) The urban heat island effect. Mean annual temperatures can be much higher in urban environments relative to nearby rural environments.

Reflection: What are specific ways that the urban heat island effect and the impacts of impervious surface cover in cities could be reduced?

Source: (A) US Environmental Protection Agency; (B) TheNewPhobia/Wikipedia

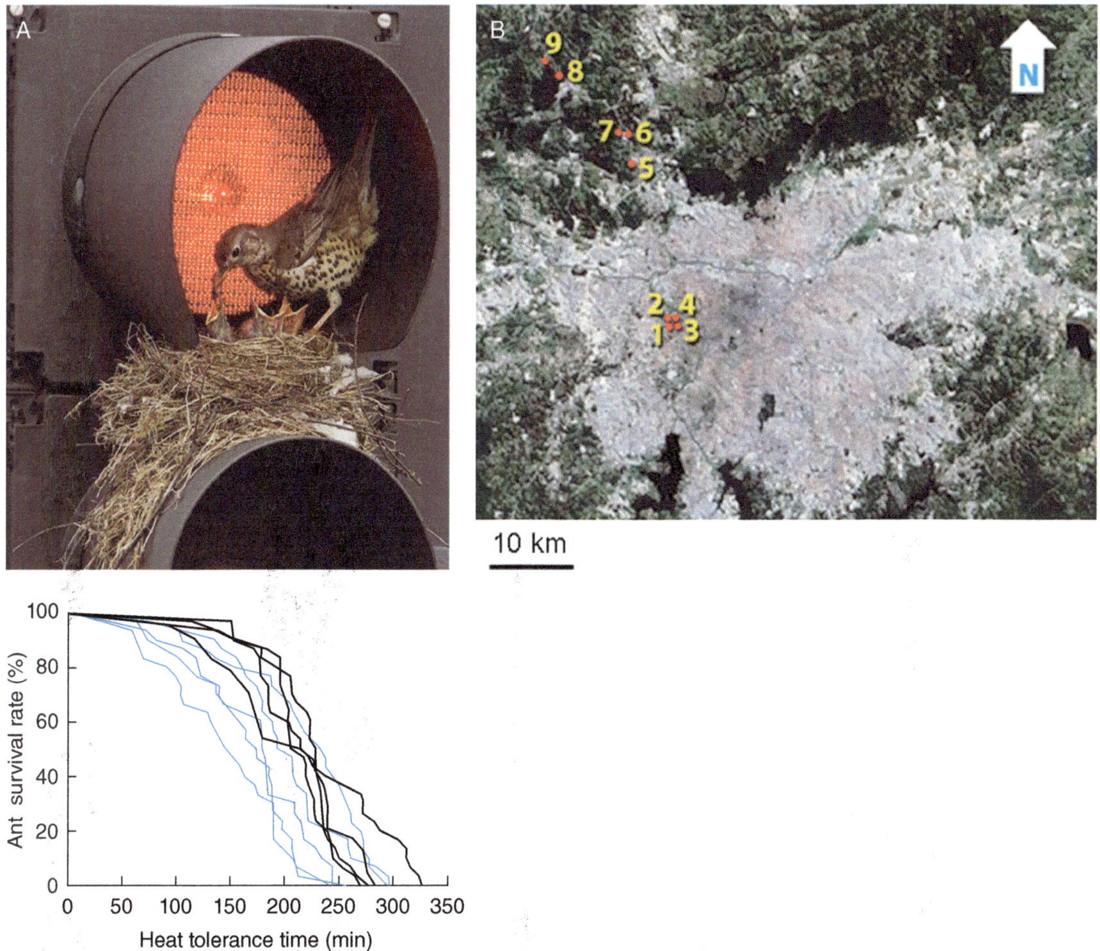

BOX FIGURE 6.6 (A) Urban environments can provide microhabitat suitable to some species, such as cliff-dwelling or cavity-nesting birds. (B) Heat tolerance in urban ants. Ants from urban areas (populations 1–4 on map and black lines on graphs) survive for longer at extreme temperatures than ants from surrounding forested areas (populations 5–9 on map and blue lines on graph).

Reflection: Generate several hypotheses for why there is variation in survivorship among ant populations, including even those collected in the same habitat type.

Source: (A) Paul Campbell/Shutterstock; (B) Angilletta MJ Jr, Wilson RS, Niehaus AC, Sears MW, Navas CA, Ribeiro PL (2007) Urban Physiology: City Ants Possess High Heat Tolerance. PLoS ONE 2(2)

also typically requires shifts in morphology, behavior, and/or physiology (for example, changes in habitat use, thermal tolerance, and diet). These phenotypic shifts often result from behavioral or developmental plasticity. However, there is also increasing evidence for genetic adaptation in urban ecosystems (Donihue & Lambert 2015). We will further explore adaptive responses to global change stressors in the next chapter.

Of course, there are ways to enrich urban environments to provide suitable habitat for a wider range of species. For example, many cities create native plant habitats, rooftop gardens, and stopovers for migratory species in an effort to enhance urban ecosystems for biodiversity (Lundholm & Richardson 2010). We will return to these conservation efforts in Chapter 11.

(Continued)

How Are Human Social Systems and Biodiversity Conservation in Cities Intertwined in Cities?

It is important to recognize the complex dynamics that link human social systems with the fates of myriad other species inhabiting cities. As we discussed in Chapter 1, numerous socioeconomic factors influence—and are influenced by—abiotic and biotic conditions in cities. For example, we saw that systemic biases and structural racism and classism have dramatic effects on the distribution of resources, pollutants, greenspaces, and even disease vectors (Schell et al. 2020). Abiotic factors then have complex feedbacks to biotic processes, such as altering habitat connectivity for nonhuman species.

Everyday issues such as how humans handle food waste, build transportation infrastructure, and apply pesticides have dramatic effects on the distribution and abundance of biodiversity in cities. One specific example is the **Luxury Effect**, where more affluent neighborhoods often have higher biological diversity than less affluent areas (e.g., Hope et al. 2003; Leong et al. 2018). Wealth in cities tends to correlate with increased vegetation and decreased air pollution, influencing habitat quality for all species. The effect of wealth on urban greenness, pollution, and species richness then has obvious feedbacks to human health and quality of life (Diaz et al. 2018; O'Neill et al. 2003).

Therefore, it is impossible to disentangle human social factors from ecological and evolutionary processes that impact all city-dwelling organisms (Des Roches et al. 2020). Thus, issues like the urban heat island effect and increased exposure to pollution in working-class neighborhoods need to be treated jointly as issues of environmental justice and topics in biodiversity conservation (e.g., Drescher 2019). Ultimately, if we do not address the grave social and economic inequities that confront our society, we cannot hope to create sustainable solutions for other species living in our built environments.

KEY CONCEPTS

What is the adjust response?

- Organisms can often track changing environmental conditions without changes at the genetic level.

What is phenotypic plasticity?

- Phenotypic plasticity is the capacity of a single genotype to exhibit different phenotypes in different environments.

Is the capacity for phenotypic plasticity consistent across traits and species?

- The degree of phenotypic plasticity can vary across traits, environments, and genotypes. When environmental conditions affect different genotypes in different ways, it is referred to as a genotype by environment interaction.

What types of plasticity occur in response to global change pressures?

- Phenotypic plasticity is observed in many different traits, in many taxonomic groups, and in response to many different anthropogenic stressors, including shifts in development, physiology, behavior, and morphology.

What mechanisms underlie phenotypic plasticity?

- Plastic responses can vary in response time (instantaneous vs. delayed), response window (inducible only in certain life stages), response magnitude (subtle or dramatic, continuous or discrete), and response permanence (reversibility).
- Phenotypic plasticity is often mediated by changes in gene expression, which typically occur only in an individual's lifetime, but occasionally can be transmitted across generations through epigenetic inheritance.

How do scientists assess and predict phenotypic plasticity?

- Experimental approaches like common garden and reciprocal transplant studies can be used to determine the relative contribution of genes and the environment to phenotypic patterns. Molecular approaches can then be used to reveal mechanisms of plasticity.

Can plasticity facilitate long-term persistence?

- Phenotypic plasticity can sometimes promote genetic adaptation through a process termed genetic assimilation.
- Phenotypic plasticity itself can be adaptive, and the degree of plasticity can itself be considered a trait that can change over time.

Core concepts: What are the mechanisms of heredity?

- DNA is the primary hereditary material for most forms of life on Earth. Genomes can be complex, with nuclear and extrachromosomal elements and coding and non-coding DNA.

Meet the data: Phenology and global warming

- Scientists used meta-analysis to show how global warming affects phenology—the timing of cyclical biological events. They found advancement of phenological events (e.g., earlier spring migration, flowering, breeding) across hundreds of diverse species: a consistent and worldwide signature of global warming on biological processes.

Taking a closer look: Urbanization

- Urbanization has dramatically accelerated in the last hundred years, with myriad effects on the abiotic environment and dramatic effects on species diversity (what species survive in urban environments) and species traits (how those species adjust or adapt to novel conditions).

CONSOLIDATE YOUR KNOWLEDGE

Answer the following questions to consolidate your knowledge, assess your progress meeting the learning outcomes, and reveal any areas in need of further exploration:

1. In your own words, write a one- or two-sentence synthesis of the big-picture take-away point of this chapter.
2. Revisit your answer to the *Blank Page* exercise in the beginning of this chapter. Would you refine your answer now based on knowledge you integrated from this chapter?
3. What is the difference between genotype and phenotype?
4. What is phenotypic plasticity? What is a genotype by environment interaction?
5. What types of phenotypic plasticity are commonly observed in response to global change stressors? What is a specific example of each?
6. How mechanistically can the same genotype produce different phenotypes?
7. What approaches can scientists use to assess whether species have exhibited (or will exhibit) phenotypic plasticity as a buffering response to global change stressors?
8. How does the adjust response relate to the adapt response? What are examples of how plasticity could increase or decrease over evolutionary timescales?
9. What are some abiotic stressors in urban environments and how have urban-dwelling species adjusted?
10. What is phenology and what is a consistent signal of phenological change with global warming?
11. In your own words, define the bolded and italicized terms in this chapter.
12. What are some questions that *you* have about the content in this chapter? If you found some content particularly challenging or particularly interesting, identify these as areas for additional reflection or reading.

LITERATURE CITED

Angilletta, Wilson, Niehaus, Sears, Navas, & Ribeiro. 2007. "Urban Physiology: City Ants Possess High Heat Tolerance." *PLoS ONE* 2: e258. doi.org/10.1371/journal.pone.0000258.

Avolio, Hoffman, & Smith. 2018. "Linking Gene Regulation, Physiology, and Plant Biomass Allocation in *Andropogon gerardii* in Response to Drought." *Plant Ecology* 219 (1): 1–15. doi.org/10.1007/s11258-017-0773-3.

Binks, Meir, Rowland, Da Costa, Vasconcelos, De Oliveira, Ferreira, & Mencuccini. 2016. "Limited Acclimation in Leaf Anatomy to Experimental Drought in Tropical Rainforest Trees." *Tree Physiology* 36 (12): 1550–61. doi.org/10.1093/treephys/tpw078.

Bossdorf, Richards, & Pigliucci. 2008. "Epigenetics for Ecologists." *Ecology Letters* 11 (2): 106–15. doi.org/10.1111/j.1461-0248.2007.01130.x.

Campbell-Staton, Winchell, Rochette, Fredette. Maayan, Schweizer, & Catchen. 2020. "Parallel Selection on Thermal Physiology Facilitates Repeated Adaptation of City Lizards to Urban Heat Islands." *Nature Ecology and Evolution* 4: 652–58. doi.org/10.1038/s41559-020-1131-8.

Chuine, Yiou, Viovy, Seguin, Daux, & Ladurie. 2004. "Grape Ripening as a Past Climate Indicator." *Nature* 432 (7015): 289–90. doi.org/10.1038/432289a.

Cobb. 2006. "Heredity Before Genetics: A History." *Nature Reviews Genetics* 7 (12): 953–58. doi.org/10.1038/nrg1948.

Cook, Wolkovich, & Parmesan. 2012. "Divergent Responses to Spring and Winter Warming Drive Community Level Flowering Trends." *Proceedings of the National Academy of Sciences* 109 (23): 9000–9005. doi.org/10.1073/pnas.1118364109.

Crews, Gillette, Scarpino, Manikkam, Savenkova, & Skinner. 2012. "Epigenetic Transgenerational Inheritance of Altered Stress Responses." *Proceedings of the National Academy of Sciences* 109 (23): 9143–48. doi.org/10.1073/pnas.1118514109.

Davidson, Jennions, & Nicotra. 2011. "Do Invasive Species Show Higher Phenotypic Plasticity than Native Species and, If So, Is It Adaptive? A Meta-Analysis." *Ecology Letters* 14 (4): 419–31. doi.org/10.1111/j.1461-0248.2011.01596.x.

Davis, Flynn, Miller, Nelson, Fangue, & Todgham. 2018. "Antarctic Emerald Rockcod Have the Capacity to Compensate for Warming When Uncoupled from CO_2 Acidification." *Global Change Biology* 24 (2): e655–70. doi.org/10.1111/gcb.13987.

DeCandido & Allen. 2006. "Nocturnal Hunting by Peregrine Falcons at the Empire State Building, New York City." *The Wilson Journal of Ornithology* 118 (1): 53–58. doi.org/10.1676/1559-4491(2006)118[0053:nhbpfa]2.0.co;2.

Des Roches, Brans, Lambert, Rivkin, Savage, Schell, et al. 2020. "Socio-Eco-Evolutionary Dynamics in Cities." *Evolutionary Applications* 00: 1-20. DOI: 10.1111/eva.13065

Diamond, Chick, Perez, Strickler, & Zhao. 2018. "Evolution of Plasticity in the City: Urban Acorn Ants can Better Tolerate more Rapid Increases in Environmental Temperature." *Conservation Physiology* 6 (1): coy030. doi.org/10.1093/conphys/coy030

Diamond & Martin. 2016. "The Interplay Between Plasticity and Evolution in Response to Human-Induced Environmental Change." *F1000Research* 5: 2835. doi.org/10.12688/f1000research.9731.1.

Díaz, Pascual, Stenseke, Martín-López, Watson, et al. 2018. "Assessing Nature's Contributions to People." *Science* 359 (6373), 270–72. doi.org/10.1126/science.aap8826.

Dodson. 1989. "The Ecological Role of Chemical Stimuli for the Zooplankton: Predator-Induced Morphology in *Daphnia*." *Oecologia* 78 (3): 361–67. doi.org/10.1007/BF00379110.

Donihue & Lambert. 2015. "Adaptive Evolution in Urban Ecosystems." *Ambio* 44 (3): 194–203. doi.org/10.1007/s13280-014-0547-2.

Drescher. 2019. "Urban Heating and Canopy Cover Need to be Considered as Matters of Environmental Justice." *Proceedings of the National Academy of Sciences* 116 (52): 26153–54. doi.org/10.1073/pnas.1917213116

Dunham, Kundaje, Aldred, Collins, Davis, Doyle, Epstein, et al. 2012. "An Integrated Encyclopedia of DNA Elements in the Human Genome." *Nature* 489 (7414): 57–74. doi.org/10.1038/nature11247.

Ehrenreich & Pfennig. 2016. "Genetic Assimilation: A Review of Its Potential Proximate Causes and Evolutionary Consequences." *Annals of Botany* 117 (5): 769–79. doi.org/10.1093/aob/mcv130.

Epperson & Clegg. 1988. "Genetics of Flower Color Polymorphism in the Common Morning Glory (*Ipomoea purpurea*)." *Journal of Heredity* 79 (1): 64–68. doi.org/10.1093/oxfordjournals.jhered.a110450.

Esbaugh. 2018. "Physiological Implications of Ocean Acidification for Marine Fish: Emerging Patterns and New Insights." *Journal of Comparative Physiology B: Biochemical, Systemic, and Environmental Physiology* 188 (1): 1–13. doi.org/10.1007/s00360-017-1105-6.

Ezzat, Maguer, Grover, Rottier, Tremblay, & Ferrier-Pagès. 2019. "Nutrient Starvation Impairs the Trophic Plasticity of Reef-Building Corals Under Ocean Warming." *Functional Ecology* 33 (4): 643–53. doi.org/10.1111/1365-2435.13285.

Forsman. 2015. "Rethinking Phenotypic Plasticity and Its Consequences for Individuals, Populations and Species." *Heredity* 15: 276–84. doi.org/10.1038/hdy.2014.92.

Fox, Donelson, Schunter, Ravasi, & Gaitán-Espitia. 2019. "Beyond Buying Time: The Role of Plasticity in Phenotypic Adaptation to Rapid Environmental Change." *Philosophical Transactions of the Royal Society B: Biological Sciences* 374: 20180174. doi.org/10.1098/rstb.2018.0174.

Fusco & Minelli. 2010. "Phenotypic Plasticity in Development and Evolution: Facts and Concepts." *Philosophical Transactions of the Royal Society B: Biological Sciences* 265: 547–56. doi.org/10.1098/rstb.2009.0267.

Gabriel, Luttbeg, Sih, & Tollrian. 2017. "Environmental Tolerance, Heterogeneity, and the Evolution of Reversible Plastic Responses." *The American Naturalist* 166 (3): 339–53. doi.org/10.2307/3473313.

Gardner, Amano, Backwell, Ikin, Sutherland, & Peters. 2014. "Temporal Patterns of Avian Body Size Reflect Linear Size Responses to Broadscale Environmental Change over the Last 50 Years." *Journal of Avian Biology* 45 (6): 529–35. doi.org/10.1111/jav.00431.

Gardner, Peters, Kearney, Joseph, & Heinsohn. 2011. "Declining Body Size: A Third Universal Response to Warming?" *Trends in Ecology and Evolution* 26 (6): 285–91. doi.org/10.1016/j.tree.2011.03.005.

Geng, Pan, Xu, Zhang, Li, Chen, Lu, & Song. 2007. "Phenotypic Plasticity Rather Than Locally Adapted Ecotypes Allows the Invasive Alligator Weed to Colonize a Wide Range of Habitats." *Biological Invasions* 9 (3): 245–56. doi.org/10.1007/s10530-006-9029-1.

Gentry, Derryberry, Danner, Danner, & Luther. 2017. "Immediate Signaling Flexibility in Response to Experimental Noise in Urban, but Not Rural, White-Crowned Sparrows." *Ecosphere* 8 (8): e01916. doi.org/10.1002/ecs2.1916.

Gormezano & Rockwell. 2013. "What to Eat Now? Shifts in Polar Bear Diet During the Ice-Free Season in Western Hudson Bay." *Ecology and Evolution* 3 (10): 3509–23. doi.org/10.1002/ece3.740.

Green. 1967. "The Distribution and Variation of *Daphnia lumholtzi* (Crustacea: Cladocera) in Relation to Fish Predation in Lake Albert, East Africa." *Journal of Zoology* 151 (2): 181–97. doi.org/10.1111/j.1469-7998.1967.tb02109.x.

Greene. 1989. "A Diet-Induced Developmental Polymorphism in a Caterpillar." *Science (New York, N.Y.)* 243 (4891): 643–46. doi.org/10.1126/science.243.4891.643.

Grimm, Grimm, Faeth, Golubiewski, Redman, Wu, Bai, et al. 2008. "Global Change and the Ecology of Cities." *Science* 319 (5864): 756–60. doi.org/10.1126/science.1150195.

Gross, Pasinelli, & Kunc. 2010. "Behavioral Plasticity Allows Short-Term Adjustment to a Novel Environment." *The American Naturalist* 176 (4): 456–64. doi.org/10.1086/655428.

Habary, Johansen, Nay, Steffensen, & Rummer. 2017. "Adapt, Move or Die—How Will Tropical Coral Reef Fishes Cope with Ocean Warming?" *Global Change Biology* 23 (2): 566–77. doi.org/10.1111/gcb.13488.

Hayes, Khoury, Narayan, Nazir, Park, Brown, Adame, et al. 2010. "Atrazine Induces Complete Feminization and Chemical Castration in Male African Clawed Frogs (*Xenopus laevis*)." *Proceedings of the National Academy of Sciences* 107 (10): 4612–17. doi.org/10.1073/pnas.0909519107.

Heuer & Grosell. 2014. "Physiological Impacts of Elevated Carbon Dioxide and Ocean Acidification on Fish." *American Journal of Physiology-Regulatory, Integrative and Comparative Physiology* 307 (9): R1061–84. doi.org/10.1152/ajpregu.00064.2014.

Hiyama, Nohara, Kinjo, Taira, Gima, Tanahara, & Otaki. 2012. "The Biological Impacts of the Fukushima Nuclear Accident on the Pale Grass Blue Butterfly." *Scientific Reports* 2: 570. doi.org/10.1038/srep00570.

Hope, Gries, Zhu, Fagan, Redman, Grimm, Nelson, Martin, Kinzig. 2003. "Socioeconomics Drive Urban Plant Diversity." *Proceedings of the National Academy of Sciences* 100: 8788–92. doi:10.1073/pnas.1537557100.

International Human Genome Sequencing Consortium. 2004. "Finishing the Euchromatic Sequence of the Human Genome International Human Genome Sequencing Consortium*." *Nature* 431 (7011): 931–45. http://www.genome.gov/10000923.

Ishizuka & Goto. 2012. "Modeling Intraspecific Adaptation of *Abies sachalinensis* to Local Altitude and Responses to Global Warming, Based on a 36-Year Reciprocal Transplant Experiment." *Evolutionary Applications* 5 (3): 229–44. doi.org/10.1111/j.1752-4571.2011.00216.x.

Iverson, Gilchrist, Smith, Gaston, & Forbes. 2014. "Longer Ice-Free Seasons Increase the Risk of Nest Depredation by Polar Bears for Colonial Breeding Birds in the Canadian Arctic." *Proceedings of the Royal Society B: Biological Sciences* 281 (1779): 20133128. doi.org/10.1098/rspb.2013.3128.

Iwanowicz, Blazer, Pinkney, Guy, Major, Munney, Mierzykowski, et al. 2016. "Evidence of Estrogenic Endocrine Disruption in Smallmouth and Largemouth Bass Inhabiting Northeast U.S. National Wildlife Refuge Waters: A Reconnaissance Study." *Ecotoxicology and Environmental Safety* 124: 50–59. doi.org/10.1016/j.ecoenv.2015.09.035.

Jablonka & Raz. 2009. "Transgenerational Epigenetic Inheritance: Prevalence, Mechanisms, and Implications for the Study of Heredity and Evolution." *The Quarterly Review of Biology* 84 (2): 131–76. doi.org/10.1086/598822.

Janzen & Phillips. 2006. "Exploring the Evolution of Environmental Sex Determination, Especially in Reptiles." *Journal of Evolutionary Biology* 19 (6): 1775–84. doi.org/10.1111/j.1420-9101.2006.01138.x.

Kelly, Panhuis, & Stoehr. 2012. "Phenotypic Plasticity: Molecular Mechanisms and Adaptive Significance." In *Comprehensive Physiology*, vol. 2, 1417–39. Hoboken, NJ: John Wiley & Sons. doi.org/10.1002/cphy.c110008.

Kowarik. 2011. "Novel Urban Ecosystems, Biodiversity, and Conservation." *Environmental Pollution* 159 (8–9): 1974–83. doi.org/10.1016/j.envpol.2011.02.022.

Kucharski, Maleszka, Foret, & Maleszka. 2008. "Nutritional Control of Reproductive Status in Honeybees via DNA Methylation." *Science* 319 (5871): 1827–30. doi.org/10.1126/science.1153069.

Kunc, Lyons, Sigwart, McLaughlin, & Houghton. 2014. "Anthropogenic Noise Affects Behavior Across Sensory Modalities." *The American Naturalist* 184 (4): E93–100. doi.org/10.1086/677545.

Lachapelle, Bell, & Colegrave. 2015. "Experimental Adaptation to Marine Conditions by a Freshwater Alga." *Evolution* 69 (10): 2662–75. doi.org/10.1111/evo.12760.

Laforsch, Beccara, & Tollrian. 2006. "Inducible Defenses: The Relevance of Chemical Alarm Cues in *Daphnia*." *Limnology and Oceanography* 51 (3): 1466–72. doi.org/10.4319/lo.2006.51.3.1466.

Lambert, Giller, Barber, Fitzgerald, & Skelly. 2015. "Suburbanization, Estrogen Contamination, and Sex Ratio in Wild Amphibian Populations." *Proceedings of the National Academy of Sciences* 112 (38): 11881–86. doi.org/10.1073/pnas.1501065112.

Leong, Dunn, & Trautwein. 2018. "Biodiversity and Socioeconomics in the City: A Review of the Luxury Effect." *Biology Letters* 14 (5): 20180082. doi.org/10.1098/rsbl.2018.0082

Liu, Wollstein, Hysi, Ankra-Badu, Spector, Park, Zhu, et al. 2010. "Digital Quantification of Human Eye Color Highlights Genetic Association of Three New Loci." Edited by Mark I. McCarthy. *PLoS Genetics* 6 (5): e1000934. doi.org/10.1371/journal.pgen.1000934.

Lundholm & Richardson. 2010. "Habitat Analogues for Reconciliation Ecology in Urban and Industrial Environments." *Journal of Applied Ecology* 47 (5): 966–75. doi.org/10.1111/j.1365-2664.2010.01857.x.

Mao, Schuler, & Berenbaum. 2015. "A Dietary Phytochemical Alters Caste-Associated Gene Expression in Honey Bees." *Science Advances* 1 (7): e1500795. doi.org/10.1126/sciadv.1500795.

Martin, Chick, Yilmaz, & Diamond. 2019. "Evolution, Not Transgenerational Plasticity, Explains the Adaptive Divergence of Acorn Ant Thermal Tolerance Across an Urban–Rural Temperature Cline." *Evolutionary Applications* 12: 1678–87. DOI: 10.1111/eva.12826

Marzluff, Bowman, & Donnelly. 2001. "A Historical Perspective on Urban Bird Research: Trends, Terms, and Approaches." In *Avian Ecology and Conservation in an Urbanizing World*, 1–17. Boston, MA: Springer. doi.org/10.1007/978-1-4615-1531-9_1.

Mitton & Ferrenberg. 2012. "Mountain Pine Beetle Develops an Unprecedented Summer Generation in Response to Climate Warming." *The American Naturalist* 179 (5): E163–71. doi.org/10.1086/665007.

Morley, Jones, & Radford. 2014. "The Importance of Invertebrates When Considering the Impacts of Anthropogenic Noise." *Proceedings of the Royal Society of London: Biological Sciences* 281: 20132683. doi.org/10.1098/rspb.2013.2683

Oates. 1934. "The Population of Rome." In *Classical Philology*. Chicago: The University of Chicago Press. Accessed August 6, 2019. doi.org/10.2307/264523.

O'Neill, Jerrett, Kawachi, Levy, Cohen, et al. 2003. Health, Wealth, and Air Pollution: Advancing Theory and Methods. *Environmental Health Perspectives* 111 (16): 1861–70. doi:10.1289/ehp.6334.

Orci, Petróczki, & Barta. 2016. "Instantaneous Song Modification in Response to Fluctuating Traffic Noise in the Tree Cricket Oecanthus Pellucens." *Animal Behaviour* 112: 187–94. doi.org/10.1016/j.anbehav.2015.12.008.

Paquette, Pelletier, Garant, & Bélisle. 2014. "Severe Recent Decrease of Adult Body Mass in a Declining Insectivorous Bird Population Is Not Related to Breeding Habitat Quality." *Proceedings of the Royal Society B: Biological Sciences* 281: 1–26.

Parmesan & Yohe. 2003. "A Globally Coherent Fingerprint of Climate Change Impacts Across Natural Systems." *Nature* 421 (6918): 37–42. doi.org/10.1038/nature01286.

Richter, Haslbeck, & Buchner. 2010. "The Heat Shock Response: Life on the Verge of Death." *Molecular Cell* 40 (2): 253–66. doi.org/10.1016/j.molcel.2010.10.006.

Robinson. 2013. "Evolution of Growth by Genetic Accommodation in Icelandic Freshwater Stickleback." *Proceedings of the Royal Society B: Biological Sciences* 280 (1772): 20132197. doi.org/10.1098/rspb.2013.2197.

Root, Price, Hall, Schneider, Rosenzweig, & Pounds. 2003. "Fingerprints of Global Warming on Wild Animals and Plants." *Nature* 421 (6918): 57–60. doi.org/10.1038/nature01333.

Roy & Sparks. 2000. "Phenology of British Butterflies and Climate Change." *Global Change Biology* 6 (4): 407–16. doi.org/10.1046/j.1365-2486.2000.00322.x.

Ryu, Veilleux, Donelson, Munday, & Ravasi. 2018. "The Epigenetic Landscape of Transgenerational Acclimation to Ocean Warming." *Nature Climate Change* 8: 504–9. doi.org/10.1038/s41558-018-0159-0.

Sahanatien & Derocher. 2012. "Monitoring Sea Ice Habitat Fragmentation for Polar Bear Conservation." *Animal Conservation* 15 (4): 397–406. doi.org/10.1111/j.1469-1795.2012.00529.x.

Sanger, Coulson, Friedmann, Air, Barrell, Brown, Fiddes, Hutchison, & Slocombe.

1978. "Nuclotide-Sequence of Bacteriophage-PHI-X174." *Journal of Molecular Biology* 125: 225–46.

Schell, Dyson, Fuentes, De Roches, Harris, Miller, Woelfle-Erskine, & Lambert. 2020. "The Ecological and Evolutionary Consequences of Systemic Racism in Urban Environments." *Science*: eaay4497. DOI: 10.1126/science/aay4497.

Schlichting & Wund. 2014. "Phenotypic Plasticity and Epigenetic Marking: An Assessment of Evidence for Genetic Accommodation." *Evolution* 68 (3): 656–72. doi.org/10.1111/evo.12348.

Schwander & Leimar. 2011. "Genes as Leaders and Followers in Evolution." *Trends in Ecology & Evolution* 26 (3): 143–51. doi.org/10.1016/j.tree.2010.12.010.

Slabbekoorn & Peet. 2003. "Birds Sing at a Higher Pitch in Urban Noise." *Nature* 424 (6946): 267–67. doi.org/10.1038/424267a.

Stamps. 2016. "Individual Differences in Behavioural Plasticities." *Biological Reviews* 91 (2): 534–67. doi.org/10.1111/brv.12186.

Sturm, Duffy, Zhao, Leite, Stark, Hayward, Martin, & Montgomery. 2008. "A Single SNP in an Evolutionary Conserved Region Within Intron 86 of the HERC2 Gene Determines Human Blue-Brown Eye Color." *American Journal of Human Genetics* 82 (2): 424–31. doi.org/10.1016/j.ajhg.2007.11.005.

Swaddle, Francis, Barber, Cooper, Kyba, Dominoni, Shannon, et al. 2015. "A Framework to Assess Evolutionary Responses to Anthropogenic Light and Sound." *Trends in Ecology & Evolution* 30 (9): 550–60. doi.org/10.1016/J.TREE.2015.06.009.

Tollrian & Dodson. 1999. "Inducible Defenses in Cladocera: Constraints, Costs, and Multipredator Environments." In *The Ecology and Evolution of Inducible Defenses*, eds. Tollrian & Harvell, 177–202. Princeton, NJ: Princeton University Press.

United Nations. 2014. "World Urbanization Prospects: The 2014 Revision, Highlights. Department of Economic and Social Affairs." population.un.org/wup/.

Verhoeven, von Holdt, & Sork. 2016. "Epigenetics in Ecology and Evolution: What We Know and What We Need to Know." *Molecular Ecology* 25 (8): 1631–38. doi.org/10.1111/mec.13617.

Verhoeven, Jansen, van Dijk, & Biere. 2010. "Stress-Induced DNA Methylation Changes and Their Heritability in Asexual Dandelions." *New Phytologist* 185 (4): 1108–18. doi.org/10.1111/j.1469-8137.2009.03121.x.

De Villemereuil, Gaggiotti, Mouterde, & Till-Bottraud. 2016. "Common Garden Experiments in the Genomic Era: New Perspectives and Opportunities." *Heredity* 116 (3): 249–54. doi.org/10.1038/hdy.2015.93.

Whitehead, Roach, Zhang, & Galvez. 2011. "Genomic Mechanisms of Evolved Physiological Plasticity in Killifish Distributed Along an Environmental Salinity Gradient." *Proceedings of the National Academy of Sciences* 108 (15): 6193–98. doi.org/10.1073/pnas.1017542108.

Winter & Holtum. 2014. "Facultative Crassulacean Acid Metabolism (CAM) Plants: Powerful Tools for Unravelling the Functional Elements of CAM Photosynthesis." *Journal of Experimental Botany* 65 (13): 3425–41. doi.org/10.1093/jxb/eru063.

Wirth. 1938. "Urbanism as a Way of Life." *American Journal of Sociology* 44 (1): 1–24. doi.org/10.1086/217913.

Zufall & Rausher. 2003. "The Genetic Basis of a Flower Color Polymorphism in the Common Morning Glory (*Ipomoea purpurea*)." *Journal of Heredity* 94 (6): 442–48. doi.org/10.1093/jhered/esg098.

Core Responses: Adapt

Learning Outcomes

After working with this chapter, you will be able to:

- Describe the conditions necessary for adaptation by natural selection.
- Analyze how organisms adapt to contemporary global change stressors.
- Evaluate the ways scientists identify adaptive evolution.
- Synthesize factors influencing the adapt response and evaluate the relationship between the adapt response and the other core responses.
- Apply your knowledge to real-world case studies and interpret data from recent scientific studies.

THE BLANK PAGE

How do human actions influence the environment of other living things? Let's take a seemingly simple choice: which soap you use to wash your hands. The use of soap by humans goes back many thousands of years (Wilcox 2000). Early soaps were fashioned from animal or vegetable oil treated with an alkaline solution. Increasingly, soaps have many other additives, including antibacterial compounds. There are currently hundreds of antibacterial soap products on the market. What do you think the consequences are of using antibacterial soaps, both for humans and for other organisms in our environment? Could you predict how species might change over time because of our pervasive use of antibacterials? Are there particular traits in particular organisms that could increase the probability of survival in an environment enriched with synthetic antibacterial compounds?

INTRODUCTION

As environmental conditions on our planet are radically altered, adaptation will be essential for the survival of myriad species. **Adaptation** is the process by which populations become better suited to the environmental conditions they experience via heritable change over generations. In contrast with the adjust response (which occurs at the individual level), adaptation is a population-level process that occurs across generations. Adaptation can happen rapidly over a few generations or slowly over geological timescales. Adaptation has been critically important in shaping broad-scale patterns of diversity across the entire tree of life. Adaptation has also shaped fine-scale patterns of trait variation within species. As species struggle to survive in anthropogenically altered habitats, adaptation will be necessary for long-term persistence. The goal of this chapter is to analyze the "adapt" response and understand what factors facilitate—and limit—the adaptive response of populations faced with global change stressors.

WHAT CONDITIONS ARE REQUIRED FOR ADAPTATION?

Before we investigate examples of contemporary adaptation to global change, let us review the fundamentals. **Natural selection** is the evolutionary process by which adaptation occurs. When environmental conditions shift, organisms that are better suited to the new conditions preferentially survive and reproduce. Thus, a subset of individuals is more likely to pass on their genes to the next generation. Differences in survival and reproduction at the individual level lead to changes at the population level. Over time, traits that facilitate survival and reproduction in the novel environment increase in frequency.

Thus, natural selection can be more accurately defined as the differential survival or reproduction of organisms based on their fit to their environment. Organisms that are better suited to local conditions preferentially survive and reproduce, leading to a change in the traits of a population over time. Biologists recognize several key conditions that facilitate evolution by natural selection, described later and summarized in Figure 7.1. These ideas trace back to Darwin's seminal work, *The Origin of Species* (1859), written more than 150 years ago.

Heredity

For adaptation to occur, traits must have a genetic basis. As we explored in the previous chapter, most traits are influenced simultaneously by genetics and by the environment. Natural selection can occur as long as some component of the trait is **heritable**. A trait that is determined only by environmental conditions cannot evolve by natural selection.

As reviewed in the *Core Concepts* feature of the previous chapter, the primary mechanism for heredity is the information encoded in macromolecules that store information and replicate: DNA and RNA. There is variation and complexity in hereditary systems across the tree of life, but the basic requirement for natural selection is any molecular mechanism that leads to a correspondence between parent and offspring phenotypes.

Variation

Another prerequisite for adaptation by natural selection is variation. Generally, individuals within a population exhibit variation at the phenotypic level. For example, some individuals may be slightly taller or shorter in height, slightly darker or lighter in color, or slightly faster or slower at processing particular environmental compounds. If these traits are heritable, observed phenotypic variation results (at least in part) from underlying genetic variation.

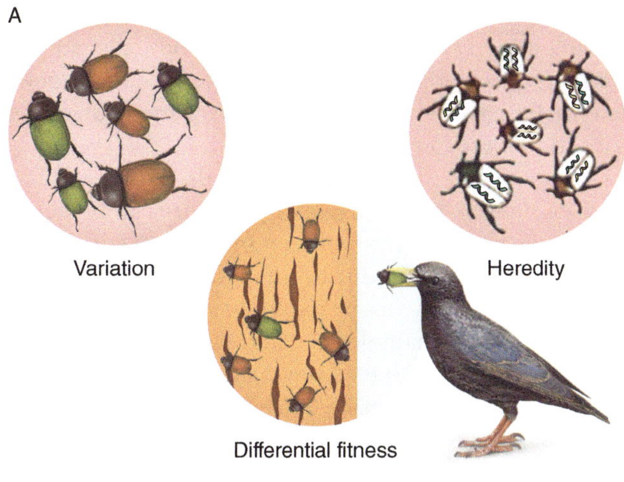

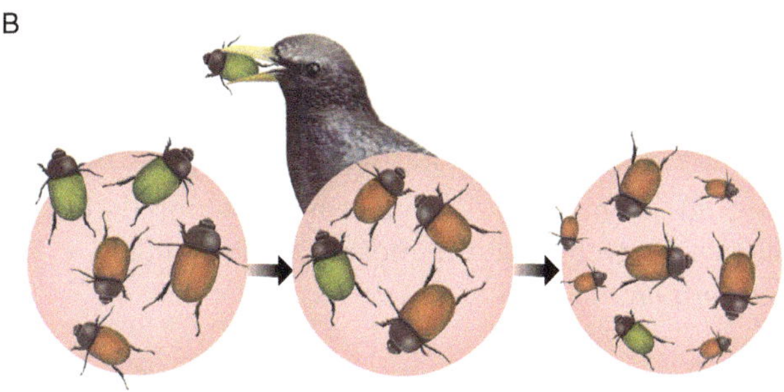

FIGURE 7.1 (A) The three fundamental conditions for evolution by natural selection. (B) Adaptation by natural selection occurs at the population level across generations.

Reflection: How can scientists determine if a trait is heritable?

Source: https://evolution.berkeley.edu/evolibrary/article/0_0_0/evo_14; https://evolution.berkeley.edu/evolibrary/search/imagedetail.php?id=281&topic_id%3D%26keywords%3D

As reviewed in the *Core Concepts* feature in this chapter, phenotypic variation is ultimately generated by **mutation**. There are many ways to generate variation at the molecular level, from alterations of a single DNA nucleotide to duplications of the entire genome. Because all molecular changes occur in a single individual, mutation ultimately creates differences among individuals and provides the raw material on which natural selection can act.

Differential Survival or Reproduction

Not all individuals in a population fare equally, and we refer to **fitness** as the ability of an organism to survive and reproduce in a particular environment. Survival and reproduction are influenced by both abiotic factors (like temperature, nutrient availability, or access to nesting sites) and biotic factors (like predation or herbivory intensity, prey availability, or access to mates). Individuals with traits that are favorable in a particular environment exhibit increased survival and/or reproductive output. In the case of environmental change, those individuals that are best suited to the new environmental conditions preferentially pass along their genes. Over the course of generations, an entire population can adapt to new conditions.

Darwin referred to the fact that not all individuals survive and reproduce equally as the "struggle for existence." This struggle for existence can be due to physiological limits; for example, rising temperatures weed out individuals with lower heat tolerance. The struggle for existence can also be due to competition. Often, in the natural world, finite resources create competition among individuals (within or between species). We often think of overt behavioral examples of competition in the animal world (such as elephant seals fighting over mates), but all organisms compete for resources, including space and nutrients. Competition is made fierce by what Darwin called the "overproduction" of gametes or offspring (for example, consider the tremendous number of human sperm that are produced but never fertilize an egg). Ultimately, the struggle for existence ensures that only a subset of individuals will ultimately contribute genes to the next generation.

The simplest mathematical formulation for predicting trait change by natural selection ($R = s*h^2$) posits that the response to selection (R) is equal to the strength of selection (s) multiplied by the heritability of the trait under consideration (h^2). There are, of course, many nuances to predicting how populations will respond to environmental change, but this equation underscores how adaptive response is jointly determined by environmental and genetic factors.

CORE CONCEPTS

WHERE DOES GENETIC VARIATION COME FROM?

Natural selection acts on genetic variation, which ultimately is generated by **mutation** and **recombination** at the molecular level. There are many ways to generate variation at the molecular level from small changes that affect only a single DNA nucleotide (called point mutations) to changes that shuffle variants within or across chromosomes (such as chromosomal rearrangements) to changes that impact the entire genome (such as whole genome duplication). Typically these changes arise from errors in DNA replication and repair, or errors during recombination. Because all molecular changes occur in a single individual, mutation ultimately creates differences among individuals.

Mutation rates vary substantially across the tree of life (Lynch 2010). For example, the flu virus has a mutation rate that is more than a million times faster than the mutation rate in humans. A higher mutation rate increases the potential for new **beneficial mutations** (but also increases the potential for new **deleterious mutations**). Organisms with faster generation times and larger population sizes also typically generate more mutations per unit time.

When a population experiences a change in environmental conditions, sometimes a genetic variation that is well suited to the new environment already exists at low frequency in the population, as shown in Box Figure 7.1. In this case, adaptation occurs from what is called **standing genetic variation**: the presence of multiple alleles at a locus in a population. Adaptation from standing genetic variation can occur more quickly because the raw material for adaptation is already present (Barrett & Schluter 2008). Conversely, adaptation can occur from **new mutation**—a genetic change that occurs after the environmental change. Because mutations occur randomly,

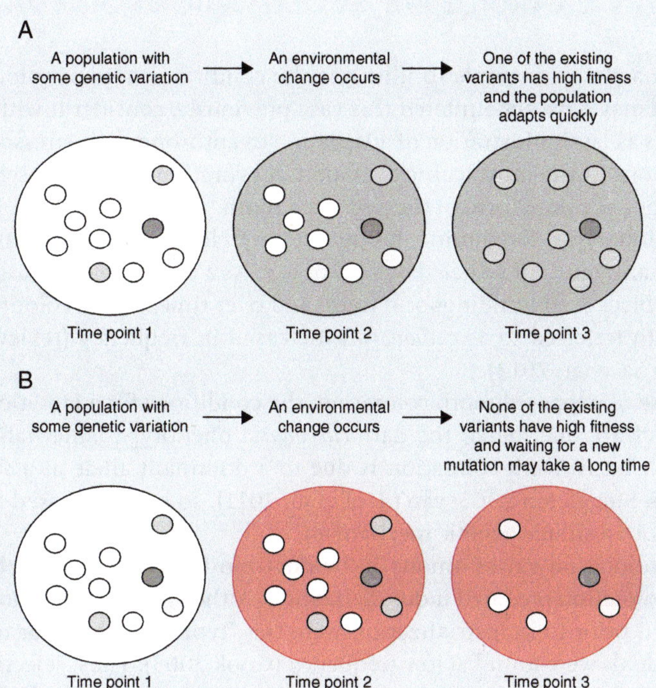

BOX FIGURE 7.1 Adaptation from standing genetic variation (A) can typically occur faster than when waiting for a new mutation (B).

Reflection: How would your expectations about levels of standing genetic variation and waiting time for new mutations change for a very large population?

Source: Courtesy of the author

the waiting time for a new mutation can be quite long. There is evidence in natural systems that both new mutation and standing genetic variation have contributed to adaptation across the tree of life. But in response to global change stressors, adaptation from standing genetic variation will be particularly important for species with smaller population sizes and longer generation times, given the more rapid pace at which selection can act.

Another potential source of adaptive alleles is **gene flow**: the movement of genetic variation between populations (or occasionally between species; Hedrick 2013). As we discussed in Chapter 5, organisms move over many spatial and temporal scales. When an individual moves from one population to another—and reproduces in the new location—new genetic variants can be introduced to the recipient population. Thus, **immigration** can be a source of adaptive alleles, and even the immigration of a single individual can change the evolutionary course of a population if that individual introduces novel genetic variation (Grant & Grant 2014).

Genetic information can also occasionally cross species boundaries, termed **introgression**. Adaptive introgression can occur via **hybridization**, when closely related species interbreed. For example, after the rodenticide warfarin was introduced in the 1950s, the European house mouse (*Mus musculus*) developed resistance via introgression of an adaptive allele from the Algerian mouse (*Mus spretus*; Song et al. 2011). Adaptive introgression can also occur via **horizontal gene transfer**. Particularly in microbes, mobile genetic elements can move genetic information across species and confer adaptation to various global change stressors like pollutants (Springael & Top 2004). Thus, mutation and recombination act to generate novel genotypes, and introgression, hybridization, and horizontal gene transfer can move these variants among populations.

WHAT IS AN EXAMPLE OF EVOLUTION BY NATURAL SELECTION?

Let us look at a classic example to illustrate the conditions necessary for adaptation. Although you may have encountered this case previously, consider it with a fresh perspective as it is a classic illustration of adaptation to anthropogenic stressors. As shown in Figure 7.2, one of the most iconic examples of evolution by natural selection is the rise in frequency of a dark form of the peppered moth (*Biston betularia*) in 19th-century Britain. The Industrial Revolution brought tremendous population and industrial growth in Britain, much of it fueled by coal power. As a result, coal pollution darkened the skies and blackened buildings and trees, and over time, a dark color morph of the peppered moth (referred to as *carbonaria*) increased in frequency (reviewed in Cook 2003; Cook & Saccheri 2013).

In the case of peppered moth coloration, the conditions for adaptation by natural selection are clearly met. First, the dark *carbonaria* phenotype is heritable and has a known genetic basis. Dark coloration is due to a dominant allele at a single genetic locus (Cook & Muggleton 2003; van't Hof et al. 2011), so color is passed from parents to offspring with a simple genetic mechanism.

Second, coloration varies among individual moths within populations. In fact, variation in coloration predated industrialization. Although the vast majority of specimens collected prior to industrialization were the "typical" light color morph, some darker individuals were found at low frequency (Cook 2003). Thus, selection could act on **standing genetic variation** as described in the *Core Concepts* box of this chapter.

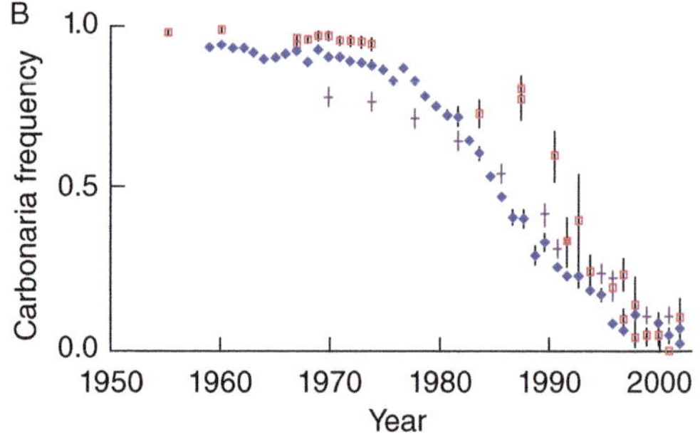

FIGURE 7.2 (A) Camouflage under different environmental conditions in the peppered moth (*Biston betularia*). The typical form is better hidden from avian predators under normal conditions, but the dark (*carbonaria*) form was better hidden when coal pollution blackened trees in industrialized Britain. (B) As coal pollution declined in the late 20th century, the frequency of the dark *carbonaria* form once again decreased in British peppered moth populations. Different shapes represent mean *carbonaria* frequencies in populations from different regions of England, and vertical lines represent standard errors.

Reflection: What expectations do you have for whether and how levels of genetic variation would change after an episode of strong natural selection like that witnessed in the peppered moth?

Source: (A) IanRedding/Shutterstock; (B) Laurence M Cook, "The Rise and Fall of the Carbonaria Form of the Peppered Moth," The Quarterly Review of Biology 78, no. 4 (December 2003): 399-417

Third, differences in fitness among color morphs resulted from differential success at avoiding predation. Moths are prey for visual hunting avian predators, and numerous studies have demonstrated that birds can act as an agent of selection in this system (reviewed in Cook 2003). Individuals that were better camouflaged were able to reduce the risk of avian predation (Cook 2003; Cook et al. 2012). Visually hunting avian predators preferentially attack color morphs that are conspicuous against their background, so smoke and soot pollution affected the relative probability of survival of different color morphs.

Ultimately, natural selection led to tremendous shifts in the *carbonaria* frequency. In some parts of Britain, the frequency of the dark *carbonaria* form increased from rarity to a height of 90% in the mid-20th century. Interestingly, the frequency of the dark *carbonaria* morph has again changed in the 20th century as coal pollution has decreased (Cook 2003; Figure 7.2).

WHAT TYPES OF ADAPTATION OCCUR IN RESPONSE TO GLOBAL CHANGE PRESSURES?

We just looked at one classic example of adaptation in moth coloration in response to coal pollution. But there are myriad examples of adaptation to diverse global change stressors across the tree of life. In Chapter 6, we organized examples of the adjust response by trait type (development, physiology, behavior, morphology). In this chapter, to promote flexibility in our thinking, we will evaluate examples instead by global change stressor. Of course, these frameworks are compatible and complementary: a full understanding of any core response requires analyzing the interplay between environmental change and trait change. Here our exemplars focus on rapid adaptation to contemporary environmental change in terrestrial ecosystems. The *Taking a Closer Look* feature at the end of this chapter focuses exclusively on marine systems.

Adaptation to Environmental Contaminants

Chemical inputs into the environment can create strong natural selection for traits that confer resistance or tolerance to these stressors. We have already looked at a number of human activities that alter the chemical composition of ecosystems, including inputs of heavy metals, pesticides, pharmaceuticals, and carbon dioxide. Simply because these stressors exist does not necessarily mean that organisms will be able to adapt to them. However, rapid adaptive responses are often observed: from plants that adapt to grow on heavy metal mining tailings (e.g., Jain and Bradshaw 1966) to bacteria that evolve resistance to pharmaceuticals like antibiotics (e.g., French 2010).

One example is the evolved resistance of insects to the use of insecticides in agricultural settings. A specific illustration is the use of a toxin from the bacterium *Bacillus thuringiensis* (Bt) as an insecticide. Because this toxin kills a number of key agricultural pests, many widespread crops like corn, cotton, and soybean are now genetically engineered to produce Bt proteins. As Figure 7.3 shows, over the last 20 years Bt crop planting has accelerated and now covers nearly 100 million hectares of agricultural fields globally.

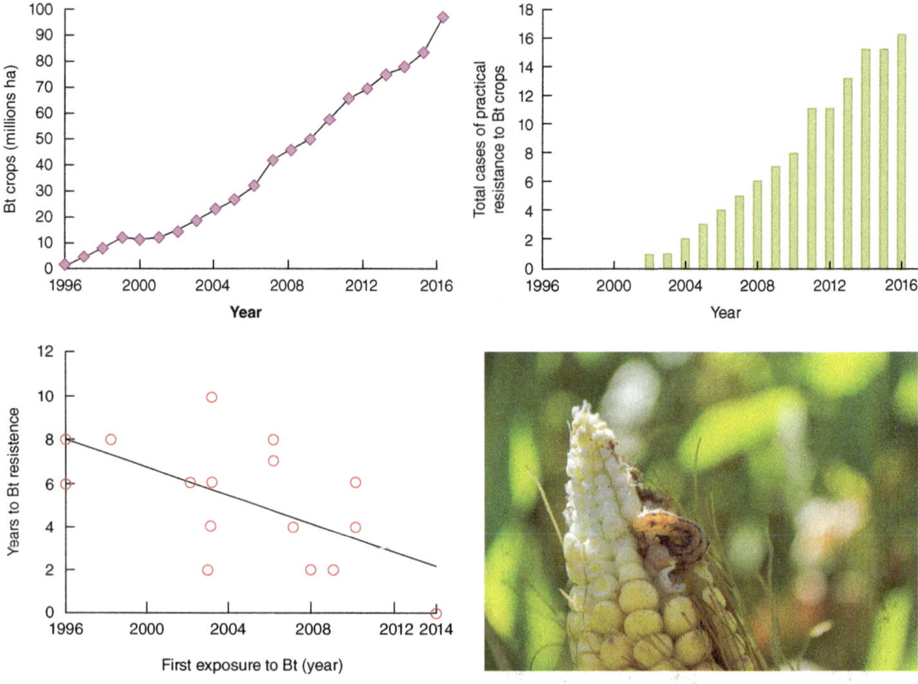

FIGURE 7.3 Adaptation to insecticides. Planting of transgenic corn and cotton crops that produce insecticidal proteins from *Bacillus thuringiensis* (Bt) has increased dramatically over the last decades. In response, a number of agricultural pest species have evolved resistance to Bt (including the armyworm, *Spodoptera frugiperdam*, whose larva is shown attacking corn). In fact, the number of years to evolve resistance after first exposure to Bt has been decreasing over time.

Reflection: What is one possible reason that the insects are evolving resistance to Bt more quickly now than they were 20 years ago? Consider the possible ways that adaptive alleles could move between populations.

Source: (A) Tabashnik & Carrière, Nature Biotechnology volume 35, pages 926–935 (2017)l; (B) Fabian Putruele/Shutterstock

However, a number of target insects—from Africa to Asia to North and South America—have rapidly evolved resistance to Bt (Tabashnik & Carrière 2017). In fact, the number of years until resistance emerges after the first planting of Bt crops appears to be decreasing, demonstrating that adaptation can occur very quickly (Tabashnik & Carrière 2017). The specific genetic mechanism of resistance has even been identified in some species. For example, the fall armyworm (*Spodoptera frugiperda*) in Puerto Rico exhibits a mutation in an ATP-binding gene that confers resistance to a toxic Bt crystalline protein (Banerjee et al. 2017). Thus, genetic changes that confer adaptation can spread through a population rapidly when natural selection is strong.

Adaptation to Introduced Species

Biotic changes can also trigger adaptive evolution, especially for species that have strong ecological interactions in their communities. One of the most common anthropogenically induced biotic changes in ecosystems is introduced species. As we discussed in Chapter 5, globalized trade facilitates the movement of species out of their native ranges at an unprecedented rate. As introduced species establish, they can dramatically change the selection environment for native species.

One classic example is insects that adapt to exploit an introduced host plant. A specific illustration is the apple maggot fly (*Rhagoletis pomonella*) which rapidly adapted to the introduction of cultivated apple trees in eastern North America in the 1800s (switching hosts from the native wild hawthorn). The life history of *Rhagoletis* flies is highly tied to their host plant because mating, oviposition, and larval growth all occur on host fruits. Apple and hawthorn trees differ in a number of ways, including their smell and the timing of fruiting. *Rhagoletis* adapted to apple trees exhibit a number of heritable changes in olfaction and life history timing (e.g., McPheron et al. 1988; Filchak et al. 2000). Contrary to the armyworm example earlier, where changes at a single gene underlie an adaptive shift, many regions throughout the genome appear to be involved in *Rhagoletis* adaptation (Egan et al. 2015).

Introduced species can provide new resources, but they can also provide new threats, as when invasive predators or competitors affect the evolutionary trajectory of native species. For example, the introduction of the ground-dwelling, predatory curly-tail lizard (*Leiocephalus carinatus*) to Caribbean islands shifted the habitat use and morphology of the local brown anole lizard (*Anolis sagrei*). As shown in Figure 7.4, on islands with the introduced predator, brown anoles exhibit a number of shifts in behavior (like perching higher off the ground) and morphology (like larger body size) that decrease exposure to predation and increase survivorship (Losos et al. 2004; Lapiedra et al. 2018). Although the specific genetic mechanisms for these shifts have not yet been identified, replicated experiments have demonstrated that natural selection on heritable variation contributes to rapid population-level trait change. Thus, biotic interactions are a key component of selection regimes, and introduced species often generate strong selection pressure in invaded ecosystems.

Adaptation to a Changing Climate

The effects of climate change generate novel—and widespread—selection pressures. Not only are organisms under pressure to survive at higher temperatures, but they experience changes in local cloud cover, rainfall, humidity, and frequency and severity of extreme weather events. Changes in abiotic climate variables can also lead to changes in biotic interactions, triggered for example by shifts in phenology. Thus, adaptation to a changing climate can include heritable changes in a wide variety of traits from thermal tolerance to osmotic balance to photoperiod response.

One classic example of rapid adaptation to climate change is found in the fruit fly *Drosophila subobscura*. Fruit flies are highly sensitive to temperature given the physiological requirements of flight, and *D. subobscura* living in colder northern latitudes exhibit a

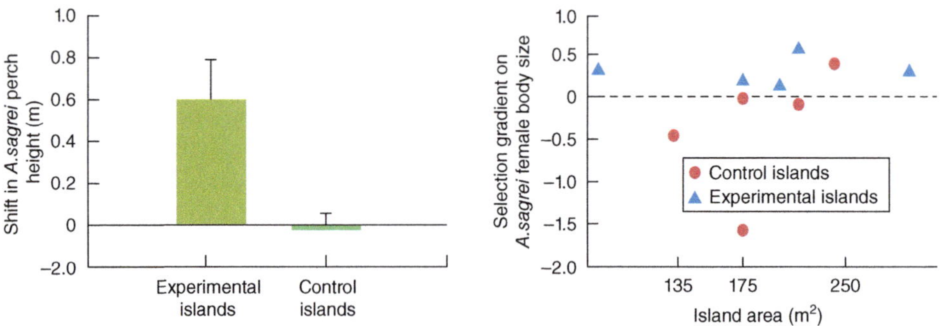

FIGURE 7.4 Rapid evolution of traits in the native brown anole (*Anolis sagrei*) in the presence of the predatory curly-tailed lizard (*Leiocephalus carinatus*). *Anolis* that perched higher and had larger body sizes survived better on experimental islands where the predatory *L. carinatus* was introduced.

Reflection: How specifically could the researchers determine whether the observed shift in perch height is due to behavioral plasticity or genetic adaptation?

Source: Losos, J., Schoener, T. & Spiller, D. Predator-induced behaviour shifts and natural selection in field-experimental lizard populations. Nature 432, 505–508 (2004)

number of morphological and behavioral adaptations to cool climates (Gilchrist et al. 2004). A specific genetic mechanism—in the form of a large-scale DNA rearrangement called a **chromosomal inversion**—appears to play a key role. The frequency of the chromosomal inversion shows latitudinal variation both in native (Europe) and introduced (North and South America) range of *D. subobscura*. Specifically, as shown in Figure 7.5, the inversion is more prevalent in colder northern climates, and this is parallel across continents. Perhaps most interesting is that the frequency of the inversion has been decreasing over the last several decades, associated with global warming (Balanyá et al. 2006; Rezende et al. 2010). Thus, adaptation to contemporary climate change may recruit pre-existing genetic variation from populations adapted to different historical climate conditions.

In addition to changes in thermal tolerance, some species exhibit heritable shifts in **phenology** in response to anthropogenic climate change. One illustration is advancement of spring breeding time by the red squirrel (*Tamiasciurus hudsonicus*), which allows offspring to better exploit the earlier spring growth of their food supply (Réale et al. 2003). Another is the pitcher-plant mosquito (*Wyeomyia smithii*), which has shifted its photoperiod response (Bradshaw and Holzapfel 2001), allowing prolonged fall activity and a longer growing season. Although there are many examples of organismal change in response to climate, the examples here are specifically attributed to genetic adaptation, not solely phenotypic plasticity.

In Sum

Of course, species may be subject to multiple stressors simultaneously, and they may show adaptive responses in numerous traits. Conversely, species may be subject

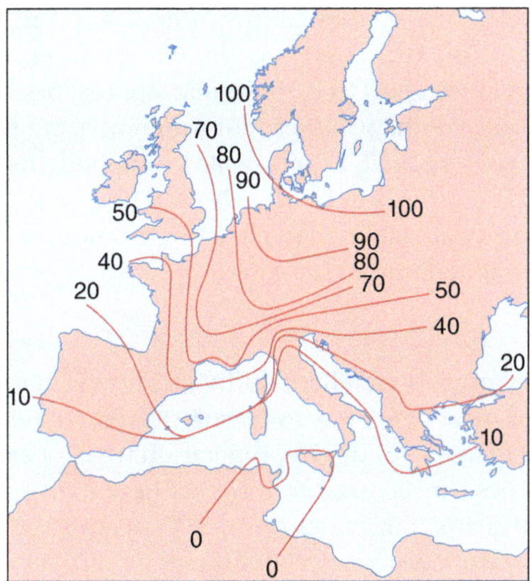

FIGURE 7.5 Adaptation to changing climates: the frequency of a chromosomal rearrangement thought to confer adaptation to colder climates in the fruit fly *Drosophila subobscura*. Lines represent breakpoints between frequencies. Historically, the chromosomal rearrangement was absent in warmer southern Europe and fixed in colder northern Europe. However, over the last 20 years, latitudinal patterns have been shifting with climate change.

Reflection: What traits do you think the species we have discussed thus far (such as moths, lizards, and flies) have in common that might facilitate a rapid evolutionary response?

Source: Rezende EL, Balanyà J, Rodríguez-Trelles F, Rego C and others (2010) Climate change and chromosomal inversions in Drosophila subobscura. Clim Res 43:103–114

to strong natural selection but fail to adapt, particularly when they lack heritable variation in relevant traits. Although adaptation is not guaranteed in response to anthropogenic stressors, the earlier examples—and numerous others—demonstrate that adaptation can occur rapidly, particularly when environmental pressures exert strong natural selection.

HOW DO SCIENTISTS IDENTIFY ADAPTATIONS AND PREDICT ADAPTIVE POTENTIAL?

Identifying adaptation in natural populations requires demonstrating the effects of natural selection on heritable genetic variation and ruling out other explanations for trait change over time. As we have previously discussed, **genetic drift** can cause heritable trait change, but simply due to the effects of chance on survival and reproduction. In addition, **phenotypic plasticity** can lead to trait changes, but plastic changes are typically not heritable and do not lead to genetic adaptation at the population level. Thus, to distinguish adaptation from drift and plasticity, scientists must demonstrate that focal traits are **heritable** and directly improve **fitness** (the ability of an organism to survive and reproduce in a particular environment).

There are many ways to assess heritability, fitness, and the signature of selection in natural populations. In the last chapter, we discussed how experimental approaches (like reciprocal transplant studies) and molecular approaches (like gene expression assays) can be used to determine whether trait differences are heritable. In addition, there are many other observational, experimental, and modeling approaches that can be used to detect selection (Merilä & Hendry 2014). Here we look at several examples in which scientists identify adaptations to changing environments and reveal mechanisms underlying these adaptations.

Field Studies

There are many ways to measure trait heritability and fitness in natural systems. Breeding experiments are the most direct way to estimate trait heritability, but many species cannot be bred in the lab due to body size, generation time, or environmental requirements. Thus, field studies can be used to estimate heritability by tracking similarity between parent and offspring trait values (Wray & Visscher 2008). Similarly, designing field experiments to measure fitness can be difficult because of variation in dynamics of natural selection over space and time, or between sexes or life stages (Hardwick et al. 2015). However, performance assays, mark-recapture studies, and population growth measurements can all be conducted in field settings to understand how particular traits affect fitness in particular environments. Thus, both observational and experimental studies in nature can be used to identify adaption.

A classic example of a long-term observational study comes from Darwin's finches on the Galapagos Islands. Rosemary and Peter Grant, shown in Figure 7.6, studied trait change in several species of ground finches (*Geospiza*) over a period of >40 years. During this time, several extreme weather events—including prolonged droughts and intense storms—triggered adaptive responses in local finch populations. For example, in 1982, strong El Niño rains brought the most severe weather event experienced in centuries. The abundance of rain altered the vegetation on the island and the types of seeds available for ground finches to eat. In response, medium ground finches (*Geospiza fortis*) with smaller, pointier beaks had increased fitness, and beak shape decreased dramatically over time, as shown in Figure 7.6 (Grant & Grant 2014). Because scientists had already identified particular genes that influence beak size and shape, the genetic mechanism responsible for the trait change could be elegantly tracked (Lamichhaney et al. 2016; Arnold & Kunte 2017). As the frequency and severity of extreme events increase with anthropogenic climate change, the potential for adaptation to climate extremes becomes increasingly relevant for numerous species.

Studies in nature can also make use of experimental approaches by directly manipulating the biotic or abiotic environment to assess adaptive response. One example comes from controlled introductions of lizards onto small islands in the Bahamas (e.g., Schoener & Schoener 1983). The effects of the predatory curly-tail lizard (*Leiocephalus carinatus*) on brown anoles (*Anolis sagrei*)—which we discussed in the last section of this chapter—were revealed using such an approach. In one recent study, scientists moved more than 250 brown anoles onto eight small islands (Lapiedra et al. 2018). They uniquely tagged each lizard, allowed them to acclimate to new conditions for

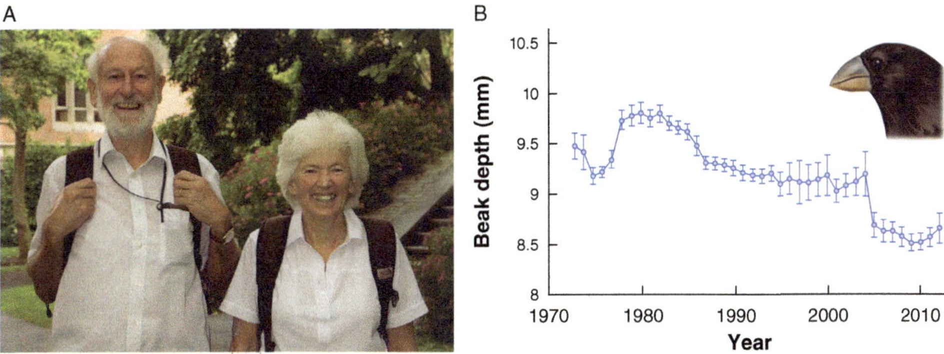

FIGURE 7.6 (A) Drs. Peter and Rosemary Grant, who have been conducting field studies of Galapogos finches for more than 45 years. (B) Changes in beak depth over time in the Galapagos ground finch (*Geospiza fortis*) are associated with extreme weather events that influence seed availability and the fitness of birds with different beak sizes.

Reflection: How specifically might beak size and shape influence different aspects of survival and reproduction?

Source: (A) Princeton University, Office of Communications, Denise Applewhite (2009); (B) "Adaptive Genetic Exchange: A Tangled History of Admixture and Evolutionary Innovation, Michael L. Arnold, Krushnamegh Kunte, Trends in Ecology & Evolution, VOLUME 32, ISSUE 8, P601-611, AUGUST 01, 2017"

1 week, and then introduced curly-tailed lizards on half of the islands. Using a mark–recapture approach and replicated study design, the scientists could compare outcomes of natural selection on control versus treatment islands. Although biotic manipulations must be done with extreme caution, studies like this can shed light on adaptation following invasion.

Lab Studies

Of course, experimental approaches can also be used in the lab under controlled conditions. Lab studies of adaptation can take many forms, but one particularly powerful approach for organisms with relatively fast generation times is **experimental evolution**. As illustrated in Figure 7.7, experimental evolution involves applying controlled selection regimes to replicated populations and tracking adaptation *in real time* over many generations (Elena & Lenski 2003). Because populations can be sampled at many time points (including the ancestral population used to found the replicates), scientists can determine the mechanisms underlying observed evolutionary responses. For example, scientists can see whether adaptation occurs in different replicates using the same genetic mechanisms. Scientists can also assess whether adaptation occurs from new mutations or from genetic variants already present in the ancestral population (a topic explored in this chapter's *Core Concepts* feature).

Perhaps the most iconic example of experimental evolution in action is a long-term experiment with the bacteria *Escherichia coli*. Richard Lenski and colleagues have been evolving *E. coli* for more than 30 years and 60,000 generations. The long-term

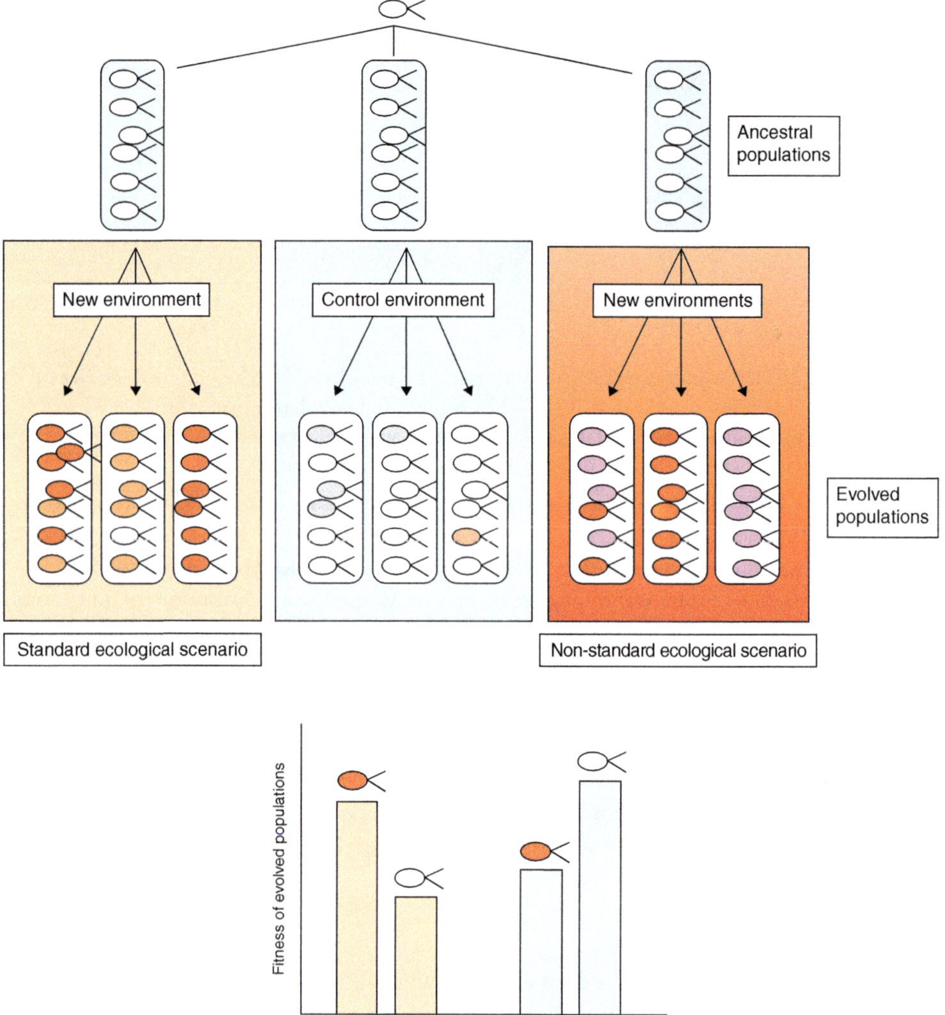

FIGURE 7.7 Experimental evolution approach. The "Standard Ecological Scenario" represents a single, abrupt environmental change, while the "Non-standard Ecological Scenario" represents using experimental evolution for more complex scenarios, such as fluctuating environmental conditions or persistent incremental environmental change over many generations.

Reflection: Describe in your own words the experimental evolution approach presented. How is an experiment started? What happens during the experiment? What is compared at the end of the experiment? How is adaptation assessed?

Source: Collins, S. Evol Biol (2011) 38: 3

study follows 12 replicate populations that originated from the same ancestral strain but have been evolving independently ever since. The approach provides incredible power to assess what aspects of evolution are predictable and what aspects are idiosyncratic over long timescales (Lenski 2017). It also generates important baseline data on evolution in a constant environment that can be compared to dynamics of adaptation in stressful environments. For example, recent studies that experimentally evolved *E. coli* under different temperature regimes found that strains adapted to elevated temperatures by converging on similar phenotypes but with different underlying genetic mechanisms (Sandberg et al. 2014; Deatherage et al. 2017). Thus, lab studies clearly demonstrate similar adaptive outcomes can occur independently in different populations.

Another example related to global change stressors is adaptation of marine invertebrates to ocean acidification. As we will see in the *Taking a Closer Look* feature at the end of this chapter, ocean acidification has the potential to disrupt marine food webs because it profoundly affects calcifying organisms. Lohbeck et al. (2012) evolved phytoplankton (*Emiliania huxleyi*) under increased carbon dioxide (CO_2) for 500 generations and then tested their performance under ocean acidification conditions. Although calcification rates were still impaired by acidification, the populations that had been preadapted to increased CO_2 fared better even after 1 year. Therefore, some micro-organisms may be able to adapt to novel conditions on a timescale relevant to contemporary marine communities.

Of course, there are also examples where experimental populations fail to adapt even under strong selection. For example, Collins and Bell (2004) showed that an experimental population of green alga did not exhibit specific adaptations to increased carbon dioxide despite 1,000 generations of selection. Thus, laboratory experiments can help scientists determine whether species are likely to have the capacity to adapt to future environmental conditions.

Molecular Approaches

Molecular approaches provide an important complement to field and laboratory studies because they can reveal the specific mechanisms underlying adaptation. Even 15 years ago, molecular approaches were applied most commonly to **model organisms**–species like yeast, fruit flies, and mice that could be kept in the lab in large numbers for detailed study. However, as shown in Figure 7.8, over the last decades, costs for genomic sequencing have exponentially decreased. With decreased cost and increased accessibility, sophisticated genetic and genomic analyses are increasingly the norm in studies of natural selection in the wild. Variation can be studied at both the DNA sequence level and also at the gene expression level. Thus, researchers can now look directly at the molecular mechanisms underlying adaptation in natural populations over space and/or time.

There are many ways to search for evidence of natural selection at the genetic level and identify specific genes or genomic regions that underlie adaptive trait change (reviewed, for example, in Nielsen 2005; Stinchcombe & Hoekstra 2008;

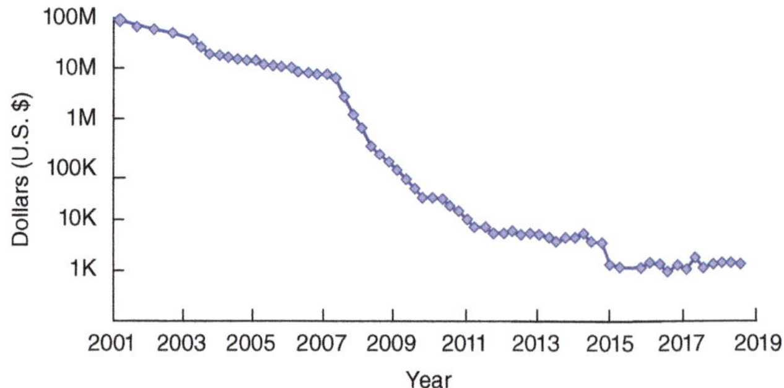

FIGURE 7.8 The cost to sequence a genome has dramatically decreased in the past decades. Here the cost per human genome is shown over time. The first human genome took more than 10 years to sequence with more than 2 billion US dollars of investment. Today, however, a human genome can be sequenced in 1 day for approximately $1,000. Declining costs mean that genome sequencing can now be conducted rapidly and inexpensively for both model and nonmodel species.

Reflection: What implications does inexpensive DNA sequencing have? What opportunities does it present? What scientific or ethical challenges might it raise?

Source: National Human Genome Research Institute

Ellegren 2014). One common approach is to conduct genome-scale sequencing to look for gene regions that show evidence of natural selection. This approach was applied to reveal the genomic signature of poleward range expansion in a damselfly species (*Coenagrion scitulum*). As we saw in Chapter 5, a consistent biotic signature of global warming is poleward range shifts. What adaptations might facilitate successful range shifts? Natural selection could favor populations with traits that increase rate of spread—particularly at the range edge—including increased dispersal capacity and thermal tolerance. For example, Swaegers et al. (2015) compared the genomes of damselflies from the center of the range to those at the northern edge. They found variation in one particular nucleotide that was associated with increased flight performance in populations at the northern edge of the range. This same variant was found in four of five independent range expansion events, suggesting that a consistent genetic mechanism may be responsible for increased dispersal capacity in this species at range boundaries.

In another recent example, Campbell-Staton et al. (2020) used RNA-sequencing to evaluate mechanisms of thermal adaptation to urban environments in the Puerto Rican crested anole (*Anolis cristatellus*). RNA-sequencing has the advantage of providing genome-scale data on both gene *expression* (changes in transcription) and gene *evolution* (changes in the coding sequence itself). The scientists compared urban to nearby forest lizards in four different regions of the country. The replicates were independent because each city was separately colonized from its nearby forest populations. The urban areas were significantly warmer than the paired forest sites, and Campbell-Staton et al. confirmed that city lizards occupied hotter microhabitats and experienced higher body temperatures than their forest counterparts. Moreover, urban lizards showed increased tolerance to high temperatures. Using a genome-scanning approach, the researchers found that a single mutation in a protein synthesis gene (RARS) was associated with heat tolerance in all replicate urban-forest pairs. This gene is known to be involved to resilience to abiotic stress in other species (Anderson et al. 2009). Parallel selection on the same gene suggests that some independent populations—and species—might find similar adaptive solutions to shared global change stressors like climate change. A key takeaway is that molecular methods are now sophisticated enough to address big picture questions about whether evolution in this era of rapid environmental change can be predictable at the genetic level.

In Sum

Of course, field observations, laboratory experiments, and molecular approaches can be combined to better understand dynamics of adaptation to global change. In addition, computational and synthesis approaches are fruitful for describing and predicting adaptive responses. Modeling approaches can be used to generate ecological and evolutionary expectations and to compare observed patterns to model predictions (Merilä & Hendry 2014; Marshall et al. 2016). Meta-analyses are also instrumental in understanding broad-scale patterns of natural selection in the wild (e.g., Hoekstra et al. 2001; Kingsolver et al. 2012).

Historically, selection studies have focused on individual traits, but selection acts on multiple traits in integrated organisms. Thus, it is increasingly important to understand complex and multidimensional phenotypes and their effect on fitness (Laughlin & Messier 2015). Moreover, it is critical to assess the likelihood that a species will exhibit an adapt response when confronted with changing conditions—sometimes referred to as **adaptive potential**. The most promising approaches for predicting adaptive potential are when projected future environmental conditions can be simulated, whether through manipulations in the field, experiments in the lab, or mathematical simulations. Thus, integrative approaches are essential for predicting whether species can adapt to novel conditions in a rapidly changing world.

CAN ADAPTATION PREVENT EXTINCTION?

Adaptation is such an important response to changing conditions that it can sometimes "save" a population from extirpation (reviewed in Gonzalez et al. 2013; Carlson et al. 2014; Bell 2017). **Evolutionary rescue** is when genetic adaptation allows a population to recover from an environmental change that would otherwise cause extirpation. As always, the environmental threat can be abiotic (like temperature rise) or biotic (like introduction of a new pathogen). As illustrated in Figure 7.9, populations undergoing evolutionary rescue show a characteristic decline and subsequent recovery fueled by adaptive genetic change (Carlson et al. 2014).

A classic example of evolutionary rescue is the case of the European rabbit, which was intentionally introduced to Australia beginning in the 18th century as a pet and game animal (Fenner 2010). The introduced rabbits proliferated and became a major pest by the mid-19th century. In the 1950s, an intentional release of the myxoma virus began to decimate the rabbit populations. As shown in Figure 7.9, there was initially

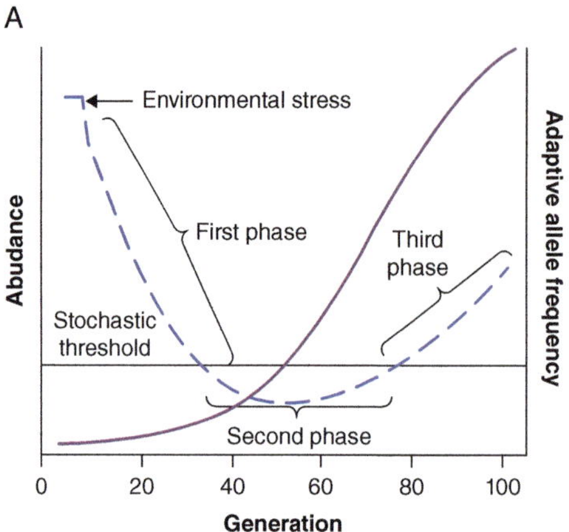

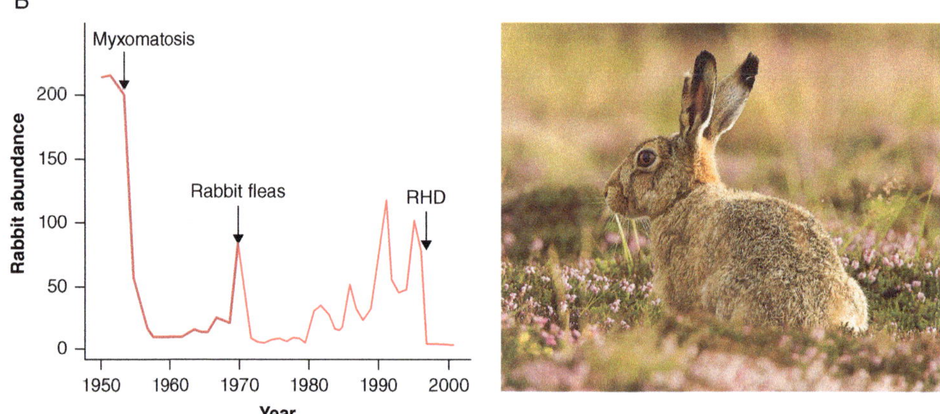

FIGURE 7.9 Evolutionary rescue in nature. (A) Evolutionary rescue is characterized by a U-shaped demographic trajectory. First, populations experience a rapid decline due to maladaptation to new environmental conditions. Second, population size falls below the threshold where extinction becomes likely simply due to stochastic fluctuations. Third, population growth increases due to spread of adaptive allele(s). (B) An empirical example showing change in introduced rabbit population size in Australia over time. Arrows indicate biological control agents that were used in attempts to eradicate rabbits. The rebound of rabbit populations in the early 1970s was due to evolved resistance to the myxoma virus.

Reflection: Describe the ways an adaptive allele can arise and spread in a population.

Source: (A) Stephanie M.Carlson, Curry J.Cunningham, Peter A.H.Westley, Trends in Ecology & Evolution, Volume 29, Issue 9, September 2014, Pages 521–530; (B) Vander Wal E., Garant D., Festa-Bianchet M. and Pelletier F. Evolutionary rescue in vertebrates: evidence, applications and uncertainty368Phil. Trans. R. Soc.; photo courtesy Mark Medcalf/Shutterstock

high susceptibility to myxomatosis, with up to 99% mortality in infected animals. However, strong selection for resistance to the virus (and likely reduction of viral virulence) eventually led to a population recovery (Fenner 2010). More recently, several other biological control agents have been used in attempts to reduce rabbit population sizes in Australia. However, it appears that rabbits may be developing resistance to these as well, including the more recently introduced rabbit hemorrhagic disease (RHD) (Saunders et al. 2010).

Examples of evolutionary rescue are relatively rare because it is difficult to demonstrate that a genetically based change conclusively saved a population from extirpation. However, both empirical and experimental studies provide examples of evolutionary rescue in action (Carlson et al. 2014). Experimental evolution studies are particularly compelling because they can more easily track the trajectory of particular adaptive mutations. For example, Bell and Gonzalez (2009) exposed experimental yeast populations to what would be typically lethal salt concentrations and found that evolutionary rescue was explained primarily by mutations in two genes and occurred relatively quickly when population sizes were large.

There are a number of demographic, genetic, and environmental factors that influence the probability of evolutionary rescue (Bell & Gonzalez 2011; Schiffers et al. 2013; Carlson et al. 2014). Critically important are factors like initial population size, availability of genetic variation, and rate of environmental change. Thus, evolutionary rescue may not occur for many threatened species, which already have low population size and reduced genetic variation (Vander Wal et al. 2013).

Evolutionary rescue also carries an implicit risk. Bouts of strong natural selection can actually—and counterintuitively—make populations *more* vulnerable to future stressors. For example, a population that experiences strong natural selection due to chemical contaminants in the environment will not automatically be better protected in the face of global warming. Strong natural selection generally depletes genetic variation (Figure 7.10), leaving populations with less adaptive potential to respond to future environmental change. Thus, evolutionary rescue will not be a magic bullet for all species. But the potential for evolutionary rescue remains important, particularly as adaptation to novel conditions becomes the best hope for survival of many species in the face of global change stressors.

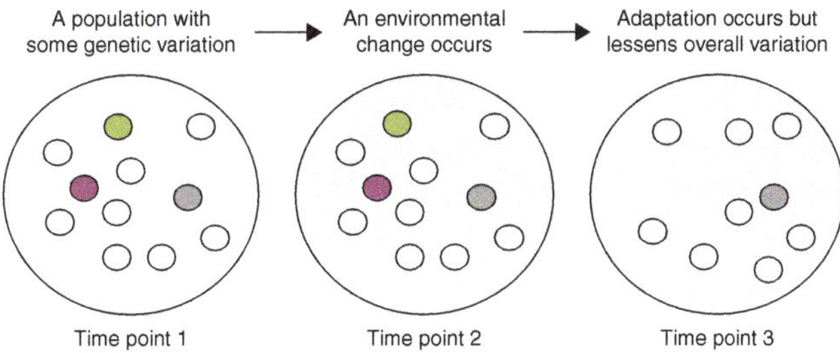

FIGURE 7.10 Strong natural selection can make a population better adapted to current conditions but can also deplete variation and create vulnerabilities to future environmental change.

Reflection: What factors influence the likelihood of evolutionary rescue?

Source: Courtesy of the author

CONCLUSION

Adaptation via natural selection has for millennia played a key role in shaping the evolutionary trajectory of life on Earth. As global and local environments change ever more quickly, adaptation will continue to be a critical response, facilitating population persistence and species survival. Although there is no one trait that can be used to predict whether a population will adapt in the face of a changing environment, a number of factors affect the probability that populations will successfully adapt when confronted with a rapidly changing environment. Typically, organisms with higher mutation rates, faster generation times, and larger population sizes will adapt more quickly because they generate more variation and are more buffered from the effects of genetic drift. However, any species can adapt, provided that natural selection is strong and there is differential survival based on variation in heritable traits.

There are many additional factors that can influence the likelihood of adaptation in response to environmental change. For example, the degree of genetic connectivity between populations can influence the probability of adaptive evolution, as genetic variants are moved between populations. As always, the magnitude and velocity of environmental change are also critically important because the rate or magnitude of change can be so great that populations cannot adapt quickly enough. As with the other core responses, the probability of adaptive evolution does not depend on just one factor but on the combination of many different organismal traits, population characteristics, and environmental factors.

Adaptive responses are clearly not mutually exclusive with the other core responses. In Chapter 5, we explored how the adaptation might facilitate—or be facilitated by—the move response. Populations adapted to human-dominated environments are more likely to be transported to other anthropogenically modified landscapes. Once species are transported to new geographic regions, they may again adapt to novel conditions. Moreover, adaptation can facilitate a move response—for example, if natural selection favors the evolution of increased dispersal capacity. In Chapter 6, we saw how the adapt and adjust responses interact. For example, plastic responses can be canalized through **genetic assimilation**, and plasticity itself can be an adaptation.

Of course, adaptation is not guaranteed. Many forecasting studies suggest that the rate of environmental change will exceed the ability of many populations to adapt (e.g., Etterson & Shaw 2001; Both et al. 2006; Sinervo et al. 2010). Thus, adaptation is far from assured when populations experience rapid environmental change, and failure to adapt can lead to the "die" response, which we turn to in the next chapter.

MEET THE DATA THE *DAPHNIA* TIME MACHINE

Who Are the Scientists and What Did They Set Out To Do?

Here we will evaluate a study that used a novel approach to understand potential for adaptive evolution on rapid timescales. "Rapid Evolution of Thermal Tolerance in the Water Flea *Daphnia*" was published by Geerts et al. in the journal *Nature Climate Change* in 2015.

When investigating adaptation in contemporary populations, researchers often wish they could access historical time points for comparison. In the *Meet the Data* feature in Chapter 5, we looked at the possibility of using museum specimens as an archival anchor to investigate organismal change over time. Although museum specimens provide a rich resource for measuring phenotypes that are well preserved, many traits cannot be measured after death.

What if researchers could have access to *live* animals from a historical time point that had somehow been frozen in time? For some organisms—those which produce dormant life stages—this is possible. Under certain conditions, seeds, eggs, cysts, and spores can be stably preserved in stratified sediments, soil, or ice and revived years, decades, or even centuries later. These resources—referred to as **propagule banks**—can be accurately dated and compared to their contemporary counterparts (Orsini et al. 2013). Sometimes called **resurrection ecology**, the ability to hatch or germinate preserved samples offers an "evolutionary time machine" (Orsini et al. 2013) and a rare opportunity to evaluate adaptation through time directly.

Freshwater invertebrates—like the water flea *Daphnia*—have provided some of the best examples of resurrection ecology in action. *Daphnia* produce dormant eggs that have been revived even 500 years after deposition (Frisch et al. 2014). Moreover, these aquatic invertebrates are easy to study in laboratory settings and lend themselves well to the **experimental evolution** approaches we discussed earlier in this chapter. Finally, genomes from several species of *Daphnia* have been well characterized (Colbourne et al. 2011; Orsini et al. 2016). Leveraging these benefits to working with *Daphnia*, Geerts et al. (2015) set out to use this model system to study genetic adaptation to climate change over the last 50 years.

What Are *Your* Predictions?

Before you read on, take a few minutes to think about how to apply what you have learned in this chapter about adaptation to the *Daphnia* system. Start by answering the following questions:

- *Do you expect adaptive changes to have occurred in Daphnia response to heat stress over the last 50 years? If so, what specific traits might show an adaptive response?*
- *Do you expect contemporary populations of Daphnia to have the potential for further evolution in response to even higher future temperatures?*
- *If so, how would you set up an experiment to determine evolutionary potential?*

Now write a summary prediction statement about whether—and how—*Daphnia* populations are adapting to climate change change.

What Were the Scientists' Predictions?

The authors did not have formalized predictions, but they had clear objectives. Geerts et al. (2015) set out to test whether *Daphnia* populations have adapted to warming water temperatures due to anthropogenic climate change

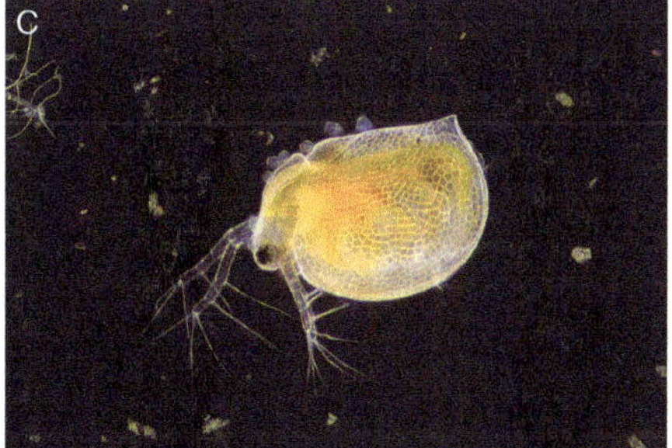

BOX FIGURE 7.2 Lead author Dr. Aurora Geerts (A), senior author Dr. Luc De Meester (B), and their study organism (*Daphnia magna*) (C).

Source: (A) Aurora Geerts; (B) Photo by Kristen Brans; (C) D. Kucharski K. Kucharska/Shutterstock

(Continued)

and to predict whether current populations have further adaptive potential.

What Data Were Collected?

Geerts and colleagues used two different approaches to understand the history of past adaptation and the potential for future adaptation in populations of *Daphnia magna*.

First, they used an experimental evolution approach to understand whether contemporary populations have the *potential* to genetically adapt to higher water temperatures. They housed *Daphnia* from contemporary populations in replicate outdoor tanks (simulating small ponds) and exposed them to one of two treatments: ambient temperature or ambient temperature + 4 degrees Celsius. After 2 years of selection, they tested thermal tolerance using a simple index called the **critical thermal maximum** (CT_{Max}). CT_{Max} is the point at which animals lose key motor function, and it is biologically relevant for climate change because it sets an upper bound on organismal survival under heat stress.

Second, they used a resurrection ecology approach to measure *actual* changes over time. They hatched *Daphnia* from historic and contemporary sediment layers representing a time series of 50 years. Now instead of comparing CT_{Max} between experimental treatment groups, they compared CT_{Max} between historical and contemporary individuals.

What Is *Your* Interpretation of the Data?

Before you read on, take a few minutes to interpret the data figure (Box Figure 7.3). Consider the following questions as you look at the figure:

- *What different components of the study are represented in the two figures?*
- *What are the overall trends observed between treatment groups in each figure?*
- *Why might there be variation within the treatment groups?*
- *Write a sentence or two describing the key findings of this study and their significance.*

What Was the Scientists' Interpretation of the Data?

The authors demonstrated that contemporary *Daphnia* populations have the capacity to respond with genetic adaptation under prolonged selection at increased temperatures. *Daphnia* from the heat treatment had a significantly higher CT_{Max} than those from the control treatment. In fact, the magnitude of the change in CT_{Max} was 3.6 degrees Celsius, closely tracking the 4 degrees Celsius environmental difference between treatments. Moreover, evolution of *Daphnia* heat tolerance occurred within a 2-year experiment, demonstrating that some species can evolve rapidly on timescales relevant to contemporary global warming.

Moreover, the authors demonstrate that adaptation to global warming has already happened in this system.

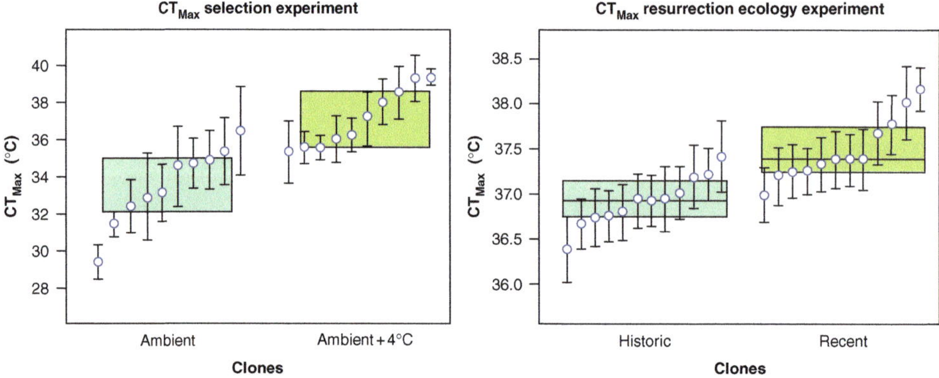

BOX FIGURE 7.3 Average CT_{Max} for *Daphnia* from the experimental evolution study (A) and the resurrection ecology study (B). Individual means (circles) and standard errors (whiskers) represent different replicates. Colored boxplots show the median and 25th and 75th percentile per population.

Source: A. N. Geerts, J. Vanoverbeke, B. Vanschoenwinkel, W. Van Doorslaer, H. Feuchtmayr, D. Atkinson, B. Moss, T. A. Davidson, C. D. Sayer & L. De Meester, Rapid evolution of thermal tolerance in the water flea Daphnia, Nature Climate Change volume 5, pages 665–668 (2015)

Over the last 40 years, *average* temperature at the focal lake has increased more than 1 degree Celsius, and temperature *extremes* (heatwaves) have increased threefold. *Daphnia* populations have closely tracked these changes with increased CT_{Max}.

It is possible that evolution for increased heat tolerance is widespread in natural populations. However, at the time, there was little direct empirical evidence. As Geerts et al. stated, *"So far, no study has documented evolutionary changes in the thermal tolerance of natural populations as a response to recent temperature increase."* Thus, their experimental demonstration of adaptation to global warming in a natural population was highly significant.

Although Geerts et al. (2015) focused on adaptation to climate change, this research group and their collaborators have used similar methods to evaluate potential for adaptation to other environmental stressors. For example, Orsini et al. (2012) applied a resurrection ecology framework to understanding the evolutionary response of *Daphnia* to predation, parasitism, and agricultural land-use intensity. These resurrection ecology approaches can be complemented with genomic approaches to identify the specific genetic changes responsible for observed adaptive response. For example, Orsini et al. (2012) identified genetic variants that were associated with general—or specific—stress responses in their treatment groups. Similarly, Jansen et al. 2017 (working with the same *Daphnia* populations as Geerts et al. 2015) identified candidate genes that show altered expression patterns between treatment groups differing in CT_{Max}. Connecting genotype, phenotype, and fitness in this way for populations undergoing rapid evolution in response to anthropogenic stressors remains a leading edge in the field of Global Change Biology.

What Are *Your* Ideas for Future Research Directions?

Given what these studies accomplished, what next steps do you envision for this research program? If you had been involved in the *Daphnia* study, how would you follow up? Start by considering the following questions:

- *How could a similar approach in this study be used to understand adaptation to additional global change stressors?*
- *What other systems could lend themselves to a similar approach and provide a useful comparison?*
- *What other approaches could provide complementary data at a broader scale?*

Now write a few sentences about future directions on this research theme.

TAKING A CLOSER LOOK CORAL REEFS

Oceans comprise >70% of Earth's surface and play a critical role in climate regulation, oxygen production, and human survival. Within the marine realm, coral reef ecosystems are some of the most biologically diverse—and anthropogenically threatened—areas of the world. Central to understanding the structure, function, and long-term fate of reef ecosystems is an ancient symbiosis between corals and algae. As we saw in Chapter 1, intracellular symbionts have played an important role in the evolution of life on Earth. Here we will take a closer look at corals, the anthropogenic stressors that threaten their symbiosis, and the potential for adaptation to play a role in the future of reef ecosystems.

What Are Coral Reefs and Why Are They Valuable?

Coral reefs are found in clear, warm ocean waters. Shallow submarine platforms provide excellent conditions for reef development, and the greatest concentration of coral reefs is in the Indo-Pacific and the Caribbean (as shown in Box Figure 7.4). There are several different types of coral reefs that exhibit structural and functional differences based on their proximity to shore, their degree of isolation, and their connection to neighboring ecosystems. However, here we will focus on cross-cutting issues that affect all coral reefs.

Coral reefs occupy <1% of the ocean floor (Spalding & Grenfell 1997) but support an incredible amount of biological diversity. For example, they harbor >25% of all species of marine fish (McAllister 1991). In addition to their inherent ecological value as hotspots for biodiversity, coral reefs provide a wealth of **ecosystem services** (as introduced in Chapter 3). Human populations make use of myriad resources from coral reefs (reviewed in Moberg & Folke 1999), including food resources (like fish and seaweed), raw materials for building (like sand and ingredients

(Continued)

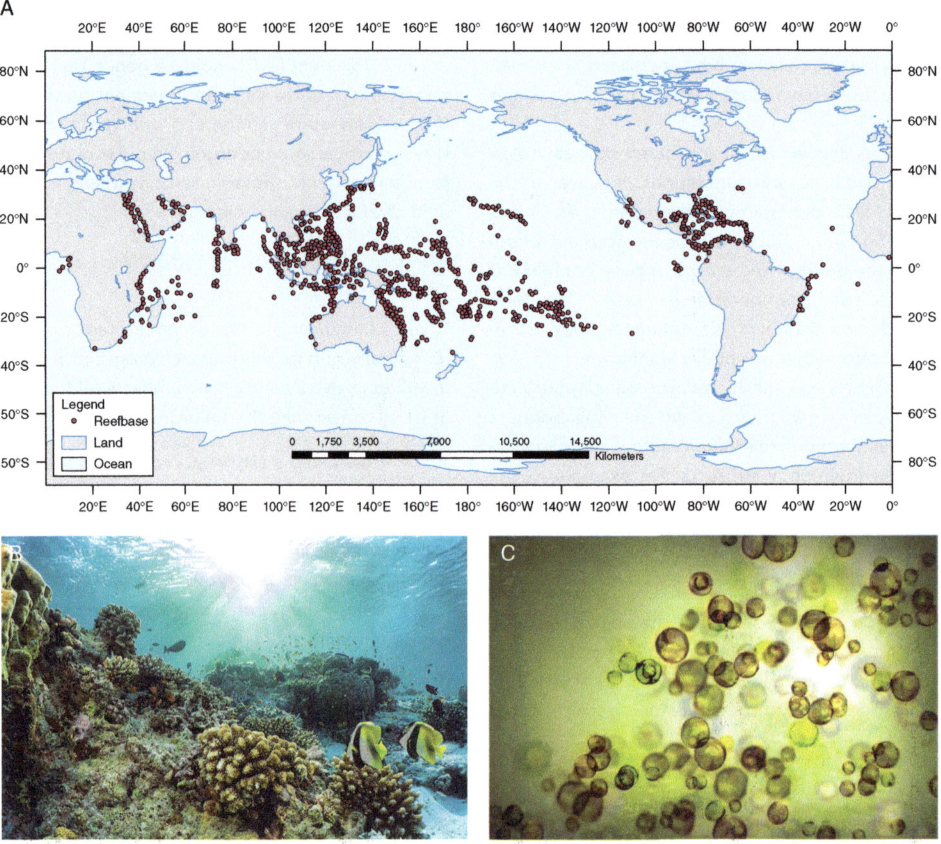

BOX FIGURE 7.4 (A) Distribution of coral reefs around the world in warm, shallow marine environments. (B) A diversity of reef-building corals contributes to reef structure by secreting calcified skeletons. (C) Symbiotic photosynthetic algae (*Symbiodinium*) are critical to coral health.

Reflection: Given the geographic distribution, physical structure, and critical interspecific interactions in coral reefs, what global change stressors do you predict will generate the most direct threats to coral ecosystems?

Source: (A) National Oceanic and Atmospheric Administration (NOAA); (B) Sven Hansche/Shutterstock; (C) Mary Alice Coffroth

for cement), and energy resources (like extracted oil and gas). In addition, the physical structure of coral reefs provides valuable coastal protection from storms and erosion. Finally, coral reefs provide livelihoods for many coastal communities from trade in reef products and in tourism from reef recreation.

The value of ecosystem services from coral reefs is estimated in the billions of dollars annually. For example, in the meta-analysis we considered in the *Meet the Data* feature of Chapter 1, de Groot et al. (2012) surveyed economic studies from 10 biomes and found that—in terms of monetary value of ecosystem services—coral reefs were the single most valuable biome per unit area. For comparison, an average hectare of coral reef was estimated to be >700 times more valuable than an average hectare of open ocean.

What Are Corals and Their Symbionts?

Corals are marine invertebrates in the phylum *Cnidaria*. There are hundreds of species of reef-building corals—and many additional non-reef-building corals. Reef-building corals secrete a hard skeleton of calcium carbonate. Secreted skeletons cement individuals together, and skeletons from dead corals form the reef structure for living corals.

In addition, many reef-building corals contain symbiotic photosynthetic algae in the genus *Symbiodinium* (often referred to as **zooxanthellae**). These symbionts confer color to the corals and their products of photosynthesis, and many corals cannot survive long term without their symbionts (Huston 1985). Symbionts can be transmitted vertically from parent to offspring or acquired from the environment (Goulet & Coffroth 2003). Some host–symbiont relationships exhibit a high degree of specificity (specialized association between particular coral species and particular symbionts), while others are more flexible (Baker 2003). Symbiont communities are quite diverse but do appear to exhibit adaptation to local conditions (Thornhill et al. 2017).

What Are the Major Threats to Coral Reefs?

As we saw in Chapter 4, there are many interacting stressors affecting marine ecosystems. Moreover, a large proportion of marine areas are strongly affected by multiple drivers simultaneously (Halpern et al. 2008). Here we consider a few examples of how anthropogenic stressors specifically affect coral ecosystems (reviewed in Cesar & Chong 2004; Harborne et al. 2017; Putnam et al. 2017).

OCEAN WARMING Global warming will have diverse effects in the ocean. Perhaps the most dramatic effect in reefs is **coral bleaching**. The coral–*Symbiodinium* symbiosis is highly sensitive to temperature. With thermal anomalies of 1 degree Celsius for just a few days, corals can lose their symbiotic algae. As shown in Box Figure 7.5, loss of symbionts—and their associated pigmentation—results in a blanched appearance and increased coral mortality. Coral bleaching can have dramatic downstream ecological effects on reef communities as species that depend on corals are impacted and macroalgae begin to take over (Hoegh-Guldberg et al. 2007). Both localized and large-scale bleaching events have been observed over the last decades (Berkelmans et al. 2004). For example, a single bleaching event in 1998 in Japan reduced coral species richness by >60% and coral cover by >80% (Loya et al. 2001). Some reefs appear to recover from localized bleaching within several years, while others are still impacted many years later (Graham et al. 2007; Diaz-Pulido et al. 2009).

In addition to warming oceanic temperatures, global warming creates additional stressors on reef ecosystems. Thermal expansion and melting of polar ice lead to sea level rise. This directly impacts corals as rising sea levels alter water depth and degrade previously suitable habitat. In addition, anthropogenic climate change alters circulation patterns in the nearshore ocean. This perturbs patterns of larval dispersal and nutrient availability, both critically important for future reef health.

OCEAN ACIDIFICATION Approximately 25% of all anthropogenic carbon emissions enter the ocean—leading to ocean acidification. Acidification can have myriad impacts on marine organisms as lower pH can alter metabolic, physiological, and stress pathways. As Box Figure 7.5 shows, acidic water can also literally dissolve the shells of many marine invertebrates. Moreover, a chemical interaction between carbon dioxide and water reduces the carbonate available to calcifying organisms (Orr et al. 2005). Reduced coral calcification can have many downstream effects as the extent and density of reef structures diminish (Hoegh-Guldberg et al. 2007). Increased storm frequency and intensity—a hallmark of anthropogenic climate change—can act synergistically with ocean acidification. Brittle skeletons are at greater risk of storm damage: as both risks increase, so will physical damage to reef ecosystems.

OVERHARVEST AND CONTAMINATION Perhaps the most direct anthropogenic impact on coral reefs is harvest. Corals are mined for ornamental use but also for raw materials for building and the production of lime. Subsistence, recreational, and commercial fishing also have dramatic impacts on reef ecosystems. The methods of harvesting reef species can also lead to structural damage of the reefs and chemical contamination. For example, in some regions of the world, cyanide is used for poison fishing and home-made bombs are used for blast fishing (McManus et al. 1997). Of course, as we reviewed in Chapter 4, other contaminants—including those from agricultural and industrial systems—can alter nutrient profiles, decrease water quality, and ultimately damage reef ecosystems.

(Continued)

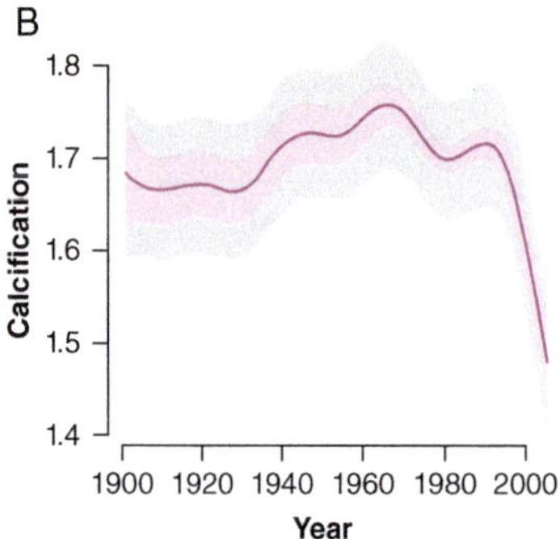

BOX FIGURE 7.5 Threats to corals. (A) Coral reef bleaching due to global warming. The coral–*Symbiodinium* association is highly temperature dependent and breaks down under thermal stress. As temperatures rise, reef ecosystems are projected to deteriorate; as corals die and key grazers are lost, macroalgae take over, fundamentally altering the ecosystem. (B) Ocean acidification can have both short and long term impacts on aquatic life. In the lab, acidic conditions can dissolve the shells of many aquatic invertebrates in just weeks. In the field, *Porites* corals exhibit a massive decrease in calcification over decades associated with acidified conditions.

Reflection: Besides the coral–*Symbiodinium* relationship, what other ecological interactions in reefs are likely stressed by global warming and ocean acidification?

Source: (A) The Ocean Agency / XL Catlin Seaview Survey; (B) "Coral Reefs Under Rapid Climate Change and Ocean Acidification. BY O. HOEGH-GULDBERG, P. J. MUMBY, A. J. HOOTEN, R. S. STENECK, P. GREENFIELD, E. GOMEZ, C. D. HARVELL, P. F. SALE, A. J. EDWARDS, K. CALDEIRA, N. KNOWLTON, C. M. EAKIN, R. IGLESIAS-PRIETO, N. MUTHIGA, R. H. BRADBURY, A. DUBI, M. E. HATZIOLOS. SCIENCE 14 DEC 2007 : 1737-1742"

Can Corals Adapt to These Threats?

Collectively, anthropogenic stressors have led to a drastic increase in the proportion of endangered coral species over the last 20 years, with more than 30% of all coral species considered to be threatened (Carpenter et al. 2008). Recent estimates suggest that coral reefs might decline by 70%–90% over the coming decades (Hoegh-Guldberg et al. 2018). However, projections for future reef health are less dismal if coral species can adapt to changing conditions (Logan et al. 2014). Because coral ecosystems are so sensitive to temperature anomalies, many studies on the potential for coral adaptation have focused on adaptation to global warming.

For example, Palumbi et al. (2014) used reciprocal transplant experiments of tabletop corals (*Acropora hyacinthus*) between cooler and hotter pools and tested transplanted colonies for thermal resistance over a 2-year period. They found differences in thermal tolerance between the two treatment groups, and these differences were associated with specific gene expression responses. Using a more comparative approach, Barshis et al. (2013) compared gene expression patterns between thermally sensitive and thermally resilient corals and found particular genes (like heat shock proteins) were upregulated in resilient corals.

In addition to intrinsic changes in coral gene expression, there is evidence that corals can adapt to thermal stress by hosting high temperature–adapted *Symbiodinium* (Rowan 2004; Howells et al. 2012). In fact, recent studies suggest that both hosts and symbionts are genetically adapting to high temperatures in regions of the world with particularly extreme summer temperatures (Howells et al. 2016). Thus, many species of corals may be able to acclimate or adapt to higher temperatures with changes in species-specific traits and symbiotic partnerships.

Adaptation or acclimatization to warmer temperatures does not ensure long-term reef health. As we explored earlier, there are many additional stressors beyond temperature, and there are numerous examples where coral colonies have not acclimated or adapted to stressors like ocean acidification (e.g., Crook et al. 2013). Moreover, even when adaptation decreases coral mortality for the near term, mortality may only be delayed if carbon emissions are not drastically reduced and other global change stressors are not mitigated (Logan et al. 2014). Thus, preserving the important biological functions of coral reefs into the future will require creative integration of biological research, conservation management, and partnerships in governance across borders (Hughes et al. 2017), topics we will return to in Unit IV.

KEY CONCEPTS

What is the adapt response?

- Adaptation is the process by which organisms become better suited to environmental conditions via heritable genetic change over time.

What conditions are required for adaptation?

- Heredity, variation, and differential survival or reproduction are requirements for evolution by natural selection.

What is an example of evolution by natural selection?

- The case of the peppered moth in 19th-century Britain provides an example of natural selection in action.

What types of adaptation occur in response to global change pressures?

- Adaptation is observed in many different traits, in many taxonomic groups, and in response to many different selection pressures, including abiotic stressors like climate change or biotic stressors like introduced species.

How do scientists identify adaptations and predict adaptive potential?

- To identify current adaptations and predict future adaptive responses, scientists can use myriad approaches, including long-term field observations, laboratory experimental evolution studies, and molecular approaches.

Can adaptation prevent extinction?

- Evolutionary rescue is when genetic adaptation prevents a population from being lost, but strong bouts of natural selection can create other risks, such as depleted genetic variation.

Core concepts: Where does genetic variation come from?

- Genetic variation is generated by mutation and recombination at the molecular level. In addition, adaptive alleles can also enter a population via gene flow.

Meet the data: The Daphnia time machine

- Scientists used resurrection ecology and experimental evolution approaches to understand the potential for the water flea *Daphnia* to adapt to global warming. They found that *Daphnia* have evolved higher thermal tolerance over the last decades and have the potential to adapt to even higher future temperatures.

Taking a closer look: Coral reefs

- Coral reefs are ecologically and economically important habitats threatened by global warming, ocean acidification, chemical pollution, and overharvest. Rising temperatures disrupt the symbiosis between reef-building corals and their photosynthetic *Symbiodinium* algae. Recent research suggests that some corals and their symbionts may be able to respond to changing conditions with a combination of phenotypic plasticity and genetic adaptation.

CONSOLIDATE YOUR KNOWLEDGE

Answer the following questions to assess your progress meeting the learning outcomes:

1. In your own words, write a one- or two-sentence synthesis of the big-picture takeaway point of this chapter.
2. Revisit your answer to the *Blank Page* exercise in the beginning of this chapter. Would you refine your answer now based on knowledge you integrated from this chapter?
3. What distinguishes adaptation from phenotypic plasticity? What distinguishes adaptation from genetic drift?
4. What conditions are required for evolution by natural selection? What is an example and how does it meet each criterion?
5. What are specific examples of different types of adaptation to global change stressors?
6. What do scientists need to do to unambiguously demonstrate that a particular trait is an adaptation? What are different methods that can be used to identify adaptation?
7. What is evolutionary rescue? Beyond demonstrating that a particular trait is adaptive, what else needs to be demonstrated to show that evolutionary rescue has occurred?
8. What traits might predict whether a species has high adaptive potential?
9. Describe how global change stressors affect species interactions in coral reefs.
10. Compare and contrast resurrection ecology and experimental evolution.
11. In your own words, define the bolded and italicized terms in this chapter.
12. What are some questions that *you* have about the content in this chapter? If you found some content particularly challenging or particularly interesting, identify these as areas for additional reflection or reading.

LITERATURE CITED

Anderson, Mao, Scott, & Crowder. 2009. "Survival from Hypoxia in *C. elegans* by Inactivation of Aminoacyl-tRNA Synthetases." *Science* 323: 630–34.

Arnold & Kunte. 2017. "Adaptive Genetic Exchange: A Tangled History of Admixture and Evolutionary Innovation." *Trends in Ecology and Evolution* 32 (8): 601–44. doi.org/10.1016/j.tree.2017.05.007.

Baker. 2003. "Flexibility and Specificity in Coral-Algal Symbiosis: Diversity, Ecology, and Biogeography of Symbiodinium." *Annual Review of Ecology, Evolution, and Systematics* 34 (1): 661–89. doi.org/10.1146/annurev.ecolsys.34.011802.132417.

Balanyá, Oller, Huey, Gilchrist, & Serra. 2006. "Global Genetic Change Tracks Global Climate Warming in *Drosophila subobscura*." *Science* 313 (5794): 1773–75. doi.org/10.1126/science.1131002.

Banerjee, Hasler, Meagher, Nagoshi, Hietala, Huang, Narva, & Jurat-Fuentes. 2017. "Mechanism and DNA-Based Detection of Field-Evolved Resistance to Transgenic Bt Corn in Fall Armyworm (*Spodoptera frugiperda*)." *Scientific Reports* 7 (1): 1–10. doi.org/10.1038/s41598-017-09866-y.

Barrett & Schluter. 2008. "Adaptation from Standing Genetic Variation." *Trends in Ecology and Evolution* 23 (1): 38–44. doi.org/10.1016/j.tree.2007.09.008.

Barshis, Ladner, Oliver, Seneca, Traylor-Knowles, & Palumbi. 2013. "Genomic Basis for Coral Resilience to Climate Change." *Proceedings of the National Academy of Sciences* 110 (4): 1387–92. doi.org/10.1073/pnas.1210224110.

Bell. 2017. "Evolutionary Rescue." *Annual Review of Ecology, Evolution, and Systematics* 48: 605–27. doi.org/10.1146/annurev-ecolsys-110316-023011.

Bell & Gonzalez. 2009. "Evolutionary Rescue Can Prevent Extinction Following Environmental Change." *Ecology Letters* 12 (9): 942–48. doi.org/10.1111/j.1461-0248.2009.01350.x.

Bell & Gonzalez. 2011. "Adaptation and Evolutionary Rescue Environmental Deterioration." *Science* 332 (June): 1327–30. doi.org/10.1126/science.1203105.

Berkelmans, De'ath, Kininmonth, & Skirving. 2004. "A Comparison of the 1998 and 2002 Coral Bleaching Events on the Great Barrier Reef: Spatial Correlation, Patterns, and Predictions." *Coral Reefs* 23 (1): 74–83. doi.org/10.1007/s00338-003-0353-y.

Both, Bouwhuis, Lessells, & Visser. 2006. "Climate Change and Population Declines in a Long-Distance Migratory Bird." *Nature* 441: 81–83. 10.1038/nature04539

Bradshaw & Holzapfel. 2001. "Genetic Shift in Photoperiodic Response Correlated with Global Warming." *Proceedings of the National Academy of Sciences* 98 (25): 14509–11. doi.org/10.1073/pnas.241391498.

Brierley & Kingsford. 2009. "Impacts of Climate Change on Marine Organisms and Ecosystems." *Current Biology* 19 (14): R602–14. doi.org/10.1016/j.cub.2009.05.046.

Campbell-Staton, Winchell, Rochette, Fredette, Maayan, Schweizer & Catchen. 2020. "Parallel Selection on Thermal Physiology Facilitates Repeated Adaptation of City Lizards to Urban Heat Islands." *Nature Ecology and Evolution* 4: 652–58. doi.org/10.1038/s41559-020-1131-8.

Carlson, Cunningham, & Westley. 2014. "Evolutionary Rescue in a Changing World." *Trends in Ecology and Evolution* 29 (9): 521–30. doi.org/10.1016/j.tree.2014.06.005.

Carpenter, Abrar, Aeby, Aronson, Banks, Bruckner, Chiriboga, et al. 2008. "One-Third of Reef-Building Corals Face Elevated Extinction Risk from Climate Change and Local Impacts." *Science* 321 (5888): 560–63. doi.org/10.1126/science.1159196.

Cesar & Chong. 2004. "Economic Valuation and Socioeconomics of Coral Reefs: Methodological Issues and Three Case Studies." In *Economic Valuation and Policy*

Priorities for Sustainable Management of Coral Reefs, edited by Ahmed, Chong, & Cesar, pages 14–215. Penang, Malaysia.

Colbourne, Pfrender, Gilbert, Thomas, Tucker, Oakley, Tokishita, et al. 2011. "The Eco-responsive Genome of *Daphnia pulex*." *Science* 331 (6017): 555–61. doi.org/10.1126/science.1197761.

Collins. 2011. "Many Possible Worlds: Expanding the Ecological Scenarios in Experimental Evolution." *Evolutionary Biology* 38: 3–14. doi.org/10.1007/s11692-010-9106-3.

Collins & Bell. 2004. "Phenotypic Consequences of 1,000 Generations of Selection at Elevated CO_2 in a Green Alga." *Nature* 431 (7008): 566–69. doi.org/10.1038/nature02945.

Cook. 2003. "The Rise and Fall of the Carbonaria Form of the Peppered Moth." *The Quarterly Review of Biology* 78 (4): 399–417. doi.org/10.1086/378925.

Cook, Grant, Saccheri, & Mallet. 2012. "Selective Bird Predation on the Peppered Moth: The Last Experiment of Michael Majerus." *Biology Letters* 8 (4): 609–12. doi.org/10.1098/rsbl.2011.1136.

Cook & Muggleton. 2003. "The Peppered Moth, *Biston betularia* (Linnaeus, 1758) (Lepidoptera: Geometridae): A Matter of Names." *Entomologist's Gazette* 54 (4): 211–21.

Cook & Saccheri. 2013. "The Peppered Moth and Industrial Melanism: Evolution of a Natural Selection Case Study." *Heredity* 110 (3): 207–12. doi.org/10.1038/hdy.2012.92.

Crook, Cohen, Rebolledo-Vieyra, Hernandez, & Paytan. 2013. "Reduced Calcification and Lack of Acclimatization by Coral Colonies Growing in Areas of Persistent Natural Acidification." *Proceedings of the National Academy of Sciences* 110 (27): 11044–49. doi.org/10.1073/pnas.1301589110.

Darwin. 1859. *On the Origin of Species*. doi.org/10.4324/9780203509104.

Deatherage, Kepner, Bennett, Lenski, & Barrick. 2017. "Specificity of Genome Evolution in Experimental Populations of *Escherichia coli* Evolved at Different Temperatures." *Proceedings of the National Academy of Sciences* 114 (10): E1904–12. doi.org/10.1073/pnas.1616132114.

Diaz-Pulido, McCook, Dove, Berkelmans, Roff, Kline, Weeks, Evans, Williamson, & Hoegh-Guldberg. 2009. "Doom and Boom on a Resilient Reef: Climate Change, Algal Overgrowth and Coral Recovery." *PLoS ONE* 4 (4): e5239. doi.org/10.1371/journal.pone.0005239.

Egan, Ragland, Assour, Powell, Hood, Emrich, Nosil, & Feder. 2015. "Experimental Evidence of Genome-Wide Impact of Ecological Selection During Early Stages of Speciation-with-Gene-Flow." *Ecology Letters* 18 (8): 817–25. doi.org/10.1111/ele.12460.

Elena & Lenski. 2003. "Evolution Experiments with Microorganisms: The Dynamics and Genetic Bases of Adaptation." *Nature Reviews Genetics* 4 (6): 457–69. doi.org/10.1038/nrg1088.

Ellegren. 2014. "Genome Sequencing and Population Genomics in Non-Model Organisms." *Trends in Ecology and Evolution* 29 (1): 51–63. doi.org/10.1016/j.tree.2013.09.008.

Etterson & Shaw. 2001. "Constraint to Adaptive Evolution in Response to Global Warming." *Science* 294 (5540): 151–54. doi.org/10.1126/science.1063656.

Fenner. 2010. "Deliberate Introduction of European Rabbit, *Oryctolagus cuniculus*, into Australia." *Revue Scientifique et Technique de l'OIE* 29 (1): 103–11. doi.org/10.20506/rst.29.1.1964.

Filchak, Roethele, & Feder. 2000. "Natural Selection and Sympatric Divergence in the Apple Maggot *Rhagoletis pomonella*." *Nature* 407 (6805): 739–42. doi.org/10.1038/35037578.

French. 2010. "The Continuing Crisis in Antibiotic Resistance." *International Journal of Antimicrobial Agents* 36 (Suppl. 3): S3–7. doi.org/10.1016/S0924-8579(10)70003-0.

Frisch, Morton, Chowdhury, Culver, Colbourne, Weider, & Jeyasingh. 2014. "A Millennial-Scale Chronicle of Evolutionary Responses to Cultural Eutrophication in *Daphnia*." *Ecology Letters* 17 (3): 360–68. doi.org/10.1111/ele.12237.

Geerts, Vanoverbeke, Vanschoenwinkel, Van Doorslaer, Feuchtmayr, Atkinson, Moss, Davidson, Sayer, & De Meester. 2015. "Rapid Evolution of Thermal Tolerance in the Water Flea *Daphnia*." *Nature Climate Change* 5 (7): 665–68. doi.org/10.1038/nclimate2628.

Gilchrist, Huey, Balanyá, Pascual, & Serra. 2004. "A Time Series of Evolution in Action: A Latitudinal Cline in Wing Size in South American *Drosophila subobscura*." *Evolution* 58 (4): 768–80. doi.org/10.1111/j.0014-3820.2004.tb00410.x.

Gonzalez, Ronce, Ferriere, & Hochberg. 2013. "Evolutionary Rescue: An Emerging Focus at the Intersection between Ecology and Evolution." *Philosophical Transactions of the Royal Society B: Biological Sciences* 368: 20120404. doi.org/10.1098/rstb.2012.0404.

Goulet & Coffroth. 2003. "Stability of an Octocoral–Algal Symbiosis over Time and Space." *Marine Ecology Progress Series* 250: 117–24. doi.org/10.3354/meps250117.

Graham, Wilson, Jennings, Polunin, Robinson, Bijoux, & Daw. 2007. "Lag Effects in the Impacts of Mass Coral Bleaching on Coral Reef Fish, Fisheries, and Ecosystems." *Conservation Biology* 21 (5): 1291–300. doi.org/10.1111/j.1523-1739.2007.00754.x.

Grant and Grant. 2014. *40 Years of Evolution: Darwin's Finches on Daphne Major Island*. Princeton, NJ: Princeton University Press.

Groot, de, Brander, van der Ploeg, Costanza, Bernard, Braat, Christie, et al. 2012. "Global Estimates of the Value of Ecosystems and Their Services in Monetary Units." *Ecosystem Services* 1 (1): 50–61. doi.org/10.1016/j.ecoser.2012.07.005.

Halpern, Walbridge, Selkoe, Kappel, Micheli, D'Agrosa, Bruno, et al. 2008. "A Global Map of Human Impact on Marine Ecosystems." *Science* 319 (5865): 948–52. doi.org/10.1126/science.1149345.

Harborne, Rogers, Bozec, & Mumby. 2017. "Multiple Stressors and the Functioning of Coral Reefs." *Annual Review of Marine Science* 9 (1): 445–68. doi.org/10.1146/annurev-marine-010816-060551.

Hardwick, Harmon, Hardwick, & Rosenblum. 2015. "When Field Experiments Yield Unexpected Results: Lessons Learned from Measuring Selection in White Sands Lizards." *PLoS ONE* 10 (2): 1–18. doi.org/10.1371/journal.pone.0118560.

Hedrick. 2013. "Adaptive Introgression in Animals: Examples and Comparison to New Mutation and Standing Variation as Sources of Adaptive Variation." *Molecular Ecology* 22 (18): 4606–18. doi.org/10.1111/mec.12415.

Hoegh-Guldberg, Kennedy, Beyer, McClennen, & Possingham. 2018. "Securing a Long-Term Future for Coral Reefs." *Trends in Ecology and Evolution* 33 (12): 936–44. doi.org/10.1016/j.tree.2018.09.006.

Hoegh-Guldberg, Mumby, Hooten, Steneck, Greenfield, Gomez, Harvell, et al. 2007. "Coral Reefs Under Rapid Climate Change and Ocean Acidification." *Science* 318 (5857): 1737–42. doi.org/10.1126/science.1152509.

Hoekstra, Hoekstra, Berrigan, Vignieri, Hoang, Hill, Beerli, & Kingsolver. 2001. "Strength and Tempo of Directional Selection in the Wild." *Proceedings of the National Academy of Sciences* 98 (16): 9157–60. doi.org/10.1073/pnas.161281098.

Howells, Abrego, Meyer, Kirk, & Burt. 2016. "Host Adaptation and Unexpected Symbiont Partners Enable Reef-Building Corals to Tolerate Extreme Temperatures." *Global Change Biology* 22 (8): 2702–14. doi.org/10.1111/gcb.13250.

Howells, Beltran, Larsen, Bay, Willis, & Van Oppen. 2012. "Coral Thermal Tolerance Shaped by Local Adaptation of Photosymbionts." *Nature Climate Change* 2 (2): 116–20. doi.org/10.1038/nclimate1330.

Hughes, Barnes, Bellwood, Cinner, Cumming, Jackson, Kleypas, et al. 2017. "Coral Reefs in the Anthropocene." *Nature* 546: 82–90. doi.org/10.1038/nature22901.

Huston. 1985. "Patterns of Species Diversity on Coral Reefs." *Annual Review of Ecology and Systematics* 16 (1): 149–77. doi.org/10.1146/annurev.es.16.110185.001053.

Jain & Bradshaw. 1966. "Evolutionary Divergence Among Adjacent Plant Populations. I. The Evidence and Its Theoretical Analysis." *Heredity* 21 (3): 407–41. doi.org/10.1038/hdy.1966.42.

Jansen, Geerts, Rago, Spanier, Denis, de Meester, & Orsini. 2017. "Thermal Tolerance in the Keystone Species *Daphnia magna*—A Candidate Gene and an Outlier Analysis Approach." *Molecular Ecology* 26 (8): 2291–305. doi.org/10.1111/mec.14040.

Kingsolver, Diamond, Siepielski, & Carlson. 2012. "Synthetic Analyses of Phenotypic Selection in Natural Populations: Lessons, Limitations and Future Directions." *Evolutionary Ecology* 26 (5): 1101–18. doi.org/10.1007/s10682-012-9563-5.

Lamichhaney, Han, Berglund, Wang, Almén, Webster, Grant, Grant, & Andersson. 2016. "A Beak Size Locus in Darwin's Finches Facilitated Character Displacement During a Drought." *Science* 352 (6284): 470–474. DOI: 10.1126/science.aad8786

Lapiedra, Schoener, Leal, Losos, & Kolbe. 2018. "Predator-Driven Natural Selection on Risk-Taking Behavior in Anole Lizards." *Science* 360 (6392): 1017–20. doi.org/10.1126/science.aap9289.

Laughlin & Messier. 2015. "Fitness of Multidimensional Phenotypes in Dynamic Adaptive Landscapes." *Trends in Ecology and Evolution* 30 (8): 487–96. doi.org/10.1016/j.tree.2015.06.003.

Lenski. 2017. "Convergence and Divergence in a Long-Term Experiment with Bacteria." *The American Naturalist* 190 (S1): S57–68. doi.org/10.1086/691209.

Logan, Dunne, Eakin, & Donner. 2014. "Incorporating Adaptive Responses into Future Projections of Coral Bleaching." *Global Change Biology* 20 (1): 125–39. doi.org/10.1111/gcb.12390.

Lohbeck, Riebesell, & Reusch. 2012. "Adaptive Evolution of a Key Phytoplankton Species to Ocean Acidification." *Nature Geoscience* 5 (5): 346–51. doi.org/10.1038/ngeo1441.

Losos, Schoener, & Spiller. 2004. "Predator-Induced Behaviour Shifts and Natural Selection in Field-Experimental Lizard Populations." *Nature* 432: 505–8. doi.org/10.1038/nature03039.

Loya, Sakai, Yamazato, Nakano, Sambali, & Van Woesik. 2001. "Coral Bleaching: The Winners and the Losers." *Ecology Letters* 4 (2): 122–31. doi.org/10.1046/j.1461-0248.2001.00203.x.

Lynch. 2010. "Evolution of the Mutation Rate." *Trends in Genetics* 26 (8): 345–52. doi.org/10.1016/j.tig.2010.05.003.Evolution.

Marshall, Burgess, & Connallon. 2016. "Global Change, Life-History Complexity and the Potential for Evolutionary Rescue." *Evolutionary Applications* 9 (9): 1189–201. doi.org/10.1111/eva.12396.

McAllister. 1991. "What Is the Status of the World's Coral Reef Fishes?" *Sea Wind* 5 (1): 14–18.

McManus, Reyes Jr., & Nañola Jr. 1997. "Effects of Some Destructive Fishing Methods on Coral Cover and Potential Rates of Recovery." *Environmental Management* 21 (1): 69–78. doi.org/10.1007/s002679900006.

McPheron, Smith, & Berlocher. 1988. "Genetic Differences Between Host Races of *Rhagoletis pomonella*." *Nature* 336 (6194): 64–66. doi.org/10.1038/336064a0.

Merilä & Hendry. 2014. "Climate Change, Adaptation, and Phenotypic Plasticity: The Problem and the Evidence." *Evolutionary Applications* 7 (1): 1–14. doi.org/10.1111/eva.12137.

Moberg & Folke. 1999. "Ecological Goods and Services of Coral Reef Ecosystems." *Ecological Economics* 29 (2): 215–33.

Nielsen. 2005. "Molecular Signatures of Natural Selection." *Annual Review of Genetics* 39 (1): 197–218. doi.org/10.1146/annurev.genet.39.073003.112420.

Orr, Fabry, Aumont, Bopp, Doney, Feely, Gnanadesikan, et al. 2005. "Anthropogenic Ocean Acidification over the Twenty-First Century and Its Impact on Calcifying Organisms." *Nature* 437 (7059): 681–86. doi.org/10.1038/nature04095.

Orsini, Gilbert, Podicheti, Jansen, Brown, Solari, Spanier, et al. 2016. "*Daphnia magna* Transcriptome by RNA-Seq Across 12 Environmental Stressors." *Scientific Data* 3: 160030. doi.org/10.1038/sdata.2016.30.

Orsini, Schwenk, de Meester, Colbourne, Pfrender, & Weider. 2013. "The Evolutionary Time Machine: Using Dormant Propagules to Forecast How Populations

Can Adapt to Changing Environments." *Trends in Ecology and Evolution* 28 (5): 274–82. doi.org/10.1016/j.tree.2013.01.009.

Orsini, Spanier, & De Meester. 2012. "Genomic Signature of Natural and Anthropogenic Stress in Wild Populations of the Waterflea *Daphnia magna*: Validation in Space, Time and Experimental Evolution." *Molecular Ecology* 21 (9): 2160–75. doi.org/10.1111/j.1365-294X.2011.05429.x.

Palumbi, Barshis, Traylor-Knowles, & Bay. 2014. "Mechanisms of Reef Coral Resistance to Future Climate Change." *Science* 344 (6186): 895–98. doi.org/10.1126/science.1251336.

Putnam, Barott, Ainsworth, & Gates. 2017. "The Vulnerability and Resilience of Reef-Building Corals." *Current Biology* 27 (11): R528–40. doi.org/10.1016/j.cub.2017.04.047.

Réale, McAdam, Boutin, & Berteaux. 2003. "Genetic and Plastic Responses of a Northern Mammal to Climate Change." *Proceedings of the Royal Society B: Biological Sciences* 270 (1515): 591–96. doi.org/10.1098/rspb.2002.2224.

Rezende, Balanyà, Rodríguez-Trelles, Rego, Fragata, Matos, Serra, & Santos. 2010. "Climate Change and Chromosomal Inversions in *Drosophila subobscura*." *Climate Research* 43 (1–2): 103–14. doi.org/10.3354/cr00869.

Rowan. 2004. "Coral Bleaching: Thermal Adaptation in Reef Coral Symbionts." *Nature* 430 (7001): 742.

Sandberg, Pedersen, Lacroix, Ebrahim, Bonde, Herrgard, Palsson, Sommer, & Feist. 2014. "Evolution of *Escherichia coli* to 42°C and Subsequent Genetic Engineering Reveals Adaptive Mechanisms and Novel Mutations." *Molecular Biology and Evolution* 31 (10): 2647–62. doi.org/10.1093/molbev/msu209.

Saunders, Cooke, McColl, Shine, & Peacock. 2010. "Modern Approaches for the Biological Control of Vertebrate Pests: An Australian Perspective." *Biological Control* 52 (3): 288–95. doi.org/10.1016/j.biocontrol.2009.06.014.

Schiffers, Bourne, Lavergne, Thuiller, & Travis. 2013. "Limited Evolutionary Rescue of Locally Adapted Populations Facing Climate Change." *Philosophical Transactions of the Royal Society B: Biological Sciences* 368 (1610): 20120083. doi.org/10.1098/rstb.2012.0083.

Schoener & Schoener. 1983. "Distribution of Vertebrates on Some Very Small Islands. II. Patterns in Species Number." *The Journal of Animal Ecology* 52 (1): 237–62. doi.org/10.2307/4598.

Sinervo, Méndez-de-la-Cruz, Miles, Heulin, Bastiaans, Cruz, Lara-Resendiz, et al. 2010. "Erosion of Lizard Diversity by Climate Change and Altered Thermal Niches." *Science* 328 (5980): 894–99. doi.org/10.1126/science.1184695.

Song, Endepols, Klemann, Richter, Matuschka, Shih, Nachman, & Kohn. 2011. "Adaptive Introgression of Anticoagulant Rodent Poison Resistance by Hybridization Between Old World Mice." *Current Biology* 21 (15): 1296–301. doi.org/10.1016/j.cub.2011.06.043.

Spalding & Grenfell. 1997. "New Estimates of Global and Regional Coral Reef Areas." *Coral Reefs* 16 (4): 225–30.

Springael & Top. 2004. "Horizontal Gene Transfer and Microbial Adaptation to Xenobiotics: New Types of Mobile Genetic Elements and Lessons from Ecological Studies." *Trends in Microbiology* 12 (2): 53–58. www.sciencedirect.com/science/article/pii/S0966842X03003378.

Stinchcombe & Hoekstra. 2008. "Combining Population Genomics and Quantitative Genetics: Finding the Genes Underlying Ecologically Important Traits." *Heredity* 100 (2): 158–70. doi.org/10.1038/sj.hdy.6800937.

Swaegers, Mergeay, Van Geystelen, Therry, Larmuseau, & Stoks. 2015. "Neutral and Adaptive Genomic Signatures of Rapid Poleward Range Expansion." *Molecular Ecology* 24 (24): 6163–76. doi.org/10.1111/mec.13462.

Tabashnik & Carrière. 2017. "Surge in Insect Resistance to Transgenic Crops and Prospects for Sustainability." *Nature Biotechnology* 35 (10): 926–35. doi.org/10.1038/nbt.3974.

Thornhill, Howells, Wham, Steury, & Santos. 2017. "Population Genetics of

Reef Coral Endosymbionts (Symbiodinium, Dinophyceae)." *Molecular Ecology* 26 (10): 2640–59. doi.org/10.1111/mec.14055.

Vander Wal, Garant, Festa-Bianchet, & Pelletier. 2013. "Evolutionary Rescue in Vertebrates: Evidence, Applications and Uncertainty." *Philosophical Transactions of the Royal Society B: Biological Sciences* 368 (1610): 20120090. doi.org/10.1098/rstb.2012.0090.

van't Hof, Edmonds, Dalíková, & Saccheri. 2011. "Industrial Melanism in British Peppered Moths Has a Singular and Recent Mutational Origin." *Science* 332 (6032): 958–60.

Wilcox. 2000. "Soap." In *Poucher's Perfumes, Cosmetics and Soaps*, edited by Hilda Butler, pages 453–65. Dordrecht: Springer Netherlands.

Wray & Visscher. 2008. "Estimating Trait Heritability." *Nature Education* 1 (1): 29.

Core Responses: Die

Learning Outcomes

After working with this chapter, you will be able to:

- Analyze how contemporary human activities can lead to losses at the individual, population, and species levels.
- Describe how scientists predict extinction risk.
- Evaluate evidence that we have entered the sixth mass extinction.
- Synthesize factors influencing the die response and evaluate the relationship between the die response and the other core responses.
- Apply your knowledge to real-world case studies and interpret data from recent scientific studies.

THE BLANK PAGE

Consider the extinction of a single species. What downstream effects might the loss of one species have? Are there some situations where the death of a single species might have greater or lesser effects? Now consider broader scale patterns of species loss. We explored the "Big Five" historical mass extinctions in Chapter 2. Do you think human activities on our planet could trigger an extinction spasm of similar proportions? What would be the ecological and evolutionary ramifications of such an event?

INTRODUCTION

The "die" response is in many ways not a response, but rather a failure to respond. When species vulnerable to global change stressors do not move, adjust, or adapt, they may be irrevocably lost. Of course, death is an inevitable part of life on Earth. The loss of individuals, populations, and species has occurred for billions of years. But in contemporary times, extinction rates are soaring, the tree of life is being pruned, and scientists caution that we have entered the sixth mass extinction on our planet (Barnosky et al. 2011). As stated by the eminent biologist E. O. Wilson (1999), *"Humanity has initiated the sixth great extinction spasm, rushing to eternity a large fraction of our fellow species in a single generation."* The goal of this chapter is to confront the reality that human activities have imperiled a large proportion of species on Earth, describe contemporary patterns of species loss, and explain how scientists evaluate extinction risk. In Unit IV, we will then evaluate ways of addressing this crisis and better support biodiversity conservation and recovery in the coming decades.

HOW IS THE SURVIVAL OF INDIVIDUALS, POPULATIONS, AND SPECIES CONNECTED?

Global change stressors can affect survival at several different levels of biological organization. Species extinctions result from population losses, and population losses result from death of individuals. Therefore, it is important to understand how the die response cuts across multiple biological levels.

Loss of Individuals

All living things eventually die; however, anthropogenic stressors have increased rates of mortality within many populations. Human activities often have direct effects on mortality rates. For example, hunting of African forest elephants led to a population size decline of >60% in just 10 years (Maisels et al. 2013). Anthropogenic activities also have indirect effects. For example, the introduction of the pathogen *Phytophthora ramorum* to California caused the death of many individual tan oak trees, resulting in a phenomenon known as "sudden oak death" (Rizzo et al. 2002; McPherson et al. 2005).

Increased mortality rates can lead to reduced population sizes. Reduced population size, in turn, can increase vulnerability to additional stressors. As shown in Figure 8.1, decreases in population size can lead to further decreases in population size, creating a positive feedback loop termed the **extinction vortex** (Gilpin & Soulé 1986; Fagan & Holmes 2006). The extinction vortex can have a number of interacting causes.

First, decreased population size or density can disrupt key aspects of survival or reproduction, a phenomenon referred to as the **Allee effect**. For example, it can be difficult to find mates, engage in cooperative feeding (e.g., pack hunting carnivores), or maintain cooperative defense strategies (e.g., schooling fish) if there are fewer individuals living at lower densities (Kramer et al. 2009).

Second, decreased population size can increase risk of population loss in chance events. Unpredictable changes in environmental conditions over space or time create **environmental stochasticity**, which can disproportionately impact small populations. For example, a small population may be less resilient to an extreme weather event than a large population (Sutton & Morgan 2009), simply because having more individuals increases the probability that some will survive.

Third, decreased population size is associated with reduced genetic variability. Genetic variation decreases in small populations for multiple reasons. Most notably, as population sizes decrease, the impact of genetic drift increases (referred to as **genetic stochasticity**). Thus, genetic variation is more likely to be lost by chance in small populations. In addition, for sexually reproducing organisms, the probability of **inbreeding** (the mating of close genetic relatives) increases as population size decreases. Mating between relatives decreases total genetic variation in a population and also metrics like **heterozygosity** (the presence of multiple alleles at a locus).

Reduced genetic variation can increase vulnerability to future stressors. In the short term, loss of genetic variation can lead to inbreeding depression, or reduced fitness due to mating between relatives. One mechanism for inbreeding depression is exposure of previously hidden deleterious recessive alleles. In the long term, loss

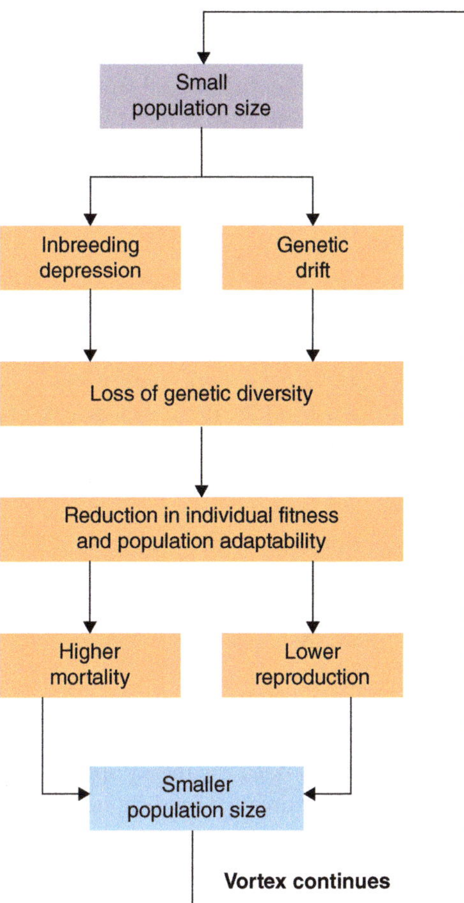

FIGURE 8.1 The extinction vortex is a positive feedback loop whereby population size reductions increase vulnerability to further population size reductions, for example by increased vulnerability to genetic and environmental stochasticity.

Reflection: How might the processes of natural selection and genetic drift relate to the extinction vortex?

Source: Oxford University Press, Life 12e, Figure 57.2

of genetic variation can decrease adaptive potential, or the ability of a population to respond to natural selection. For example, if a population has lost genetic variation for key traits (e.g., immune traits), there is less variation on which selection can act, and populations may not be able to adapt to future threats (Sommer 2005).

In one specific example, the endangered Tasmanian devil (*Sarcophilus harrisii*) is a marsupial with a tiny geographic range (the island of Tasmania). This species was severely impacted by an infectious cancer (devil facial tumor disease), which can be transmitted across individuals and killed more than 85% of Tasmanian devils in the last several decades. Several studies have shown that this species has extremely low genomic diversity, including low diversity in immune genes (Miller et al. 2011; Morris et al. 2015). Lack of genetic diversity could help explain vulnerability to disease epidemics in species that are vulnerable to the extinction vortex.

It is important to note that loss of individuals does not necessarily lead to loss of populations and species. In fact, loss of individuals can be a component of an adaptive response. Natural selection requires the differential survival of individuals in a population. Therefore, mortality can promote adaptation if individuals better suited to new conditions preferentially survive and reproduce. However, after a bout of increased mortality, the remaining individuals—even if they are better adapted—must still contend with the risks of reduced population size.

Loss of Populations

It is important to distinguish between the loss of a species from a particular geographic area and loss of a species from its entire range. The loss of an entire population is referred to as **extirpation** and the loss of an entire species as **extinction**. Some authors interchangeably use the alternate terminology of *local extinction* (for loss of a population) versus *global extinction* (for loss of a species). However, for clarity we use *extirpation* versus *extinction* throughout.

Population loss is not always a cause for alarm. Many species are adapted to have high rates of extirpation, particularly in highly variable environments. However, these species also generally exist in a **metapopulation**, where geographically separated subpopulations interact via frequent migration (Levins 1969; Hanski 1998). Therefore, species with naturally high rates of extirpation also generally have high rates of recolonization, where migration reestablishes lost populations periodically. Many butterfly species provide classic examples of metapopulations. Figure 8.2 provides an illustration from the Glanville fritillary butterfly (*Melitaea cinxia*): populations at specific locations may not survive year to year, but the species persists at a regional scale due to high rates of migration (Ojanen et al. 2013).

Even species without the safety net of a metapopulation structure can recover from local extirpation if there is a source for seeding new populations. For example,

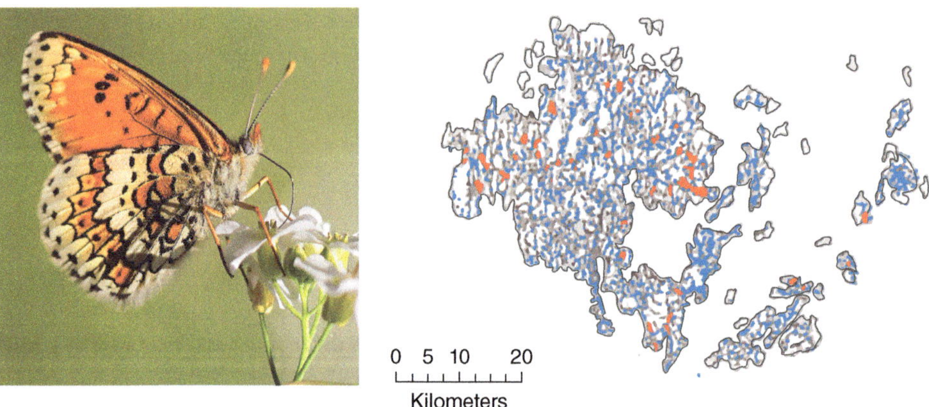

FIGURE 8.2 Metapopulation structure for the Glanville fritillary butterfly (*Melitaea cinxia*) in Finland. Locations of 4,000 meadows known to be used by this species. High natural rates of extirpation and recolonization mean that only a subset of meadows is occupied each year. For example, meadows occupied in 2012 are shown in red and those unoccupied are shown in blue.

Reflection: How might anthropogenic stressors alter the dynamics of metapopulation recolonization?

Source: Ojanen et al. 2013 (https://onlinelibrary.wiley.com/doi/full/10.1002/ece3.733)

the fire poppy (*Papaver californicum*) is a California endemic annual herb that is common in recently burned environments. During postfire succession, the fire poppy is quickly replaced by other plant species and appears to be extirpated. However, storage of seeds in the seed bank and fire-induced germination allow the fire poppy to quickly reappear when conditions are right for its growth and reproduction (Keeley & Keeley 1987).

However, some populations are not naturally adapted to boom-and-bust dynamics. When these species experience high extirpation rates, it can jeopardize the survival of the entire species. Even species that commonly experience extirpation and recolonization can be threatened by increased extirpation or increased barriers to recolonization. In these cases, population loss can cause range contraction (explored in Chapter 5) or, ultimately, loss of entire species.

Loss of Species

Extinction occurs when all individuals of all populations have been lost. Using the umbrella term *extinction* can give the false impression that species losses result from shared underlying processes. However, the dynamics of extinction are highly variable. As we explored in Chapter 2, extinction rates have varied through time and across different groups of organisms. Extinction rates are currently accelerating due to anthropogenic impacts on biodiversity. Yet even these contemporary extinctions have diverse causes and diverse consequences. Thus, it is important to consider that just as each species on Earth is unique, each extinction story is also unique.

WHAT ARE EXAMPLES OF EXTINCTION IN RESPONSE TO GLOBAL CHANGE PRESSURES?

To foster our ability to evaluate global change problems from different perspectives, each core response chapter has presented examples using a slightly different organizational framing. We evaluated examples of the move response by type of range change (i.e., range contractions, expansions, and shifts), the adjust response by trait type (i.e., developmental, physiological, behavioral, and morphological), and the adapt response by global change stressor (i.e., contaminants, introduced species, and climate change). Here we will look at just a few examples of human-mediated extinctions from across the tree of life. The short obituaries for the species shown in Figure 8.3 highlight the fact that many different global change stressors—and their interactions—can contribute to species losses.

Pinta Island Tortoise

Historically, tortoises were plentiful on the Galapagos Islands. However, beginning in the 1800s, the Pinta Island tortoise (Chelonodis nigra abingdoni) was hunted by humans extensively for food. A further stressor was introduced species, such as goats, which decimated their habitat. Ultimately, the last known Pinta Island tortoise, named "Lonesome George," was located in 1972 and taken into captivity (Jones 2012). While in captivity, unsuccessful attempts were made to mate Lonesome George with other closely related species. Pinta Island tortoises were declared extinct in June 2012, when Lonesome George died in captivity.

FIGURE 8.3 Example of anthropogenically induced extinctions: (A) Pinta Island tortoise, (B) passenger pigeon, (C) Polynesian tree snail, (D) superb cyanea, (E) Yangtze River dolphin, (F) smallpox virus.

Reflection: These are taxonomically diverse examples of recent extinctions, but not all contemporary extinctions are well-documented. Are there taxonomic groups, geographic regions, or global change stressors for which extinctions might be more difficult to document?

Source: (A) Christopher Werner, MyShot; (B) Rick Wicker photo ©DENVER MUSEUM OF NATURE & SCIENCE; (C) valentinrussanov/iStockphoto; (D) US Army Environmental Command; (E) AFP/Sixth Tone/Imagine China; (F) Everett Historical/Shutterstock

Passenger Pigeon

The passenger pigeon (*Ectopistes migratorius*) was once one of the most abundant birds in North America (Ehrlich et al. 1988). Flocks of these birds could reach more than a million-strong, and the entire population was estimated at 3 to 5 billion. However, these birds were hunted extensively for food and suffered from habitat loss during the deforestation of the eastern United States (Ehrlich et al. 1988). The passenger pigeon went extinct in 1914, when the last individual bird died in a zoo.

Polynesian Tree Snail

The Polynesian tree snail (*Partula nodosa*) was a small tree snail endemic to Tahiti. Like many of its close relatives, this species declined dramatically after the introduction of a carnivorous snail species, *Euglandina rosea* (Tonge & Bloxam 1991; Coote 2009). *Euglandina rosea* was intentionally introduced by humans to help control another invasive species, but the unintended consequence was the devastation of many native Polynesian tree snails. Due to predation from this introduced species, at least 10 species of Polynesian tree snails, including *Partula nodosa*, are now extinct in the wild (Coote 2009).

Superb Cyanea

The superb cyanea (*Cyanea superba*) was a small tree native to O'ahu, Hawaii, that was lost to the synergistic effects of many different invasive species. *Cyanea superba* was affected by competition from many different invasive plant species. It was also heavily impacted by predation of its seedlings and fruits by non-native slugs, introduced rats, and feral pigs (USFWS 2007). The effects of introduced species were compounded by human-caused wildfires. *Cyanea superba* is now considered extinct in the wild (Bruegmann & Caraway 2003).

Yangtze River Dolphin

The Yangtze River dolphin (*Lipotes vexillifer*) was one of the few species of freshwater dolphins in the world. This species suffered a dramatic decline due to overharvest, entanglement in fishing gear, collisions with boats, habitat degradation, and pollution (Smith et al. 2008). Although *Lipotes vexillifer* is yet to be declared extinct by the IUCN, the last known sighting was in 2002, and a comprehensive survey failed to find any surviving individuals (Turvey et al. 2007; Smith et al. 2008).

Smallpox Virus

Loss of pathogens is not often considered in the extinction literature, but humans are intentionally and unintentionally altering the fate of many microbes (Carlson et al. 2017). One example comes from pox viruses, which are complex viruses with large double-stranded DNA genomes. Pox viruses infect a wide range of vertebrate and arthropod hosts, but perhaps the most commonly known is smallpox. Smallpox (*Variola virus major and minor*) was responsible for the deaths of millions of humans worldwide through the 18th century. A smallpox vaccine was developed in 1796, and by 1980 the virus was eradicated worldwide. All known lab stocks were destroyed, and neither strain of the smallpox virus has appeared in natural populations for decades (Melamed et al. 2018).

In Sum

Before we dive more formally into the science of contemporary extinction, these short obituaries serve to illustrate several key points. First, human-mediated extinctions result from myriad global change stressors. From overharvest to habitat loss and from introduced species to pollution, many anthropogenic activities put the very survival of other species at risk. Second, the contemporary extinction spasm impacts all major lineages on the tree of life and all major biomes. From microbes to mammals and from marine to montane habitats, examples of extinction are far-reaching. Third, some human-mediated extinctions are intentional (like the eradication of pathogens). However, all extinctions represent a loss of evolutionary potential on our planet and potentially have ramifications for other species. Finally, human-mediated extinction rates are accelerating. We evaluated the role of *Homo sapiens* in Quaternary megafaunal extinctions in Chapter 3, but our impact on biodiversity has only increased in the last several hundred years. Let us now evaluate in more detail how scientists study extinction and extinction risk across the tree of life.

HOW DO SCIENTISTS ESTIMATE EXTINCTION RISK?

We can learn important lessons from the extinction stories of the recent past, but we must also look at **extinction risk** for species that are currently on the verge of being lost. There are many ways to study extinction risk. Observational approaches are critical to determine the trajectory of populations in the wild. For example, long-term monitoring studies can help determine whether population sizes are decreasing or ranges are contracting. Experimental approaches can be used to assess species response to changing conditions. For example, physiological tolerance to warmer temperatures can be measured in a laboratory setting to assess whether populations are likely to survive in future environmental conditions. Finally, quantitative modeling studies can be used to assess current threats and to predict the fate of species under alternative future scenarios. Because modeling studies are so widely used to understand extinction risk, here we will look in more detail at two examples of quantitative approaches.

Species Distribution Modeling

We have already discussed **species distribution models** (SDMs) in Chapter 5. SDMs are most commonly used to evaluate the move response, but they also play an important role in assessing probability of extinction. SDMs are used to estimate extinction risk by projecting changes in suitable habitat. By comparing current and projected future distributions, scientists can assess whether sufficient suitable habitat will remain to support species persistence. In addition, scientists can identify where suitable habitat may occur on the landscape and if there is spatial overlap between current and future patches of suitable habitat. As with the move response, SDMs allow scientists to evaluate extinction risk over different timescales and under different future environmental scenarios. An example for California endemic plants is shown in Figure 8.4.

Extinction is clearly predicted if 100% of suitable habitat is lost; however, extinction can occur even without the loss of all suitable habitat, as illustrated in Figure 8.4. Projections from SDMs suggest that there will be complete separation between the current distribution of many species and the geographic regions that will be suitable for them in the future (e.g., Thomas et al. 2004). Without geographic overlap between current and future patches of suitable habitat, dispersal-limited species will be at particular risk because they may go extinct before colonizing new patches of suitable habitat. In fact, the proportion of species at risk of extinction almost doubles if dispersal limitation is taken into account (Thomas et al. 2004). As we discussed in Chapter 5, dispersal limitation can be imposed by intrinsic organismal traits (e.g., lack of a mobile life stage) or by landscape features (e.g., highly fragmented habitats).

Population Viability Analysis

Population viability analysis (PVA) is an alternative approach for modeling extinction risk. PVA describes a broad category of quantitative modeling methods that allow researchers to evaluate the likelihood that a population will persist for a given amount of time into the future based on particular threats (Reed et al. 2003). Rather than focusing on the environmental envelope of a species, PVA focuses on the demographic characteristics of a population (e.g., birth and death rates, life stages, population sizes). PVA models can address relatively simple or highly complex systems, as illustrated in Figure 8.5.

Ultimately, PVA is a flexible, scenarios-based modeling framework that can be used to evaluate different questions about extinction risk. Most typically, PVA is used to assess the probability of a particular outcome (e.g., population decline, extinction, recovery) over a specific timeframe. For example, what is the risk of extirpation of stellar sea lions in the Gulf of Alaska over the next 100 years (Winship & Trites 2006)? PVA can also be used to predict a **minimum viable population** (MVP) size, a threshold under which populations

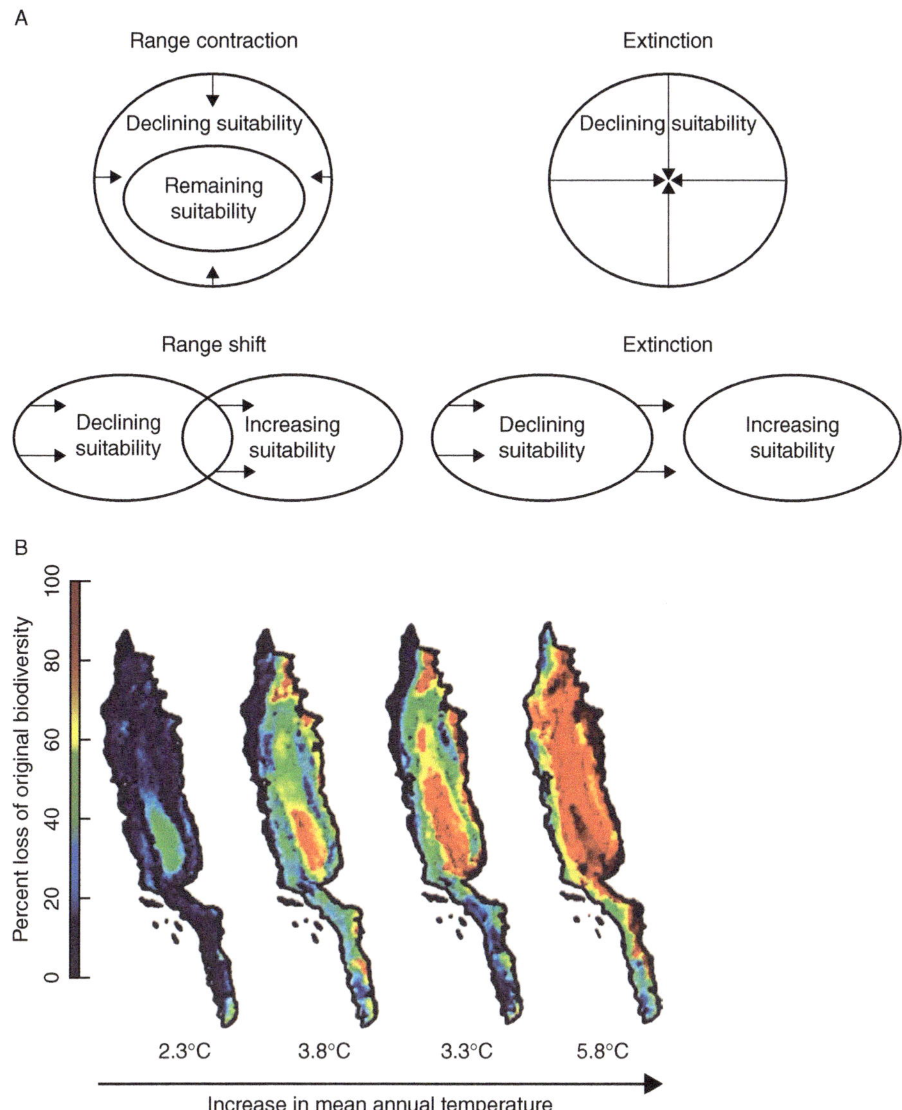

FIGURE 8.4 The use of species distribution modeling (SDM) to predict extinction. (A) Conceptual diagram of how SDMs can be used to forecast range changes versus extinction. Extinction probability increases as amount of projected future suitable habitat decreases. Similarly, extinction probability increases as the overlap between current and future suitable habitat decreases, particularly for species with low dispersal capacity. (B) Example of the use of SDMs to predict species loss. Projected species loss of California endemic plants in 80 years under different global warming scenarios. The percent of current floristic biodiversity lost will increase with rising global temperatures.

Reflection: What are some reasons why the proportion of species projected to be lost may be uneven across the landscape?

Source: (A) Courtesy of author (modified from Thomas 2012); (B) Loarie SR, Carter BE, Hayhoe K, McMahon S, Moe R, Knight CA, et al. (2008) Climate Change and the Future of California's Endemic Flora. PLoS ONE 3(6)

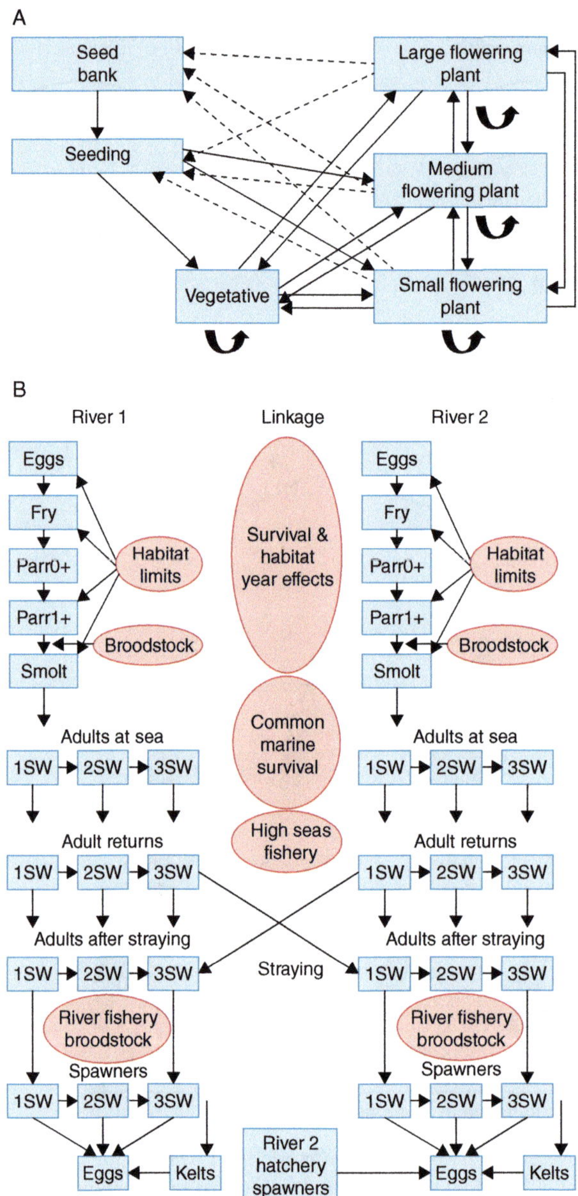

FIGURE 8.5 Examples of demographic parameters that are incorporated into population viability analyses. PVA models generally track different age/stage classes and the transition probabilities among them. Demographic scenarios can be simple, as in the plant example (A), or complex, as in the the fish example (B).

Reflection: In what different ways might "straying" fish—those that were born in one river but then return to the other river to breed—impact the survival of either population?

Source: (A) Eric S. Menges, Pedro F. Quintana Ascencio, Carl W. Weekley, Orou G. Gaouec, Population viability analysis and fire return intervals for an endemic Florida scrub mint, Biological Conservation Volume 127, Issue 1, January 2006, Pages 115–127; (B) Salmon PVA: A Population Viability Analysis Model for Atlantic Salmon in the Maine Distinct Population Segment, Christopher M. Legault, U.S. DEPARTMENT OF COMMERCE, National Oceanic and Atmospheric Administration

are likely to be lost over a specified time period. For example, what is the minimum population size of the grey wolf to ensure its persistence over the next 40 generations (Reed et al. 2003)? Finally, PVA can be used in a comparative framework to determine the sensitivity to different threats or rank different management options. For example, is it better to focus biodiversity conservation efforts within one large habitat patch or two small habitat patches (Simberloff & Abele 1982)? Answering these kinds of comparative questions is the most powerful and appropriate use of PVA (Reed et al. 2003).

A specific illustration is explored in Figure 8.6. The greater glider (*Petauroides volans*) is a nocturnal gliding Australian marsupial. It requires relatively large stands of old-growth Eucalyptus for food and habitat and has low reproductive output, with females raising a maximum of a single pup each year. A number of different PVAs have been conducted for this species evaluating probability of extinction under different demographic, environmental, and conservation scenarios, including alternate logging and fire management approaches (e.g., Possingham et al. 1993; Goldingay & Possingham 1995). Findings suggest that the survival of this species is particularly sensitive to the amount of habitat protected and the presence of a key predator: the powerful owl (*Ninox sterna*). In addition, more recent efforts have integrated field data from years of monitoring at more than 150 sites with PVA approaches. Models recalibrated by field data suggest that extinction risk is slightly higher than previously thought, but overall the PVA results were quite robust (Lindenmayer & McCarthy 2006).

Using data from lab or field studies to parametrize and refine models is particularly useful for obtaining biologically relevant results. This can be done with either a PVA or

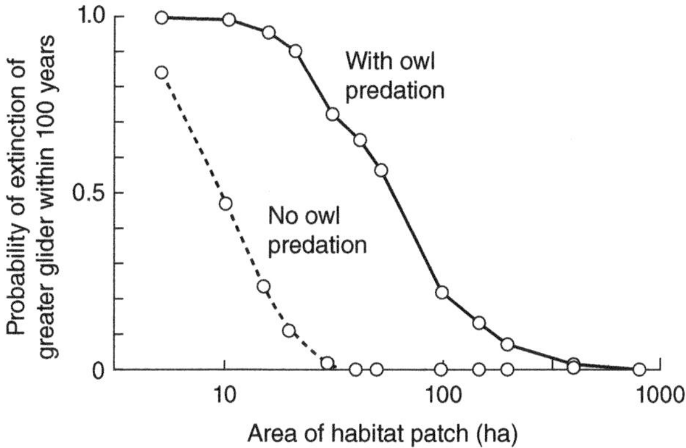

FIGURE 8.6 Example of PVA for assessing the likely outcomes of different management alternatives. Here the probability of extinction of the greater glider (*Petauroides volans*) is evaluated under different conditions. The survival of this species is sensitive to the amount of habitat protected and whether a key predator is present.

Reflection: Based on the data presented, what is the probability of *extinction* for the greater glider in the next 100 years if 10 hectares are protected and owl predators are not removed? What if they are removed? How many hectares need to be protected to ensure the greater glider has a <20% probability of extinction regardless of owl predation?

Source: Reproduced from P. Possingham H. , B. Lindenmayer D. W. Norton T. (1994) A framework for the improved management of threatened species based on Population Viability Analysis (PVA). Pacific Conservation Biology 1, 39-45. https://doi.org/10.1071/PC930039 with permission from CSIRO Publishing

SDM approach. One example is a study that set out to understand the impact of climate change on lizard biodiversity. Sinervo et al. (2010) modeled expected survival under future climate conditions for 34 different lizard families around the world. They evaluated multiple different climate change scenarios and included detailed information about lizard physiology in their model. For example, they used a critical temperature threshold for survival for each family and incorporated biologically realistic information about which lizard species are thermoregulators (and can behaviorally regulate their own temperature by basking) versus which species are thermoconformers (and simply track ambient temperature).

The integrative modeling approach used by Sinervo et al. suggested that many lizard species are unlikely to survive in warmer conditions. Specifically, they found that globally we may lose 20% of lizard species by 2080. The authors then grounded their extinction predictions with detailed observational studies across 200 sites in Mexico. By comparing contemporary and historical data, they identified sites where particular species had disappeared and calculated rates of species loss over time. They then used data loggers to record local temperatures at sites where particular species had persisted through time versus those where species had gone locally extinct. Thus, they had high confidence that their modeling results were biologically relevant.

In Sum

In addition to the two methods we just evaluated in detail, there are a number of additional modeling approaches that can be used to estimate extinction risk over different spatial, temporal, or taxonomic scales (e.g., Botkin et al. 2007; Harfoot et al. 2014). Some of these focus on the fate of individual species, while others consider entire species assemblages. Ultimately, integrating multiple approaches and diverse datasets (on complex organismal traits and dynamics of environmental change) remains at the forefront of predicting extinction risk from global change stressors.

HOW DO SCIENTISTS SUMMARIZE GLOBAL PATTERNS OF EXTINCTION RISK?

Understanding extinction risk for individual species and regional species aggregates is important, but documenting patterns of extinction at the global scale is also essential. There are a number of ways to summarize global patterns of extinction and extinction risk. Here we look at two examples. We will return to evaluate other conservation and policy tools—like the **Endangered Species Act**—that aim to assess and ameliorate extinction risk in Unit IV.

Biodiversity Databases

There are several global consortiums of scientists working to standardize the way we classify extinction risk and understand large-scale patterns of extinction across the tree of life. Some maintain online databases which include information on tens of thousands of species and their vulnerability to extinction (e.g., NatureServe [explorer.natureserve.org] and RedList [www.iucnredlist.org]).

Perhaps the most widely used database regarding extinction risk is the **Red List of Threatened Species** maintained by the International Union for Conservation of

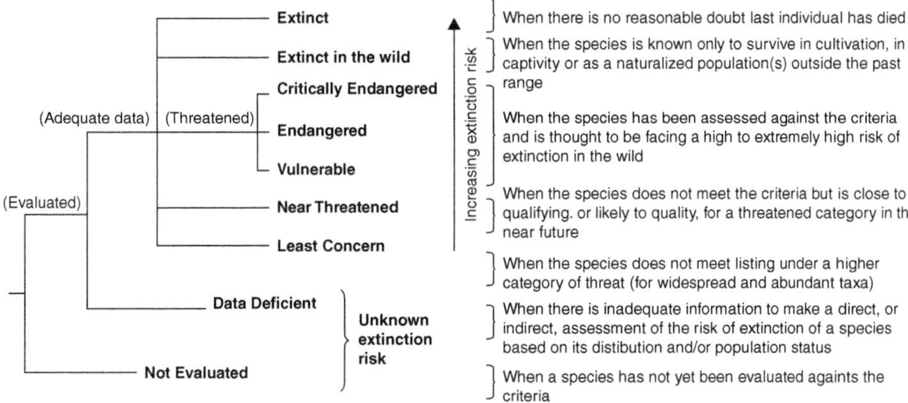

FIGURE 8.7 IUCN Red List threat categories. While a considerable number of species remain unevaluated or data deficient, many species have been categorized along a gradient of extinction risk, from least concern to critically endangered.

Reflection: Do you think it is useful to have multiple subcategories for threatened species? Why or why not? If you were designing a database to assess extinction risk, would you use similar categories as shown here?

Source: Ana S.L.Rodrigues, John D.Pilgrim, John F.Lamoreux, Michael Hoffmann, Thomas M.Brooks, The value of the IUCN Red List for conservation, Trends in Ecology & Evolution Volume 21, Issue 2, February 2006, Pages 71–76

Nature (IUCN). The IUCN has maintained this database of species at high risk of extinction for more than 50 years. Over the decades, the IUCN assessment process has become more rigorous, the number of species evaluated has increased, and the list has influenced conservation practices (Rodrigues et al. 2006; Mace et al. 2008). At its core, the list provides an assessment of the level of threat to individual species, as illustrated in Figure 8.7. Threat level is based on different categories of information, including species range area, population size, decline rate, and mathematical analysis of extinction risk (Mace et al. 2008). However, the sheer number of species included allows us to understand broader patterns of extinction risk across the tree of life.

The Red List underscores several important points about overall extinction risk. First, a large number of species are currently at risk. Taxonomic groups that have thousands of species assessed provide the best examples, as shown in Figure 8.8. For example, 25% of all assessed mammals, 14% of birds, 40% of amphibians, 31% of sharks and rays, and 34% of conifer plants are listed as threatened (i.e., facing a high risk of extinction in the wild; IUCN 2016). Second, there is tremendous variation in extinction risk across taxonomic groups. For example, more than 60% of cycad plants assessed are threatened, whereas only ~1% of lobsters are threatened (IUCN 2016). Third, extinction risk can change over even short periods of time. For example, over the last 50 years, extinction risk has increased for many groups of organisms. One particularly bleak example is the change in status of the corals, which have experienced precipitous declines in the last decades, as shown in Figure 8.8 and discussed in the last chapter. The *Taking a Closer Look* feature in this chapter provides an additional perspective from amphibians.

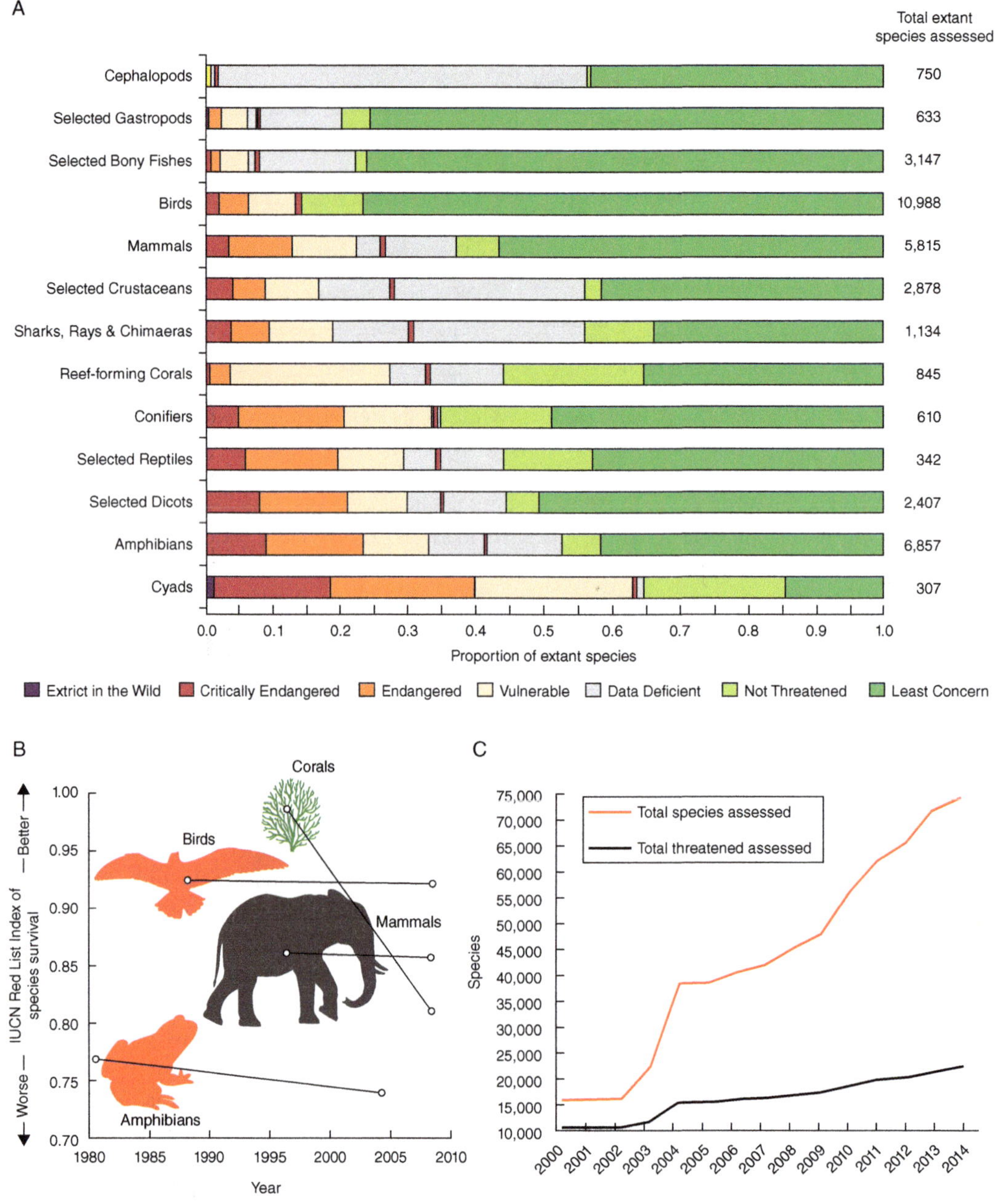

FIGURE 8.8 Data from the IUCN Red List. (A) Proportion of assessed species in each extinction risk category varies across taxonomic groups. (B) Trends in biodiversity endangerment for exemplar groups from 1980 to 2010. Over very recent timescales, some groups have declined more precipitously than others. (C) The number of total species assessed has increased dramatically in the last decade and, with it, the total number of threatened species.

Reflection: Compare and contrast the endangerment patterns for amphibians and corals. If you had limited funds for conservation, which group would you preferentially target for conservation intervention and why?

Source: IUCN. 2019. Summary Statistics Figure 2. The IUCN Red List of Threatened Species. Version 2019-2. https://www.iucnredlist.org/resources/summary-statistics

Although the number of species evaluated in the Red List has increased more than fourfold (to a current total of >70,000 species), the Red List currently only includes a small fraction (<5%) of the total species on Earth, as shown in Figure 8.8. Moreover, there are taxonomic biases in our current understanding of extinction risk, given that some groups are more difficult to study than others. For example, the IUCN estimates that it has evaluated 100% of all mammals, but only <6% of plants and <1% of all insects (IUCN 2016). Further, some highly diverse and ecologically important branches of the tree of life (such as archaea and eubacteria) are not assessed at all.

The limitations to the Red List are understandable given the tremendous effort involved in conducting comprehensive assessments, and it is unrealistic to expect that every species on Earth will be studied in detail. For example, just considering arthropods (a phyla that includes insects, arachnids, and crustaceans), there are more than 1 million currently described species, but potentially millions more undescribed species (Ødegaard 2000). Moreover, it is difficult to even define distinct species for some parts of the tree of life (e.g., microbes). Thus, there are a vast number of species that are listed as "data deficient" in the Red List and an enormous amount of biodiversity on our planet that remains unknown to science. Therefore, we are likely losing species to extinction before we even know they exist.

Meta-analyses

Given that extinction risk varies widely across lineages and habitats, meta-analyses can be used to synthesize data from different regions of the world and branches of the tree of life. Many of these global studies focus on extinction risk due to global warming. An early influential study that used both SDM and species-area modeling predicted that we may have committed between 18% and 35% of terrestrial species on Earth to extinction by 2050 (Thomas et al. 2004). Studies on the marine realm have been similarly bleak. For example, more than 30% of reef-building coral species currently face elevated risk of extinction (Carpenter et al. 2008).

Several recent meta-analyses have also aimed to develop an average extinction probability across previously published studies. For example, Maclean and Wilson (2011) integrated more than 300 records across different taxonomic groups (plants, invertebrates, and vertebrates) and biomes (polar, temperate, tropical, and marine) and found an average extinction risk by 2100 of 10%–14%. Similarly, Urban (2015) synthesized more than 130 published studies and found an average estimated extinction risk of 8% (with a wide range from 0% to 54%).

Of course, there is no single correct estimate given that extinction risk will vary depending on taxa, timescale, and region. However, whether the total number of species at risk of extinction in the next 50–100 years due to global warming is closer to 8% or closer to 35%, this degree of loss could have catastrophic effects on global ecosystems. Moreover, many recent meta-analyses focus exclusively on impacts from climate change, which is only one of many interacting global change stressors. Many modeling approaches also omit the role of important biotic factors such as dispersal, acclimation, adaptation, and species interactions. Therefore, meta-analyses based solely on climate may significantly underestimate true extinction risk.

In Sum

Global databases and large-scale analyses are essential for revealing broad patterns of extinction and extinction risk. However, continued empirical research is also urgently needed. Not only is there ongoing debate over how many species currently live on planet Earth, but estimates of extinction risk vary depend on the taxonomic, spatial, and temporal scale considered. Therefore, analytical advances must be paired with on-the-ground studies so that robust data are available to seed databases and meta-analyses.

WHAT IS THE SIXTH MASS EXTINCTION?

Overall, when we evaluate global patterns, we see that current extinction rates far exceed historical background extinction rates. In fact, the magnitude of the current extinction crisis is approaching that of the mass extinctions that occurred prior to our species' evolution. The ultimate fate of all species on our planet is extinction. However, this is the only time in Earth's history that a single species has precipitated rapid and widespread extinctions across the tree of life.

Scientists have termed the contemporary anthropogenically induced extinction pulse the **sixth mass extinction** (Wake & Vredenburg 2008; Barnosky et al. 2011). As the name implies, we are currently at risk of losing a large proportion of species on Earth. As we evaluated in the last section, specific projections of how much biodiversity will be lost over the next century vary, but broad-scale studies have all sounded the alarm bell. Not only have many species already been lost, but ongoing anthropogenic stressors have committed a large number of species to future extinction. The *Core Concepts* feature in this chapter further discusses the concept of **extinction debt**. The *Meet the Data* feature provides a more in-depth comparison between historical and contemporary extinction rates, demonstrating the role of modern humans in this contemporary extinction spasm.

CORE CONCEPTS

WHAT IS EXTINCTION DEBT?

Global change stressors can have immediate effects and also **time lag** effects. The lag effects of environmental disturbances on biodiversity are particularly important, because it can take a substantial amount of time for impacted populations to finally disappear. These delayed—but largely inevitable—extinctions are referred to as **extinction debt**. In other words, extinction debt is the future extinction of a species caused by past or current events.

Predictions about extinction debt are usually formalized as the number of species in a community that are predicted to go extinct, as shown in Box Figure 8.1. Prior to a disturbance, a community has a particular equilibrium **species richness** (number of total species). After a disturbance (like habitat loss), species may be lost, and it may take time for a new, lower equilibrium species richness to be reached. The difference in predisturbance and postdisturbance species richness is the extinction debt. The amount of time it takes to reach the new equilibrium is termed the **relaxation time**.

The term *extinction debt* was popularized by Tilman et al. (1994), who studied delayed extinction in the context

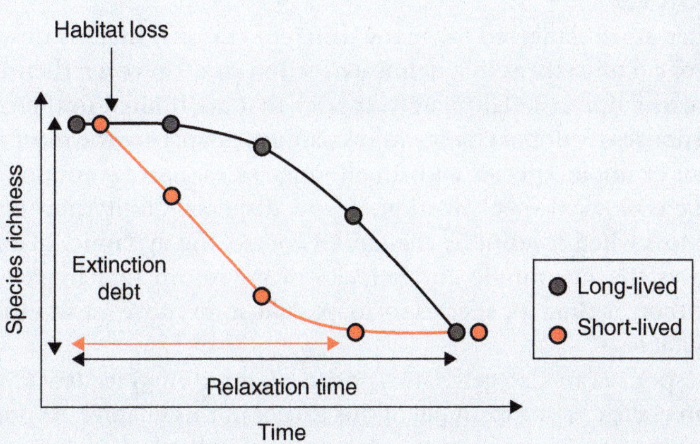

BOX FIGURE 8.1 Hypothetical model of extinction debt. Any number of global change stressors can lead to species extinctions and decreased species richness in a community. The difference in species richness before and after the impact is termed *extinction debt*, and the amount of time it takes to reach a new equilibrium is termed *the relaxation time*. Species that are short-lived tend to have shorter relaxation times than long-lived species. For example, adults of a long-lived species with slow generation intervals might persist on the landscape longer than that of a short-lived species.

Reflection: If stressors (like habitat loss) were reversed, what factors would determine whether or not species richness recovers?

Source: Mikko Kuussaari, Riccardo Bommarco, Risto K. Heikkinen, Aveliina Helm, Jochen Krauss, Regina Lindborg, Erik Öckinger, Meelis Pärtel, Joan Pino, Ferran Rodà, Constantí Stefanescu, Tiit Teder, Martin Zobel, Ingolf Steffan-Dewenter, Extinction debt: a challenge for biodiversity conservation, Trends in Ecology & Evolution Volume 24, Issue 10, October 2009, Pages 564–571

of habitat destruction. Tilman et al. used mathematical models to demonstrate how current habitat destruction and habitat fragmentation can have future ecological costs that may not be realized for generations. Since that seminal paper, other studies have demonstrated that extinction debts from anthropogenic stressors can last decades or centuries (e.g., Vellend et al. 2006).

Why do past or current stressors have lag effects on biodiversity? There are many possible causes of extinction debts (e.g., Hylander & Ehrlén 2013). First, some but not all life stages may be able to survive in an altered environment. For example, if adults can survive an initial perturbation but there is no suitable habitat for reproduction or juvenile survival, populations may persist for some time, especially for long-lived species. Second, once population size or population density has been reduced, populations are at increased risk of extinction. For example, Allee effects (which we explored earlier in this chapter) can increase extinction debt (Labrum 2011). If population size decreases after an initial environmental perturbation, the remaining individuals may lack genetic diversity, be more vulnerable to environmental stochasticity, and have more difficulty finding mates. Third, some populations may persist for a time even if the species as a whole is headed toward extinction. This is particularly true if colonization dynamics have been irreversibly disrupted in a metapopulation.

Ultimately, any anthropogenic stressor that has time-delayed effects on population persistence can contribute to extinction debt. The concept of extinction debt is highly analogous to the concept of **climate change commitment** that we will discuss in Chapter 10. Past human actions often have legacy effects that cannot be avoided even if current stressors are abated. Of course, reducing anthropogenic pressures on ecosystems is still critically important because some impacts can be reversed so that new lagged effects will not be set in motion.

CONCLUSION

Because species are endangered for many different reasons, there is no single characteristic or set of circumstances that define extinction risk. However, there are traits that can increase extinction risk. Ultimately, species that are limited in their potential for other core responses to global change (move, adjust, adapt) are the most vulnerable to extinction. For example, species with small range sizes, narrow environmental tolerances, extreme ecological specialization, or low dispersal ability may be particularly vulnerable to loss when conditions change. Of course, the dynamics of anthropogenic change (such as the magnitude and velocity of environmental impacts) will influence whether there is time for species to adapt, adjust, or move, or whether extinction becomes inevitable.

The die response can also generate a positive (amplifying) feedback. We discussed the **extinction vortex** as one example of this earlier in this chapter. As population size decreases, populations become more vulnerable to additional stressors. Extirpations and extinctions can also have a positive feedback at the ecosystem level. The complex relationships in ecological communities have evolved over millions of years. When one species is lost, the species it interacts with ecologically—its predators, prey, competitors, and mutualists—are also impacted. When species disappear in great numbers and at great speed, the structure and function of communities can rapidly deteriorate. Thus, even the loss of an individual species can have dramatic cascading effects, a topic we will turn to next.

In the next two chapters, we will address the consequences of extinction on ecosystems and how global change stressors affect species interactions and entire communities. Then in the final unit, we will address conservation and policy actions that can be taken to reduce biodiversity loss in this era of rapid global change.

MEET THE DATA THE SIXTH MASS EXTINCTION

Who Are the Scientists and What Did They Set Out To Do?

Here we examine a study conducted by an international team of scientists from Mexico and the United States of America (Ceballos et al. 2015). The study was entitled "Accelerated Modern Human-Induced Species Losses: Entering the Sixth Mass Extinction" and was published in the journal *Science Advances* in 2015. Led by Dr. Gerardo Ceballos (Box Figure 8.2), the collaborating scientists straddle a number of fields, including ecology, evolutionary biology, paleontology, and conservation biology, and include leading voices in the study of large-scale patterns of extinction. The study appealed to a wide audience, reaching an estimated 100 million people in the first month after publication.

As we explored in Chapter 2, scientists recognize five historical mass extinction events. The "Big Five" mass extinctions had different causes, but shared a dramatic outcome: the loss of more than 75% of species alive at the time. Currently, anthropogenic stressors are driving many species to the brink of extinction. But can humans precipitate a mass extinction of these proportions?

Prior studies suggest that the answer to this question is yes. Nearly 1,000 recent species extinctions cataloged by the IUCN have been directly caused by humans (IUCN 2016). This number only considers extinction of described species and is weighted toward human effects of the last century. The total number of human-mediated extinctions is much higher when we consider the loss of undescribed species and the effects of humans on biodiversity prior to the last century. Moreover, meta-analyses suggest

BOX FIGURE 8.2 Dr. Gerardo Ceballos led an international team of scientists from Mexico and the United States to study contemporary extinction rates.

Source: Gerardo Ceballos

that current extinction rates exceed typical historical "background" extinction rates in many diverse groups of organisms (e.g., Pimm et al. 2006; Barnosky et al. 2011; Burkhead 2012).

However, there are also challenges in assessing the magnitude of human-induced extinctions because both background extinction rates and modern extinction rates must be measured accurately. For example, if background extinction rates are estimated poorly, modern extinction patterns might appear more dramatic—or less dramatic—than they really are.

Ceballos and colleagues set out to understand whether modern rates of extinction are higher than background rates of extinction for vertebrates. The researchers avoided assumptions that are often used in meta-analyses of extinction rates and used a particularly conservative estimate of background extinction rates. Thus, any conclusions drawn from the data would not overestimate modern extinction rates. In the authors' own words: *"The analysis we present here avoids using assumptions such as loss of species predicted from species–area relationships, which can suggest very high extinction rates, and which have raised the possibility that scientists are 'alarmists' seeking to exaggerate the impact of humans on the biosphere."*

What Are *Your* Predictions?

Before you read on, take a few minutes to apply what you have learned about extinction to the analysis of extinction rates through time:

- *Do you predict that modern extinction rates will be different from background extinction rates?*
- *If so, do you expect the magnitude of current extinction rates to be similar to those during historical mass extinctions?*
- *Do you expect results to be similar across all vertebrate groups? Which groups do you think might have the highest rates of extinction and why?*

Now write a summary prediction statement about whether—and how—historical and modern extinction rates are likely to differ.

What Were the Scientists' Predictions?

The authors did not present a formal prediction in their study, but rather an objective: *"Here, we ascertain whether even the lowest estimates of the difference between background and contemporary extinction rates still justify the conclusion that people are precipitating a global spasm of biodiversity loss."* Rephrased as a prediction, this could read: *We predict that even very conservative estimates of extinction rates indicate that anthropogenic stressors are precipitating a global spasm of biodiversity loss.*

What Data Were Collected?

The authors compared background and contemporary extinction rates for vertebrates. For the contemporary extinction rates, the authors used the IUCN database (which we explored earlier in this chapter) to identify the number of anthropogenic extinctions in the recent past (since ~1500 CE). The authors estimated a *highly conservative* modern extinction rate (based only on species known to be extinct) and a *conservative* modern extinction rate (based on species known to be extinct, species thought to be extinct but not confirmed, and species extinct in the wild). A *conservative* estimate is one that is intentionally moderate, and both

(*Continued*)

estimates used are conservative because true modern extinction rates are likely higher than can be inferred from the IUCN database (because there are many species without formal IUCN assessments).

For the background extinction rates, the authors used an estimate of prevailing extinction rates over the last 2 million years: 2 extinctions per million species years (E/MSY). E/MSY is a commonly used metric that reports extinction rates per million species per year, and 2 E/MSY is 2 species extinctions per 10,000 species per 100 years. The 2 E/MSY rate was obtained by rounding-up an estimate from an influential meta-analysis (Barnosky et al. 2011) that evaluated the stratigraphic occurrence of thousands of mammal species in the fossil record over multiple time intervals. This 2 E/MSY is also conservative because it is higher than what is typically used as a background extinction rate in similar studies.

In sum, the authors use a conservative estimate of background extinction rates (one that is on the high end of scientific expectations) and a conservative estimate of modern extinction rates (one that is on the low end of scientific expectations). Therefore, they intentionally minimize the difference between background and contemporary extinction rates, providing a stringent test of whether contemporary extinction rates are exceptionally elevated.

What Is *Your* Interpretation of the Data?

Before you read on, take a few minutes to interpret the data in Box Figure 8.3. Consider the following questions as you look at the figure:

- *What are the overall trends observed?*
- *How consistent are the trends across species?*
- *What are possible explanations for similarities and differences across major lineages?*

Write a sentence or two describing the *key findings* of this study and their *significance*.

What Was the Scientists' Interpretation of the Data?

The data in Box Figure 8.3 demonstrate that modern rates of extinction are exceptionally high relative to historical background rates of extinction. Modern rates of extinction are elevated for all major groups of vertebrates. Depending on the lineage and time period evaluated, modern extinction rates are between 5 and 100 times higher than the background extinction rate of 2 E/MSY. Extinctions have accelerated dramatically since 1900.

Another way to look at the data is to consider how long it would have taken for extinctions observed over the last century to have occurred under the background extinction rate. The number of extinctions that have occurred since 1900 would have taken between 800 and 10,000 years to occur under a constant background extinction rate of 2 E/MSY (Box Figure 8.4). The authors emphasize that because they used conservative estimates for background and modern extinction rates, their results likely represent a *lower bound* on contemporary extinction dynamics.

The significance of this study is that—even using very conservative estimates—anthropogenically induced extinctions are occurring at an alarming rate. Anthropogenic stressors are precipitating a "global spasm of biodiversity loss," suggesting that the sixth mass extinction is already underway. The reality that humans are reshaping the biosphere can seem quite frightening. However, there are countless ways that individuals, communities, and organizations can help turn the tide. Unit IV is devoted to exploring ways to reduce biodiversity loss and realign the needs of our species and the rest of the natural world.

What Are *Your* Ideas for Future Research Directions?

Given what this study accomplished, what next steps do you envision for this research program? If you had been involved in this study, how would you follow up? Start by considering the following questions:

- *What new questions does this study raise?*
- *Can the causes or consequences of the patterns revealed be investigated?*
- *How could the study be broadened to achieve more generality?*

Now write a few sentences about future directions on this research theme.

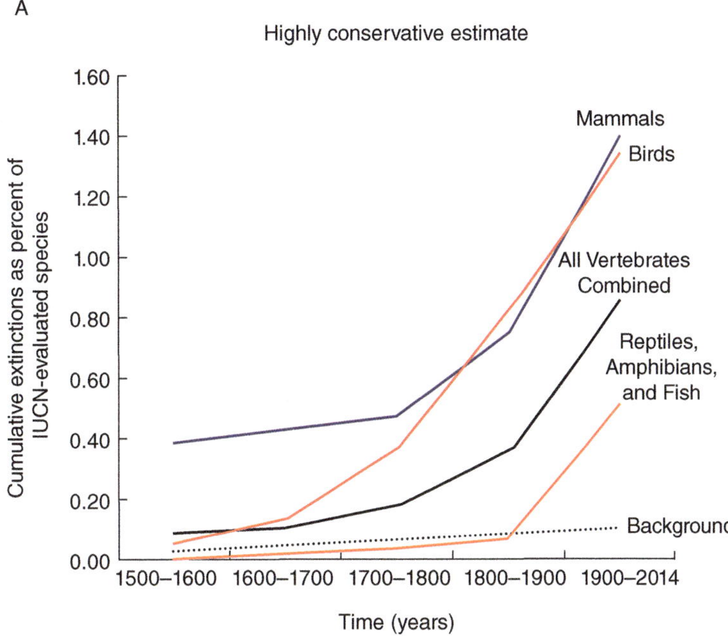

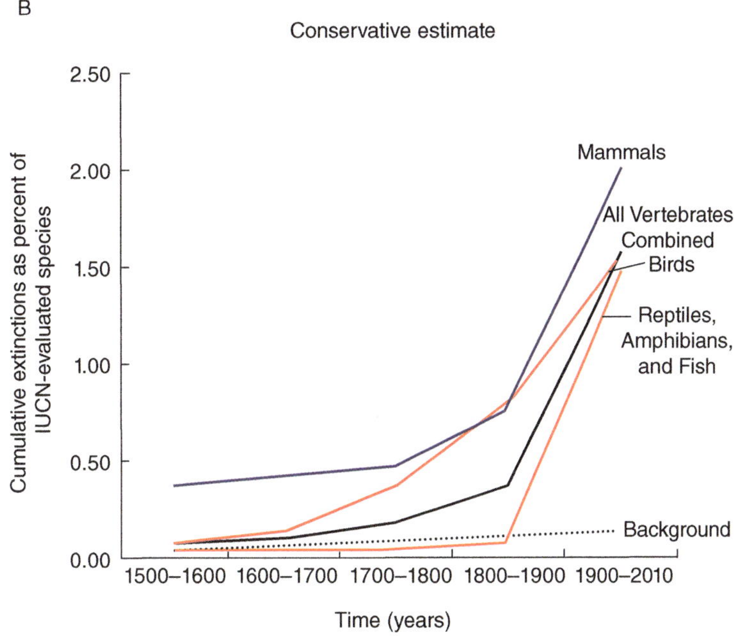

BOX FIGURE 8.3 Cumulative percentage of extinct species for all vertebrates combined and for mammals, birds, and fish, amphibians, and reptiles. (A) and (B) show results under highly conservative and conservative estimates of modern extinction, respectively. The dashed line represents the number of extinctions expected under a background extinction rate of 2 E/MSY.

Source: Accelerated modern human–induced species losses: Entering the sixth mass extinction. BY GERARDO CEBALLOS, PAUL R. EHRLICH, ANTHONY D. BARNOSKY, ANDRÉS GARCÍA, ROBERT M. PRINGLE, TODD M. PALMER. SCIENCE ADVANCES 19 JUN 2015 : E1400253

(Continued)

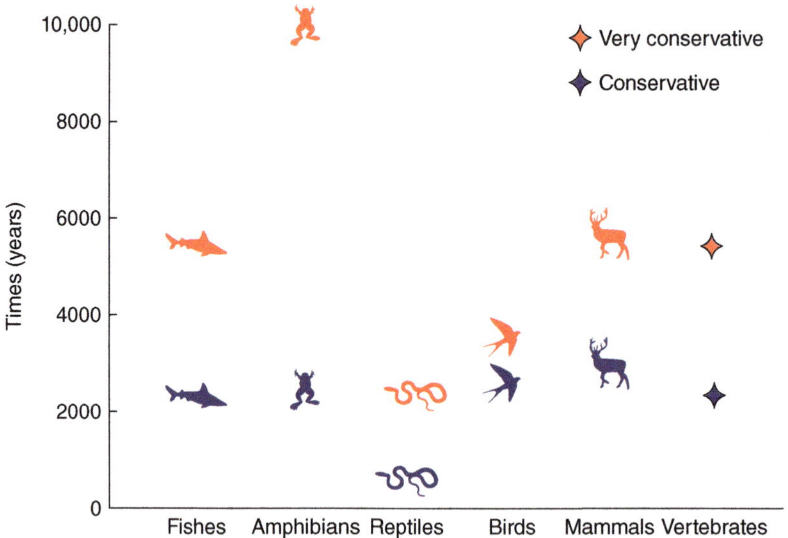

BOX FIGURE 8.4 Number of years it would take for the vertebrate extinctions since 1900 to occur under the background extinction rate of 2 E/MSY.

Source: Accelerated modern human–induced species losses: Entering the sixth mass extinction. BY GERARDO CEBALLOS, PAUL R. EHRLICH, ANTHONY D. BARNOSKY, ANDRÉS GARCÍA, ROBERT M. PRINGLE, TODD M. PALMER. SCIENCE ADVANCES 19 JUN 2015 : E1400253

TAKING A CLOSER LOOK AMPHIBIAN DECLINES

Many lineages across the tree of life have experienced population losses and species extinctions on anthropogenic timescales. Here we will look at amphibians as a case study because they are one of the most imperiled lineages on Earth. Amphibians are an ancient group of animals, first evolving around 350 million years ago. They survived the last three major mass extinction events on our planet, but many amphibian species are now on the brink of extinction.

What Are Amphibians?

Amphibians are a remarkably diverse group of organisms. Amphibians are divided into three major lineages: Anura (frogs and toads), Caudata (salamanders and newts), and Gymnophiona (caecilians). These lineages contain a striking amount of diversity in morphology, behavior, and life history, as illustrated in Box Figure 8.5. There are currently more than 7,000 described species of amphibians. However, new species are still being discovered and described. Over the last 10 years, more than 100 new species have been added per year to amphibian databases (amphibiaweb.org/amphiban/newspecies.html).

Amphibians have great ecological, economic, and aesthetic value. In natural populations, amphibians play important roles in food webs as both predators and prey and keep invertebrate populations in check. Amphibians also are considered important environmental sentinels. Because many amphibians live in aquatic ecosystems and have very porous skin, amphibians can provide an early warning of environmental contaminants that may be harmful to other species. Amphibians also can provide direct benefits to humans. For example, secretion-rich amphibian skin has been an inspiration for the development of biomedicines (such as new antimicrobial and antiviral drugs). Finally, amphibians are revered around the world; they are used as food sources and are viewed as agents of rain, fertility, and luck in many cultures.

What Are the Current Threats to Amphibians?

As illustrated in Box Figure 8.6, the IUCN estimates that more than one-third of all amphibians are threatened with extinction. Amphibians are considered the most imperiled group of vertebrates and one of the most imperiled branches on the tree of life (Ceballos et al. 2015). As with other lineages on our planet, there is a diversity of interacting threats to amphibians.

HABITAT ALTERATION Habitat loss and habitat fragmentation are often critical drivers of extinction. Amphibians have complex habitat needs, especially when different life stages require contrasting—but adjacent—habitats (e.g., larvae and adults requiring ponds and forested areas, respectively). Deforestation, urban development, and roads are just some of the stressors that can affect amphibian survival, juvenile dispersal, and the

BOX FIGURE 8.5 Amphibian biodiversity. (A) Representatives of the major lineages of amphibians: Anura (top), Caudata (center), and Gymnophiona (bottom). Frogs and salamanders are relatively common, but most people have never heard of the limbless, burrowing caecilians. (B) Amphibian diversity in size. Adult amphibians can be as large as the Chinese giant salamander (top) or as small as the newly discovered microhylid frog from Papua New Guinea (bottom). (C) Diversity in reproductive mode. Some amphibians lay eggs, others give birth to live young, some exhibit paternal care (as with the egg-guarding behavior of the midwife toad shown), and still others undergo metamorphosis in usual places like in the gut (as with the gastric brooding frog). Many species with these unusual traits are critically endangered.

Reflection: What factors likely contribute to the dramatic diversity in the amphibian lineage?

Source: (A) Dirk Ercken/Shutterstock; Matt Jeppson/Shutterstock; https://www.wired.com/2015/02/absurd-creature-week-caecilian/; (B) Matt Jeppson/Shutterstock; Audrey Wilson1/Shutterstock; (C) Geoff Gallice/Wikipedia (CC BY 2.0); COULANGES/Shutterstock

(Continued)

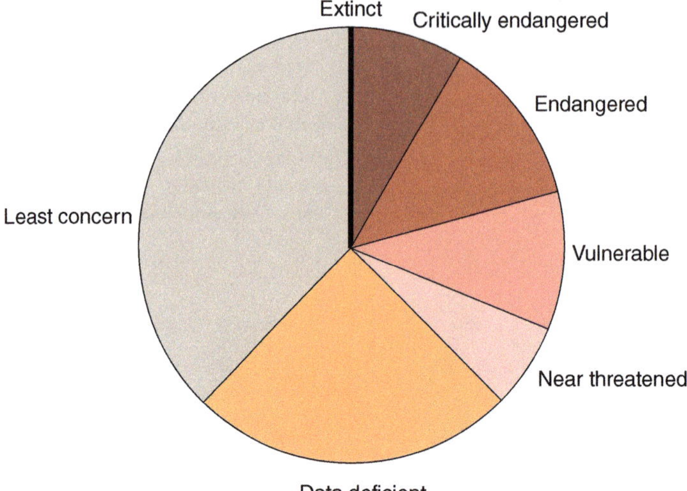

BOX FIGURE 8.6 IUCN Red List assessment for amphibians. Nearly one-third of all amphibians are threatened with extinction.

Reflection: Why might nearly one-third of all amphibians be considered data deficient?

Source: Courtesy of the author (using data from http://www.iucnredlist.org)

probability of recolonization in metapopulations (Cushman 2006).

OVEREXPLOITATION There are dozens of amphibian species in the food and pet trade. Commerce in meat (e.g., frog legs for consumption) and live animals (e.g., frogs for pets) moves millions of animals around the world every year (Carpenter et al. 2014). Trade of the three most popular genera of frogs in pet stores is worth more than 10 million US dollars annually. Some pet store amphibians are reared in captivity, but many are taken from the wild. Recent analyses show that a number of critically endangered species are found regularly in the international trade market (Carpenter et al. 2014), a practice that can lead to population declines.

INTRODUCED SPECIES Introduced species have caused population declines of native amphibians in many ecosystems around the world (Bucciarelli et al. 2014). Invasive plants can alter the habitat of native amphibians, affecting amphibian behavior, survival, development, and breeding success. Invasive fish often become predators and competitors of native amphibians, leading to population-level impacts. Finally, invasive amphibians have had some of the most devastating effects because they can prey on, compete with, hybridize with, and/or vector disease to native amphibians.

CONTAMINANTS Chemical pollutants are increasingly common in amphibian habitats, whether from direct application, run-off from agricultural settings, industrial sewage, or atmospheric deposition. Amphibians are often highly exposed and extremely sensitive to chemical stressors given their close association with aquatic habitats and their permeable skin. Meta-analyses suggest that chemical pollutants have broad detrimental effects on amphibian survival, life cycle, and developmental abnormalities (Egea-Serrano et al. 2012). There are many high-profile examples of particular pollutants that have dramatic effects on amphibians. For example, the agricultural herbicide atrazine is an endocrine disruptor in amphibians, and it can alter the sex of frogs even at low environmental concentrations (Hayes et al. 2010).

CLIMATE CHANGE Global warming has affected the phenology, elevational distribution, and body size in some species of amphibians (reviewed in Li et al. 2013). However, there is little direct evidence that climate alone has caused species extinctions. The direct role of climate change on amphibian declines has been difficult to assess given the challenge of disentangling the effects of global warming from those of other global change pressures. However, even if climate change is not directly responsible for current amphibian declines, climate change can have indirect and sublethal effects that make amphibian populations more vulnerable to other stressors (Li et al. 2013).

DISEASE Amphibians are susceptible to a number of disease threats. Bacterial, viral, fungal, and animal pathogens and parasites can all impact amphibian populations (reviewed in Blaustein et al. 2012). Disease threats are on the rise and contributing directly to catastrophic amphibian population losses. We now turn to the impact of these emerging infectious diseases and their role in contemporary amphibian declines.

Why Are Emerging Infectious Diseases on the Rise?

One of the most pressing threats to amphibian biodiversity is emerging infectious disease. Emerging infectious diseases (EIDs) are diseases that have recently appeared or increased in incidence, severity, geography, or host range. The impact of EIDs has risen dramatically over the last century, as shown in Box Figure 8.7 (Jones et al. 2008). The increased incidence and severity of EIDs are at least partially explained by an increasingly globalized transportation network, as shown in Box Figure 8.4. As people and goods move rapidly around the world, pathogens have an opportunity to invade new regions of the world and are exposed to new host species.

In addition, as human livelihoods increasingly rely on products from domesticated and wildlife species, the opportunity for cross-species transmission and **zoonoses** (diseases that can be transmitted to humans from other animals) increases. EIDs have had dramatic effects on human populations (for example, HIV, Ebola, Covid-19), domesticated and agricultural species (like avian influenza and potato blight), and natural populations across the tree of life (such as sudden oak death, sea star wasting disease, and bat white nose syndrome). One such EID has had catastrophic effects on amphibians around the world: **chytridiomycosis**.

What Is Chytridiomycosis?

Chytridiomycosis is a skin disease of amphibians caused by the chytrid fungus *Batrachochytrium dendrobatidis* (Bd). Bd is an aquatic fungus that attacks amphibian skin and compromises vital skin functions such as osmoregulation, electrolyte balance, and protection from other pathogens. An example of a Bd die-off and an illustration of the lifecycle of Bd are shown in Box Figure 8.8. Bd was discovered

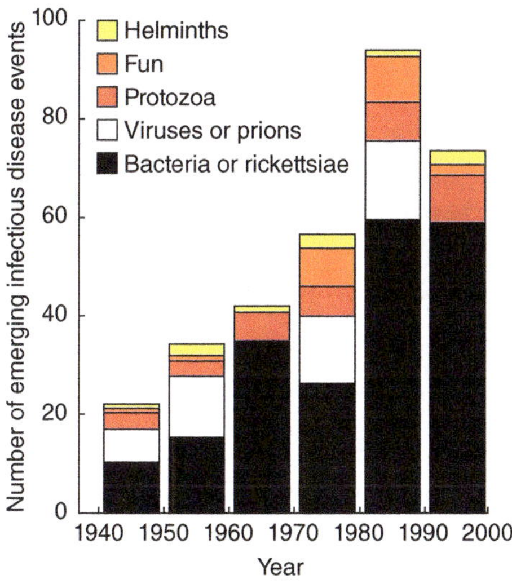

BOX FIGURE 8.7 The incidence of emerging infectious disease (EID) events has increased dramatically over the last century as globalization facilitates the movement of pathogens to new geographic regions and new hosts.

Reflection: Do you expect diseases of humans, agricultural species, and natural populations to spread via the same or different channels?

Source: (A) Jones, K., Patel, N., Levy, M. et al. Global trends in emerging infectious diseases. Nature 451, 990–993 (2008); (B) globaia.org

(Continued)

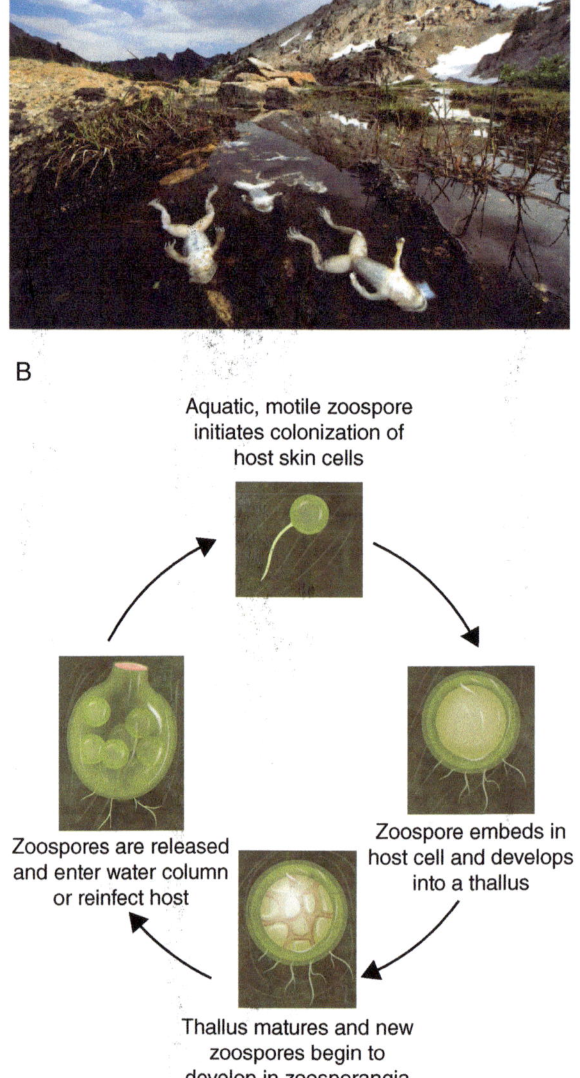

BOX FIGURE 8.8 The deadly fungal pathogen *Batrachochytrium dendrobatidis* (Bd). (A) A Bd-induced die-off of mountain yellow-legged frogs in the Sierra Nevada Mountains of California. (B) The life cycle of Bd. Zoospores with flagella find amphibian hosts and embed in the skin. The zoospore then matures into the reproductive zoosporangia, which releases more zoospores.

Reflection: Bd is an aquatic fungus—what hypotheses could explain its transportation and establishment around the world?

Source: (A) Joel Sartore / www.joelsartore.com; (B) courtesy of author

in 1998 (Berger et al. 1998; Longcore et al. 1999) and now infects hundreds of amphibians species on all continents where amphibians occur. Recently, a closely relative of Bd has been discovered, *Batrachochytrium salamandrivorans* (Bsal; Martel et al. 2013), which thus far appears to be primarily affecting European salamanders. Not all species are equally susceptible to chytridiomycosis. Some species experience rapid mortality while others are more resistant and can serve as disease reservoirs. Globally, chytridiomycosis has had widespread effects on amphibians and has been called "the most spectacular loss of vertebrate biodiversity due to disease in recorded history" (Skerratt et al. 2007).

Chytrids are ancient fungi, and thousands of species of chytrids are found around the world (Longcore & Simmons 2012). Bd and Bsal are the only chytrids known to be animal pathogens. When and why did these species evolve the ability to kill vertebrates? Despite decades of research, these questions are not entirely resolved. It is clear that these chytrids have specialized adaptations that allow them to invade amphibian skin. They also have rapidly evolving genomes and several specific classes of genes that may serve as virulence factors (Farrer et al. 2011; Rosenblum et al. 2013). At least one Bd strain has spread around the world relatively recently (Farrer et al. 2011; Rosenblum et al. 2013; O'Hanlon et al. 2018). The exact mechanisms of spread are unknown, but international trade in amphibians is likely to play a role (e.g., Fisher & Garner 2007).

What Is Being Done to Confront Amphibian Declines?

There are a number of conservation actions that can be undertaken to ameliorate the effects of chytridiomycosis. Examples include treatment with antifungal drugs to reduce Bd prevalence in populations in the wild, selection for disease resistance in captive animals that can then be released to the wild; reducing habitat suitability for Bd by altering temperature or moisture regimes. Many of these conservation actions are labor intensive and focused on the short-term goal of protecting the most at-risk populations from chytridiomycosis. In addition, many address disease but not other global change stressors contributing to amphibian declines. Ultimately, to ensure long-term persistence of amphibian biodiversity, integrative approaches will be needed, a topic we will explore in Unit IV.

KEY CONCEPTS

What is the die response?

- When species do not adapt, adjust, or move, they may be irrevocably lost via the process of extinction.

How is the survival of individuals, populations, and species connected?

- By increasing mortality rates, global change stressors can lead to losses at the population (extirpation) and species (extinction) levels.
- Decreases in population size can create a positive feedback loop called the extinction vortex.

What are examples of extinction in response to global change pressures?

- Extinction can result from many different anthropogenic pressures and their interactions.

How do scientists estimate extinction risk?

- Species distribution modeling (SDM) and population viability analysis (PVA) can be used to assess extinction risk under different future environmental scenarios.

How do scientists summarize global patterns of extinction risk?

- Biodiversity databases and meta-analyses are used to understand global extinction patterns across the tree of life.
- There is a variation in estimates of extinction risk across taxonomic groups, geographic regions, and temporal scales, but all suggest elevated extinction rates in the Anthropocene.

What is the sixth mass extinction?

- Extinction rates caused by human activities far exceed historical background extinction rates, suggesting that we have entered a sixth mass extinction.

Core concepts: What is extinction debt?

- Time lag effects of environmental stressors on species richness are called extinction debts.

Meet the data: The sixth mass extinction

- Scientists set out to assess whether we are entering a sixth mass extinction. They compared modern and background extinction rates and found that modern extinction rates are exceptionally high. They concluded that anthropogenic stressors are precipitating species losses of mass extinction proportions.

Taking a closer look: Amphibian declines

- More than one-third of all amphibians are currently threatened with extinction. There are diverse threats to amphibians, including habitat alteration, overexploitation, introduced species, contaminants, climate change, and disease. With globalization, emerging infectious diseases like chytridiomycosis play an increasingly devastating role.

CONSOLIDATE YOUR KNOWLEDGE

Answer the following questions to assess your progress meeting the learning outcomes:

1. In your own words, write a one- or two-sentence synthesis of the big-picture takeaway point of this chapter.
2. Revisit your answer to the Blank Page exercise in the beginning of this chapter. Would you refine your answer now based on knowledge you integrated from this chapter?
3. How are losses at the individual, population, and species level connected?
4. How does the die response precipitate positive feedbacks at both the population and community levels?
5. What human activities contribute to the die response?
6. What approaches can scientists use to predict extirpation and extinction rates under global change pressures?
7. What are some lineages that have an elevated number of species at risk of extinction? What are some lineages where we know very little about relative extinction risk?
8. What specific factors make it more or less likely that a species will undergo human-mediated extinction?
9. What factors and processes likely contribute to the increase in emerging infectious diseases over the last century?
10. Have we entered the sixth mass extinction? What evidence can you give to support your position?
11. In your own words, define the bolded and italicized terms in this chapter.
12. What are some questions that you have about the content in this chapter? If you found some content particularly challenging or particularly interesting, identify these as areas for additional reflection or reading.

LITERATURE CITED

Barnosky, Matzke, Tomiya, Wogan, Swartz, Quental, Marshall, et al. 2011. "Has the Earth's Sixth Mass Extinction Already Arrived?" *Nature* 471 (7336): 51–57. doi.org/10.1038/nature09678.

Berger, Speare, Daszak, Green, Cunningham, Goggin, Slocombe, et al. 1998. "Chytridiomycosis Causes Amphibian Mortality Associated with Population Declines in the Rain Forests of Australia and Central America." *Proceedings of the National Academy of Sciences* 95 (15): 9031–36. doi.org/10.1073/pnas.95.15.9031.

Blaustein, Gervasi, Johnson, Hoverman, Belden, Bradley, & Xie. 2012. "Ecophysiology Meets Conservation: Understanding the Role of Disease in Amphibian Population Declines." *Philosophical Transactions of the Royal Society B: Biological Sciences* 367 (1596): 1688–707. doi.org/10.1098/rstb.2012.0011.

Botkin, Saxe, Araújo, Betts, Bradshaw, Cedhagen, Chesson, et al. 2007. "Forecasting the Effects of Global Warming on Biodiversity." *BioScience* 57 (3): 227–36. doi.org/10.1641/B570306.

Bruegmann & Caraway. 2003. "Cyanea Superba." The IUCN Red List of Threatened Species. http://www.iucnredlist.org.

Bucciarelli, Blaustein, Garcia, & Kats. 2014. "Invasion Complexities: The Diverse Impacts of Nonnative Species on Amphibians." *Copeia* 2014 (4): 611–32. doi.org/10.1643/ot-14-014.

Burkhead. 2012. "Extinction Rates in North American Freshwater Fishes, 1900–2010." *BioScience* 62 (9): 798–808. doi.org/10.1525/bio.2012.62.9.5.

Carlson, Burgio, Dougherty, Phillips, Bueno, Clements, Castaldo, et al. 2017. "Parasite Biodiversity Faces Extinction and Redistribution in a Changing Climate." *Science Advances* 3 (9): e1602422. doi.org/10.1126/sciadv.1602422.

Carpenter, Abrar, Aeby, Aronson, Banks, Bruckner, Chiriboga, et al. 2008. "One-Third of Reef-Building Corals Face Elevated Extinction Risk from Climate Change and Local Impacts." *Science* 321 (5888): 560–63. doi.org/10.1126/science.1159196.

Carpenter, Andreone, Moore, & Griffiths. 2014. "A Review of the International Trade in Amphibians: The Types, Levels and Dynamics of Trade in CITES-Listed Species." *Oryx* 48 (4): 565–74. doi.org/10.1017/S0030605312001627.

Ceballos, Ehrlich, Barnosky, García, Pringle, & Palmer. 2015. "Accelerated Modern Human-Induced Species Losses: Entering the Sixth Mass Extinction." *Science Advances* 1 (5): e1400253. doi.org/10.1126/sciadv.1400253.

Coote. 2009. "*Partula nodosa*." The IUCN Red List of Threatened Species. http://www.iucnredlist.org.

Cushman. 2006. "Effects of Habitat Loss and Fragmentation on Amphibians: A Review and Prospectus." *Biological Conservation* 128 (2): 231–40. doi.org/10.1016/j.biocon.2005.09.031.

Egea-Serrano, Relyea, Tejedo, & Torralva. 2012. "Understanding of the Impact of Chemicals on Amphibians: A Meta-Analytic Review." *Ecology and Evolution* 2 (7): 1382–97. doi.org/10.1002/ece3.249.

Ehrlich, Dobkin, & Wheye. 1988. "The Passenger Pigeon." Stanford University. http://web.stanford.edu/group/stanfordbirds/text/essays/Passenger_Pigeon.html.

Fagan & Holmes. 2006. "Quantifying the Extinction Vortex." *Ecology Letters* 9 (1): 51–60. doi.org/10.1111/j.1461-0248.2005.00845.x.

Farrer, Weinert, Bielby, Garner, Balloux, Clare, Bosch, et al. 2011. "Multiple Emergences of Genetically Diverse Amphibian-Infecting Chytrids Include a Globalized Hypervirulent Recombinant Lineage." *Proceedings of the National Academy of Sciences* 108 (46): 18732–36. doi.org/10.1073/pnas.1111915108.

Fisher & Garner. 2007. "The Relationship between the Emergence of *Batrachochytrium dendrobatidis*, the International Trade in Amphibians and Introduced Amphibian Species." *Fungal Biology Reviews* 21 (1): 2–9. doi.org/10.1016/j.fbr.2007.02.002.

Gilpin & Soulé. 1986. "Minimum Viable Populations: Processes of Species Extinction." In

Conservation Biology: The Science of Scarcity and Diversity, pages 19–34. Sunderland, MA: Sinaeur Associates.

Goldingay & Possingham. 1995. "Area Requirements for Viable Populations of the Australian Gliding Marsupial Petaurus Australis." *Biological Conservation* 73 (2): 161–67. doi.org/10.1016/0006-3207(95)90043-8.

Hanski. 1998. "Metapolulation Dynamics." *Nature* 396: 41–49.

Harfoot, Newbold, Tittensor, Emmott, Hutton, Lyutsarev, Smith, Scharlemann, & Purves. 2014. "Emergent Global Patterns of Ecosystem Structure and Function from a Mechanistic General Ecosystem Model." *PLoS Biology* 12 (4): e1001841. doi.org/10.1371/journal.pbio.1001841.

Hayes, Khoury, Narayan, Nazir, Park, Brown, Adame, et al. 2010. "Atrazine Induces Complete Feminization and Chemical Castration in Male African Clawed Frogs (*Xenopus Laevis*)." *Proceedings of the National Academy of Sciences* 107 (10): 4612–17. doi.org/10.1073/pnas.0909519107.

Hylander & Ehrlén. 2013. "The Mechanisms Causing Extinction Debts." *Trends in Ecology and Evolution* 28 (6): 341–346. doi.org/10.1016/j.tree.2013.01.010.

Internatoinal Union for Consevation of Nature (IUCN). 2016. International Union for Conservation of Nature Annual Report 2016. portals.iucn.org/library/node/46619

Jones. 2012. "Lonesome George, Last of the Pinta Island Tortoises, Dies." *CNN*. http://www.cnn.com/2012/06/25/world/americas/lonesome-george-giant-tortoise-dies/index.html.

Jones, Patel, Levy, Storeygard, Balk, Gittleman, & Daszak. 2008. "Global Trends in Emerging Infectious Diseases." *Nature* 251 (7181): 900–93. doi.org/10.1038/nature06536.

Keeley & Keeley. 1987. "Role of Fire in the Germination of Chaparral Herbs and Suffrutescents." *California Botanical Society* 34 (3): 240–49.

Kramer, Dennis, Liebhold, & Drake. 2009. "The Evidence for Allee Effects." *Population Ecology* 51 (3): 341–54. doi.org/10.1007/s10144-009-0152-6.

Labrum. 2011. "Allee Effects and Extinction Debt." *Ecological Modelling* 222 (5): 1205–7. doi.org/10.1016/j.ecolmodel.2010.12.013.

Levins. 1969. "Some Demographic and Genetic Consequences of Environmental Heterogeneity for Biological Control." *Bulletin of the Entomological Society of America* 15 (3): 237–40.

Li, Cohen, & Rohr. 2013. "Review and Synthesis of the Effects of Climate Change on Amphibians." *Integrative Zoology* 8 (2): 145–61. doi.org/10.1111/1749-4877.12001.

Lindenmayer & McCarthy. 2006. "Evaluation of PVA Models of Arboreal Marsupials: Coupling Models with Long-Term Monitoring Data." *Biodiversity and Conservation* 15 (13): 4079–96. doi.org/10.1007/s10531-005-3367-7.

Longcore & Simmons. 2012. "Chytridiomycota." In *ELS*. Chichester, UK: John Wiley & Sons. doi.org/DOI: 10.1002/9780470015902.a0000349.pub3.

Longcore, Pessier, & Nichols. 1999. "*Batrachochytrium dendrobatidis* gen. et sp. nov., a Chytrid Pathogenic to Amphibians." *Mycologia* 91 (2): 219. doi.org/10.2307/3761366.

Mace, Collar, Gaston, Hilton-Taylor, Akçakaya, Leader-Williams, Milner-Gulland, & Stuart. 2008. "Quantification of Extinction Risk: IUCN's System for Classifying Threatened Species." *Conservation Biology* 22 (6): 1424–42. doi.org/10.1111/j.1523-1739.2008.01044.x.

Maclean & Wilson. 2011. "Recent Ecological Responses to Climate Change Support Predictions of High Extinction Risk." *Proceedings of the National Academy of Sciences* 108 (30): 12337–42. doi.org/10.1073/pnas.1017352108.

Maisels, Strindberg, Blake, Wittemyer, Hart, Williamson, Aba'a, et al. 2013. "Devastating Decline of Forest Elephants in Central Africa." *PLoS ONE* 8 (3): e59469. doi.org/10.1371/journal.pone.0059469.

Martel, Spitzen-van der Sluijs, Blooi, Bert, Ducatelle, Fisher, Woeltjes, et al. 2013. "*Batrachochytrium salamandrivorans* sp. nov. Causes Lethal Chytridiomycosis in Amphibians." *Proceedings of the National Academy of Sciences* 110 (38): 15325–15329. doi.org/10.1073/pnas.1307356110.

McPherson, Mori, Wood, Storer, Svihra, Kelly, & Standiford. 2005. "Sudden Oak Death in California: Disease Progression in Oaks and Tanoaks." *Forest Ecology and Management* 213 (1–3): 71–89. doi.org/10.1016/j.foreco.2005.03.048.

Melamed, Israely, & Paran. 2018. "Challenges and Achievements in Prevention and Treatment of Smallpox." *Vaccines* 6 (1): 8. doi .org/10.3390/vaccines6010008.

Miller, Hayes, Ratan, Peterson, Wittekindt, et al. 2011. Causes Lethal Chytridiomycosis in Amphibians." *Proceedings of the National Academy of Sciences* 108 (3): 12348–53. rg /cgi/doi/10.1073/pnas.1102838108

Morris, Wright, Grueber, Hogg, & Belov. 2015. "Lack of Genetic Diversity Across Diverse Immune Genes in an Endangered Mammal, the Tasmanian Devil (*Sarcophilus harrisii*)." *Molecular Ecology* 24 (15): 3860–72. doi.org /10.1111/mec.13291.

Ødegaard. 2000. "How Many Species of Arthropods? Erwin's Estimate Revised." *Biological Journal of the Linnean Society* 71 (4): 583–97. doi.org/10.1006/bijl.2000.0468.

O'Hanlon, Rieux, Farrer, Rosa, Waldman, Bataille, Kosch, et al. 2018. "Recent Asian Origin of Chytrid Fungi Causing Global Amphibian Declines." *Science* 360 (6389): 621–27. doi.org/10.1126/science.aar1965.

Ojanen, Nieminen, Meyke, Pöyry, & Hanski. 2013. "Long-Term Metapopulation Study of the Glanville Fritillary Butterfly (*Melitaea cinxia*): Survey Methods, Data Management, and Long-Term Population Trends." *Ecology and Evolution* 3 (11): 3713–37. doi.org /10.1002/ece3.733.

Pimm, Raven, Peterson, Sekercioglu, & Ehrlich. 2006. "Human Impacts on the Rates of Recent, Present, and Future Bird Extinctions." *Proceedings of the National Academy of Sciences* 103 (29): 10941–46. doi.org/10.1073 /pnas.0604181103.

Possingham, Lindenmayer, & Norton. 1993. "A Framework for the Improved Management of Threatened Species Based on Population Viability Analysis (PVA)." *Pacific Conservation Biology* 1 (1): 39–45.

Reed, Grady, Brook, Ballou, & Frankham. 2003. "Estimates of Minimum Viable Population Sizes for Vertebrates and Factors Influencing Those Estimates." *Biological Conservation* 113: 23–34.

Rizzo, Garbelotto, Davidson, Slaughter, & Koike. 2002. "*Phytophthora ramorum* as the Cause of Extensive Mortality of *Quercus* spp. and *Lithocarpus densiflorus* in California." *Plant Disease* 86 (3): 205–14. doi.org/10 .1094/pdis.2002.86.3.205.

Rodrigues, Pilgrim, Lamoreux, Hoffmann, & Brooks. 2006. "The Value of the IUCN Red List for Conservation." *Trends in Ecology and Evolution* 21 (2): 71–76. doi.org/10.1016/j .tree.2005.10.010.

Rosenblum, James, Zamudio, Poorten, Ilut, Rodriguez, Eastman, et al. 2013. "Complex History of the Amphibian-Killing Chytrid Fungus Revealed with Genome Resequencing Data." *Proceedings of the National Academy of Sciences* 110 (23): 9385–90. doi.org /10.1073/pnas.1300130110.

Simberloff & Abele. 1982. "Refuge Design and Island Biogeographic Theory: Effects of Fragmentation" 120 (1): 41–50.

Sinervo, Méndez-de-la-Cruz, Miles, Heulin, Bastiaans, Cruz, Lara-Resendiz, et al. 2010. "Erosion of Lizard Diversity by Climate Change and Altered Thermal Niches." *Science* 328 (5980): 894–99. doi.org/10.1126 /science.1184695.

Skerratt, Berger, Speare, Cashins, McDonald, Phillott, Hines, & Kenyon. 2007. "Spread of Chytridiomycosis Has Caused the Rapid Global Decline and Extinction of Frogs." *EcoHealth* 4 (2): 125–34. doi.org/10.1007 /s10393-007-0093-5.

Smith, Zhou, Reeves, Barlow, Taylor, & Pitman. 2008. "*Lipotes vexillifer.*" The IUCN Red List of Threatened Species. http://www .iucnredlist.org.

Sommer. 2005. "The Importance of Immune Gene Variability (MHC) in Evolutionary Ecology and Conservation." *Frontiers in Zoology* 2: 16. doi.org/10.1186/1742-9994-2-16.

Sutton & Morgan. 2009. "Functional Traits and Prior Abundance Explain Native Plant Extirpation in a Fragmented Woodland Landscape." *Journal of Ecology* 97 (4): 718–27. doi .org/10.1111/j.1365-2745.2009.01517.x.

Thomas, Cameron, Green, Bakkenes, Beaumont, Collingham, Erasmus, et al. 2004. "Extinction Risk from Climate Change." *Nature* 427 (6970): 145–48. doi.org/10.1038 /nature02121.

Tilman, May, Lehman, & Nowak. 1994. "Habitat Destruction and the Extinction Debt." *Nature* 371: 65–66. DOI: 10.1038/371065a0

Tonge & Bloxam. 1991. "A Review of the Captive-Breeding Programme for Polynesian Tree Snails *Partula* spp." *International Zoo*

Yearbook 30 (1): 51–59. doi.org/10.1111/j.1748-1090.1991.tb03465.x.

Turvey, Pitman, Taylor, Barlow, Akamatsu, Barrett, Zhao, et al. 2007. "First Human-Caused Extinction of a Cetacean Species?" *Biology Letters* 3 (5): 537–40. doi.org/10.1098/rsbl.2007.0292.

United States Fish and Wildlife Service. 2007. "*Cyanea superba* (Haha): 5-Year Review, Summary and Evaluation." http://ecos.fws.gov/docs/five_year_review/doc1131.pdf.

Urban. 2015. "Accelerating Extinction Risk from Climate Change." *Science* 348 (6234): 571–73.

Vellend, Verheyen, Jacquemyn, Kolb, Calster, Peterken, & Hermy. 2006. "Extinction Debt of Forest Plants Persists for More than a Century Following Habitat Fragmentation." *Ecology* 87 (3): 542–48.

Wake & Vredenburg. 2008. "Are We in the Midst of the Sixth Mass Extinction? A View from the World of Amphibians." *Proceedings of the National Academy of Sciences* 105 (supplement 1): 11466–73. doi.org/10.1073/pnas.0801921105.

Wilson. 1999. *The Diversity of Life*. New York: W.W. Norton.

Winship & Trites. 2006. "Risk of Extirpation of Steller Sea Lions in the Gulf of Alaska and Aleutian Islands: A Population Viability Analysis Based on Alternative Hypotheses for Why Sea Lions Declined in Western Alaska." *Marine Mammal Science* 22 (January): 124–55.

Community-Level Responses

Learning Outcomes

After working with this chapter, you will be able to:

- Distinguish among different types of species interactions.
- Evaluate how global change stressors can disrupt species interactions.
- Assess the significance of coextinction and cascading effects.
- Apply your knowledge to real-world case studies and interpret data from recent scientific studies.

THE BLANK PAGE

When individual species are impacted by global change pressures, the surrounding biological community can experience ripple effects. For each core response (move, adjust, adapt, die), brainstorm one way that changes in one species could impact interacting species. Be as specific as possible about the types of biological interactions that might be perturbed.

INTRODUCTION

The response of individual species is not divorced from factors at the community, ecosystem, and biosphere levels. In fact, species responses influence—and are influenced by—higher-level processes. In Unit II, we focused primarily on factors at the species level and below. We saw how molecular, organismal, population, and species traits influenced different core responses. However, the response an individual species exhibits—whether to move, adjust, adapt, or die—will have community- and ecosystem-level reverberations. Similarly, characteristics of the community, ecosystem, and global biosphere will help determine species-level responses. The reciprocal interaction of biotic responses across levels is a hallmark of Global Change Biology.

As a reminder from our earlier exploration of biological levels: a **community** is a set of interacting species in a particular area; an **ecosystem** is comprised of the biological community and the surrounding abiotic environment; and the **biosphere** is the sum total of all ecosystems on the planet. In Unit III, we will evaluate signatures of global change at these broader scales to see how systems as a whole respond to global change perturbations. The goal of this chapter is to evaluate how global change pressures impact biological communities—primarily through altering species interactions. The following chapter will then address shifts at ecosystem, biome, and biosphere levels. Ultimately, integration across all biotic levels leads to a holistic understanding of global change.

WHAT ARE KEY TYPES OF BIOLOGICAL INTERACTIONS?

Before we evaluate how global change pressures alter species interactions, let us review the fundamental ways that organisms interact. All organisms are embedded in complex webs of biological interaction. Individuals *within* a species interact in numerous ways, and these **intraspecific** interactions can be cooperative (like cooperative hunting, breeding, or territory defense) or competitive (like competing for nutrient resources or mates). Organisms also interact *across* species boundaries. These **interspecific** interactions also take many forms and are the focus of this chapter.

Six major classes of biological interaction are illustrated in Table 9.1. These classes are defined by whether the interaction has a positive (beneficial), negative (detrimental), or neutral (negligible) effect for each partner in the interaction. For example, negative effects from predation, competition, or parasitism could include reduced body condition, biomass, reproductive output, or survivorship. However, even negative effects are extremely important both ecologically and evolutionarily as they contribute to population size regulation and the biotic context for natural selection. Therefore, "negative" and "positive" are not used as value judgements in this context.

It is important to recognize that these six categories of biotic interaction are idealized, and in reality, the dynamics of species interaction can be more complex. The named biotic interactions are extremes on a continuum and interactions can be of different strengths. There are several additional nuances when categorizing biotic interactions. Let us look at a few examples.

Facultative Versus Obligate Interactions

On the extreme end of interaction strength, some species rely on an interacting partner for their very survival. Examples of these **obligate** interactions include trophic specialists (like an insect that can feed only on a single plant) and organisms that live inside

TABLE 9.1 Key Categories of Species Interactions

Type of Interaction	Effect on Species 1	Effect on Species 2
Mutualism	positive	positive
Commensalism	positive	neutral
Neutralism	neutral	neutral
Exploitation	positive	negative
Amensalism	negative	neutral
Competition	negative	negative

Question: Can you think of a specific example of each of the six basic types of species interactions?

Source: Adapted from Holland & DeAngelis 2009

other organisms (like a parasite that can reproduce only in a single host species). Other species interactions are not essential but rather occur optionally. Examples of these **facultative** interactions include trophic generalists (like omnivores that can exploit many different food resources or parasites that can make use of many different hosts). Of course, biotic interactions can be facultative for both partners, obligate for both partners, or facultative for one and obligate for the other.

Multispecies Interactions

Categorizing biological interactions in a simple pairwise framework overlooks the fact that species interact in complex multispecies webs. For example, some interactions have more than two key players. One classic example is the interaction between three-toed sloths, algae, and moths, which is illustrated in Figure 9.1. The green algae that grows on the sloth's fur provides food for the sloth, the sloth's fur and feces provide habitat for different life stages of the moth, and decomposing moths provide nutrients

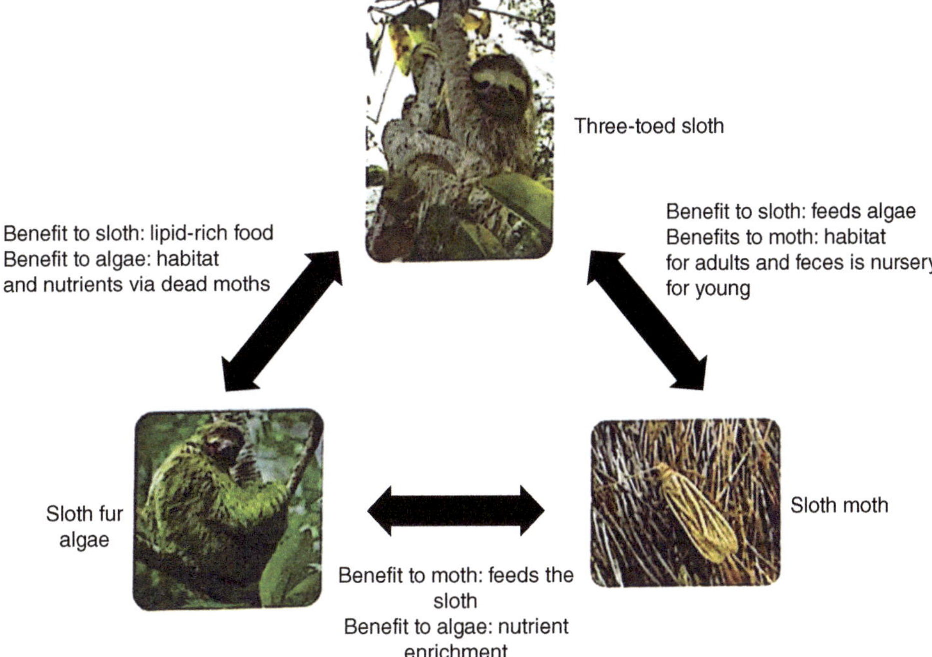

FIGURE 9.1 Three-way mutualism between three-toed sloths, sloth moths, and algae. The sloth moth (*Cryposes* spp.) appears to be an obligate mutualist, with all life stages interacting with the sloth host. Adult moths live and mate in sloth fur, eggs are deposited in sloth dung, and larvae develop entirely in these feces. The moths contribute to high nutrient resources in sloth fur, which then facilitate algal growth. Sloths in turn eat these algae to supplement their diet.

Reflection: Would you predict that facultative or obligate interactors would be more impacted by global change stressors? Why?

Source: Pauli Jonathan N., Mendoza Jorge E., Steffan Shawn A., Carey Cayelan C., Weimer Paul J. and Peery M. Zachariah A syndrome of mutualism reinforces the lifestyle of a sloth281Proc. R. Soc. B

for the algae (Pauli et al. 2014). Thus, the strength and consequences of an interaction between two species may depend on whether other interacting partners are present.

Direct Versus Indirect Interactions

One species may affect many other species, either because it has many direct interaction partners or due to indirect effects. A **direct effect** is the impact one species has on another without being mediated by any other species. The effect of a predator on its prey and an herbivore on its host plant are examples of direct effects. An **indirect effect** is the impact of one species on another despite not being direct interaction partners. For example, chains of biotic interaction can link multiple species, even if they do not all interact directly. A classic illustration of this is a **trophic cascade**, where a predator has a negative effect on its prey and thus a positive effect on plant primary productivity (by suppressing herbivory). Indirect effects can also occur if one species modifies the behavior of a second species, which impacts a third species. For example, a prey species might change its foraging habits when a predator is present.

In Sum

Ultimately, it is the aggregate of all biotic interactions that defines an ecological community. However, species interactions are not static through space or time. For example, the strength and impact of interactions can vary geographically, seasonally, or by life stage. Moreover, species interactions can be perturbed by global change stressors, a topic we will turn to next.

HOW DO GLOBAL CHANGE PRESSURES AFFECT BIOLOGICAL INTERACTIONS?

Global change stressors impact biological interactions in numerous ways. Environmental change can alter which species are present in an ecosystem and how those species interact over space and time. These changes in community structure can perturb ecological interactions in any number of fashions (Kiers et al. 2010; Traveset & Richardson 2014). For example, the type of interaction can be altered (for example, a previously mutualistic interaction becomes antagonistic). Alternatively, species can change their interaction partners (for example, to exploit different food resources or hosts). Finally, interactions can collapse entirely, threatening the persistence of one or more interacting partners. Here we will look at three fundamental causes of altered species interactions.

Effects of Species Loss

The die response can lead to the subtraction of species from a community. Population declines reduce the abundance of a species in a community, and extirpation removes a species entirely from its historical interaction network. We have already looked at some examples of how species declines have disrupted biotic interactions, for example in the *Meet the Data* feature in Chapter 4 on bee declines. In fact, parallel declines in pollinators and the plants they pollinate provide a clear demonstration of how loss of a key mutualist can negatively impact other species in a community (e.g., Biesmeijer et al.

2006; Portman et al. 2018). We will further evaluate the consequences of extinction on interaction partners in the next section of this chapter.

Species loss can occasionally have positive effects on some interaction partners, for example if species are released from predation or competition. For example, when cougar (*Puma concolor*) populations were reduced in Zion National Park due to increased tourism, mule deers (*Odocoileus hemionus*) were released from predation (Ripple & Beschta 2006). Although this had a positive impact on deer populations, it had catastrophic effects on other members of the community. Without deer populations controlled, intense herbivory altered vegetation landscape dramatically. Ultimately, this decreased the abundance of numerous other species, including butterflies, frogs, and lizards (Ripple & Beschta 2006). Thus, even positive growth of some species can be detrimental to a community as a whole.

Effects of Species Gain

The move response can lead to the addition of species to a community. Over evolutionary timescales, dispersal has always provided opportunities for species additions to local communities. However, as we evaluated in Chapter 5, human-mediated transport of species outside of their native range has dramatically accelerated the rate of biological invasions. Invasive species can have a variety of direct negative effects on native species, whether they serve as herbivores, predators, or competitors (Traveset & Richardson 2014). Of course, species gains can also have positive effects on some species if they provide new resources that can be exploited or new mutualistic partnerships.

Invasive species also have numerous indirect effects. For example, invaders often contribute to the movement and transmission of pathogens (e.g., Young et al. 2017). Invasive species can bring new pathogens and parasites into an ecosystem, which can then begin to infect native species (referred to as spillover). Invasive species can also amplify the effects of existing pathogens if they increase rates of pathogen transmission by providing additional hosts (referred to as amplification). Of course, invasive species can also potentially lessen disease risk for native species, for example if they interfere with pathogen transmission (referred to as the dilution effect; Civitello et al. 2015).

One example of the role of invasive species in disease spread is the invasive bullfrog (*Lithobates catesbeianus*), which can serve as a vector of the chytrid fungus (*Batrachochytrium dendrobatidis*). As we evaluated in Chapter 8, the chytrid fungus has been responsible for amphibian declines worldwide. Bullfrogs are highly competent reservoirs for Bd as they typically carry and transmit the pathogen without succumbing to the disease (e.g., Miaud et al. 2016). Because infected bullfrogs are farmed and traded globally, they may have played an important role in introducing Bd to previously naive amphibian communities (e.g., Garner et al. 2006).

The arrival of an invasive species also can disrupt existing species interactions. For example, many native ant species play important ecological roles as mutualists, tending and dispersing native plant partners (Ness & Bronstein 2004). As we saw in Chapter 5, invasive ant species often displace native ants, which can dismantle these key mutualisms (Rodriguez-Cabal et al. 2009, 2012). One specific example—a meta-analysis describing the effects of the invasive Argentine ant (*Linepithema humile*)—is shown in Figure 9.2. Sites with invasive Argentine ants had 92% fewer native ant seed

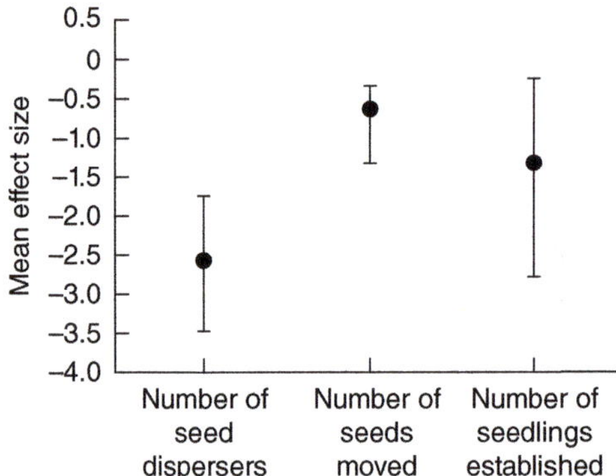

FIGURE 9.2 Invasive species can disrupt existing mutualisms. Shown here are results from a meta-analysis looking at the effects of the invasive Argentine ant (*L. humile*) on native ant species and their role as seed dispersers. Many native ants disperse seeds and contribute to seedling establishment. Argentine ants outcompete many of these native species without replacing their seed dispersal services. The graph shows mean effect size (and confidence intervals) for three aspects of the plant–ant mutualism: number of seed dispersers (left), number of seeds moved (middle), and number of seedlings established (right).

Reflection: All of the response ratios are significantly negative, and none of the confidence intervals overlaps zero. What does this mean about the overall effect of the invasive Argentine ant on seed dispersal mutualisms?

Source: Rodriguez-Cabal Mariano A., Stuble Katharine L., Nuñez Martin A. and Sanders Nathan J. Quantitative analysis of the effects of the exotic Argentine ant on seed-dispersal mutualisms5Biol. Lett.

dispersers and correspondingly lower metrics of seeds moved and seedlings established (Rodriguez-Cabal et al. 2009). Thus, invasive species do not always perform similar ecological functions as the native species they displace, altering species interactions and community structure.

Effects of Species Change

The adjust and adapt responses can lead to changes in physiological, developmental, morphological, or behavioral traits that impact where and when species are found. For example, as we evaluated in Chapter 6, a common signature of global warming is temporal shifts in development or activity season. One recent large-scale study looked at >10,000 long-term phenological datasets (more than 20 years) across both aquatic and terrestrial species (Thackeray et al. 2016). As shown in Figure 9.3, the scientists found that phenological events are expected to happen earlier in the year, as expected. However, they also found significant variation across taxonomic groups and trophic levels. For example, phenological shifts were of greater magnitude for primary consumers compared to primary producers.

If consumers in a particular community are active earlier in the season than their food resources, this could lead to severe **trophic mismatch**. Numerous observational studies have demonstrated trophic mismatches in biomes around the world (reviewed in Renner & Zohner 2018). One example comes from Arctic geese where mean gosling

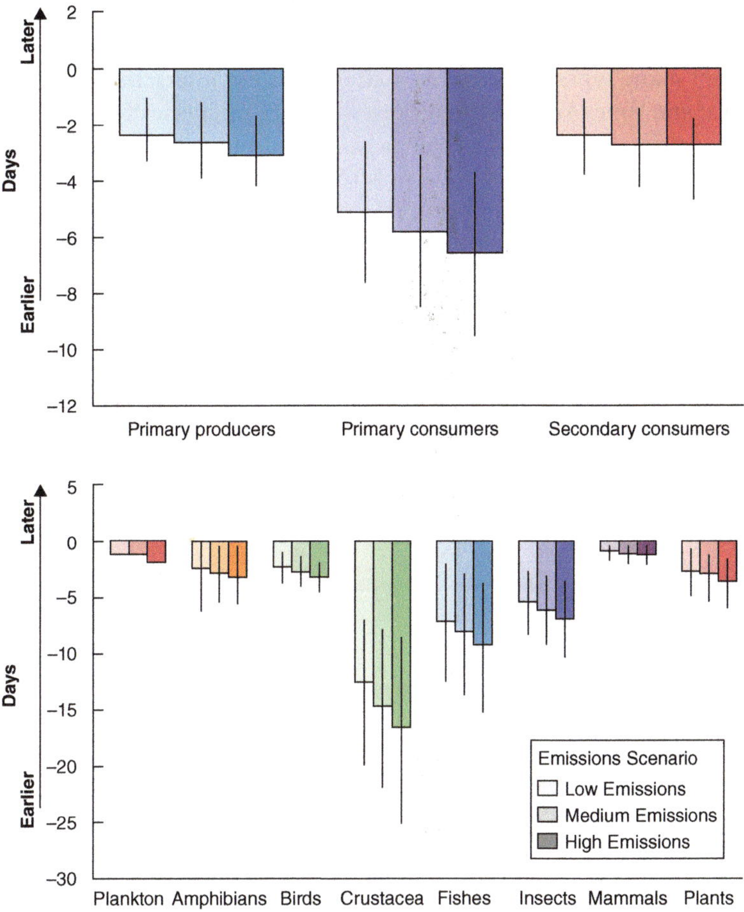

FIGURE 9.3 Results from a large-scale study with estimates of phenological shifts predicted by 2050. Estimates are for different trophic levels (top) and different taxonomic groups (bottom). For each category, three different climate change projections were used based on alternate emissions scenarios. Bars are mean predicted responses, and whiskers represent 90th percentile extremes.

Reflection: Why are some trophic levels and taxonomic groups more susceptible to phenological shifts than others? What is a specific scenario whereby a modest phenological shift of a few days could still have grave consequences?

Source: Stephen J. Thackeray, et. al, Phenological sensitivity to climate across taxa and trophic levels, Nature volume535, pages241–245 (14 July 2016)

hatch date is increasingly mismatched with the peak of their food resources (Ross et al. 2017). Both plant and bird phenologies are advancing to occur earlier in the year. However, the advancement of goose phenology appears to be constrained by the increasing predation of eggs by polar bears (which we discussed in Chapter 6). Therefore, the goose hatch date is not advancing as fast as vegetation green-up, leading to a trophic mismatch and declines in gosling production (Ross et al. 2017).

Experimental studies have also shown that even a few days of trophic mismatch can have grave fitness consequences (e.g., Schenk et al. 2016). Ultimately, changes in phenology can disrupt many types of biotic interactions because interacting partners must overlap in space and time. Any shift that causes interacting species to lose their temporal or spatial synchrony can alter the ecological community as a whole.

Changes in phenology do not always have negative impacts on species survival. Some phenological shifts might actually bring interacting species into greater synchrony (Kharouba et al. 2018). Other phenological changes might release pressure on an interaction partner. One intriguing example comes from brown bears (*Ursus arctos*) on Kodiak Island in Alaska, illustrated in Figure 9.4 (Deacy et al. 2017). These bears rely

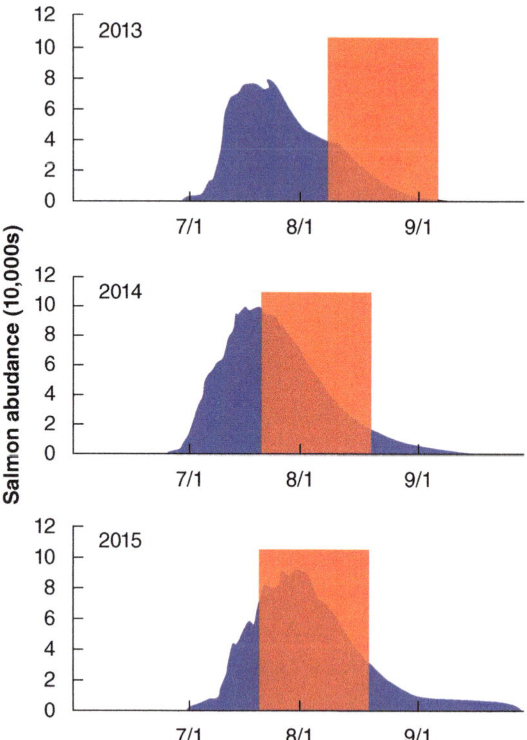

FIGURE 9.4 Phenological shifts can have unexpected effects on species interactions. The availability of one key food resource for brown bears on Kodiak Island—red elderberry (shown by red bars)—is shifting earlier in the year. It now overlaps with peak salmon availability (blue shaded regions). As a result, brown bears leave streams to forage on red elderberry, thus reducing predation on salmon.

Reflection: Both increased and decreased synchrony between interacting species can have broader effects in an ecological community. Provide one specific example of each.

Source: Resource synchronization disrupts predation William W. Deacy, Jonathan B. Armstrong, William B. Leacock, Charles T. Robbins, David D. Gustine, Eric J. Ward, Joy A. Erlenbach, Jack A. Stanford Proceedings of the National Academy of Sciences Sep 2017, 114 (39) 10432–10437

on two key resources, which historically have been seasonally complementary. Sockeye salmon are typically most available in July and August, while red elderberry is typically most abundant in late August and September. However, in recent anomalously warm years, elderberry fruited weeks earlier than usual. In response, bears altered their foraging behavior and left salmon streams to feed on berries. Thus, a phenological shift in one resource (elderberry) indirectly reduced predation on another (spawning salmon). Whether phenological shifts have positive or negative impacts on individual species, it is clear that they are currently dramatically altering food webs around the world (Bartley et al. 2019).

In Sum

Ultimately, every change in an individual species has the potential to cascade and impact other species in its community. Data increasingly suggest that it may not be hard physiological limits that put species at risk in an era of global change; rather, disruption of species interactions is one of the gravest threats (e.g., Cahill et al. 2012).

HOW DOES EXTINCTION AFFECT COMMUNITIES?

As we have just seen, the loss, gain, or change of individual species can impact entire ecological communities. Here we will evaluate an extreme example: where the loss of a single species can generate cascading effects on the species that depend on it. This phenomena is termed coextinction (or secondary extinction; Dunn et al. 2009; Brodie et al. 2014). In the extreme, loss of a single species can drive the extinction of numerous other affiliate species. In practice, it can be difficult to detect coextinction (Colwell et al. 2012; Brodie et al. 2014), but considering the effect of extinction on different types of biological interactions is essential given the interconnectedness of species within ecological networks.

Coextinction of Mutualists

Many ecological interactions are mutualistic, where both interacting species benefit. As we discussed earlier in the chapter, obligate mutualists depend on each other for their very survival. Darwin famously ruminated on the possibility that the extinction of mutualists—like pollinators—could endanger their host plants. More than 150 years ago, Darwin described an unusual orchid (*Angræcum sesquipedale*) in Madagascar with nectar stored nearly a foot deep in a specialized tubular structure. He hypothesized that there must be moths with probosces long enough to pollinate these orchids. As shown in Figure 9.5, these moths do exist (*Xanthopan morganii*) and were discovered after Darwin's death. Darwin conjectured: *If such great moths were to become extinct in Madagascar, assuredly the* Angraecum *would become extinct (Darwin 1862).*

Although decline in pollinators or seed dispersers has not unambiguously been tied to global extinction of their ecological partners, it is clear that anthropogenic stressors can disrupt these mutualistic interactions and lead to local declines. For example, in a meta-analysis, Markl et al. (2012) integrated >80 comparisons of animal-mediated seed dispersal between disturbed and undisturbed forests. The

FIGURE 9.5 (A) The orchid *Angræcum sesquipedale* and (B) its sphinx moth pollinator *Xanthopan morganii*. These species have not gone extinct but are iconic examples of pollinator–plant coevolution and led Darwin to speculate about the possibility of coextinction.

Reflection: What general factors are likely to increase the risk of coextinction?

Source: (A) Credit: Wellcome Collection. CC BY; (B) Nick Garbutt/NaturePL/Science Source

authors demonstrated that hunting and logging were associated with decreased seed dispersal. Moreover, they showed that there were different effects on large-seeded and small-seeded plants. Animals large enough to eat large fruits are often disproportionately affected by hunting and habitat fragmentation. Thus, certain plant species (like those with large seeds) may be disproportionately at risk of coextinction. We will look at another clear example of the impacts of disrupting pollination and seed dispersal in this chapter's *Meet the Data* feature.

Coextinction of Parasites

Largely unrecognized until recently, extinction of host species also can have catastrophic effects on parasites (Dunn et al. 2009; Strona 2015; Cizauskas et al. 2017). For example, we do not often consider the number of lice species that might be lost with their endangered vertebrate hosts. However, both micro- and macroparasites and pathogens can exhibit high host specificity, and many parasite species often exploit a single host species. Therefore, as shown in Figure 9.6, parasite coextinctions can represent significant loss of biodiversity (Koh et al. 2004).

An elegant example comes from Wood et al. (2013), who used ancient DNA and microscopic analysis of preserved feces to identify gastrointestinal parasites from four species of extinct moa birds. As illustrated in Figure 9.6, the moa birds were large, flightless birds in New Zealand, some of which grew to more than 200 kilograms (nearly 500 pounds). They were rapidly extirpated in the late 13th century, likely by hunting and habitat loss following colonization by the Polynesians (Allentoft et al. 2014). By identifying host-specific parasites, Wood et al. inferred that Late Quaternary megafaunal extinctions likely had cascading effects on parasites. Given high rates of contemporary extinction, the loss of parasites and pathogens has likely accelerated dramatically. Although they are often considered

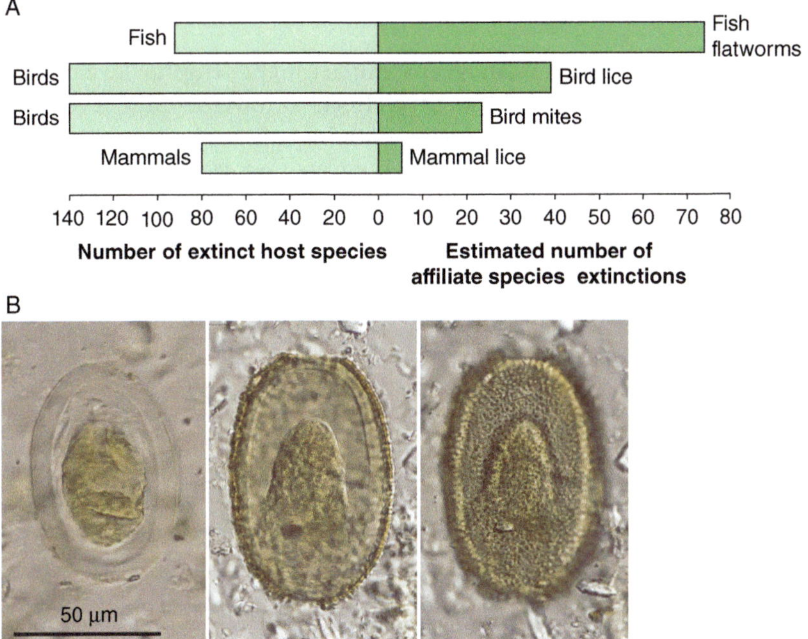

FIGURE 9.6 Coextinction of parasites and pathogens. (A) Predictions of extinctions of affiliate species based on the number of recorded contemporary host species extinctions. (B) Scientists have been able to identify now extinct parasites using ancient DNA and analysis of fossilized materials. For example, helminth parasite eggs were recovered from feces naturally preserved for hundreds of years since the extinction of New Zealand moa birds.

Reflection: Why should we care from ecological and evolutionary perspectives about the extinction of parasites and pathogens?

Source: (A) Species Coextinctions and the Biodiversity Crisis. BY LIAN PIN KOH, ROBERT R. DUNN, NAVJOT S. SODHI, ROBERT K. COLWELL, HEATHER C. PROCTOR, VINCENT S. SMITH. SCIENCE 10 SEP 2004 : 1632-1634; (B) Wood JR, Wilmshurst JM, Rawlence NJ, Bonner KI, Worthy TH, Kinsella JM, et al. (2013) A Megafauna's Microfauna: Gastrointestinal Parasites of New Zealand's Extinct Moa (Aves: Dinornithiformes). PLoS ONE 8(2)

to be simply nuisance species, the ecological impacts of losing parasites and pathogens can be far-reaching.

Coextinction of Predators and Herbivores

Trophic specialization is a key risk factor for coextinction. Consumers at any trophic level that feed on one or few food resources are particularly vulnerable if conditions change and their nutrient source is impacted. One example is the koala (*Phascolarctos cinereus*), which is not only a trophic specialist on *Eucalyptus* foliage but also exhibits low dispersal and high site fidelity. In one study, scientists demonstrated that in a year where the local eucalyptus trees were defoliated, >70% of the koala population died (Whisson et al. 2016). Thus, trophic specialists—both herbivores and predators—can be especially sensitive to disturbances within their ecological communities.

Loss of nutrient resources can clearly impact higher trophic levels, but loss of a key herbivore or predator can also have effects at lower trophic levels (e.g., Gilljam et al. 2015). We will evaluate the potential for these ripple effects in the next section on cascading effects. In addition, the *Core Concepts* feature of this chapter provides a review of **trophic levels** and **top-down effects** versus **bottom-up effects**. Ultimately, with any one extinction, dynamics of predation, herbivory, and competition among remaining species change, potentially leading to additional species losses.

CORE CONCEPTS

WHAT ARE ABOVE- AND BELOW-GROUND FOOD WEBS?

The **trophic structure** of a community refers to how energy and matter flow between hierarchical levels. These **trophic levels** describe the position of an organism in a local food web. A simple illustration is provided in Box Figure 9.1. **Producers** convert energy from the sun into biomass via photosynthesis. Then **consumers** obtain this energy by feeding on producers or other consumers at lower trophic levels.

Simplistic depictions of trophic structure or food webs in terrestrial ecosystems typically highlight energy flows in the community living above ground. However, there is a rich below-ground community, including bacteria, fungi, lice, worms, moles, and many other lineages (examples shown in Box Figure 9.1). Many members of this below-ground community serve as **decomposers** and play an essential role in breaking down organic matter and nutrient cycling.

Recognizing the distinct processes that occur above and below ground, ecologists sometimes distinguish between the **green food web** (which is anchored by photosynthetic plants and occurs primarily above ground) and the **brown food web** (which is anchored by decomposers and occurs primarily below ground). The green food web produces dead organic matter (such as feces, dead organisms, shed skin, and leaf litter), which is then decomposed in the brown food web. The brown food web releases nutrients essential for primary producer growth. Moreover, microbivores (typically arthropods) feed directly on decomposers and then can be eaten by both above- and below-ground predators. Ultimately, the green and brown food webs (and their above- and below-ground processes) are inextricably linked (Zou et al. 2016).

All ecological communities—whether above or below ground—are structured by numerous factors (e.g., Oksanen et al. 1981). **Top-down effects** are those imposed by higher trophic levels (for example, a predator limiting prey population size and keeping herbivory in check). **Bottom-up effects** are those imposed by resource availability (for example, nitrogen limiting primary productivity and thus keeping consumer populations in check). Ecologists have long debated whether resource limitation or predation pressure is most important (e.g., Wilkinson & Sherratt 2016), but it is clear that both top-down and bottom-up effects interact to structure ecological communities (e.g., Lynam et al. 2017). Box Figure 9.2 illustrates top-down and bottom-up effects in a marine food web.

Of course, changes to ecological communities can occur when any element is perturbed. When a key community member is lost—whether a producer, consumer, or decomposer—the impacts can ripple throughout the food web. For example, as we will see in the *Taking a Closer Look* feature, human activities like hunting have removed apex predators from many terrestrial and marine ecosystems. Simultaneously, hunting has established humans as a new top predator in many ecosystems, radically altering community structure and composition. Ultimately, to understand the impacts of anthropogenic stressors on biological communities, it is essential to consider complex interactions between trophic levels and processes occurring both above and below ground.

BOX FIGURE 9.1 Trophic structure in ecological communities. (A) Basic illustration of trophic levels within a terrestrial ecological community. (B) Often terrestrial trophic diagrams omit below-ground species. However, there is a wealth of micro- and macroscopic organisms in the soil (from bacteria to fungi to tardigrades to worms to centipedes to moles) contributing to decomposition and nutrient cycling in food webs.

Reflection: Create an analogous illustration of trophic levels for an aquatic ecosystem.

Source: (A) http://cesonoma.ucanr.edu/4H/Clubs/4-H_Projects/Wildlife_Project;

(B): 3Dstock/Shutterstock; https://www.natgeoimagecollection.com/archive/-2KWGDN1LJ82V.html; Somluck Rungaree/Shutterstock; Neil Bromhall/Shutterstock

(Continued)

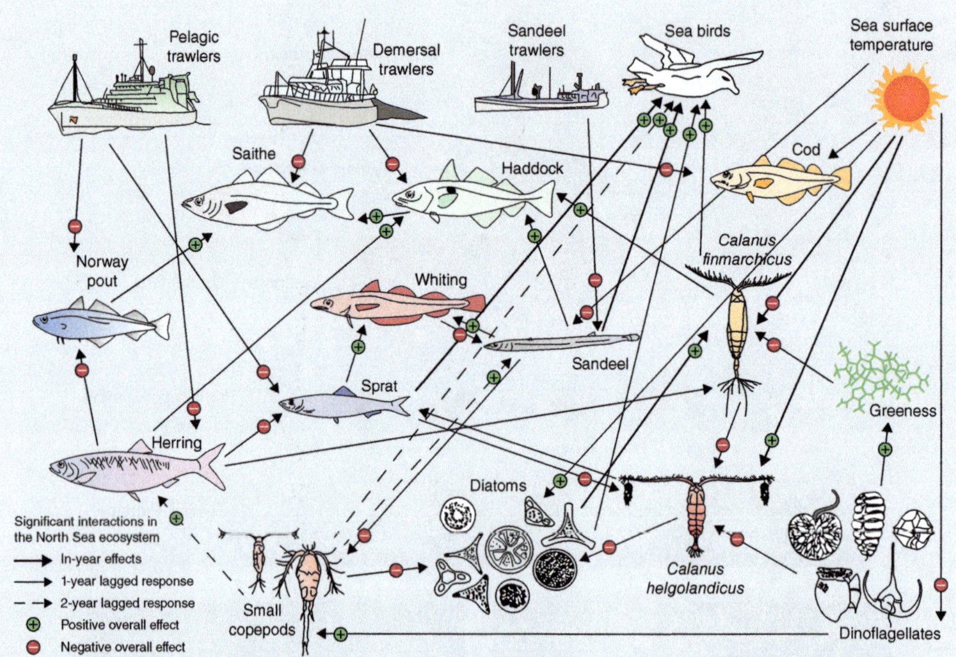

BOX FIGURE 9.2 Illustration of the complex interactions in a marine food web in the North Sea. Researchers used >40 years of data to model both top-down and bottom-up effects. Some impacts occur immediately (in-year), while others exhibit a lag effect.

Reflection: Humans function as a top predator in the North Sea. If human fishing pressure was halted, what specific prediction could you make about the impact on a species at least two trophic links away?

Source: Trophic and environmental control in the North Sea Christopher Philip Lynam, Marcos Llope, Christian Möllmann, Pierre Helaouët, Georgia Anne Bayliss-Brown, Nils C. Stenseth Proceedings of the National Academy of Sciences Feb 2017, 114 (8) 1952–1957

In Sum

Whether species are at risk of coextinction depends on a number of factors (Brodie et al. 2014). Critically, the more specialized and dependent the affiliate species is on the host, the greater the risk of coextinction. The potential for host switching can ameliorate this risk if novel partnerships are possible once a primary host has been lost. The development of new interaction partnerships depends on many factors, including the richness of the species pool and the potential for phenotypic plasticity or rapid evolution. Ultimately, it can be difficult to predict the effects of coextinction because secondary extinctions can themselves precipitate further losses, altering the structure and function of ecological systems. Yet it is critical that we consider not only the risk of extinction of individual species but also the ecological interactions they support (Valiente-Banuet et al. 2015).

WHAT ARE CASCADING EFFECTS?

As we have been discussing, often an initial global change perturbation leads to a chain of events. These **cascading effects** are often unforeseen. For example, deforestation has direct negative effects like removing habitat for organisms, but it can also have cascading effects by facilitating invasions, allowing increased access to hunting and harvesting, and changing fire regimes. One key cause of cascading effects is **biotic interaction chains**. In the previous sections, we evaluated how changes in the presence, abundance, and activity level of one species can have direct effects on immediate interaction partners. However, changes in one species can also ripple across trophic levels and have indirect effects on many additional species.

A classic example of cascading effects due to biotic interaction chains is the decline of keystone predators. When top predators are removed from a community, this directly impacts prey populations, which in turn affects rates of herbivory, plant primary productivity, and even nutrient cycling. We already saw one illustration of this with cougars in Zion National Park. In another example, Terborgh et al. (2001) studied the effects of decreased predation as apex predators (e.g., jaguars and cougars) were removed from tropical forest fragments. Forests without apex predators exhibited decreased plant recruitment and survival. The connection between top predators and plant community structure was mediated by herbivores released from predation pressure. As habitat loss and hunting disproportionately impact large-bodied carnivores, similar cascading effects can be observed in terrestrial and marine communities around the world.

Another example comes from marine rocky shoreline communities. In an experimental study, Donohue et al. (2017) tested whether loss of individual predator species would trigger coextinctions at distant trophic levels. The authors conducted a replicated experiment with a fully crossed design. They removed one of two predators (whelk and crab) and two different classes of primary consumers (grazers and mussels). With appropriate controls, they followed nine different treatment groups for more than a year and measured loss of macroalgae species. Figure 9.7 illustrates their key finding: removing a single predator led to the loss of numerous producer species several trophic links away.

Cascading effects can be triggered not only by species losses but also by species additions. Moreover, cascading effects can be caused by changes at other trophic levels, not only by changes in predation. Figure 9.7 illustrates changes in forest understory caused by introduced earthworms in a temperate forest (Wardle et al. 2011). Although earthworms are often beneficial for plant growth, invasive earthworms alter nutrient availability, water infiltration, and root exposure (Hale et al. 2008; Craven et al. 2017). Ultimately, the effects of this one introduced decomposer have extensive cascading effects.

We will evaluate additional examples of biotic interaction chains in the special features of this chapter. A classic example of losing a keystone predator in marine kelp forests is highlighted in the *Taking a Closer Look* feature. The role of introduced species in disrupting mutualisms in tropic forests is analyzed in the *Meet the Data* feature.

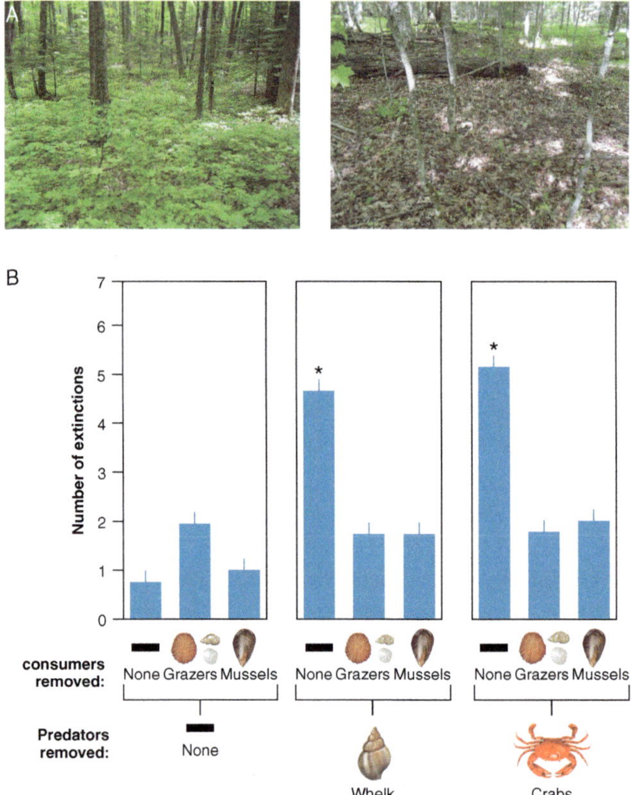

FIGURE 9.7 Cascading effects precipitated by the loss or gain of key species. (A) Forests that have been colonized by an invasive earthworm (*Lumbricus terrestris*) show reduced understory (right) compared to the typically lush understory in native oak forests (left). Dramatic shifts in vegetation are also observed in tropical forest fragments that have lost apex predators where increased herbivory has led to decreased plant productivity. (B) Number of extinctions of macroalgal taxa in nine different pairwise treatment groups where predators and consumers were experimentally removed from a rocky intertidal community. Scientists removed different combinations of primary consumers (both filter feeders and grazers) and predators (like whelk and crabs). When primary consumers alone were removed, there was no significant effect. When predators alone were removed, many producers went extinct. When predators and primary consumers were removed, the impact on producers was alleviated.

Reflection: Why is the marine intertidal experiment an example of a cascading effect?

Source: (A) Trophic Downgrading of Planet Earth. BY JAMES A. ESTES, JOHN TERBORGH, JUSTIN S. BRASHARES, MARY E. POWER, JOEL BERGER, WILLIAM J. BOND, STEPHEN R. CARPENTER, TIMOTHY E. ESSINGTON, ROBERT D. HOLT, JEREMY B. C. JACKSON, ROBERT J. MARQUIS, LAURI OKSANEN, TARJA OKSANEN, ROBERT T. PAINE, ELLEN K. PIKITCH, WILLIAM J. RIPPLE, STUART A. SANDIN, MARTEN SCHEFFER, THOMAS W. SCHOENER, JONATHAN B. SHURIN, ANTHONY R. E. SINCLAIR, MICHAEL E. SOULÉ, RISTO VIRTANEN, DAVID A. WARDLE. SCIENCE 15 JUL 2011: 301–306; (B) © 2017 John Wiley & Sons Ltd

CONCLUSION

Ecological communities are governed by complex webs of biotic interactions. Dynamics of competition, predation, parasitism, mimicry, and other interactions among species play a critical role in the response of individual species to environmental change. Moreover, the response of any one species to global change stressors can reverberate throughout the community. Gains or losses of species—and changes in the spatial or temporal overlap between interacting partners—can disrupt long-standing ecological relationships. Moreover, changes in any one ecological partnership can have cascading effects that ripple through the community.

Of course, environmental changes can also lead to new opportunities and new interaction partnerships, as some species exploit new resources due to range changes or host switching. However, most species interactions are the result of millennia of coevolution. Thus, disassembly of species interactions is not likely to be quickly offset by the development of new interaction partnerships. Ultimately, changes in species interactions can have dramatic effects not only throughout a community but can also impact ecosystem-level processes, a topic we will explore in the next chapter.

MEET THE DATA THE COLLAPSE OF MUTUALISMS

Who Are the Scientists and What Did They Set Out To Do?

Here we examine a collaborative study conducted by scientists from Europe, North America, and South America. This study, entitled "Node-by-Node Disassembly of a Mutualistic Interaction Web Driven by Species Introductions," was published by Rodriguez-Cabal et al. in the journal *Proceedings of the National Academy of Sciences* in 2013 (Box Figure 9.3). As we explored earlier in this chapter, gaining or losing species in an ecosystem can disrupt key **mutualisms**. Moreover, changes in any one species can have cascading effects throughout an entire biological community. This study set out to understand the complex direct and indirect effects of invasive species in the forests of Patagonia, Argentina.

The story centers on two interacting plant species. The first is an evergreen shrub endemic to Chile and Argentina (*Aristotelia chilensis*). The second is a mistletoe species (*Tristerix corymbosus*), which grows as an aerial parasite, rooting into the branches of the host shrub. These two interacting plant species in turn support at least three other key mutualisms. The white-crested elaenia bird (*Elaenia albiceps*) eats the berries of the host shrub and disperses their seeds. A small endemic opossum (*Dromiciops gliroides*) is the only known disperser for the mistletoe plant and appears to rely on mistletoe fruits as a food resource. An endemic hummingbird (*Sephanoides sephaniodes*) also has an obligate mutualism with the mistletoe: the hummingbird pollinates the mistletoe, and the mistletoe nectar is a key food resource for the hummingbird. These interactions are shown in Box Figure 9.4.

In the late 18th century, non-native herbivores (such as cows and deer) were introduced by humans to this region. The introduced herbivores foraged intensively on *A. chilensis*, reducing not only the host plant but also the mistletoe. In addition, another exotic species has more recently invaded the ecosystem: a wasp (*Vespula germanica*) that feeds on the fruits on the host shrub. These introduced herbivores and competitors impact the resources upon which the native hummingbird, opossum, and elaenia bird rely. On this backdrop, Rodriguez-Cabal and colleagues set out to quantify the impact of introduced species on endemic pollinator and seed-dispersal mutualisms.

What Are *Your* Predictions?

Before you read on, take a few minutes to apply what you have learned about species interactions and cascading effects to this case study.

- *With the addition of non-native deer and wasps, do you expect populations of the three native*

(Continued)

BOX FIGURE 9.3 Dr. Mariano Rodriguez-Cabal led a collaborative international team of researchers to understand how invasive species could alter mutualistic interactions between native plants (like mistletoe) and endemic animals (like opossums) in the forests of Patagonia.

Source: Mariano Rodriguez-Cabal

vertebrates (hummingbird, elaenia bird, opossum) to increase or decrease?

- *What specific direct and indirect effects do you expect the introduced species to have on the native vertebrates?*
- *Do you expect direct or indirect effects of the introduced species to be stronger? Why?*

Now write a summary prediction statement about how the introduced species are expected to perturb this local interaction web.

What Were the Scientists' Predictions?

The authors did not present a formal prediction in their study, but rather an objectives statement: "*Examining the effects of species gains and losses is fundamental to understanding the assembly and disassembly of ecological communities in a changing world. However, field-based empirical studies that demonstrate the disassembly of mutualistic webs are exceedingly rare. In this study, we take advantage of an ongoing natural experiment that links the gain of invasive species (introduced large mammals and wasps), the loss of a native keystone species (a mistletoe), and the ensuing node-by-node disassembly of an interaction web in Patagonia, Argentina.*"

What Data Were Collected?

The authors selected 26 sites in national parks to study. To understand the effects of exotic herbivores, they compared sites with introduced herbivores (invaded sites) to sites without (intact sites). In addition they conducted an exclosure experiment where fencing prevented browsing by the introduced herbivores. To understand the effects of exotic wasps, the scientists used a toxin with high specificity to remove invasive wasps in a subset of sites without impacting native arthropods. They measured a variety of outcomes in all of the plots, including the number of reproductive and seedling host plants and mistletoes, abundance of focal bird species, and opossum presence. The scientists used a variety of statistical approaches to assess

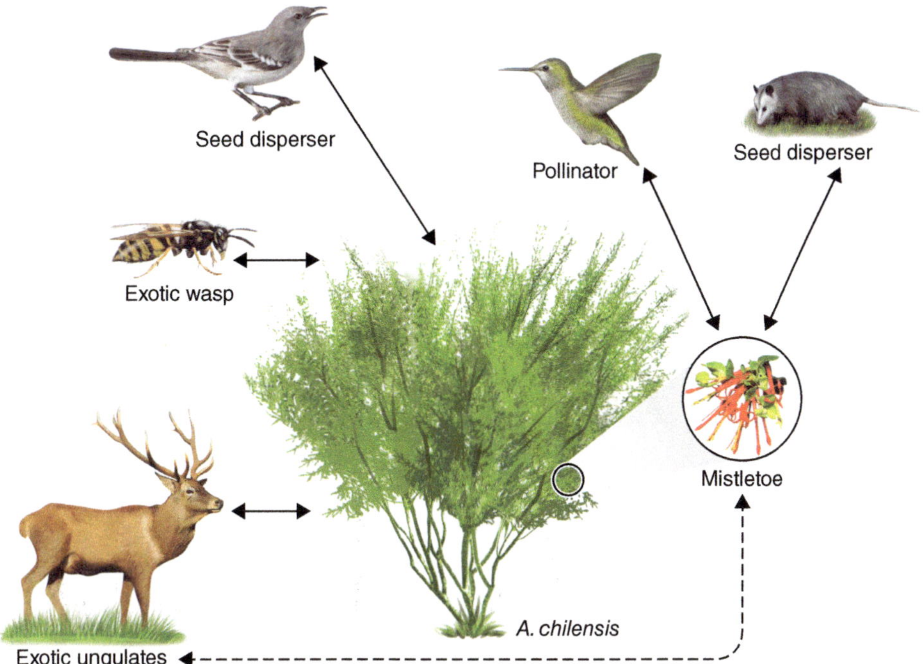

BOX FIGURE 9.4 Interactions in the Patagonia forest. Keystone plant species include an evergreen shrub (*Aristotelia chilensis*) and the mistletoe (*Tristerix corymbosus*) that grows on it. Several endemic species are key mutualists of the plants. The elaenia bird (*Elaenia albiceps*), opossum (*Dromiciops gliroides*), and hummingbird (*Sephanoides sephaniodes*) are important for pollination or seed dispersal. These mutualisms are threatened by introduced species, including exotic ungulates (like deer and cow) and invasive wasps.

Source: Disassembly of a mutualistic web Mariano A. Rodriguez-Cabal, M. Noelia Barrios-Garcia, Guillermo C. Amico, Marcelo A. Aizen, Nathan J. Sanders Proceedings of the National Academy of Sciences Oct 2013, 110 (41) 16503–16507

differences between control and treatment plots and to estimate the strength of the effect of the invasive species on the endemic species.

The authors also shared that they overcame one significant setback during their research. During the winter of 2011, they had to select a new set of sites because the eruption of a local volcano covered all of their previous sites with up to 0.3 meters of ash. This illustrates the perseverance often required to conduct large-scale field studies.

What Is *Your* Interpretation of the Data?

Before you read on, take a few minutes to interpret the data in Box Figure 9.5. Consider the following questions as you look at the figure:

- *What direct effect do introduced herbivores have on plants in Patagonia forests?*
- *What indirect effect do introduced herbivores have on hummingbirds?*
- *What impact do introduced wasps have?*

Write a sentence or two describing the *key findings* of this study and their *significance*.

What Was the Scientists' Interpretation of the Data?

Several key mutualisms in Patagonian forests were disassembled by the arrival of two exotic species. First, the exotic herbivores had dramatic *direct effects* on *A. chilensis*. The herbivores foraged intensively on *A. chilensis*, reducing both the host plant and the associated mistletoe. Second, the invasive wasps had direct effects on *A. chilensis* by eating the fruits of the host plant without dispersing the seeds.

(*Continued*)

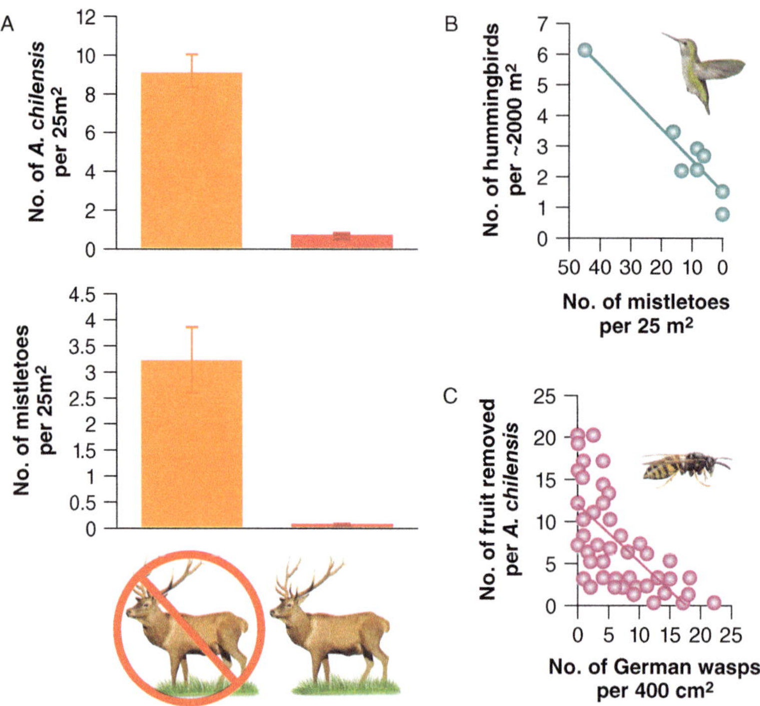

BOX FIGURE 9.5 (A) Direct effects of introduced herbivores. Comparison of shrub and mistletoe density between intact and invaded sites. (B) Association between mistletoe and hummingbird abundances. (C) Association between invasive wasp abundance and shrub seed dispersal.

Source: Disassembly of a mutualistic web Mariano A. Rodriguez-Cabal, M. Noelia Barrios-Garcia, Guillermo C. Amico, Marcelo A. Aizen, Nathan J. Sanders Proceedings of the National Academy of Sciences Oct 2013, 110 (41) 16503–16507

These direct effects on the two key plant species led to a *cascade* of *indirect effects*. For example, there was a strong impact of introduced herbivores on hummingbird populations. Herbivores browsed on the host plant, reducing the number of mistletoes. This in turn reduced the nectar resources on which the hummingbird relies. Similarly, the invasive wasps fed on the fruits of the host shrub, reducing the food resources for the native elaenia bird. As Box Figure 9.6 shows, some of these indirect effects were even stronger than the direct effects.

It is also important to note that once native pollinators and seed dispersers have declined, this could have further impacts on Patagonian forests. For example, the reduction in hummingbird abundance reduces pollination services, potentially leading to further loss of mistletoe. Similarly, declines in elaenia birds, which disperse seeds for a large number of plants, could lead to reductions of other plant species that were not measured in this study.

Ultimately, this study shows how invasive species can disrupt entire interaction webs. In the authors' own words: *"Here we show how the addition of novel species from disparate taxa and the ensuing loss of a keystone species leads to the node-by-node disassembly of an interaction web in Patagonia, Argentina. . . . Our results demonstrate that the gains and losses of species are both consequences and drivers of global change that can lead to under-appreciated cascading coextinctions through the disruption of mutualisms."*

What Are *Your* Ideas for Future Research Directions?

Given what this study accomplished, what next steps do you envision for this research program? If you had been involved in this study, how would you follow up? Start by considering the following questions:

- *What new questions does this study raise?*
- *Can the consequences of the patterns revealed be further investigated?*
- *How could the study be broadened over spatial, temporal, or taxonomic scales to achieve more generality?*

Now write a few sentences about future directions on this research theme.

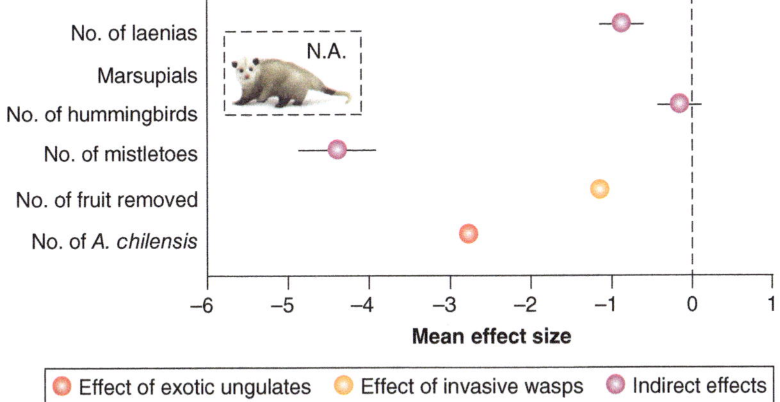

BOX FIGURE 9.6 Mean effect size of the introduced species on different response variables. The effect of exotic species is stastically significant if 95% confidence interval bars do not overlap with zero. Marsupial presence/absence data cannot be displayed in this format, but endemic marsupials were found in all intact sites and none of the invaded sites.

Source: Disassembly of a mutualistic web Mariano A. Rodriguez-Cabal, M. Noelia Barrios-Garcia, Guillermo C. Amico, Marcelo A. Aizen, Nathan J. Sanders Proceedings of the National Academy of Sciences Oct 2013, 110 (41) 16503–16507

TAKING A CLOSER LOOK KELP FORESTS AND TROPHIC CASCADES

What Are Kelp Forests?

Kelp forests are among the most productive and biodiverse habitats on Earth. These ecosystems—characterized by elaborate underwater structures produced by brown algae—are found in shallow, typically cold-water coastal regions around the world, as illustrated in Box Figure 9.7. A variety of algal species in the order Laminariales contribute to the structure of kelp forests, and different species of kelp dominate in different parts of the world. Some kelps—like the giant kelps in the genus *Macrocystis*—can grow up to 45 meters tall and produce floating canopies. Others have low fronds that blanket the ocean floor.

The multitiered structures are home to an incredible diversity of marine species (Steneck et al. 2002). Darwin himself mused on the importance of kelp forests during his oceanic voyage aboard the *Beagle*. After visiting kelp forests off the coast of Chile in 1834, he remarked: *"If in any country a forest was destroyed, I do not believe nearly so many species of animals would perish as would here, from the destruction of the kelp."*

Why Are Kelp Forests Important?

Kelp forests provide cover, nutrients, and nursery grounds for a variety of invertebrates and vertebrates specialized to thrive in kelp habitats. Kelp forests are also important players in marine–terrestrial nutrient and energy flows as they provide biomass and nutrients for coastal ecosystems. Kelp ecosystems also contribute ecosystem services to human populations. The sequestration of carbon in kelp forests is valued at hundreds of millions of dollars per year. In addition, not only do they provide protection against coastal erosion by dampening waves and storm surges, but they also provide direct economic and nutritional benefit to coastal communities that depend on kelp ecosystems for their livelihoods.

How Do Ecological Interactions Drive Kelp Forest Dynamics?

Kelp forests are host to complex food webs, and top predators have cascading effects on other biotic interactions and ecosystem processes in these systems. A key player

(*Continued*)

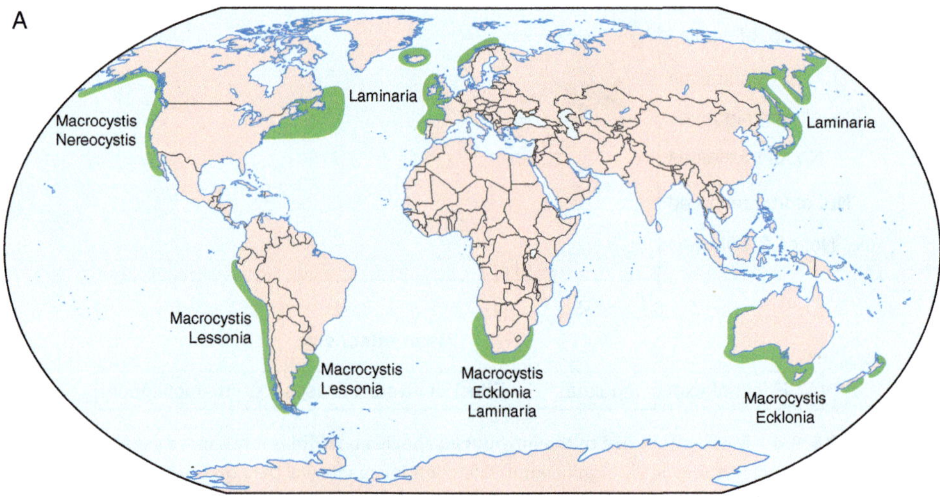

BOX FIGURE 9.7 (A) Distribution of kelp forests around the world with dominant forms of kelp identified. (B) Kelp forests are home to a diversity of marine life, including otters, which function as a keystone species in this ecosystem.

Reflection: Although otters do not feed on kelp, they are key engineers of kelp ecosystems. Brainstorm all the direct and indirect effects that otters might have on kelp forests.

Source: (A) Maximilian Dörrbecker (Chumwa)/Wikimedia Commons; https://www.natgeoimagecollection.com/archive/-2KWGDN31BRO1.html; (B) Chase Dekker/Shutterstock

in kelp forest trophic cascades is the sea otter (*Enhydra lutris*), shown in Box Figure 9.7. Sea otters predate on sea urchins (species in the genus *Strongylocentrotus*), strongly suppressing urchin population and limiting their herbivory on kelp. As shown in Box Figure 9.8, the impact of urchin herbivory on kelp forests is dramatic. In one study of kelp forests in coastal British Columbia, forests with otters were 3.7 times deeper and 18.8 times larger than kelp forests without otters (Markel & Shurin 2015). Otters mediate kelp growth in this system by eliminating red sea urchins (*Strongylocentrotus franciscanus*).

The ecological effects of sea otters as a **keystone species** reach beyond the kelp forest and have indirect effects on marine, terrestrial, and even atmospheric systems. For example, otters affect food webs, not only in the kelp community but also in nearby shoreline habitats. When sea

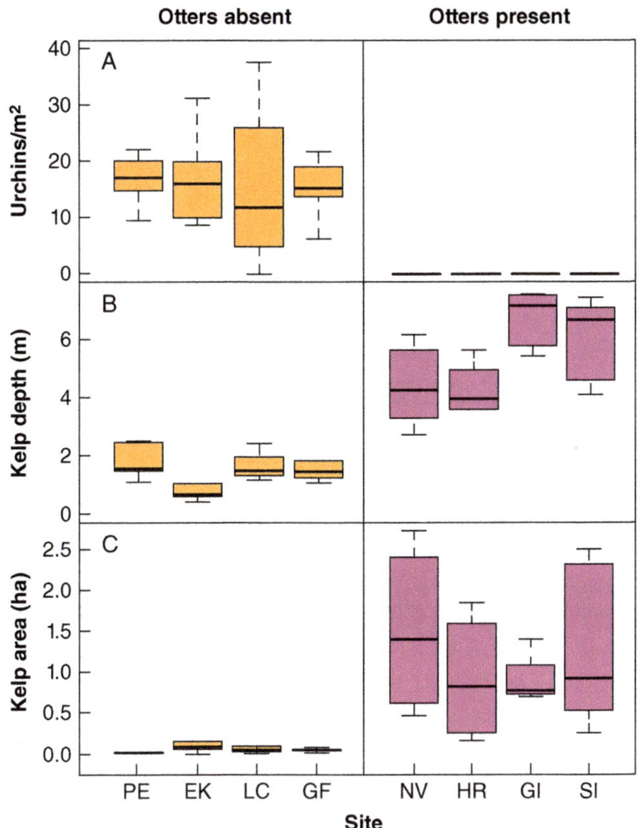

BOX FIGURE 9.8 For multiple sites off the coast of British Columbia, sea otter presence in kelp forests is associated with (A) significantly reduced urchin density, (B) deeper kelp forests, and (C) larger forest area. Horizontal lines in the center of the boxes show the median, and error bars show the full range of data.

Reflection: The box plots in panel A show strong differences *between* treatment groups (otters present versus otters absent). But is there also variation among sites *within* treatment groups? What factors other than otter presence might explain variation within treatment groups?

Source: Russell W. Markel Jonathan B. Shurin, Indirect effects of sea otters on rockfish (Sebastes spp.) in giant kelp forests, Ecology, Volume 96, Issue 11, November 2015, Pages 2877–2890

otters are present and kelp forests are healthy, predatory birds like the glaucous-winged gull (*Larus glaucescens*) and bald eagle (*Haliaeetus leucocephalus*) consume a higher proportion of fish in their diet (Irons et al. 1986; Anthony et al. 2008). Otter abundance also impacts bacterial community structure. Kelp patches with otters have 2.7 times higher bacterial abundance because more organic matter leads to more microbial growth and activity (Clasen & Shurin 2015). Amazingly, otter abundance even can affect atmospheric carbon flows. In one study, researchers found that otters—by suppressing urchins—facilitated kelp ecosystems to develop more than 10 times more biomass and 10 times more primary productivity (per square meter per year; Wilmers et al. 2012). At the landscape level, this translates into an incredible amount of carbon sequestration, an essential ecosystem service.

What Are the Current Threats to Kelp Forests?

Kelp ecosystems are subject to many of the same stressors that we have evaluated for aquatic systems in previous

(Continued)

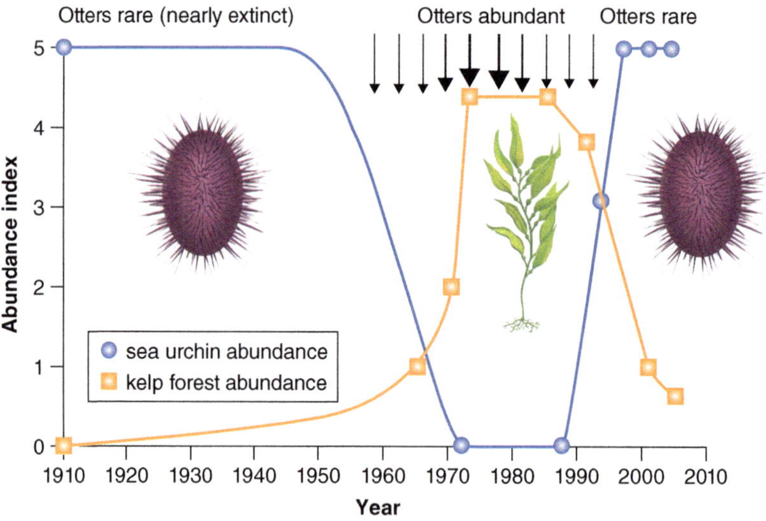

BOX FIGURE 9.9 Trends over time in sea urchin abundance (blue line) and kelp forest abundance (orange line) in a kelp forest off the coast of Alaska. At this location it is only when sea otters are abundant—and sea urchin populations controlled—that kelp forests thrive.

Reflection: What types of conservation or policy efforts could help protect sea otters and kelp forests?

Source: Steneck, R., Graham, M., Bourque, B., Corbett, D., Erlandson, J., Estes, J., & Tegner, M. (2002). Kelp forest ecosystems: Biodiversity, stability, resilience and future. Environmental Conservation, 29(4), 436–459

chapters. For example, direct harvest and overfishing can impact kelp forests (Dayton et al. 1998). Pollutants in the form of chemical contaminants, sewage effluents, nutrient inputs, and oil spills also have particularly detrimental effects in coastal kelp ecosystems (Steneck et al. 2002). Additionally, climate change can have devastating effects on kelp forests, given how vulnerable these ecosystems are to physical and physiological stress. Increased storm frequency and severity are predicted to increasingly compromise the structure of kelp forests. Rising ocean temperatures also have negative impacts, given how dependent kelp forests are on cool water temperatures (Dayton et al. 1999).

However, altered ecological interactions are perhaps the most pressing threat to kelp forests. Hunting and fishing both remove important predators that keep populations of herbivorous sea urchins in check. Without regulation by predators, sea urchins can decimate kelp forests. Take the famous example of sea otter population losses in coastal Alaskan kelp forests. Although indigenous communities reduced sea otter populations beginning several thousand years ago (Simenstad et al. 1978), it was not until European and North American fur traders arrived in the 1700 and 1800s that otters were hunted to near extinction. As otter populations declined, sea urchin populations exponentially increased.

The 20th century brought new environmental protection laws. As shown in Box Figure 9.9, sea otter populations began to rebound and kelp forests began to recover. But in the 1990s, otter populations began to decline again. There are a number of hypotheses for these subsequent declines, including contaminants, disease, and increased predation pressure from orcas (reviewed in Kuker & Barrett-Lennard 2010).

Not all kelp forests exhibit the same patterns seen in Alaskan kelp forests, but most have seen changes in sea urchin abundance and kelp forest health driven by changes in trophic interactions (Steneck et al. 2002). As in other marine ecosystems, conservation of kelp forests will require mitigation of numerous threats, a topic we will explore in more detail in Unit IV. However, ecosystem structure and function cannot be conserved without detailed attention to maintaining key ecological interactions. Ultimately, changes in one species can have dramatic downstream and indirect effects on entire ecosystems.

KEY CONCEPTS

What are key types of biological interactions?

- Interspecific interactions can take many forms from, mutualism (like symbiosis) to exploitation (like parasitism). Some interactions are obligate, while others are facultative. Species can have indirect effects on each other even if they do not interact directly.

How do global change pressures affect biological interactions?

- Environmental change can alter how species interact by changing community composition (species loss or gain) or by changing the timing, strength, or type of interaction between interaction partners.

How does extinction affect communities?

- Coextinction is when loss of one species endangers affiliate species (like mutualists, parasites, predators, and herbivores).

What are cascading effects?

- Cascading effects are those where an initial perturbation leads to a chain of downstream effects. Cascading effects in ecological communities are often caused by disrupted biotic interaction chains.

Core concepts: What are above- and below-ground food webs?

- The above-ground food web (anchored by photosynthetic plants as primary producers) and the below-ground food web (anchored by decomposers) are inextricably linked by nutrient and energy flows.

Meet the data: The collapse of mutualisms

- Scientists studied the effects of introduced species on a complex interaction web in Argentina. They found that both direct and indirect effects of exotic species disrupted key mutualisms and ultimately led to dramatic declines of local species.

Taking a closer look: Kelp forests and trophic cascades

- Kelp forests are among the most productive habitats on Earth and illustrate the cascading effects of disrupting ecological interactions; in this example, overhunting otters caused a trophic cascade that threatened large swaths of kelp forest.

CONSOLIDATE YOUR KNOWLEDGE

Answer the following questions to assess your progress in meeting the learning outcomes:

1. In your own words, write a one- or two-sentence synthesis of the big-picture takeaway point of this chapter.
2. Revisit your answer to the *Blank Page* exercise in the beginning of this chapter. Would you refine your answer now based on knowledge you integrated from this chapter?
3. Give clear examples of both direct and indirect species interactions.
4. What are three ways that invasive species can disrupt species interactions?

5. What are trophic mismatches and why are they a concern in this era of rapid global change?
6. If you wanted to predict whether a species was at high risk for coextinction, what would you want to know about the species and its situation?
7. Can the extinction of one species impact species several trophic levels away? If so, provide a clear example.
8. Do you think we should care about parasite extinctions? Why or why not?
9. What is the difference between a top-down and a bottom-up effect in ecology?
10. Describe how disruption of species interactions threatens kelp forests around the world.
11. In your own words, define the bolded and italicized terms in this chapter.
12. What are some questions that *you* have about the content in this chapter? If you found some content particularly challenging or particularly interesting, identify these as areas for additional reflection or reading.

LITERATURE CITED

Allentoft, Heller, Oskam, Lorenzen, Hale, Gilbert, Jacomb, Holdaway, & Bunce. 2014. "Extinct New Zealand Megafauna Were Not in Decline Before Human Colonization." *Proceedings of the National Academy of Sciences* 111 (13): 4922–27. doi.org/10.1073/pnas.1314972111.

Anthony, Estes, Ricca, Miles, & Forsman. 2008. "Bald Eagles and Sea Otters in the Aleutian Archipelago: Indirect Effects of Trophic Cascades." *Ecology* 89 (10): 2725–35. doi.org/10.1890/07-1818.1.

Bartley, McCann, Bieg, Cazelles, Granados, Guzzo, MacDougall, Tunney, & McMeans. 2019. "Food Web Rewiring in a Changing World." *Nature Ecology and Evolution* 3: 345–54. doi.org/10.1038/s41559-018-0772-3.

Biesmeijer, Roberts, Reemer, Ohlemüller, Edward, et al. 2006. "Parallel Declines in Pollinators and Insect-Pollinated Plants in the Britain and the Netherlands" *Science* 313 (5785): 351–54. DOI: 10.1126/science.1127863.

Brodie, Aslan, Rogers, Redford, Maron, Bronstein, & Groves. 2014. "Secondary Extinctions of Biodiversity." *Trends in Ecology and Evolution* 29 (12): 664–72. doi.org/10.1016/j.tree.2014.09.012.

Cahill, Llimona, Cabaneros, & Calomardo. 2012. "The Increasing Dilemma of Wild Boar (*Sus scrofa*) Habituation to Urban Areas: Traits from Collserola Park (Barcelona) and Comparison with This Problem in Other Cities." *Animal Biodiversity and Conservation* 35 (2): 221–23.

Civitello, Cohen, Fatima, Halstead, Liriano, McMahon, Ortega, et al. 2015. "Biodiversity Inhibits Parasites: Broad Evidence for the Dilution Effect." *Proceedings of the National Academy of Sciences of the United States of America* 112 (28): 8667–71. doi.org/10.1073/pnas.1506279112.

Cizauskas, Carlson, Burgio, Clements, Dougherty, Harris, & Phillips. 2017. "Parasite Vulnerability to Climate Change: An Evidence-Based Functional Trait Approach." *Royal Society Open Science* 4 (1): 160535. doi.org/10.1098/rsos.160535.

Clasen & Shurin. 2015. "Kelp Forest Size Alters Microbial Community Structure and Function on Vancouver Island, Canada." *Ecology* 96 (3): 862–72. doi.org/10.1890/13-2147.1.sm.

Colwell, Dunn, & Harris. 2012. "Coextinction and Persistence of Dependent Species in a Changing World." *Annual Review of Ecology, Evolution, and Systematics* 43 (1): 183–203. doi.org/10.1146/annurev-ecolsys-110411-160304.

Craven, Thakur, Cameron, Frelich, Beauséjour, Blair, Blossey, et al. 2017. "The Unseen

Invaders: Introduced Earthworms as Drivers of Change in Plant Communities in North American Forests (a Meta-Analysis)." *Global Change Biology* 23 (3): 1065–74. doi.org/10.1111/gcb.13446.

Darwin. 1862. *On the Various Contrivances by Which British and Foreign Orchids Are Fertilised by Insects, and on the Good Effects of Intercrossing.* London: John Murray.

Dayton, Tegner, Edwards, & Riser. 1999. "Temporal and Spatial Scales of Kelp Demography: The Role of Oceanographic Climate." *Ecological Monographs* 69 (2): 219–50.

Dayton, Tegner, Edwards, & Riser. 1998. "Sliding Baselines, Ghosts, and Reduced Expectations in Kelp Forest Communities." *Ecological Applications* 2: 309–22. doi.org/10.1890/1051-0761(1998)008[0309:SBGARE]2.0.CO;2.

Deacy, Armstrong, Leacock, Robbins, Gustine, Ward, Erlenbach, & Stanford. 2017. "Phenological Synchronization Disrupts Trophic Interactions Between Kodiak Brown Bears and Salmon." *Proceedings of the National Academy of Sciences* 114 (39): 10432–37. doi.org/10.1073/pnas.1705248114.

Donohue, Petchey, Kéfi, Génin, Jackson, Yang, & O'Connor. 2017. "Loss of Predator Species, Not Intermediate Consumers, Triggers Rapid and Dramatic Extinction Cascades." *Global Change Biology* 23 (8): 2962–72. doi.org/10.1111/gcb.13703.

Dunn, Harris, Colwell, Koh, & Sodhi. 2009. "The Sixth Mass Coextinction: Are Most Endangered Species Parasites and Mutualists?" *Proceedings of the Royal Society B: Biological Sciences* 276 (1670): 3037–45. doi.org/10.1098/rspb.2009.0413.

Estes, Estes, Terborgh, Brashares, Power, Berger, Bond, et al. 2011. "Trophic Downgrading of Planet Earth." *Science* 333 (6040): 301–6. doi.org/10.1126/science.1205106.

Garner, Perkins, Govindarajulu, Seglie, Walker, Cunningham, & Fisher. 2006. "The Emerging Amphibian Pathogen *Batrachochytrium dendrobatidis* Globally Infects Introduced Populations of the North American Bullfrog, *Rana Catesbeiana*." *Biology Letters* 2 (3): 455–59. doi.org/10.1098/rsbl.2006.0494.

Gilljam, Curtsdotter, & Ebenman. 2015. "Adaptive Rewiring Aggravates the Effects of Species Loss in Ecosystems." *Nature Communications* 6: 1–10. doi.org/10.1038/ncomms9412.

Hale, Frelich, Reich, & Pastor. 2008. "Exotic Earthworm Effects on Hardwood Forest Floor, Nutrient Availability and Native Plants: A Mesocosm Study." *Oecologia* 155 (3): 509–18. doi.org/10.1007/s00442-007-0925-6.

Irons, Anthony, & Estes. 1986. "Foraging Strategies of Glaucous-Winged Gulls in a Rocky Intertidal Community." *Ecology* 67 (6): 1460–74.

Kharouba, Ehrlén, Gelman, Bolmgren, Allen, Travers, & Wolkovich. 2018. "Global Shifts in the Phenological Synchrony of Species Interactions over Recent Decades." *Proceedings of the National Academy of Sciences* 115 (20): 5211–16. doi.org/10.1073/pnas.1714511115.

Kiers, Palmer, Ives, Bruno, & Bronstein. 2010. "Mutualisms in a Changing World: An Evolutionary Perspective." *Ecology Letters* 13 (12): 1459–74. doi.org/10.1111/j.1461-0248.2010.01538.x.

Koh, Dunn, Sodhi, Colwell, Proctor, & Smith. 2004. "Species Coextinctions and the Biodiversity Crisis." *Science* 305 (5690): 1632–34. doi.org/10.1126/science.1101101.

Kuker & Barrett-Lennard. 2010. "A Re-Evaluation of the Role of Killer Whales Orcinus Orca in a Population Decline of Sea Otters *Enhydra lutris* in the Aleutian Islands and a Review of Alternative Hypotheses." *Mammal Review* 40 (2): 103–24. doi.org/10.1111/j.1365-2907.2009.00156.x.

Lynam, Llope, Möllmann, Helaouët, Bayliss-Brown, & Stenseth. 2017. "Interaction Between Top-down and Bottom-up Control in Marine Food Webs." *Proceedings of the National Academy of Sciences* 114 (8): 1952–57. doi.org/10.1073/pnas.1621037114.

Markel & Shurin. 2015. "Indirect Effects of Sea Otters on Rockfish (*Sebastes* spp.) in Giant Kelp Forests." *Ecology* 96 (11): 2877–90.

Markl, Schleuning, Forget, Jordano, Lambert, Traveset, Wright, & Böhning-Gaese. 2012. "Meta-Analysis of the Effects of Human Disturbance on Seed Dispersal by Animals."

Conservation Biology 26 (6): 1072–81. doi.org/10.1111/j.1523-1739.2012.01927.x.

Miaud, Dejean, Savard, Millery-Vigues, Valentini, Gaudin, & Garner. 2016. "Invasive North American Bullfrogs Transmit Lethal Fungus *Batrachochytrium dendrobatidis* Infections to Native Amphibian Host Species." *Biological Invasions* 18 (8): 2299–308. doi.org/10.1007/s10530-016-1161-y.

Ness & Bronstein. 2004. "The Effects of Invasive Ants on Prospective Ant Mutualists." *Biological Invasions* 6 (4): 445–61. doi.org/10.1023/B:BINV.0000041556.88920.dd.

Oksanen, Fretwell, Arruda, & Niemela. 1981. "Exploitation Ecosystems in Gradients of Primary Productivity." *The American Naturalist* 18 (2): 240–61. doi.org/10.1086/283817.

Pauli, Peery, Carey, Mendoza, Steffan, & Weimer. 2014. "A Syndrome of Mutualism Reinforces the Lifestyle of a Sloth." *Proceedings of the Royal Society B: Biological Sciences* 281 (1778): 20133006. doi.org/10.1098/rspb.2013.3006.

Portman, Tepedino, Tripodi, Szalanski, & Durham. 2018. "Local Extinction of a Rare Plant Pollinator in Southern Utah (USA) Associated with Invasion by Africanized Honey Bees." *Biological Invasions* 20 (3): 593–606. doi.org/10.1007/s10530-017-1559-1.

Renner & Zohner. 2018. "Climate Change and Phenological Mismatch in Trophic Interactions Among Plants, Insects, and Vertebrates." *Annual Review of Ecology, Evolution, and Systematics* 49: 165–82. doi.org/10.1146/annurev-ecolsys-110617-062535.

Ripple & Beschta. 2006. "Linking a Cougar Decline, Trophic Cascade, and Catastrophic Regime Shift in Zion National Park." *Biological Conservation* 133 (4): 397–408. doi.org/10.1016/j.biocon.2006.07.002.

Rodriguez-Cabal, Barrios-Garcia, Amico, Aizen, & Sanders. 2013. "Node-by-Node Disassembly of a Mutualistic Interaction Web Driven by Species Introductions." *Proceedings of the National Academy of Sciences* 110 (41): 16503–7. doi.org/10.1073/pnas.1300131110.

Rodriguez-Cabal, Stuble, Guénard, Dunn, & Sanders. 2012. "Disruption of Ant-Seed Dispersal Mutualisms by the Invasive Asian Needle Ant (*Pachycondyla chinensis*)." *Biological Invasions* 14 (3): 557–65. doi.org/10.1007/s10530-011-0097-5.

Rodriguez-Cabal, Stuble, Nuñez, & Sanders. 2009. "Quantitative Analysis of the Effects of the Exotic Argentine Ant on Seed-Dispersal Mutualisms." *Biology Letters* 5 (4): 499–502. doi.org/10.1098/rsbl.2009.0297.

Ross, Alisauskas, Douglas, & Kellett. 2017. "Decadal Declines in Avian Herbivore Reproduction: Density-Dependent Nutrition and Phenological Mismatch in the Arctic." *Ecology* 98 (7): 1869–83. doi.org/10.1002/ecy.1856.

Simenstad, Estes, & Kenyon. 1978. "Aleuts, Sea Otters, and Alternate Stable-State Communities." *Science* 200 (4340): 403–11. doi.org/10.2307/1746443.

Steneck, Graham, Bourque, Corbett, Erlandson, Estes, & Tegner. 2002. "Kelp Forest Ecosystems: Biodiversity, Stability, Resilience and Future." *Marine Sciences Faculty Scholarship* 29 (4): 436–59. doi.org/10.1017/S0376892902000322.

Strona. 2015. "Past, Present and Future of Host-Parasite Co-Extinctions." *International Journal for Parasitology: Parasites and Wildlife* 4 (3): 431–41. doi.org/10.1016/j.ijppaw.2015.08.007.

Terborgh, Lopez, Nuñez, Rao, Shahabuddin, et al. 2001. "Ecological Meltdown in Predator-Free Forest Fragments." *Science* 294 (5548): 1923–26. science.sciencemag.org/content/294/5548/1923.short.

Thackeray, Henrys, Hemming, Bell, Botham, Burthe, Helaouet, et al. 2016. "Phenological Sensitivity to Climate Across Taxa and Trophic Levels." *Nature* 535: 241–45. doi.org/10.1038/nature18608.

Traveset & Richardson. 2014. "Mutualistic Interactions and Biological Invasions." *Annual Review of Ecology, Evolution, and Systematics* 45: 89–113. doi.org/10.1146/annurev-ecolsys-120213-091857.

Valiente-Banuet, Aizen, Alcántara, Arroyo, Cocucci, Galetti, García, et al. 2015. "Beyond Species Loss: The Extinction of Ecological Interactions in a Changing World." *Functional Ecology* 29 (3): 299–307. doi.org/10.1111/1365-2435.12356.

Wardle, Bardgett, Callaway, & Van der Putten. 2011. "Terrestrial Ecosystem Responses to Species Gains and Losses." *Science* 332 (6035): 1273–77.

Whisson, Dixon, Taylor, & Melzer. 2016. "Failure to Respond to Food Resource Decline Has Catastrophic Consequences for Koalas in a High-Density Population in Southern Australia." *PLoS ONE* 11 (1): 1–12. doi.org/10.1371/journal.pone.0144348.

Wilkinson & Sherratt. 2016. "Why Is the World Green? The Interactions of Top–down and Bottom–up Processes in Terrestrial Vegetation Ecology." *Plant Ecology and Diversity* 9 (2): 127–40. doi.org/10.1080/17550874.2016.1178353.

Wilmers, Estes, Edwards, Laidre, & Konar. 2012. "Do Trophic Cascades Affect the Storage and Flux of Atmospheric Carbon? An Analysis of Sea Otters and Kelp Forests." *Frontiers in Ecology and the Environment* 10 (8): 409–15. doi.org/10.1890/110176.

Wood, Wilmshurst, Rawlence, Bonner, Worthy, Kinsella, & Cooper. 2013. "A Megafauna's Microfauna: Gastrointestinal Parasites of New Zealand's Extinct Moa (Aves: Dinornithiformes)." *PLoS ONE* 8 (2): e57315. doi.org/10.1371/journal.pone.0057315.

Young, Parker, Gilbert, Sofia Guerra, & Nunn. 2017. "Introduced Species, Disease Ecology, and Biodiversity–Disease Relationships." *Trends in Ecology and Evolution* 32 (1): 41–54. doi.org/10.1016/j.tree.2016.09.008.

Zou, Thébault, Lacroix, & Barot. 2016. "Interactions between the Green and Brown Food Web Determine Ecosystem Functioning." *Functional Ecology* 30 (8): 1454–65. doi.org/10.1111/1365-2435.12626.

Ecosystem-Level Responses

Learning Outcomes

After working with this chapter, you will be able to:

- Assess the different components of ecosystems that can be altered by global change stressors.
- Evaluate the complex effects of anthropogenic activities on large-scale Earth systems and cycles.
- Summarize factors that contribute to ecosystem collapse and ecosystem resilience.
- Analyze the general factors that affect the response of biological systems across hierarchical levels to anthropogenic inputs.
- Apply your knowledge to real-world case studies and interpret data from recent scientific studies.

THE BLANK PAGE

Imagine that you are a contestant on a strange dystopian game show where you will win a million dollars if you correctly predict the response of a particular population, species, community, or ecosystem to global change. To assist you in making your prediction, you are allowed to ask for five pieces of information about the situation. Brainstorm a list of general factors that influence how living things respond to global change stressors. Consider each hierarchical level and identify key determinants of whether populations, species, communities, and ecosystems will survive under altered conditions. Then highlight five pieces of information that you would like from the game show host.

INTRODUCTION

Ecosystems are interdependent networks. If one biotic or abiotic element is altered, these changes can have far-reaching consequences. In a simple world, we might expect a direct relationship between an act and its consequence: cause and effect. But in complex systems, there can be nuanced relationships among global change stressors, their downstream effects, and their interrelated outcomes for the biosphere. The goal of this chapter is to assess the impact of anthropogenic perturbations on large-scale Earth systems and to develop an integrated understanding of the factors that influence response to environmental change across biological levels.

WHAT ARE BIOGEOCHEMICAL CYCLES?

An **ecosystem** is an interacting community of living things and the abiotic elements of the surrounding environment. One hallmark of ecosystem-level change is the interplay between biotic and abiotic processes. Physical, chemical, geological, and biological processes constantly interface, and these interactions ultimately determine how ecosystems will respond to global change stressors. Let us briefly review how energy and nutrients flow through ecosystems and how these flows can be impacted by anthropogenic activities.

Life on Earth is composed of energy and matter. Most of the energy that moves through living systems comes from the sun, although some ecosystems rely on geothermal energy released from the mantle and core of the planet. Matter—in the form of chemical building blocks of life, such as carbon, hydrogen, oxygen, nitrogen, phosphorus, and sulfur—also flows through ecosystems. The interlinked cycles that govern the movement of energy and matter on Earth are referred to as **biogeochemical cycles**.

Water Cycle

Water is essential for life on earth and is stored in numerous reservoirs, including surface waters (like rivers and lakes), groundwater, oceans, glaciers, soils, the atmosphere, and living organisms. Individual water molecules can move quickly through less stable reservoirs like living organisms or spend thousands of years in more stable reservoirs like glaciers. Thus, water can cycle quickly or slowly through the processes of evaporation, sublimation, condensation, precipitation, streamflow, run-off, and snowmelt. The water cycle is integrally linked to many other biogeochemical cycles as water carries nutrients and minerals across biomes. Numerous human activities—such as irrigation, dam building, groundwater extraction, and pollution—directly alter the dynamics of the water cycle and its role in nutrient cycling (Zhou et al. 2016).

Carbon Cycle

Carbon is a key building block of life and found in all living organisms on Earth. The carbon cycle is complex. Carbon dioxide is found in gas form in the atmosphere. This inorganic form of carbon is transformed into its organic form through photosynthesis (both terrestrial and marine) and travels through trophic levels. Ultimately, carbon is released back to the atmosphere by both biotic and abiotic processes. From the biosphere, carbon is released through respiration and decomposition. Carbon can also be stored in soils, rocks, fossils, groundwater, and fossil fuels, and numerous abiotic processes like volcanic activity, erosion, and fire can release carbon back into the atmosphere. Of course, human activities—like burning fossil fuels—are dramatically altering the global carbon cycle. In addition to large-scale perturbations of atmospheric carbon, human activities like deforestation also impact local and regional carbon sequestration.

Nitrogen Cycle

Although nitrogen is the most abundant molecule in the Earth's atmosphere, atmospheric nitrogen is not easily transformed into the nutrient form necessary for synthesizing proteins. Therefore, nitrogen is often a limiting nutrient in ecosystems.

Nitrogen-fixing microbes play an essential role in providing organic nitrogen to terrestrial and marine ecosystems. They supply nitrogen to primary producers by converting nitrogen into a biologically available form that can be absorbed from water or soil and through symbiotic associations with plant roots. After moving through trophic levels, organic nitrogen is processed back into its gas form through a multistep process by fungal and bacterial decomposers. Again, human activities—like the application of nitrogenous fertilizers in agriculture—can have direct effects on the nitrogen profile.

In Sum

Of course, other cycles like oxygen, phosphorus, and sulfur cycles are also essential for life on Earth. Similar principles apply whereby the flow of energy and matter between biotic and abiotic reservoirs can be disrupted by global change stressors. We will further evaluate the impact of global change pressures on biogeochemical cycles in the *Meet the Data* feature of this chapter. Ultimately, it is biogeochemistry that determines the conditions for different biomes (illustrated in Figure 10.1) and the structure of ecological communities. At local, regional, and global scales, biogeochemical cycles influence the structure and function of ecosystems, biomes, and the biosphere as a whole. Thus, perturbations of biogeochemical cycles have far-reaching consequences.

HOW DO GLOBAL CHANGE PRESSURES IMPACT ECOSYSTEMS?

As anthropogenic pressures alter fundamental Earth processes, different ecosystem characteristics can be affected. Here we will introduce three core aspects of ecosystems that can be modified by human enterprises. Then, in the remainder of this chapter, we will consider specific examples of ecosystem-level change around the globe.

Ecosystem Structure

An **ecosystem's structure** is described by its basic biotic and abiotic components. The abundance and distribution of abiotic resources (like light, moisture, and nutrients) and biotic resources (like producers, consumers, and decomposers) are interacting elements of ecosystem structure. Human activities can affect any aspect of ecosystem structure. Changes in land use—including agricultural, industrial, or urban development—alter fundamental biotic and abiotic elements of ecosystem structure. Changes in the distribution and abundance of native, introduced, and domesticated species also alter the structure of the biotic community, as we evaluated in Chapter 5. Any change in biotic or abiotic components of an ecosystem can have rippling effects on ecosystem structure, a theme that we will explore throughout this chapter.

Ecosystem Functions

Ecosystem functions—or ecosystem processes—are the physical, chemical, and biological processes that occur in a given ecosystem. The structural elements of an ecosystem—and their interactions—shape which processes occur. For example, decomposers contribute to the process of soil formation; in turn, soil contributes to the retention and storage of water; water cycling then contributes to temperature regulation; and so on. Thus, there are myriad processes—from nutrient cycling to carbon

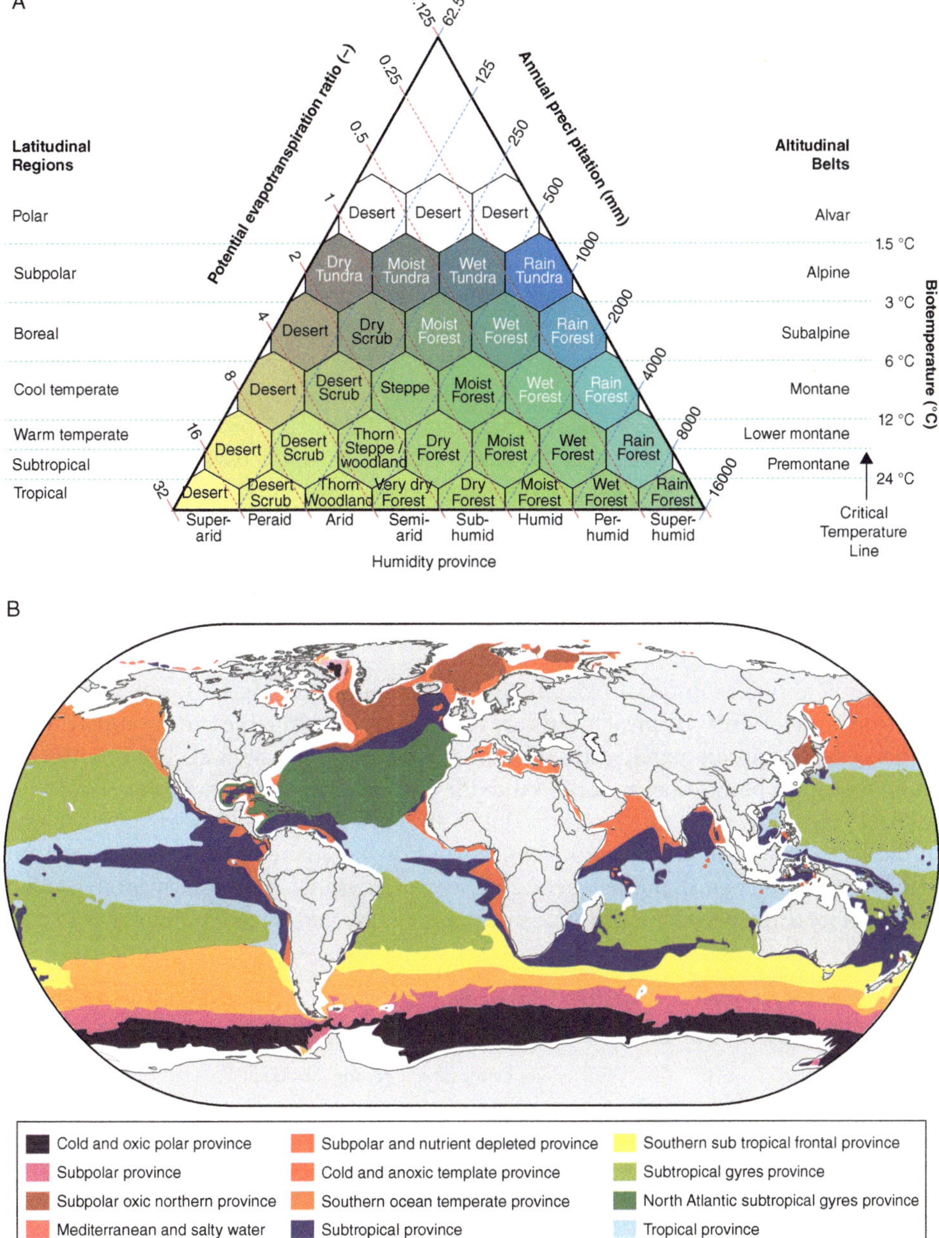

FIGURE 10.1 Biogeochemistry as the foundation of biome classification. (A) One classic classification scheme for terrestrial biomes. (B) One recent classification scheme for marine biogeochemical provinces.

Reflection: Pick one terrestrial and one marine biome and for each identify a specific global change stressor that you expect to have disproportionate effects in that biome.

Source: (A) Wikipedia (modified from Holdridge 1947), https://en.wikipedia.org/wiki/Holdridge_life_zones; (B) Reygondeau, G, Guidi, L, Beaugrand, G, et al. Global biogeochemical provinces of the mesopelagic zone. J Biogeogr. 2018; 45: 500– 514

sequestration—occurring at any given time in any given ecosystem. As human activities alter ecosystem structure, ecosystem functions are also perturbed. For example, urban land use (including roads and buildings) decreases soil respiration and water retention and increases water and nutrient run-off (Alberti 2005), as we evaluated in Chapter 6. Reciprocally, perturbations of ecosystem functions can lead to further alteration of ecosystem structure.

Ecosystem Services

When ecosystem functions benefit humans, they are referred to as **ecosystem services**. Typically, ecosystem services are grouped into four general categories—supporting services (like nutrient cycling and primary production), provisioning services (like food and lumber production), regulating services (like climate regulation and water purification), and cultural services (like recreational and therapeutic opportunities) (Millennium Ecosystem Assessment Report 2005). Table 10.1 provides some specific examples of ecosystem services. As human activities disrupt ecosystem structure and ecosystem function, so too are ecosystem services threatened, as we evaluated in Chapter 1. For example, climate change has led to dramatic melting of Arctic sea ice, snow cover, and permafrost, and the lost climate regulation services is estimated to cost more than 7 trillion US dollars per year (Euskirchen et al. 2013).

Links Between Ecosystem Properties

Let us look at two examples to clearly identify the links among ecosystem structure, function, and services. Tropical forests are one of the most valuable carbon sinks on the planet, storing approximately 40% of terrestrial carbon (Dixon et al. 1994). Through photosynthesis, plants take in carbon dioxide and store this carbon as biomass. Large tropical trees play a key role in terrestrial carbon storage given their size. As we saw in the last chapter, many tropical trees depend on frugivores (animals that feed on fruits) for seed dispersal and regeneration. Seed dispersers—particularly large-bodied species—are threatened by anthropogenic habitat loss and hunting (Dirzo et al. 2014).

TABLE 10.1 Examples of Ecosystem Services

Ecosystem Service	Example
Climate regulation	Regulation of atmospheric carbon dioxide by forests
Water supply	Provisioning of water by watersheds
Erosion control	Prevention of soil loss (through wind and water run-off) by vegetation
Soil formation	Production of soil by microbial, fungal, and animal decomposers
Pollination	Pollination by bees necessary for reproduction of flowering plants
Population regulation	Control of population size of prey species by keystone predators
Food production	Production of fish, game, fruit, and nuts for consumption
Raw materials	Production of lumber and fuel
Genetic resources	Production of compounds that have medicinal use
Nutrient cycling	Fixation of nitrogen
Disturbance protection	Protection from floods and storm surges by natural levees

Question: Which of the four broad categories of ecosystem services does each entry in the table above best fit?

Source: Modified from Costanza et al. 1997

In one specific study, Bello and collaborators (2015) used data from tropical forest communities to understand the effects of **defaunation** (the loss of animal species) on ecosystem structure, function, and services. They used data from 31 field sites in the Atlantic Forest of Brazil to parameterize a mathematical model simulating the effects of frugivore (fruit-eating organism) extinction. They found that larger trees—which store more carbon—tend to have larger seeds that rely on larger frugivores to disperse. When these frugivores are removed, it dramatically impacts overall ecosystem structure. As shown in Figure 10.2, the loss of large-bodied seed dispersers leads to declines in large-seeded tree species. Changes in the biotic community in turn impact ecosystem functions such as seed dispersal and carbon storage. Bello et al. found that most of their forest sites had a carbon deficit (lost carbon storage) once seed dispersers were removed, as illustrated in Figure 10.2.

In addition to disrupting carbon sequestration, defaunation can impact myriad ecosystem functions like pollination, nutrient cycling, water quality, and soil erosion (Dirzo et al. 2014). Many of these ecosystem processes provide important ecosystem services to humans. For example, decreased carbon storage can negatively impact human populations that rely on tropical forests for climate regulation. Thus, any change in ecosystem structure can impact numerous ecosystem processes and ecosystem services.

In the earlier example, anthropogenic activities (hunting) impacted ecosystem structure (loss of large frugivores), which had rippling effects on ecosystem function (carbon sequestration) and ecosystem services (climate regulation). However, human pressures can also impact ecosystem processes directly, and these impacts can then alter ecosystem structure. One specific example comes from anthropogenic perturbations of the carbon cycle. As we saw in Chapter 6, global warming has increased the number of generations per year of many insects, including the mountain pine beetle (*Dendroctonous ponderosae*). Beetle infestations have decimated the structure of lodgepole pine (*Pinus contorta*) forests across the western United States, with some regions losing nearly all adult trees (Raffa et al. 2008), as shown in Figure 10.3.

Loss of a keystone pine species in turn precipitated further disruption of ecosystem functions and services. Brouillard and colleagues surveyed sites in Colorado with healthy versus killed trees. The scientists found extensive changes in soil pH, water content, rates of evapotranspiration, carbon–nitrogen ratios, and soil respiration (carbon dioxide released from the soil) at sites with high tree mortality (Brouillard et al. 2017). Beetle-impacted watersheds also exhibit decreased water quality, even at municipal water treatment facilities (Brouillard et al. 2016). Thus, ecosystem structure, function, and services can all be impacted directly or indirectly by global change pressures.

HOW DO GLOBAL CHANGE PRESSURES IMPACT LARGE-SCALE EARTH SYSTEMS?

Ecosystem-level change ultimately impacts entire biomes and large-scale Earth systems. Here we will briefly review how human activities have modified key systems, focusing on broad-scale changes to terrestrial, atmospheric, aquatic, and frozen environments. We have evaluated examples of global change from different habitat types and regions of the world throughout the book, but this is an opportunity to consolidate prior knowledge and consider anthropogenic impacts at a broad level.

Although we can evaluate impacts on specific Earth systems, these systems are not independent. Many global change stressors affect multiple systems simultaneously, and

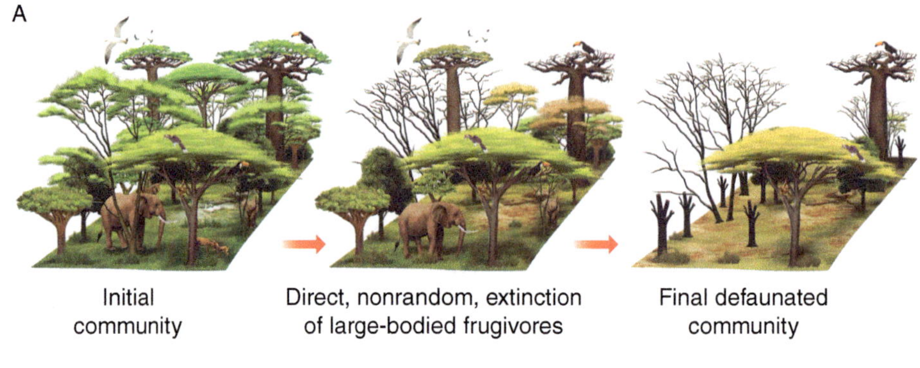

B

<0.1
0.1–0.25
0.25–1.134
1.134–2.0
2.0–2.89
2.89–3.77

FIGURE 10.2 Changes in ecosystem structure can have dramatic effects on ecosystem function and ecosystem services. (A) Defaunation in tropical communities. Loss of large-bodied seed dispersers is predicted to lead to preferential declines of large trees. (B) Changes in ecosystem structure lead to changes in ecosystem functions like carbon storage. Size of points indicates amount of carbon lost (Mg/ha) at 31 field sites in the Atlantic Forest after loss of frugivores.

Reflection: How might lost carbon storage in tropical rainforests impact human societies both in the short and long term?

Source: Defaunation affects carbon storage in tropical forests. BY CAROLINA BELLO, MAURO GALETTI, MARCO A. PIZO, LUIZ FERNANDO S. MAGNAGO, MARIANA F. ROCHA, RENATO A. F. LIMA, CARLOS A. PERES, OTSO OVASKAINEN, PEDRO JORDANO. SCIENCE ADVANCES 18 DEC 2015 : E1501105

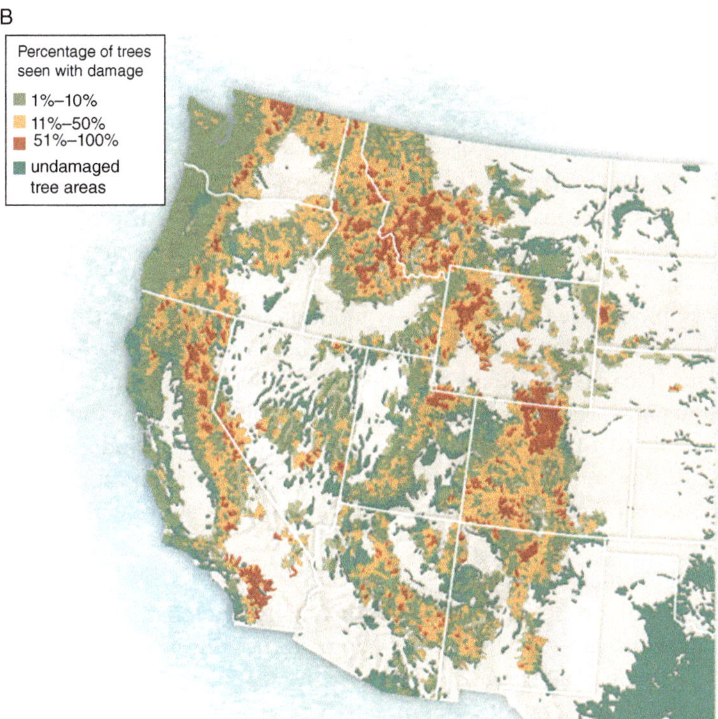

FIGURE 10.3 Impacts of bark beetles on western pine forests. (A) Example of pine forest with dead and dying trees. (B) Degree of damage across the western U.S. Impacted ecosystems show changes in structure (e.g., loss of keystone plant species), function (e.g., decreased evapotranspiration and soil respiration), and services (e.g., drinking water quality).

Reflection: Besides water quality, how specifically might bark beetle infestations impact human health or livelihoods?

Source: Dezene Huber/SFU/flickr; US Forest Service/Karen Minot

changes in any one system can have rippling effects on others. Moreover, the systems that follow represent the broadest possible classification, and many contain multiple biomes (as illustrated previously in Figure 10.1). For example, in the terrestrial realm, deserts, tundra, temperate forests, tropical forests, and grasslands will all experience unique stressors and responses. This is also true within the aquatic realm where freshwater and marine systems experience different threats. Even within marine ecosystems, nearshore regions will respond differently to global change pressures than the deep sea. Therefore, it is important to remember the diversity represented within each of these key Earth systems.

Terrestrial Systems

As we have seen throughout our studies, terrestrial systems have undergone dramatic transformations due to anthropogenic activities. Terrestrial environments have experienced significant land-use change through urban growth, infrastructure development, food production, and resource extraction. Many of these activities involve land clearing and alter habitat structure in obvious—and often irreversible—ways. Even activities that do not involve land clearing can affect myriad biotic and abiotic factors. For example, conditions in croplands can be dramatically changed by irrigation and/or drainage, altered fire regimes, chemical contamination, pest suppression, and introduction of non-native species. Thus, even a single land-use category—like agriculture—can alter the surrounding biotic and abiotic environment in numerous ways.

Overall, approximately 75% of terrestrial Earth has been impacted by human activities, and more than 50% experiences heavy use (Ellis 2011). However, anthropogenic impacts on terrestrial systems are not even across habitat types or geography, as illustrated in Figure 10.4. For example, over the last several hundred years, grasslands, shrublands, savannas, and temperate woodlands have disproportionately been exploited to create crop and rangelands. Even within biomes, impacts are not evenly distributed. The *Core Concepts* feature of this chapter reviews the concept of **biodiversity hotspots**: biodiverse regions of the world that are under grave threat.

Of course, human activities on terrestrial lands generate downstream effects in other systems. Changes to the terrestrial landscape affect both above-ground and below-ground processes, including water, gas, and nutrient cycling (Sterling et al. 2013; Cavagnaro et al. 2016). In addition, run-off from terrestrial landscapes can directly impact aquatic environments. Moreover, fossil fuel use and forest clear-cutting have had far-reaching repercussions on the climate system, which we will turn to next.

Atmospheric Systems

Contemporary human activities have dramatically altered the composition of the Earth's atmosphere. As we evaluated in Chapter 4, the concentration of carbon dioxide in the atmosphere has increased rapidly over the last hundred years as deforestation and fossil fuel use accelerated. Human activities also have altered concentrations of other gases and particles in the atmosphere, including methane, nitrous oxide, ozone, chlorofluorocarbon, and aerosols. Changes in greenhouse gases have unambiguously perturbed our planet's climate system (Oreskes 2004; Intergovernmental Panel on Climate Change [IPCC] 2014).

Climate change has many facets, including shifts in mean and variability of both temperature and precipitation. As we have seen throughout this book, one key directional

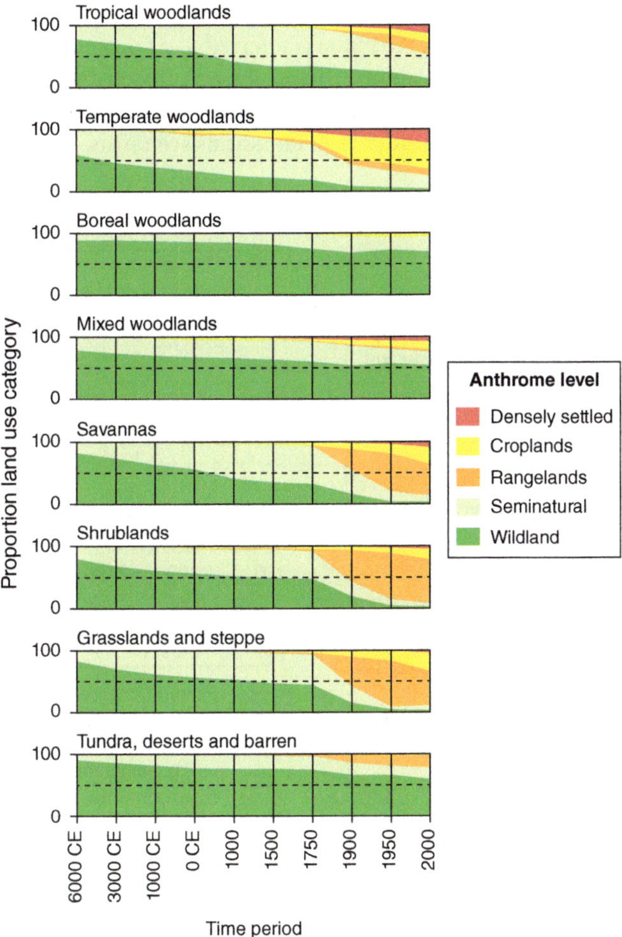

FIGURE 10.4 Changes in "anthrome level" in different terrestrial environments over the last 6,000 years. All terrestrial ecosystems have been impacted by human activities, but some disproportionately so.

Reflection: What were the key points in the last 6,000 years that human impacts on terrestrial ecosystems accelerated? What explains differences across major biomes?

Source: Ellis Erle C. Anthropogenic transformation of the terrestrial biosphere, 369, Philosophical Transactions of the Royal Society A: Mathematical, Physical and Engineering Sciences

change with widespread impacts is global warming. Global mean surface temperatures have risen approximately 1 degree Celsius over the last century. This amount of global warming that has already occurred is referred to as **realized climate change**.

In addition to measuring past change, scientists also predict future climate trends using sophisticated mathematical climate models that track numerous climate variables and incorporate realistic biogeochemical processes. Climate models generally evaluate different possible future climate change scenarios. A scenario-based approach allows scientists to explicitly incorporate uncertainty about future carbon dioxide emissions and climate response so that scientists can project into a future range with high confidence. The amount of global warming yet to come is referred to as **projected climate change**.

Figure 10.5 summarizes this approach and shows projected future warming based on different scenarios. Under low emissions scenarios, we may realize less than 1 degree Celsius additional warming, but under high emissions scenarios, we could realize more

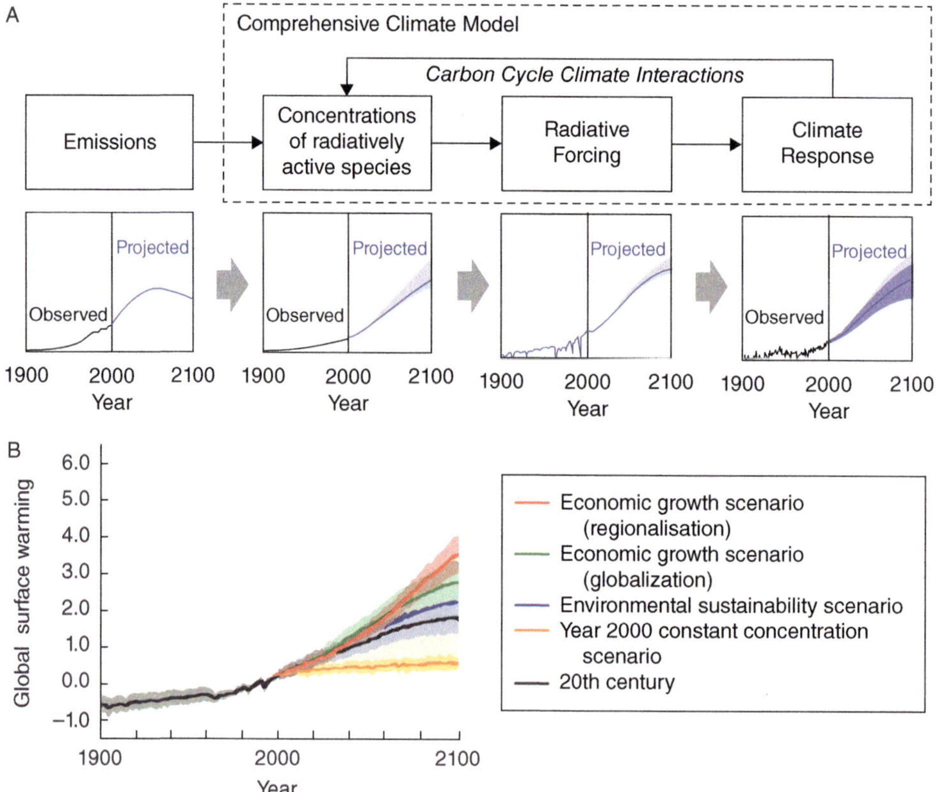

FIGURE 10.5 Projecting the future climate. (A) Climate modeling approaches are now extremely sophisticated and can make future predictions based not only on different emissions scenarios but also on feedbacks in the climate system. As different factors are incorporated, the uncertainty (or probable range) around the mean can be explicitly considered. (B) Projected climate change based on four different emissions scenarios. Solid bars represent projected means and shaded areas represent uncertainty around the mean.

Reflection: What does uncertainty mean? Why does uncertainty about global warming increase as we project further into the future? What general factors influence how much further warming will occur in the next hundred years?

Source: (A) Figure 10.1 from Meehl, G.A., T.F. Stocker, W.D. Collins, P. Friedlingstein, A.T. Gaye, J.M. Gregory, A. Kitoh, R. Knutti, J.M. Murphy, A. Noda, S.C.B. Raper, I.G. Watterson, A.J. Weaver and Z.-C. Zhao, 2007: Global Climate Projections. In: Climate Change 2007: The Physical Science Basis. Contribution of Working Group I to the Fourth Assessment Report of the Intergovernmental Panel on Climate Change [Solomon, S., D. Qin, M. Manning, Z. Chen, M. Marquis, K.B. Averyt, M. Tignor and H.L. Miller (eds.)]. Cambridge University Press, Cambridge, United Kingdom and New York, NY, USA.; (B) IPCC

than 6 degrees additional warming by the end of the 21st century. For context, ~5 degrees Celsius of past warming ended the last glaciation, and 5 degrees Celsius of future warming would create an environment hotter than humans have ever experienced in our evolutionary history. It is important to recognize that under *any* emissions scenarios, we have a **global warming commitment** (Wigley 2005). The planet will continue to warm for at least some period of time even under unrealistically stringent conditions.

Why will the climate continue to change even if human activities that promote climate change are reduced? There is inertia in the climate system, and even if emissions were curtailed entirely, the climate system would be slow to respond. This is partially explained by the fact that greenhouse gases have a long lifespan in our atmosphere. In addition, inertial warming of the oceans will also persist for centuries after emissions have been reduced. Moreover, there are **feedbacks** in the climate system that magnify anthropogenic forcings, a topic we will discuss later in this chapter.

Aquatic Systems

As we have evaluated throughout this book, both freshwater and marine environments have been highly modified by anthropogenic activities. Ponds, lakes, streams, rivers, wetlands, coastal, and marine environments are subject to a variety of direct and indirect impacts (Meybeck 2003). The physical structure of aquatic ecosystems is modified by dredging, channelization, construction of levees, irrigation, and drainage systems. Both chemical and nutrient composition of aquatic ecosystems are also altered by point source pollution and run-off from adjacent terrestrial landscapes. Concentrations of naturally occurring compounds in aquatic environments—like salts and nutrients—are also perturbed by human activities. These anthropogenic inputs alter water quality and productivity in aquatic habitats. For example, nutrient and chemical run-off can lead to eutrophication (excess nutrients), hypoxia (decreased oxygen), algal blooms, and dead zones (Rabalais et al. 2010). Ultimately, more than 40% of marine areas are strongly affected by human stressors, as illustrated in Figure 10.6 (Halpern et al. 2008).

Aquatic systems are also influenced by carbon dioxide emissions and climate change. As global temperature rises, so does the temperature of the Earth's water bodies. In fact, >90% of the heat associated with global warming is stored in the ocean (Levitus et al. 2005; Durack et al. 2014). Recent research shows warming is occurring not only in the upper ocean but also thousands of meters below the surface (Gleckler et al. 2016). Ocean warming also leads to sea level rise due to thermal expansion (as sea water warms, it expands) and the melting of polar ice (as ice melts more, water is added to oceans) (Milne et al. 2009). Changes in temperature gradients alter ocean currents, which in turn perturb circulation patterns, nutrient availability, and dispersal corridors. Finally, carbon dioxide emissions also lead to changes in ocean chemistry. Most notably, as the oceans absorb increasing amounts of carbon dioxide, the pH of the water drops. As we have seen throughout, these changes have dramatic downstream effects on marine ecosystems (Guinotte & Fabry 2008).

Cryospheric Systems

The cryosphere is comprised of all the frozen water on the planet and has also undergone rapid shifts in the last century. The vast majority (~75%) of freshwater on Earth is locked in a solid state as land-based glacial ice. Ninety-nine percent of glacial ice is

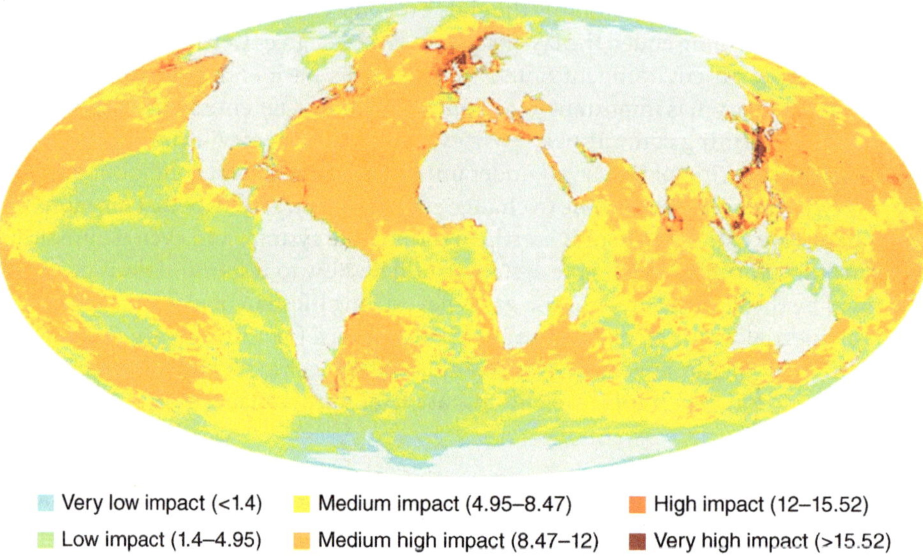

FIGURE 10.6 Global map of human impact in the oceans of the world.

Reflection: What general trends do you observe in the figure? What drivers do you think explain these trends?

Source: A Global Map of Human Impact on Marine Ecosystems. BY BENJAMIN S. HALPERN, SHAUN WALBRIDGE, KIMBERLY A. SELKOE, CARRIE V. KAPPEL, FIORENZA MICHELI, CATERINA D'AGROSA, JOHN F. BRUNO, KENNETH S. CASEY, COLIN EBERT, HELEN E. FOX, ROD FUJITA, DENNIS HEINEMANN, HUNTER S. LENIHAN, ELIZABETH M. P. MADIN, MATTHEW T. PERRY, ELIZABETH R. SELIG, MARK SPALDING, ROBERT STENECK, REG WATSON. SCIENCE 15 FEB 2008 : 948–952

found in polar regions, where ice masses can be hundreds of meters deep and tens of thousands of kilometers wide (Ohmura 2004). In addition to land-based ice (which covers ~10% of the Earth's surface), floating sea ice covers ~12% of the surface of the oceans (Weeks 2010). Ice, snow, and permafrost (frozen soil) characterize the polar regions of the world. However, mountain glaciers are found on nearly all continents—where freezing temperatures occur for at least part of the year at high elevations.

The cryosphere naturally undergoes fluctuations over multiple timescales. Seasonal temperature variation and longer term change in climatic conditions both affect the extent of glacial and sea ice on Earth. For example, during the Pleistocene, an estimated 30% of the Earth's surface was covered by ice (Clark & Mix 2002), contrasted with ~10% today.

Contemporary climate change has led to rapid—and human-mediated—thawing of the cryosphere. In frozen environments around the world, sea ice is melting, glaciers are retreating, snow cover is declining, and permafrost environments are degrading (Ohmura 2004; Fountain et al. 2012; Kang et al. 2010). Figure 10.7 illustrates cryospheric loss in different regions of the world. For example, the surface air temperature in the Arctic is increasing nearly 2 times faster than the rest of the world, and Antarctic sea ice is being lost at a rate of nearly 1% per decade (Curran et al. 2003; Johannessen et al. 2004).

Degradation of the cryosphere has reverberations in the climate system. In addition to being a reservoir of freshwater, the cryosphere plays a critical role in water, gas,

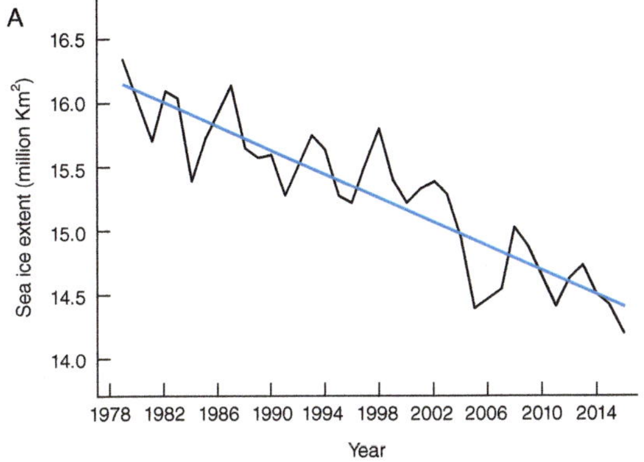

FIGURE 10.7 Thawing ice. Examples of changes in (A) average monthly arctic sea ice extent from 1979–2016 (black line represents actual sea ice extent and blue line represents linear trend over time) and (B) mountain glaciers over the last century. The glacial photographs are from the Upsala Glacier in Patagonia, Argentina, comparing 1928 (top photo) to 2004 (bottom photo).

Reflection: What are two downstream effects of the melting of the cryosphere?

Source: (A) National Snow and Ice Data Center; (B) © Daniel Beltrá / Archivo Museo Salesiano / Greenpeace

and energy cycles. Notably, the cryosphere is critical for climate regulation and stabilization, and it is often referred to as Earth's "air conditioner." Thus, thawing of the cryosphere contributes not only to sea level rise but also to continued global warming, another feedback, which we will revisit later in this chapter.

In Sum

Although we have considered each major Earth system independently, perturbations are all closely coupled. For example, terrestrial deforestation causes increased atmospheric carbon, which induces melting of polar ice, which leads to sea level rise. In addition, a hallmark of ecosystem-level change is the interaction between biotic and abiotic processes. Physical, chemical, geological, and biological processes interdependently determine how ecosystems will respond to global change stressors. In fact, many changes observed in large-scale biomes result from perturbations of biogeochemical cycles, which we reviewed earlier in this chapter. Given the complex interactions and feedbacks between Earth systems and cycles, a holistic perspective is required to understand and predict the effects of global change stressors on the biosphere.

CORE CONCEPTS

WHAT IS A BIODIVERSITY HOTSPOT?

As we have seen, all regions of the planet are under threat from anthropogenic activities. However, some biogeographic regions are especially vulnerable to global change stressors. These regions—termed **biodiversity hotspots**—are cradles of biodiversity that have already experienced significant loss.

The term *biodiversity hotspot* was originally proposed to describe biologically diverse and heavily impacted terrestrial ecosystems (Myers et al. 2000; Brooks et al. 2002). These terrestrial hotspots cover ~2% of the Earth's surface but contain a disproportionate number of endemics (>50% of endemic plants and >40% of endemic vertebrates).

(Continued)

They have also already experienced dramatic impact, with loss of at least 70% of their native vegetation. When the concept of biodiversity hotspots was first proposed, 25 regions of the Earth were identified as hotspots (Myers et al. 2000; Brooks et al. 2002). Now, as Box Figure 10.1 shows, ~36 terrestrial hotspots are typically recognized.

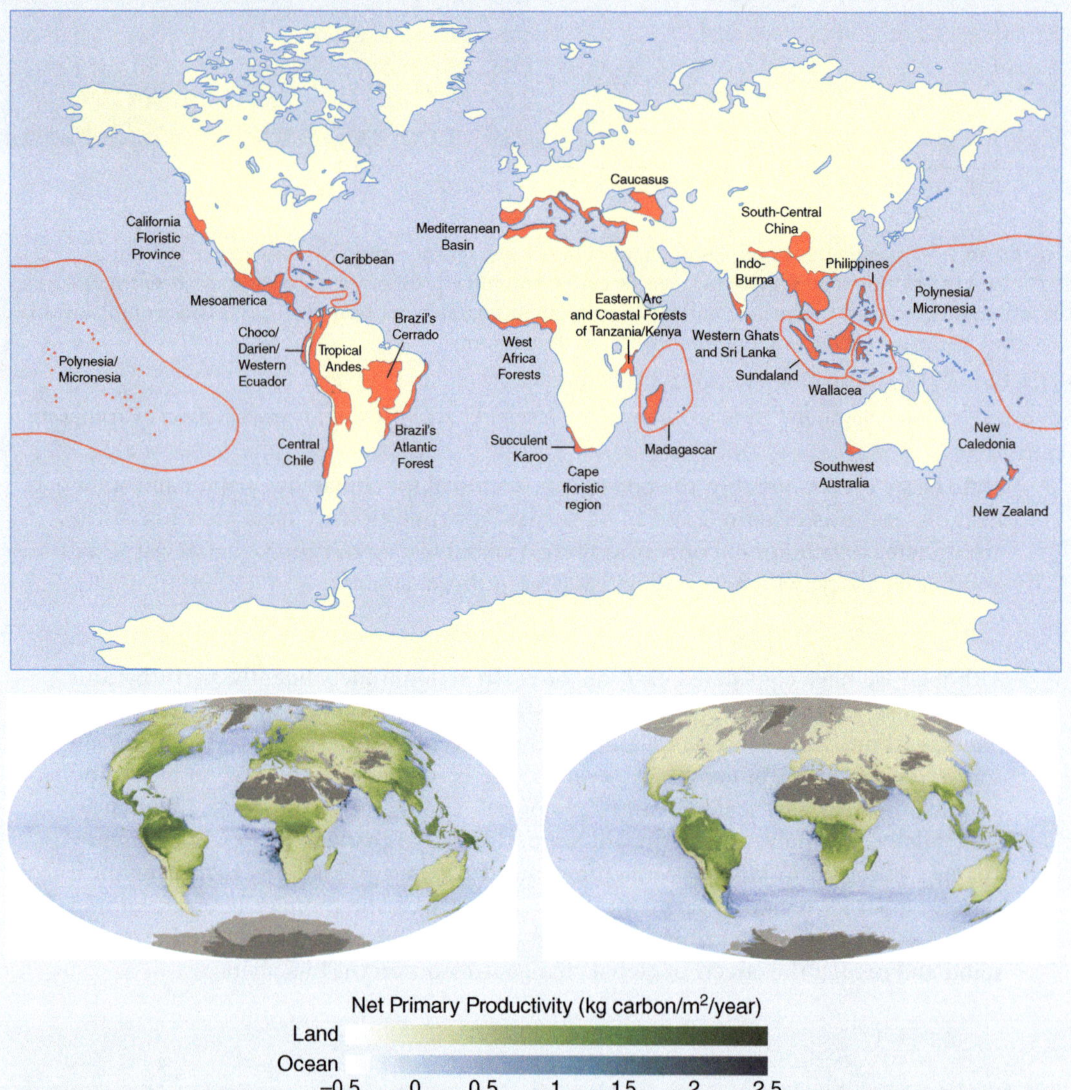

BOX FIGURE 10.1 (A) Global distribution of terrestrial biodiversity hotspots, regions of the world that are particularly biodiverse and under threat. (B) Patterns of species diversity and ecosystem processes fluctuate temporally as well as spatially. Here global patterns of net primary productivity are contrasted between summer (left) and winter (right). Dark green and dark blue represent areas of highest productivity in terrestrial and marine systems, respectively.

Reflection: What special conservation considerations are required for species or ecosystems that exhibit strong seasonal dynamics?

Source: (A) "Christian Marchese, Biodiversity hotspots: A shortcut for a more complicated concept, Global Ecology and Conservation Volume 3, January 2015, Pages 297–309"; (B) NASA Earth Observatory

Of course, there are also many aquatic ecosystems that are centers of endemism and threatened by human activities (Roberts et al. 2002), for example the coral reef ecosystems we evaluated in Chapter 7. Of the ~10 marine hotspots typically recognized, many are close to terrestrial biodiversity hotspots, suggesting that coordinated strategies could address terrestrial and marine biodiversity conservation jointly (Roberts et al. 2002; Marchese 2015).

There are other classification schemes that have been proposed to identify biologically important regions of the planet. For example, the International Union for Conservation of Nature (IUCN) defines **key biodiversity areas** as "sites contributing significantly to the global persistence of biodiversity" (IUCN 2016). To be listed as a key biodiversity area, a site needs to meet specific criteria, for example particular thresholds for threatened species, geographically restricted species, and ecological irreplaceability.

In addition, the IUCN has recently created a **Red List of Ecosystems.** In contrast with the Red List of Species (which we discussed in Chapter 8), the Red List of Ecosystems provides regional assessments of whether particular ecosystems should be considered vulnerable, endangered, or critically endangered. Ultimately, this effort could provide a new global standard for assessing the threat status of terrestrial ecosystems.

Regardless of how they are defined (e.g., as biodiversity hotspots, key biodiversity areas, or endangered ecosystems), it is clear that certain ecoregions are particularly important to the biosphere. Not only do they contain large numbers of threatened species, but also contribute disproportionately to key ecosystem processes and ecosystem services, including photosynthesis, pollination, water filtration, and biomedicines. They also likely house most of the world's remaining undiscovered species (Joppa et al. 2011).

The concept of biodiversity hotspots draws attention to spatial patterns, but it is important to remember that there are also temporal fluctuations in biodiversity and key ecosystem processes. As illustrated in Box Figure 10.1, there are seasonal patterns in the distribution of biodiversity for migratory species like birds (Somveille et al. 2013) and in the distribution of ecosystem services like oxygen production and carbon storage. Similar dynamics are at play in marine systems as the activity of photosynthetic plankton track seasonal changes in temperature and nutrient availability. These spatial and temporal dynamics are key factors in how biodiverse regions will respond to anthropogenic stressors over longer timescales.

WHAT IS A FEEDBACK?

As we have just seen, anthropogenic activities alter ecosystems, biomes, and large-scale Earth processes in extensive and interdependent ways. One hallmark of changes at these higher levels of biological organization is feedbacks. Feedbacks occur when a process or system is modified by its own effects.

There are two basic types of feedbacks. **Negative feedback loops** are self-regulatory because the response decreases the stimulus. Take the simple example of body temperature: when we are too cold, we begin to shiver, which warms us, which decreases shivering. **Positive feedback loops** are amplifying because the response increases the stimulus. Take the example of fatigue: disrupted sleep increases fatigue, which in turn reduces our ability to cope with stress, and increased stress can further disrupt our sleep.

Examples of global change feedbacks abound. Positive feedbacks are of particular concern because they lead to amplified effects. Figure 10.8 illustrates several examples in the climate system, which we have already touched upon. For example, snow and ice have high **albedo**—they are particularly reflective to incoming solar radiation. As frozen environments melt, darker terrestrial and oceanic surfaces are revealed, which absorb more light and energy. This is a feedback where increased energy absorption leads to higher temperatures. Warmer temperatures cause continued ice melt, which further accelerates global warming (Flanner et al. 2011).

FIGURE 10.8 Positive feedbacks in the climate system. Global warming leads to a variety of ecosystem-level impacts from loss of polar ice to increased drought. Many of these *effects* of global warming then become *causes* of further warming.

Reflection: Before you read on, think of one example of a non-climate positive feedback in Global Change Biology.

Source: Climate Emergency Institute

Another positive feedback in the climate system is caused by accelerated loss of terrestrial carbon stores. Human activities like deforestation and biomass burning lead to direct losses of terrestrial carbon storage. As excess carbon is released into the atmosphere, global warming occurs. As global temperatures rise, rates of decomposition in the soil increase. Increased rates of decomposition lead to increased release of soil carbon into the atmosphere, which again can further accelerate global warming (Davidson & Janssens 2006; Heimann & Reichstein 2008).

Of course, there are many examples of non-climate feedbacks in Global Change Biology. For example, habitat loss and hunting have led to dramatic tropical bird losses over the last century. As bird populations decline, so do their ecosystem services such as seed dispersal. With decreased seed dispersal, forest stands can collapse, ultimately leading to more tropical bird losses (Brook et al. 2008).

Another illustration comes from invasive species. As we saw in Chapter 5, invasive species tend to decrease the biotic diversity of the ecosystems they invade. In turn, less biologically diverse ecosystems are more likely to be invaded by non-native species (e.g., Stachowicz et al. 1999). Thus, ecosystem disturbance often generates opportunities for further disturbance, magnifying the effects of global change stressors on biological diversity and ecosystem processes.

Positive feedbacks with global warming, biodiversity loss, and ecosystem disturbance are particularly concerning because they can lead to magnified future impacts. In fact, escalating feedbacks are one factor bringing some ecosystems to the brink of collapse, a topic we turn to next.

WHAT IS ECOSYSTEM COLLAPSE?

At the extreme end of the spectrum, global change perturbations can be so extreme as to cause ecosystem collapse. **Ecosystem collapse** typically refers to a rapid, substantial, and lasting loss of ecosystem structure and function. In the extreme, an ecosystem could shift regimes entirely.

Over geological timescales, ecological regimes change regularly. For example, northern Africa shifted from a "green Sahara" to a "desert Sahara" ecosystem ~5,500 years ago (Foley et al. 2003). This shift was due to changes in the Earth's orbit (and thus incoming solar radiation) and also vegetation–climate feedbacks (which altered monsoon rainfall over the Sahara). However, anthropogenic stressors can also lead to dramatic ecological regime shifts and ecosystem collapse on contemporary timescales. For example, a similar transition from wet to dry conditions occurred over the last 50 years in the adjacent Sahel region of North Africa. This transition was due to complex feedbacks between land, ocean, and atmosphere, with anthropogenic climate change and habitat degradation both contributing (Foley et al. 2003).

Abrupt transitions to a new state are called threshold effects or **tipping points**. An ecosystem may initially exhibit resilience to perturbation, but beyond a certain threshold can no longer recover its previous state. There are many illustrations of tipping point responses to global change stressors in natural systems. For instance, climate change is radically reshaping polar ecosystems. With global warming, sea ice melts earlier in the year, and one specific consequence is increased sunlight penetration into aquatic ecosystems. Some plants and algae flourish under these new high-light conditions and outcompete endemic invertebrates (Clark et al. 2013). These changes in species composition precipitate biodiversity losses, alter ecosystem function, and can entirely transform polar ecosystems.

Ecosystems can collapse for numerous reasons. Ecosystems that are particularly at risk tend to experience increased frequency, severity, and/or duration of global change

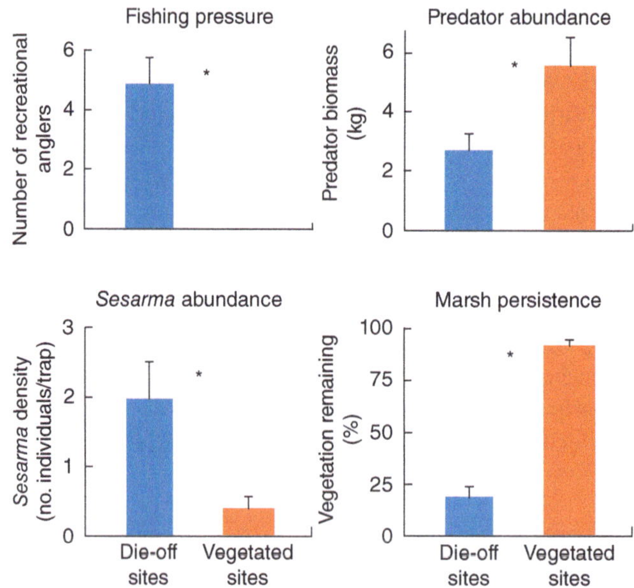

FIGURE 10.9 Ecosystems at the brink. Collapse of Atlantic salt marshes. Overfishing led to a depletion of apex predators. This in turn led to a dramatic increase of *Sesarma* crabs, which then decimated native vegetation. Asterisks indicate statistically significant differences.

Reflection: The depletion of native salt marsh habitat resulted from a top-down trophic cascade. Once marsh vegetation is eroded, what bottom-up effects do you predict might occur?

Source: Andrew H. Altieri, Mark D. Bertness, Tyler C. Coverdale, Nicholas C. Herrmann, Christine Angelini, A trophic cascade triggers collapse of a salt-marsh ecosystem with intensive recreational fishing, Ecology, Volume 93, Issue 6, June 2012, Pages 1402–1410

pressures. Increased disturbance can lead to degradation of abiotic conditions, altered biological interactions, and loss of native biota. Ecosystems are particularly vulnerable to collapse when entire functional groups or trophic levels are perturbed. We evaluated the effects of trophic cascades on the biotic community in Chapter 9, but trophic cascades can also have catastrophic ecosystem-wide effects. One example comes from Atlantic salt-marsh ecosystems where fishing pressure depleted endemic apex predators. As illustrated in Figure 10.9, this led to a cascade of interactions and ultimately the loss of entire salt marshes (Altieri et al. 2012). Ecosystems with strong positive feedbacks are particularly vulnerable to collapse as escalating impacts carry the risk of reaching a tipping point.

Scientists are intensely interested in predicting approaching tipping points (e.g., Scheffer et al. 2012). There are numerous qualitative factors that can indicate ecosystems may be at risk. There are also quantitative methods for anticipating regime shifts (Boettiger et al. 2013). Of course, no one metric will be able to predict ecosystem

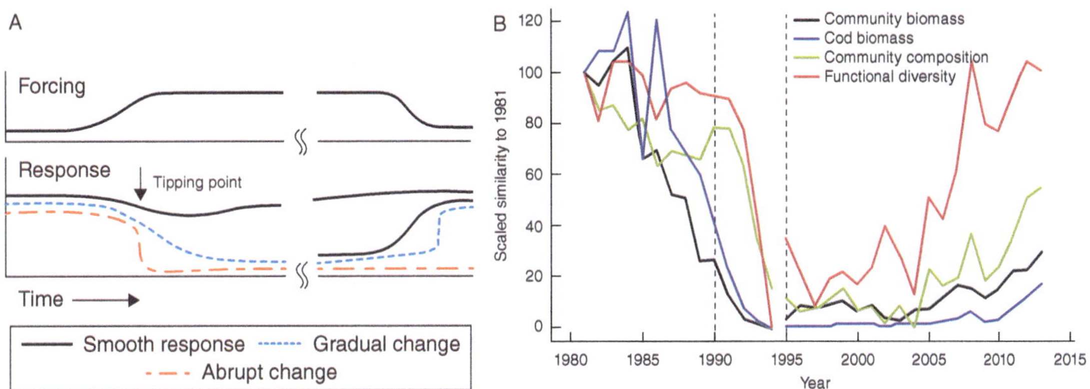

FIGURE 10.10 Tipping points. (A) Conceptual illustration of how ecological systems can exhibit gradual or abrupt tipping point responses and that these responses can be reversible, partially reversible, or irreversible. (B) Empirical tipping point example from cod fisheries, which collapsed in the 1990s. Some metrics of collapse have been at least partially reversed over the last two decades.

Reflection: How can functional diversity recover at a faster rate than species diversity?

Source: (A) Figure 10.1 from Meehl, G.A., T.F. Stocker, W.D. Collins, P. Friedlingstein, A.T. Gaye, J.M. Gregory, A. Kitoh, R. Knutti, J.M. Murphy, A. Noda, S.C.B. Raper, I.G. Watterson, A.J. Weaver and Z.-C. Zhao, 2007: Global Climate Projections. In: Climate Change 2007: The Physical Science Basis. Contribution of Working Group I to the Fourth Assessment Report of the Intergovernmental Panel on Climate Change [Solomon, S., D. Qin, M. Manning, Z. Chen, M. Marquis, K.B. Averyt, M. Tignor and H.L. Miller (eds.)]. Cambridge University Press, Cambridge, United Kingdom and New York, NY, USA; (B) Figure 10.10b Pedersen Eric J., Thompson Patrick L., Ball R. Aaron, Fortin Marie-Josée, Gouhier Tarik C., Link Heike, Moritz Charlotte, Nenzen Hedvig, Stanley Ryan R. E., Taranu Zofia E., Gonzalez Andrew, Guichard Frédéric and Pepin Pierre Signatures of the collapse and incipient recovery of an overexploited marine ecosystem4. R. Soc. open sci.

collapse across different systems. Recent studies have called for a more uniform set of indictor metrics to promote generality and synthesis across studies (e.g., Sato & Lindenmayer 2018; Rowland et al. 2018).

By definition, ecological collapse is dramatic and long lasting. However, regime shifts are not always irreversible. With removal of the initial stressor(s), a system could recover from even a large perturbation, as illustrated in Figure 10.10. For example, overfishing has decimated many aquatic ecosystems. However, when decisive conservation actions are taken, many fisheries can recover on decadal timescales (Costello et al. 2016). Of course, not all metrics of ecosystem health may recover on the same timescales. One illustration comes from Atlantic cod (*Gadus morhua*), which collapsed in the 1990s from overexploitation. Once a fishing moratorium was established, a number of ecosystem health metrics have slowly increased (Pedersen et al. 2017). However, cod biomass itself has increased the most slowly. Thus, recovery is never assured, and some recoveries may exhibit **lag effects** in which there is a time delay between when an action is taken and when the effect is observed in the ecosystem.

WHAT IS ECOSYSTEM RESILIENCE?

The flip side of ecosystem collapse is **ecosystem resilience**. The term *resilience* is used differently in different subfields, and there is a rich debate on best uses of the term in ecology (e.g., Holling 1996; Miller et al. 2010; Mori 2016). Here we define resilience as the ability of a system to absorb—or recover from—a perturbation while maintaining its overall structure and function. In the words of Cumming and Peterson (2017): *"Collapse and resilience are two sides of the same coin; collapse occurs when resilience is lost, and resilient systems are less likely to collapse."*

What contributes to ecosystem resilience? As we evaluated in Chapter 4 and review in Figure 10.11, there are many factors that influence overall vulnerability to global change stressors. At the ecosystem level, several key elements provide increased resilience (e.g., Biggs et al. 2012; Standish et al. 2014). First, **species diversity** is an important component of resilience. For example, more diverse communities are less likely to be invaded by non-native species. Second, **functional redundancy** is a key factor. For example, if important ecological processes are supported by multiple species that can compensate for each other, key functions are less likely to be lost during perturbations. Third, **connectivity** can enhance resilience so that species and resources can flow among habitat patches. For example, connectivity can provide opportunities for more rapid recovery after disturbance events. Resilience can be an inherent property of an ecosystem. However, in this era of rapid global change, humans can also intentionally manage ecosystems to promote resilience, a topic we will turn to in the next chapter.

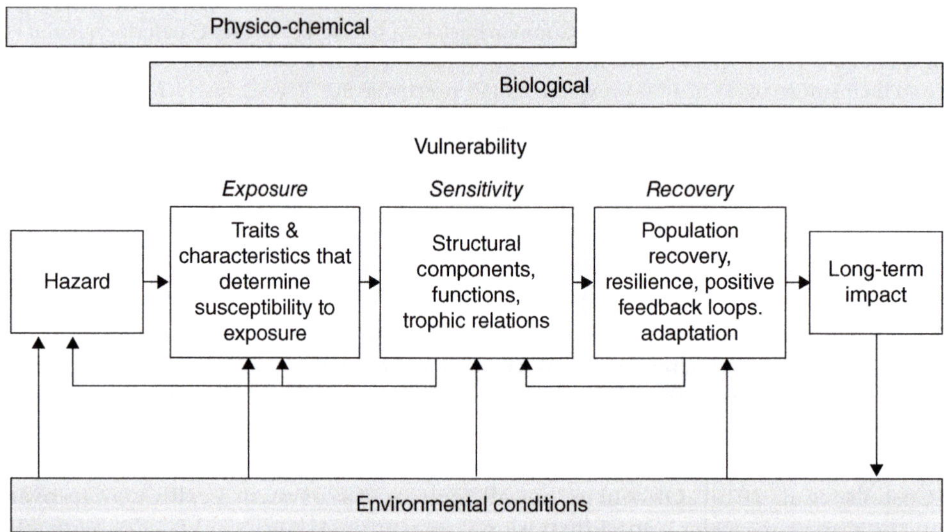

FIGURE 10.11 Ecological vulnerability is determined by the interaction of a number of factors, including exposure, sensitivity, and response. Both physico-chemical and biological factors affect vulnerability. Environmental conditions can both impact—and be impacted by—these factors.

Reflection: What is one specific way that a conservation practitioner could help decrease vulnerability and increase resilience of an ecosystem?

Source: H.J. De Lange, S. Sala, M.Vighi, J.H.Faber, Ecological vulnerability in risk assessment — A review and perspectives, Science of The Total Environment, Volume 408, Issue 18, 15 August 2010, Pages 3871–3879

CONCLUSION

Ultimately, the response of species, communities, and ecosystems to global change stressors depends on many different factors from the molecular to the planetary. The response of living systems to human activities is not a simple linear process. Communities and ecosystems exhibit complex responses, interactions, and feedbacks. Moreover, biotic responses can themselves shape dynamics of ongoing change as biological systems modulate—and sometimes even modify—anthropogenic inputs. Thus, it is essential to understand interactions and feedbacks across biological levels. Only then can we better understand and predict how the biosphere will respond to current and future human impacts.

Anthropogenic activities have already altered conditions on our planet from the depths of the oceans to the upper reaches of the atmosphere. However, humans can more intentionally influence the trajectory of our planetary system by managing ecosystems for resilience. As we conclude Unit III, our journey now turns to addressing more directly how human societies can prioritize conservation actions to preserve and protect the biological wealth of planet Earth.

MEET THE DATA GREENHOUSE GASES IN THE SOIL

Who Are the Scientists and What Did They Set Out To Do?

Here we will look at a recent study by Carolina Voigt and collaborators published in the journal *Global Change Biology* in 2017: "Warming of Subarctic Tundra Increases Emissions of All Three Important Greenhouse Gases—Carbon Dioxide, Methane, and Nitrous Oxide." The study evaluated the important interplay between anthropogenic climate change, greenhouse gas fluxes, and soil microbes. Dr. Voigt and the study area are shown in Box Figure 10.2.

Microbes are the most biodiverse life forms on the planet. Bacteria and Archaea are ubiquitous, abundant, and diverse in nearly all habitats. Take, for example, the skin of a single adult human, which is home to 300 million microbial cells, or 1 cubic meter of soil, which can contain more than a billion microbial cells (Whitman et al. 1998). It is difficult

BOX FIGURE 10.2 Lead author Dr. Carolina Voigt (A) and the study area of subarctic tundra (B).

Source: Carolina Voigt.

(Continued)

to apply traditional species concepts to microbial diversity (Rosselló-Mora & Amann 2001), but there is no doubt that microbial types vastly outnumber other life forms. Recent analyses suggest that there may be more than 1 trillion microbial species on Earth (Locey & Lennon 2016). Microbes perform an incredible array of ecosystem functions from decomposition to nutrient cycling.

In the context of global change, one key role of microbes is below-ground carbon sequestration. Bacteria and Archaea found in the soil contain a large proportion of the total carbon stored in life of Earth. Permafrost—permanently frozen soil—in the northern latitudes contains the largest quantities of soil organic carbon of any terrestrial ecosystem (as shown in Box Figure 10.3; Jackson et al. 2017). These high-latitude regions are under disproportionate pressure from anthropogenic climate change, with rates of warming approximately twice as fast as the global average (IPCC 2014). With global warming, permafrost regions have begun to thaw, increasing rates of decomposition and releasing previously sequestered carbon into the atmosphere. The release of below-ground carbon can have dramatic effects—and create complex feedbacks—on the climate system (e.g., Davidson & Janssens 2006; Schuur et al. 2015: Crowther et al. 2016).

On this backdrop, Voigt et al. set out to quantify the effects of warming on different permafrost habitats in the subarctic. Their study area was in northwestern Russia, near the southern extent of tundra habitat. In this region, permafrost temperatures are increasing, making thaw an imminent threat (Romanovsky et al. 2010). Voigt et al. compared results of warming across three local habitat types: upland tundra (with a shallow organic layer of >9 centimeters dominated by lichens, graminoids, mosses, and shrubs), peat plateau (with thick peat deposits up to 4 meters and dominated by bog vegetation and mosses), and bare peat surfaces (with exposed and eroded peat and no vascular plants).

What Are *Your* Predictions?

Before you read on, take a few minutes to apply what you have learned in this chapter about vulnerability, feedbacks, and tipping points to the permafrost system. Start by considering the following questions:

- *What direct and indirect effects do you expect global warming to have on permafrost habitats?*
- *What positive or negative feedbacks might occur?*
- *What factors might lead to different responses in different microhabitats?*

Now write a summary statement about how global warming might affect sequestered soil organic carbon and atmospheric greenhouse gases.

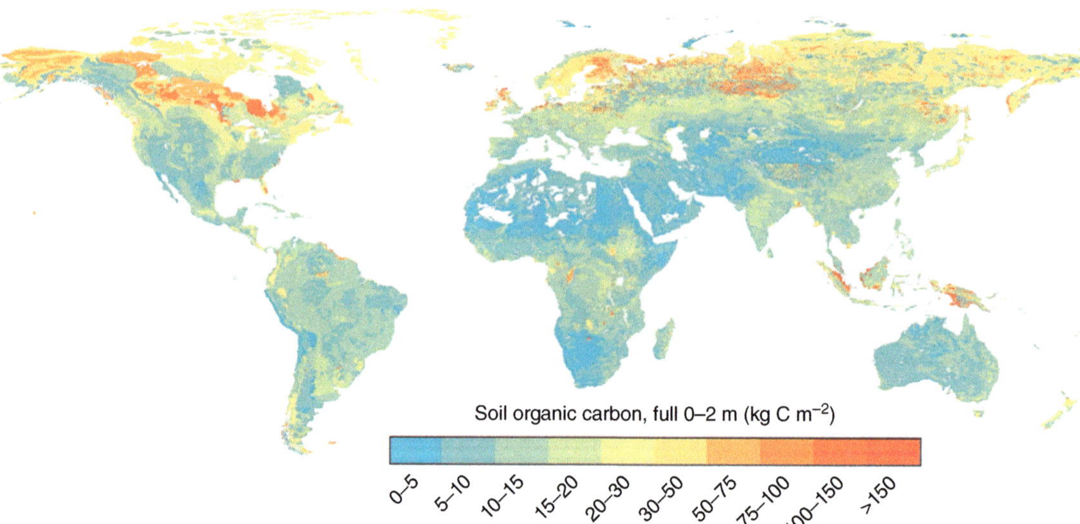

BOX FIGURE 10.3 Global estimates of soil organic carbon (in the top 2 meters of soil) showing the vast stores of carbon in northern latitudes.

Source: The Ecology of Soil Carbon: Pools, Vulnerabilities, and Biotic and Abiotic Controls. Robert B. Jackson, Kate Lajtha, Susan E. Crow, Gustaf Hugelius, Marc G. Kramer, Gervasio Piñeiro. Annual Review of Ecology, Evolution, and Systematics 2017 48:1, 419–445

BOX FIGURE 10.4 Example of one study replicate with paired control and treatment chambers. The open top chambers were used to raise the temperature of treatment plots by an average of 1 degree Celsius.

Source: Carolina Voigt

What Were the Scientists' Predictions?

The authors articulated a clear expectation of how global warming would impact gas release from different subarctic habitats. The authors *"expected that the warming response would strongly depend on the surface type and that warming would not only increase C emissions, but also promote arctic N_2O release."*

What Data Were Collected?

The authors chose a permafrost zone to study in Russia, one that contained three dominant surface types: upland tundra, peat plateau, and bare peat. They conducted a warming experiment to simulate a realistic and moderate summer warming scenario. Using open top chambers (shown in Box Figure 10.4), they raised the surface temperature of five replicated plots for each surface type. Each transparent warming chamber covered approximately 1.5 meters of habitat in diameter, and the study employed a paired design with control plots within 3 meters of warmed plots. The treatment increased air temperature by an average of ~1 degree Celsius relative to the control plots, a biologically relevant amount of warming that this region could experience in the near term.

The authors compared a number of outcomes for control and treatment sites, measured weekly over two consecutive summers. They measured basic environmental parameters like soil and air temperature with small electronic data loggers. They also measured soil gas profiles at different soil depths and nitrous oxide (N_2O) and methane (CH_4) fluxes using collection chambers placed on the soil and later analyzed with a gas chromatograph. In addition, they measured carbon dioxide (CO_2) fluxes and soil microbial respiration with chambers and infrared gas analyzer. Gas fluxes are simply the rate of flow or exchange per unit area. A flux is a change of concentration over time, so it can be positive or negative.

What Is *Your* Interpretation of the Data?

Before you read on, take a few minutes to interpret the results from the study shown in Box Figure 10.5. Consider the following questions as you look at the figures:

- *How do gas fluxes differ for control and treatment plots?*
- *What surface types show the greatest effect of warming on gas fluxes?*
- *Is there substantial variation between the two years of the study?*

Write a sentence or two describing the *key findings* of this study and their *significance*.

What Was the Scientists' Interpretation of the Data?

The warming treatment increased release of all three greenhouse gases studied. Net CO_2 emissions increased for all three surface types, and N_2O and CH_4 emissions increased for the two peat surface types. Soil microbial respiration (not shown in Box Figure 10.5) also increased significantly in the tundra sites. Results were largely consistent across the two years of study.

The observed increase in greenhouse gas emissions represents a shift across a key threshold. Under control conditions, the tundra and peat plateau habitats were greenhouse gas sinks, reservoirs for sequestration of carbon dioxide and other greenhouse gases. However, with just 1 degree Celsius of warming, all sinks were turned into greenhouse gas sources, releasing greenhouse gases into the atmosphere.

This study demonstrates that even low levels of global warming can have critical downstream effects on below-ground processes. As the authors indicate: *"Warming of the air and immediate soil surface affected plant functioning and microbial processes in the litter layer and surface soil; in this way, the warming effect propagated downwards, indirectly affecting deeper soil layers."* These changes in the below-ground community can then have important

(Continued)

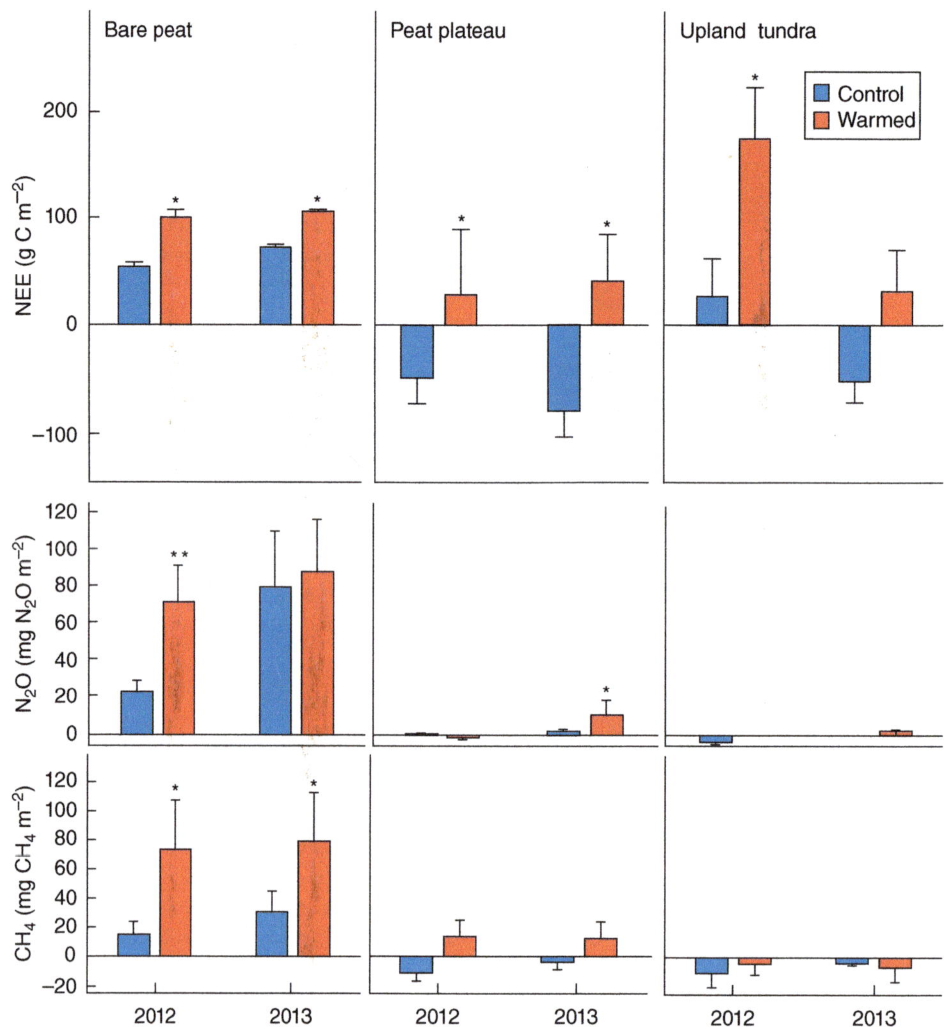

BOX FIGURE 10.5 Carbon dioxide (CO_2), nitrous oxide (N_2O), and methane (CH_4) fluxes in control versus treatment plots in three surface types. CO_2 data are shown in the top row as "net ecosystem exchange (NEE)," which is a composite of CO_2 lost through microbial and plant respiration and CO_2 uptake through photosynthesis. N_2O and CH_4 data are shown in the middle and bottom rows. Asterisks indicate statistically significant differences between treatment groups.

Source: Carolina Voigt, Richard E. Lamprecht Maija E. Marushchak Saara E. Lind Alexander Novakovskiy Mika Aurela Pertti J. Martikainen Christina Biasi, Warming of subarctic tundra increases emissions of all three important greenhouse gases – carbon dioxide, methane, and nitrous oxide, Global Change Biology, Volume 23, Issue 8, August 2017, Pages 3121–3138

positive feedbacks on the climate system as greenhouse gas concentrations increase, accelerating global warming and causing further thawing in permafrost habitats.

All three greenhouse gases measured by Voigt et al. are all critically important to climate regulation. Most attention focuses on CO_2 emissions because it accounts for the majority of emissions by volume, as shown in Box

Figure 10.6. However, nitrous oxide and methane also contribute significantly to global warming and are actually more potent than carbon dioxide. Scientists measure the global warming potential of greenhouse gases as the amount of energy 1 ton of gas will absorb over 100 years relative to CO_2. The global warming potential of N_2O is more than 250 times that of CO_2 and CH_4 and more than 25 times that of CO_2. Thus, the attention to other greenhouse gases in this study was significant. In fact, this paper provides *"the first in situ evidence for the production and release of N_2O also from vegetated tundra surfaces as a consequence of warming."* Thus, the vast stocks of below-ground carbon and nitrogen in permafrost biomes provide a tremendous ecosystem service as a greenhouse gas sink—but also are at risk of becoming a greenhouse gas source.

What Are *Your* Ideas for Future Research Directions?

Given the findings of Voigt et al. about the effects of global warming on soil processes in permafrost habitat, what next steps do you envision for this research program? If you had been involved in this study, how would you follow up? Start by considering the following questions:

- *Are there other structural or functional aspects of below-ground microbial processes that could be studied?*
- *Are there broader regional or global patterns that could be evaluated?*
- *Are there longer term predictions about feedbacks that you could make and later test?*

Now write a few sentences about future directions on this research theme.

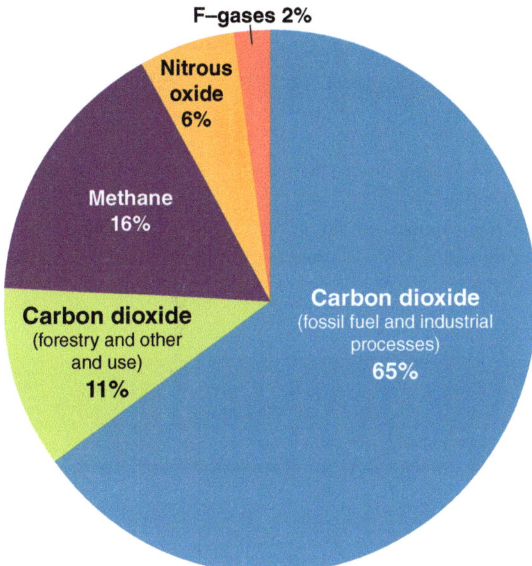

BOX FIGURE 10.6 Proportion of anthropogenic greenhouse gas emissions for carbon dioxide (CO_2), methane (CH_4), nitrous oxide (N_2O), and fluorinated (F) gases (like hydrofluorocarbons).

Source: EPA (with IPCC data)

TAKING A CLOSER LOOK FACTORS INFLUENCING RESPONSE TO GLOBAL CHANGE

As elicited from this chapter's *Blank Page* exercise, there are many factors that play a role in determining the response of living systems to global change stressors. In Unit II, we evaluated factors at the species level, and in Unit III, we studied factors at the community and ecosystem levels. Here we will review examples of key determinants, organized by the biological level at which they operate. We introduced nested and interdependent levels of biological organization in Unit I, but revisit them here as a way to synthesize our understanding of responses to global change.

What Factors Matter at the Molecular Level?

Many factors at the molecular level affect organismal response to changing environments. One example is the mutation rate, the frequency of new DNA variants introduced by errors during DNA replication or repair. Mutation rates vary substantially across the tree of life (Lynch 2010). For example, the flu virus has a mutation rate that is more than a million times faster than the mutation rate in humans. A higher mutation rate produces more genetic variation, the raw material on which selection can act. Thus, when environments change rapidly, populations will be more likely to adapt if they have more genetic variation.

Meta-analyses lend support to the supposition that extinction risk is correlated with low genetic diversity. For example, Spielman et al. (2004) compared estimates of genetic diversity between endangered species and their nonendangered relatives for hundreds of plant and animal species. As shown in Box Figure 10.7, they found that threatened species had significantly lower genetic diversity than nonthreatened relatives. Thus, reduced genetic diversity can be both a cause and a consequence of endangerment. At the community and ecosystem levels, genetic diversity also matters because genetic diversity typically correlates with functional diversity in ecosystem processes.

What Factors Matter at the Organismal Level?

A wide variety of organismal traits—including body size, growth rates, reproductive output, trophic level, and generation time—affect response to global change stressors. One key example is intrinsic dispersal ability. As local conditions become less suitable, organisms with limited dispersal capability will be less likely to move, even if suitable

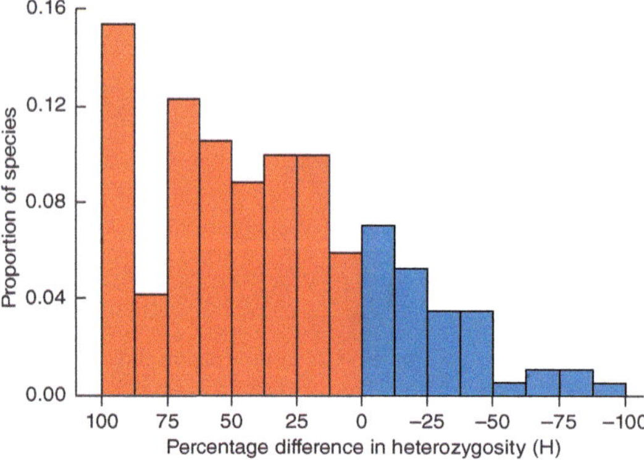

BOX FIGURE 10.7 Reduced genetic diversity (measured as heterozygosity) in threatened species when compared to nonthreatened relatives. Red shading indicates species pairs where genetic diversity is lower in the threatened species compared to the nonthreatened species. Seventy-seven percent of species pairs studied fit this pattern of reduced genetic diversity in threatened species.

Reflection: In the 23% of comparisons that did not fit the expected pattern, what other factors might explain species endangerment?

Source: Most species are not driven to extinction before genetic factors impact them, Derek Spielman, Barry W. Brook, Richard Frankham, Proceedings of the National Academy of Sciences Oct 2004, 101 (42) 15261–15264. Copyright (2004) National Academy of Sciences, U.S.A.

habitat exists in another location. On the other hand, organisms with high dispersal ability may find opportunities to exploit altered conditions outside their native range.

Like other organismal traits, dispersal ability is not fixed. Many species exhibit phenotypic plasticity for dispersal capacity. For example, environmental conditions like crowding, diet quality, photoperiod, and temperature can regulate wing development in insects (e.g., Harrison 1980; Zera & Denno 1997). Many dispersal-related traits can also undergo rapid evolution. Thus, species with strong intrinsic dispersal ability, high plasticity for dispersal traits, and/or the capacity for dispersal traits to rapidly evolve are more likely to successfully move to suitable habitat patches when environmental conditions change.

What Factors Matter at the Population Level?

Many population-level characteristics—such as population density, connectivity, growth rate, age class structure, and carrying capacity—influence response to global change stressors and are in turn influenced by global change stressors. However, one key predictive factor for persistence is population size. Myriad studies demonstrate that probability of population persistence increases with local population size. An example is shown in Box Figure 10.8.

There are a number of reasons why small populations are more imminently threatened with extirpation than large populations. First, small populations are more likely to be wiped out by local catastrophic events (like heat waves, severe winters, floods, and introduced disease). Second, the stochastic effects of genetic drift are more pronounced in small populations. Third, small populations contain less genetic variation. Conversely, in large populations, genetic drift is weaker and genetic variation is higher. Thus, natural selection can more quickly—and predictably—lead to adaptation in new conditions. Thus, larger populations tend to have greater adaptive potential and decreased vulnerability to stochastic events—key factors that increase population persistence over time.

What Factors Matter at the Species Level?

A key species-level attribute that affects response to global change stressors is range size. Species with narrow geographic ranges tend to be more specialized to particular environmental conditions, making them intrinsically more vulnerable to anthropogenic stressors. In addition, they are more vulnerable to stochastic disturbance given that local extreme events can impact a large proportion of their range. Numerous studies have shown that both terrestrial and aquatic species with small geographic ranges are more likely to fare poorly in the face of both local and global anthropogenic impacts (e.g., Harris & Pimm 2008; Sunday et al. 2015; Böhm et al. 2016). Results of one recent meta-analysis showing that range size predicts probability of endangerment is shown in Box Figure 10.8 (Ripple et al. 2017).

Like many other factors, geographic range size is not a fixed attribute and can change through time. In fact, small range size can be a predictor of extinction risk but also a response to anthropogenic stressors. As global change stressors cause range contractions, species become even more vulnerable to future declines. Thus, declining range sizes can create a feedback loop similar to the demographic feedback of the extinction vortex.

What Factors Matter at the Community Level?

At the community level, the strength and specificity of biotic interactions modulate organismal response to global change stressors—and can be altered by anthropogenic threats. Species with obligate interaction partners are particularly vulnerable because threats to their partners will impact their own survival. Close ecological interactions can be of all types (e.g., parasites and their hosts, predators and their prey) and can be disrupted by many changes (e.g., loss of an interaction partner, invasion of a new predator or competitor, or mismatches in activity times). Changes in the distribution or abundance of any species can then have cascading effects on the rest of the biological community. These cascading effects can be top-down impacts of losing an apex predator or bottom-up effects of disturbance on producers or decomposers.

What Factors Matter at the Ecosystem Level?

Cascading effects of species loss are not limited to the biological community. In fact, many community-level changes impact the surrounding abiotic environment. Numerous ecosystem processes from rates of photosynthesis to dynamics of decomposition can affect—and be affected by—global change stressors. Complex interactions and feedbacks between the biotic community and biogeochemical cycles determine whether an ecosystem will exhibit resilience in the face of global change perturbations.

One important ecosystem-level consideration is prior history of anthropogenic change. Once an ecosystem has been altered, it is more vulnerable to future impacts. For example, habitat loss is often paired with development of transportation infrastructure. In turn, roads generate increased access for harvesting. Similarly, disturbed ecosystems tend to generate opportunities for invasion of non-native species. These impacts can generate feedbacks whereby more disturbed ecosystems generate opportunities for further disturbance, magnifying the effect on biological diversity and ecosystem processes.

(Continued)

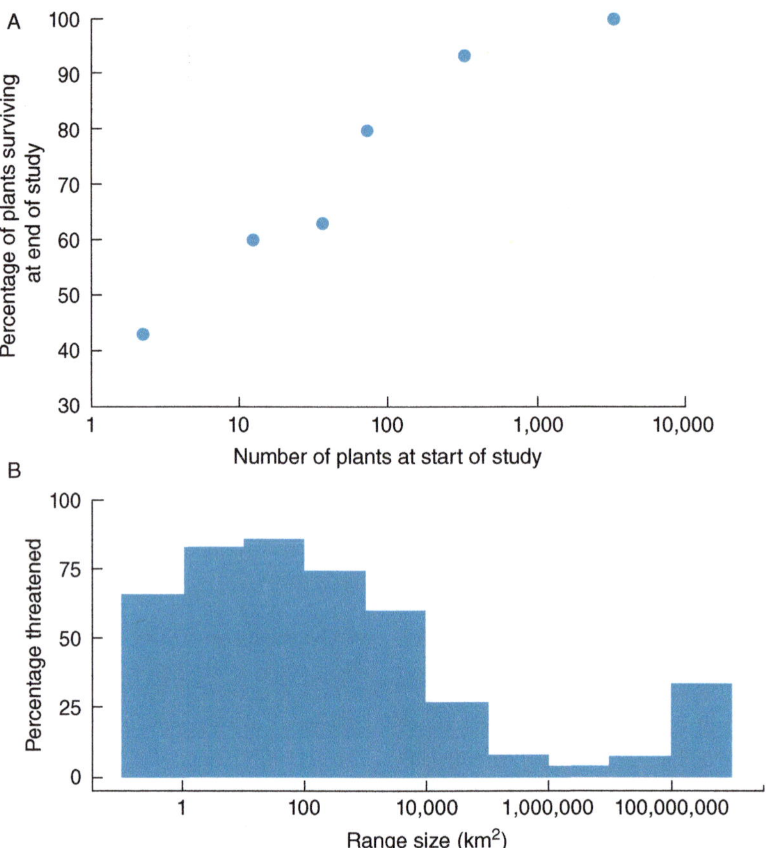

BOX FIGURE 10.8 Importance of population size and range size. (A) Relationship between population size and probability of survival over a 10-year empirical study of >350 populations of flowering plants. (B) Relationship between range size and probability of endangerment from a meta-analysis of >20,000 vertebrate species.

Reflection: How does population size interact with range size to influence risk of endangerment? All else being equal, which do you think would be at greater risk: a geographically widespread species with very low population densities throughout its range or a geographically restricted species with high local population density?

Source: (A) Diethart Matthies, Ingo Bräuer, Wiebke Maibom, Teja Tscharntke, Population size and the risk of local extinction: empirical evidence from rare plants, Oikos, Volume 105, Issue 3, June 2004, Pages 481–488; (B) "Vertebrate species extinction risk William J. Ripple, Christopher Wolf, Thomas M. Newsome, Michael Hoffmann, Aaron J. Wirsing, Douglas J. McCauley Proceedings of the National Academy of Sciences Oct 2017, 114 (40) 10678–10683"

What Factors Matter at the Biosphere Level?

At the global level, biodiversity responses are driven by complex feedbacks and the dynamics of the global change stressors themselves. For example, the magnitude, velocity, and reversibility of anthropogenic change are fundamental considerations because rates of change can be so great that species and ecosystems cannot respond quickly enough. In fact, a number of studies suggest that contribution of organismal, population, and species-level factors can simply be dwarfed by the magnitude of anthropogenic threats (e.g., Purvis et al. 2000; Brook et al. 2006). Take the example of marine sea cucumbers that are harvested for the luxury seafood market (Box Figure 10.9). Extinction risk in this group is correlated with market value, much more than with other organismal (e.g., body size) or species (e.g., range size) traits (Purcell et al. 2014). Thus, for many species and ecosystems, the key consideration will be dynamics of the stressors themselves.

In Sum

It is important to remember there are many additional factors at play besides the examples provided here and that factors interact across all levels. For example, when we consider the magnitude or velocity of environmental change, organismal traits modulate how global change pressures are experienced. For example, the *relative* velocity of environmental change is slower for organisms with fast generation times. Consider a global change stressor (e.g., global warming) over a fixed time period (e.g., the next 50 years). The response of species with generation times of minutes or hours (e.g., many viruses and bacteria) will be very different than those with long generation times (e.g., African elephants or giant sequoias). Thus, it is critical to consider factors across levels—and their interaction—to gain a holistic perspective on organismal response to global change stressors.

BOX FIGURE 10.9 Endangered sea cucumbers harvested for the food trade.

Reflection: How might other anthropogenic stressors (e.g., climate change, ocean acidification) interact with the pressures on species from direct take through hunting, fishing, or harvesting? As species become more endangered, is there typically more or less economic incentive to continue harvesting them?

Source: Purcell Steven W., Polidoro Beth A., Hamel Jean-François, Gamboa Ruth U. and Mercier Annie The cost of being valuable: predictors of extinction risk in marine invertebrates exploited as luxury seafood, 281, Proc. R. Soc. B

KEY CONCEPTS

What are biogeochemical cycles

- Biotic and abiotic processs are intimately liinked through biogeochemical cycles, and global change stressors impact the flow of energy and matter through these cycles.

How do global change pressures impact ecosystems?

- Human activities can impact ecosystem structure (abundance and distribution of biotic and abiotic elements), ecosystem function (physical, chemical, and biological processes), and ecosystem services (processes that specifically benefit human societies).

How do global change pressures impact large-scale Earth systems?

- Anthropogenic activities have dramatic impacts on terrestrial, atmospheric, aquatic, and cryospheric systems. These perturbations are all closely coupled with complex interactions and feedbacks.

What is a feedback?

- Feedbacks occur when a process or system is modified by its own effects. Positive feedback loops (where the response increases the stimulus) are particularly common in global change biology and can lead to amplification of climate change and biodiversity loss.

What is ecosystem collapse?

- Ecosystem collapse is a rapid and lasting loss of ecosystem structure and function. Tipping points that cause ecosystem collapse are difficult to predict, but many ecological regime shifts involve dramatic trophic cascades.

What is ecosystem resilience?

- Ecosystem resilience is the ability of a system to absorb or recover from a perturbation without losing its overall structure or function. Many factors contribute to ecosystem resilience, including species diversity, functional redundancy, and connectivity.

Core concepts: What is a biodiversity hotspot?

- Biodiversity hotspots are cradles of biological diversity that house a disproportionate number of endemics and have experienced significant habitat loss.

Meet the data: Greenhouse gases in the soil

- Below-ground processes—including microbial carbon sequestration—play an important role in climate regulation; in this study, researchers demonstrate that global warming can shift permafrost habitats from greenhouse gas sinks to sources, causing positive feedbacks in the climate system.

Taking a closer look: Factors influencing response to global change

- Characteristics of anthropogenic stressors and characteristics of biological systems interactively influence response to global change. Factors from the molecular to the planetary determine how populations, species, communities, and ecosystems will respond to environmental perturbation.

CONSOLIDATE YOUR KNOWLEDGE

Answer the following questions to assess your progress meeting the learning outcomes:

1. In your own words, write a one- or two-sentence synthesis of the big picture takeaway point of this chapter.
2. Revisit your answer to the *Blank Page* exercise in the beginning of this chapter. Would you refine your answer now based on knowledge you integrated from this chapter?
3. Compare and contrast the concepts of ecosystem structure, ecosystem function, and ecosystem services. Provide a clear example of how each might be altered by a global change stressor.
4. In your opinion, what are the top three threats to each major Earth system (terrestrial, atmospheric, aquatic, cryospheric)?
5. Describe one change in each of the major Earth systems that has ripple effects on the others.
6. Explain the basics of climate change science. How do scientists predict future climate change? How much climate change is expected to occur in the next 100 years? Why do we have a global warming commitment even if all carbon emissions were curtailed today?
7. Describe in detail two different positive feedback loops in Global Change Biology.
8. Compare and contrast ecosystem collapse and ecosystem resilience. What makes ecosystems vulnerable to collapse and what makes ecosystems resilient in the face of global change?
9. How do perturbations to biogeochemical cycles relate to biodiversity loss? Provide an example of how altered biogeochemical cycles can lead to biodiversity loss and reciprocally how biodiversity loss can alter biogeochemical cycles.
10. What are the implications of the thawing Arctic permafrost?
11. In your own words, define the bolded and italicized terms in this chapter.
12. What are some questions that *you* have about the content in this chapter? If you found some content particularly challenging or particularly interesting, identify these as areas for additional reflection or reading.

LITERATURE CITED

Alberti. 2005. "The Effects of Urban Patterns on Ecosystem Function." *International Regional Science Review* 28 (2): 168–92. doi.org/10.1177/0160017605275160.

Altieri, Bertness, Coverdale, Herrmann, Angelini, Ecology, June, et al. 2012. "A Trophic Cascade Triggers Collapse of a Salt-Marsh Ecosystem with Intensive Recreational Fishing." *Ecology* 93 (6): 1402–10.

Bello, Galetti, Pizo, Magnago, Rocha, Lima, Peres, Ovaskainen, & Jordano. 2015. "Defaunation Affects Carbon Storage in Tropical Forests." *Science Advances* 1 (11): e1501105. doi.org/10.1126/sciadv.1501105.

Biggs, Schlüter, Biggs, Bohensky, BurnSilver, Cundill, Dakos, et al. 2012. "Toward Principles for Enhancing the Resilience of Ecosystem Services." *Annual Review of Environmental Resources* 37: 421–48. doi.org/10.1146/annurev-environ-051211-123836.

Boettiger, Ross, & Hastings. 2013. "Early Warning Signals: The Charted and Uncharted Territories." *Theoretical Ecology* 6 (3): 255–64. doi.org/10.1007/s12080-013-0192-6.

Böhm, Williams, Bramhall, Mcmillan, Davidson, Garcia, Bland, Bielby, & Collen. 2016. "Correlates of Extinction Risk in Squamate Reptiles: The Relative Importance of Biology, Geography, Threat and Range Size." *Global Ecology and Biogeography* 25 (4): 391–405. doi.org/10.1111/geb.12419.

Brook, Bradshaw, Koh, & Sodhi. 2006. "Momentum Drives the Crash: Mass Extinction in the Tropics." *Biotropica* 38 (3): 302–5. doi.org/10.1111/j.1744-7429.2006.00141.x.

Brook, Sodhi, & Bradshaw. 2008. "Synergies among Extinction Drivers Under Global Change." *Trends in Ecology and Evolution* 23 (8): 453–60. doi.org/10.1016/j.tree.2008.03.011.

Brooks, Mittermeier, Mittermeier, da Fonseca, Rylands, et al. 2002. "Habitat Loss and Extinction in the Hotspots of Biodiversity." *Conservation Biology* 16 (4): 909–23.

Brouillard, Dickenson, Mikkelson, & Sharp. 2016. "Water Quality Following Extensive Beetle-Induced Tree Mortality: Interplay of Aromatic Carbon Loading, Disinfection Byproducts, and Hydrologic Drivers." *Science of the Total Environment* 572: 649–59. doi.org/10.1016/j.scitotenv.2016.06.106.

Brouillard, Mikkelson, Bokman, Berryman, & Sharp. 2017. "Extent of Localized Tree Mortality Influences Soil Biogeochemical Response in a Beetle-Infested Coniferous Forest." *Soil Biology and Biochemistry* 114: 309–18. doi.org/10.1016/j.soilbio.2017.06.016.

Cavagnaro, Cunningham, & Fitzpatrick. 2016. "Pastures to Woodlands: Changes in Soil Microbial Communities and Carbon Following Reforestation." *Applied Soil Ecology* 107: 24–32. doi.org/10.1016/j.apsoil.2016.05.003.

Clark & Mix. 2002. "Ice Sheets and Sea Level of the Last Glacial Maximum." *Quaternary Science Reviews* 21 (1–3): 1–7. doi.org/10.1016/S0277-3791(01)00118-4.

Clark, Stark, Johnston, Runcie, Goldsworthy, Raymond & Riddle. 2013. Light-driven tipping points in polar ecosystems. *Global Change Biology* 19: 3749–3761. doi: 10.1111/gcb/12337

Costello, Ovando, Clavelle, Strauss, Hilborn, Melnychuk, Branch, et al. 2016. "Global Fishery Prospects Under Contrasting Management Regimes." *Proceedings of the National Academy of Sciences* 113 (18): 5125–29. doi.org/10.1073/pnas.1520420113.

Crowther, Todd-Brown, Rowe, Wieder, Carey, MacHmuller, Snoek, et al. 2016. "Quantifying Global Soil Carbon Losses in Response to Warming." *Nature* 540 (7631): 104–8. doi.org/10.1038/nature20150.

Cumming & Peterson. 2017. "Unifying Research on Social-Ecological Resilience and Collapse." *Trends in Ecology and Evolution* 32 (9): 695–713. doi.org/10.1016/j.tree.2017.06.014.

Curran, Van Ommen, Morgan, Phillips, & Palmer. 2003. "Ice Core Evidence for Antarctic Sea Ice Decline Since the 1950s." *Science* 302 (5648): 1203–6. doi.org/10.1126/science.1087888.

Davidson & Janssens. 2006. "Temperature Sensitivity of Soil Carbon Decomposition and Feedbacks to Climate Change." *Nature* 440 (7081): 165–73. doi.org/10.1038/nature04514.

De Lange, De, Sala, Vighi, & Faber. 2010. "Ecological Vulnerability in Risk Assessment: A Review and Perspectives." *Science of the Total Environment* 408 (18): 3871–79. doi.org/10.1016/j.scitotenv.2009.11.009.

Dirzo, Young, Galetti, Ceballos, Isaac, & Collen. 2014. "Defaunation in the Anthropocene." *Science* 345 (6195): 401–6. doi.org/10.1126/science.1251817.

Dixon, Brown, Houghton, Solomon, Trexler, & Wisniewski. 1994. "Carbon Pools and Flux of Global Forest Ecosystems." *Science* 263 (5144): 185–90. doi.org/10.1126/science.263.5144.185.

Durack, Wijffels, & Gleckler. 2014. "Long-Term Sea-Level Change Revisited: The Role of Salinity." *Environmental Research Letters* 9 (11): 114017. doi.org/10.1088/1748-9326/9/11/114017.

Ellis. 2011. "Anthropogenic Transformation of the Terrestrial Biosphere." *Philosophical Transactions of the Royal Society A: Mathematical, Physical and Engineering Sciences* 369 (1938): 1010–35. doi.org/10.1098/rsta.2010.0331.

Euskirchen, Goodstein, & Huntington. 2013. "An Estimated Cost of Lost Climate Regulation Services Caused by Thawing of the Arctic Cryosphere." *Ecological Applications* 23 (8): 1869–80. doi.org/10.1890/11-0858.1.

Flanner, Shell, Barlage, Perovich, & Tschudi. 2011. "Radiative Forcing and Albedo Feedback from the Northern Hemisphere Cryosphere Between 1979 and 2008." *Nature Geoscience* 4 (3): 151–55. doi.org/10.1038/ngeo1062.

Foley, Coe, Scheffer, & Wang. 2003. "Regime Shifts in the Sahara and Sahel: Interactions between Ecological and Climatic Systems in Northern Africa." *Ecosystems* 6 (6): 524–39. doi.org/10.1007/s10021-002-0227-0.

Fountain, Campbell, Schuur, Stammerjohn, Williams, & Ducklow. 2012. "The Disappearing Cryosphere: Impacts and Ecosystem Responses to Rapid Cryosphere Loss." *BioScience* 62 (4): 405–15. doi.org/10.1525/bio.2012.62.4.11.

Gleckler, Durack, Stouffer, Johnson, & Forest. 2016. "Industrial-Era Global Ocean Heat Uptake Doubles in Recent Decades." *Nature Climate Change* 6 (4): 394–98. doi.org/10.1038/nclimate2915.

Guinotte & Fabry. 2008. "Ocean Acidification and Its Potential Effects on Marine Ecosystems." *Annals of the New York Academy of Sciences* 1134: 320–42. doi.org/10.1196/annals.1439.013.

Halpern, Walbridge, Selkoe, Kappel, Micheli, et al. 2008. "A Global Map of Human Impact on Marine Ecosystems." *Science* 319 (5865): 948–52. doi.org/10.1126/science.1149345.

Harris & Pimm. 2008. "Range Size and Extinction Risk in Forest Birds." *Conservation Biology* 22 (1): 163–71. doi.org/10.1111/j.1523-1739.2007.00798.x.

Harrison. 1980. "Dispersal Polymorphisms in Insects." *Annual Review of Ecology and Systematics* 11 (1): 95–118. doi.org/10.1146/annurev.es.11.110180.000523.

Heimann & Reichstein. 2008. "Terrestrial Ecosystem Carbon Dynamics and Climate Feedbacks." *Nature* 451 (7176): 289–92. doi.org/10.1038/nature06591.

Holdridge. 1947. "Determination of World Plant Formations from Simple Climatic Data." *Science* 105 (2727): 367–68. doi.org/10.1126/science.105.2727.367.

Holling. 1996. "Engineering Resilience Versus Ecological Resilience." In *Engineering Within Ecological Constraints*, pages 31–43. Washington, DC: National Academies Press. www.nap.edu/read/4919/chapter/4

Intergovernmental Panel on Climate Change (IPCC). 2014. *Climate Change 2014: Synthesis Report*. Geneva, Switzerland.

International Union for Conservation of Nature (IUCN). 2016. International Union for Conservation of Nature Annual Report 2016. portals.iucn.org/library/node/46619

Jackson, Lajtha, Crow, Hugelius, Kramer, & Piñeiro. 2017. "The Ecology of Soil Carbon: Pools, Vulnerabilities, and Biotic and Abiotic Controls." *Annual Review of Ecology, Evolution, and Systematics* 48 (1): 419–45. doi.org/10.1146/annurev-ecolsys-112414-054234.

Johannessen, Bengtsson, Miles, Kuzmina, Semenov, Alekseev, Nagurnyi, et al. 2004. "Arctic Climate Change: Observed and Modelled Temperature and Sea-Ice

Variability." *Tellus, Series A: Dynamic Meteorology and Oceanography* 56 (4): 328–41. doi.org/10.1111/j.1600-0870.2004.00060.x.

Joppa, Roberts, Myers, & Pimm. 2011. "Biodiversity Hotspots House Most Undiscovered Plant Species." *Proceedings of the National Academy of Sciences* 108 (32): 13171–76. doi.org/10.1073/pnas.1109389108.

Kang, Xu, You, Flügel, Pepin, & Yao. 2010. "Review of Climate and Cryospheric Change in the Tibetan Plateau." *Environmental Research Letters* 5 (2010): 015101. doi.org/10.1088/1748-9326/5/1/015101.

Levitus, Antonov, & Boyer. 2005. "Warming of the World Ocean, 1955–2003." *Geophysical Research Letters* 32 (2): 1–4. doi.org/10.1029/2004GL021592.

Locey & Lennon. 2016. "Scaling Laws Predict Global Microbial Diversity." *Proceedings of the National Academy of Sciences* 113 (21): 5970–75. doi.org/10.1073/pnas.1521291113.

Lynch. 2010. "Evolution of the Mutation Rate." *Trends in Genetics* 26 (8): 345–52. doi.org/10.1016/j.tig.2010.05.003.Evolution.

Marchese. 2015. "Biodiversity Hotspots: A Shortcut for a More Complicated Concept." *Global Ecology and Conservation* 3: 297–309. doi.org/10.1016/j.gecco.2014.12.008.

Matthies, Brauer, Maibom, Matthies, & Tscharntke. 2004. "Population Size and Risk of Extinction: Empirical Evidence from Rare Plants." *Oikos* 105: 481–88.

Meybeck. 2003. "Global Analysis of River Systems: From Earth System Controls to Anthropocene Syndromes." *Philosophical Transactions of the Royal Society B: Biological Sciences* 358 (1440): 1935–55. doi.org/10.1098/rstb.2003.1379.

Millennium Ecosystem Assessment Report. 2005. *Ecosystems and Human Well-Being.* Washington, DC: Island Press.

Miller, Osbahr, Boyd, Thomalla, Bharwani, Ziervogel, Walker, et al. 2010. "Resilience and Vulnerability: Complementary or Conflicting Concepts?" *Ecology and Society* 15 (3): 11. http://www.ecologyandsociety.org/vol15/iss3/art11/

Milne, Gehrels, Hughes, & Tamisiea. 2009. "Identifying the Causes of Sea-Level Change." *Nature Geoscience* 2 (7): 471–78. doi.org/10.1038/ngeo544.

Mori. 2016. "Resilience in the Studies of Biodiversity-Ecosystem Functioning." *Trends in Ecology and Evolution* 31 (2): 87–89. doi.org/10.1016/j.tree.2015.12.010.

Myers, Mittermeier, Mittermeier, Fonseca, & Kent. 2000. "Biodiversity Hotspots for Conservation Priorities." *Nature* 403 (February): 853–58. doi.org/10.1038/35002501.

Ohmura. 2004. "Cryosphere During the Twentieth Century." In *Geophysical Monograph Series.* doi.org/10.1029/150GM19.

Oreskes. 2004. "The Scientific Consensus on Climate Change." *Science* 306 (January): 1686. doi.org/10.1126/science.1103618.

Pedersen, Link, Thompson, Taranu, Gonzalez, Ball, Moritz, et al. 2017. "Signatures of the Collapse and Incipient Recovery of an Overexploited Marine Ecosystem." *Royal Society Open Science* 4 (7): 170215. doi.org/10.1098/rsos.170215.

Purcell, Polidoro, Hamel, Gamboa, & Mercier. 2014. "The Cost of Being Valuable: Predictors of Extinction Risk in Marine Invertebrates Exploited as Luxury Seafood." *Proceedings of the Royal Society B: Biological Sciences* 281 (1781): 20133296. doi.org/10.1098/rspb.2013.3296.

Purvis, Gittleman, Cowlishaw, & Mace. 2000. "Predicting Extinction Risk in Declining Species." *Proceedings of the Royal Society B: Biological Sciences* 267 (1456): 1947–52. doi.org/10.1098/rspb.2000.1234.

Rabalais, Diaz, & Gilbert. 2010. "Dynamics and Distribution of Natural and Human-Caused Hypoxia Recommended Citation." *Biogeosciences* 7: 585–619. doi.org/10.5194/bg-7-585-2010.

Raffa, Aukema, Bentz, Carroll, Hicke, Turner, & Romme. 2008. "Cross-Scale Drivers of Natural Disturbances Prone to Anthropogenic Amplification: The Dynamics of Bark Beetle Eruptions." *BioScience* 58 (6): 501–17. doi.org/10.1641/B580607.

Reygondeau, Guidi, Beaugrand, Henson, Koubbi, MacKenzie, Sutton, Fioroni, & Maury. 2018. "Global Biogeochemical Provinces of the Mesopelagic Zone." *Journal of Biogeography.* doi.org/10.1111/jbi.13149.

Ripple, Wolf, Newsome, Hoffmann, Wirsing, & McCauley. 2017. "Extinction Risk Is Most Acute for the World's Largest and Smallest Vertebrates." *Proceedings of the National Academy of Sciences* 114 (40): 10678–83. doi.org/10.1073/pnas.1702078114.

Roberts, McClean, Veron, Hawkins, Allen, McAllister, Mittermeier, et al. 2002. "Marine Biodiversity Hotspots and Conservation Priorities for Tropical Reefs." *Science* 295 (5558): 1280–84. doi.org/10.1126/science.1067728.

Romanovsky, Drozdov, Oberman, Malkova, Kholodov, Marchenko, Moskalenko, et al. 2010. "Thermal State of Permafrost in Russia." *Permafrost and Periglacial Processes* 21 (2): 136–55. doi.org/10.1002/ppp.683.

Rosselló-Mora & Amann. 2001. "The Species Concept for Prokaryotes." *FEMS Microbiology Reviews* 25 (1): 39–67. doi.org/10.1016/s0168-6445(00)00040-1.

Rowland, Nicholson, Murray, Keith, Lester, & Bland. 2018. "Selecting and Applying Indicators of Ecosystem Collapse for Risk Assessments." *Conservation Biology* 32 (6): 1233–45. doi.org/10.1111/cobi.13107.

Sato & Lindenmayer. 2018. "Meeting the Global Ecosystem Collapse Challenge." *Conservation Letters* 11 (1): 1–7. doi.org/10.1111/conl.12348.

Scheffer, Bascompte, Pascual, Brock, Carpenter, van Nes, van de Leemput, et al. 2012. "Anticipating Critical Transitions." *Science* 338 (6105): 344–48. doi.org/10.1126/science.1225244.

Schuur, McGuire, Schädel, Grosse, Harden, Hayes, Hugelius, et al. 2015. "Climate Change and the Permafrost Carbon Feedback." *Nature* 520 (7546): 171–79. doi.org/10.1038/nature14338.

Somveille, Manica, Butchart, & Rodrigues. 2013. "Mapping Global Diversity Patterns for Migratory Birds." *PLoS ONE* 8 (8): e70907. doi.org/10.1371/journal.pone.0070907.

Spielman, Brook, & Frankham. 2004. "Most Species Are Not Driven to Extinction before Genetic Factors Impact Them." *Proceedings of the National Academy of Sciences* 101 (42): 15261–64. doi.org/10.1073/pnas.0403809101.

Stachowicz, Whitlatch, & Osman. 1999. "Species Diversity and Invasion Resistance in a Marine Ecosystem." *Science* 286: 1577–80. doi.org/10.1126/science.286.5444.1577.

Standish, Hobbs, Mayfield, Bestelmeyer, Suding, Battaglia, Eviner, et al. 2014. "Resilience in Ecology: Abstraction, Distraction, or Where the Action Is?" *Biological Conservation* 177: 43–51. doi.org/10.1016/j.biocon.2014.06.008.

Sterling, Ducharne, & Polcher. 2013. "The Impact of Global Land-Cover Change on the Terrestrial Water Cycle." *Nature Climate Change* 3 (4): 385–90. doi.org/10.1038/nclimate1690.

Sunday, Pecl, Frusher, Hobday, Hill, Holbrook, Edgar, et al. 2015. "Species Traits and Climate Velocity Explain Geographic Range Shifts in an Ocean-Warming Hotspot." *Ecology Letters* 18 (9): 944–53. doi.org/10.1111/ele.12474.

Vasilakopoulos & Marshall. 2015. "Resilience and Tipping Points of an Exploited Fish Population over Six Decades." *Global Change Biology* 21 (5): 1834–47. doi.org/10.1111/gcb.12845.

Voigt, Lamprecht, Marushchak, Lind, Novakovskiy, Aurela, Martikainen, & Biasi. 2017. "Warming of Subarctic Tundra Increases Emissions of All Three Important Greenhouse Gases—Carbon Dioxide, Methane, and Nitrous Oxide." *Global Change Biology* 23: 3121–38. doi.org/10.1111/gcb.13563.

Weeks. 2010. *On Sea Ice*. Fairbanks: University of Alaska Press.

Whitman, Coleman, & Wiebe. 1998. "Prokaryotes: The Unseen Majority." *Proceedings of the National Academy of Sciences* 95 (12): 6578–83. doi.org/10.1073/pnas.95.12.6578.

Wigley. 2005. "The Climate Change Commitment." *Science* 307 (5716): 1766–69

Zera & Denno. 1997. "Physiology and Ecology of Dispersal Polymorphism in Insects." *Annual Review of Entomology* 42 (1): 207–30. doi.org/10.1146/annurev.ento.42.1.207.

Zhou, Li, Zhao, Gao, Yun, Saino, & Wang. 2016. "Experimental Comparisons of Three Submerged Plants for Reclaimed Water Purification through Nutrient Removal." *Desalination and Water Treatment* 57 (26): 12037–46. doi.org/10.1080/19443994.2015.1048736.

Conservation in an Era of Global Change

Learning Outcomes

After working with this chapter, you will be able to:

- Classify methods for defining conservation priorities in a changing world.
- Analyze and illustrate conservation strategies appropriate at different scales.
- Identify challenges to conservation in an era of global change.
- Consider practical and ethical issues raised by the use of emerging technologies in conservation.
- Apply your knowledge to real-world case studies and interpret data from recent scientific studies.

THE BLANK PAGE

What do you think are the most effective strategies for supporting biological diversity on Earth? Brainstorm a list of specific conservation strategies considering different scales. What strategies are effective at fine scales (focusing on a specific species or location)? What strategies are effective at broad scales (focusing on regional or global issues)? Generate a few ideas in each category. Then pick one idea to explore more deeply and consider whether there may be added challenges for implementing this conservation action in an era of continued rapid global change.

INTRODUCTION

Thus far, we have evaluated how trends in human society—like population growth, resource extraction, and waste production—have unintentional and complex effects on biological systems. Over the last century, these impacts have become more widespread, more visible, and more likely to impact the survival of our own species. Thus, intentional interventions to protect the biological heritage of our planet have become increasingly urgent.

Conservation is the act of preserving, protecting, or restoring ecological systems. The goal of this chapter is to analyze key conservation interventions used to protect biological diversity in an era of global change. Conservation strategies can range from local efforts that protect ecologically valuable species to global efforts that confront the root causes of endangerment. Conservation in the face of rapid global change presents particular challenges, and our exploration focuses on interventions that address the complex issues raised by ongoing environmental change.

WHY IS IT IMPORTANT TO EXPLICITLY DEFINE CONSERVATION PRIORITIES?

Conservation biologists, managers, and policymakers face myriad scientific, ethical, economic, cultural, political, and practical challenges. Foremost in a period of rapid change is deciding how to allocate limited resources. As Pressey and Bottrill (2008) said: *"Conservation efforts and emergency medicine face comparable problems: how to use scarce resources wisely to conserve valuable assets."* Sometimes referred to as **conservation triage**, conservation planning involves decisions about which conservation targets will be given resources—and which will not. Developing an explicit way to rank conservation targets by importance or priority is imperative when resources are scare and conservation decisions are time-sensitive.

There are many ways of explicitly defining conservation priorities. Historically, conservation planners have considered specific attributes of species or ecosystems. For example, if we value all species equally, conservation efforts could focus on geographic areas with the greatest number of species. This is why biodiversity hotspots (introduced in the previous chapter) initially received so much attention as a "silver bullet" strategy for conserving biological diversity (e.g., Myers et al. 2000). However, we might care not only about raw numbers of species but also about the characteristics of those species. Perhaps individual species are of particular value based on their rarity, their economic value, or their cultural significance. Or perhaps particular species contribute disproportionately to the genetic or functional diversity of an ecosystem. Thus, it is essential to consider different attributes explicitly in conservation planning.

Increasingly, scientists use quantitative decision-making tools to select and apply criteria for conservation planning (reviewed in Bower et al. 2018). For example, Belote et al. (2017) evaluated alternative priorities for expanding the reach of protected lands in the United States. As illustrated in Figure 11.1, different criteria lead to different recommendations for which lands to propose for additional protection. However, the criteria can also be applied compositely to identify areas with high combined conservation value. Our *Meet the Data* feature in this chapter provides another more in-depth example of how specific criteria can be operationalized to guide conservation planning.

Of course, scientists and decision makers need to be careful in how they select, apply, and communicate conservation priorities. For example, prioritization schemes can appear arbitrary, contain hidden value judgments, or signal that some populations, species, or landscapes are expendable (e.g., Game et al. 2013; Buckley 2016). They can also neglect important aspects of social context and alienate *human* populations that will be impacted by different conservation actions and policies. Therefore, regardless of whether quantitative or qualitative metrics are used to define conservation priorities, it is important for practitioners to be explicit and transparent about their chosen approach and have clear community-based processes that honor the priorities of local and indigenous people.

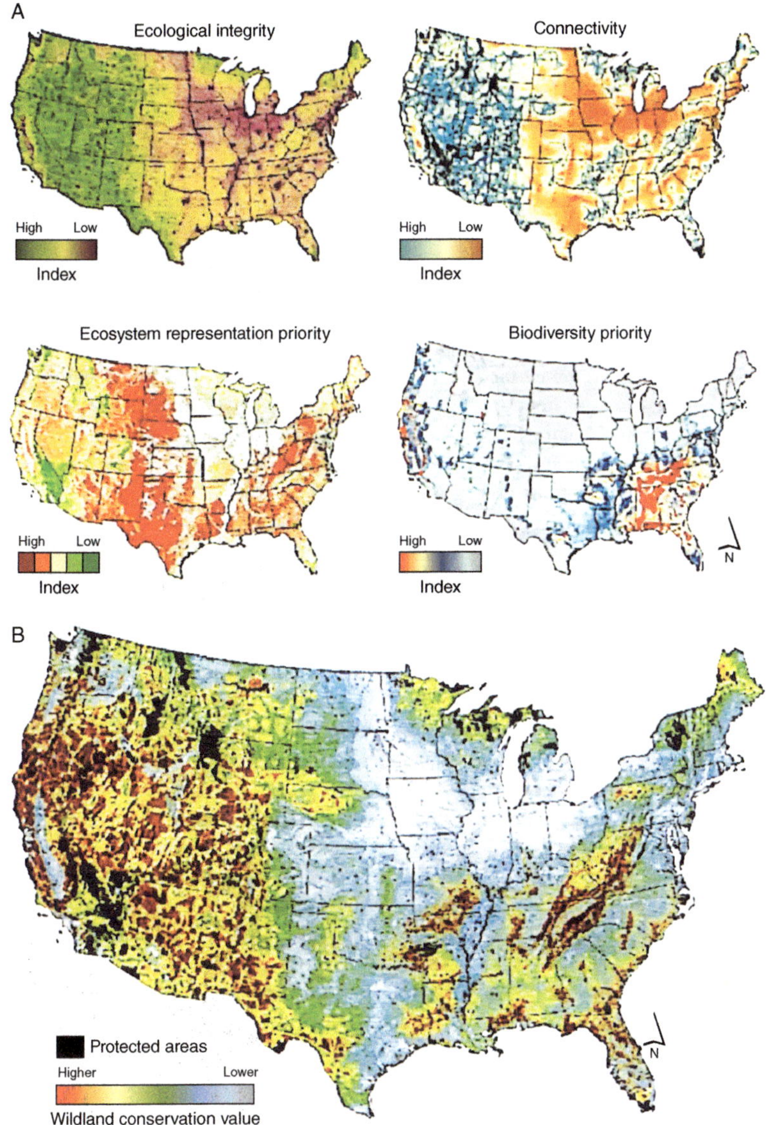

FIGURE 11.1 (A) Alternative prioritization schemes for expanding the network of protected lands in the continental United States. The *ecological integrity* criterion focuses on protecting lands with a low degree of anthropogenic modification. The *connectivity* criterion emphasizes establishing corridors among existing protected areas. The *ecosystem representation priority* and *biodiversity priority* criteria select ecosystems or species (respectively) that are under-represented in existing protected areas. (B) A composite map showing existing protected areas (in black) and areas with high conservation value based on the combined weight of the four aforementioned criteria.

Reflection: How would you rank the four criteria in order of their importance to you?

Source: R. Travis Belote, Matthew S. Dietz, Peter S. McKinley, Anne A. Carlson, Carlos Carroll, Clinton N. Jenkins, Dean L. Urban, Timothy J. Fullman, Jason C. Leppi, Gregory H. Aplet, Mapping Conservation Strategies under a Changing Climate, BioScience, Volume 67, Issue 6, June 2017, Pages 494–497

CORE CONCEPTS

WHAT IS CLIMATE MITIGATION?

The most effective conservation strategies focus on reducing the root causes of biodiversity declines. Take climate change as an example. Although global warming is only one of myriad anthropogenic pressures, it is a backdrop against which all others play out. As we discussed earlier, some degree of future global warming is assured based on feedbacks and inertia in the climate system. However, the extent of future warming depends on whether anthropogenic emissions are curtailed.

Reducing the threat of climate change on living systems will require reducing the accumulation of anthropogenic greenhouse gases in the atmosphere. **Climate change mitigation** refers to actions that reduce radiative forcing on the climate system. Mitigation can lower concentrations of greenhouse gases by reducing carbon emissions and/or enhancing carbon sinks. Here we briefly review fundamental strategies for climate mitigation.

REDUCE ENERGY USE AND USE CLEANER ENERGY SOURCES We can trim energy use by reducing energy demand (e.g., biking to work, turning off appliances when not in use) and by increasing energy efficiency (e.g., replacing traditional incandescent lightbulbs, insulating heated buildings). We can also decrease carbon emissions by using cleaner fossil fuels (e.g., natural gas emits ~50% as much carbon dioxide as coal when burned) and by using renewable energy sources (e.g., solar, wind, and hydro power). All of these strategies can be applied in residential, agricultural, and industrial sectors.

SEQUESTER CARBON We can reduce greenhouse gases via **carbon sequestration**—the process of removing carbon from the atmosphere and depositing it into a reservoir for long-term storage. Carbon is sequestered naturally in biological systems, and we can facilitate increased biological carbon storage by protecting and restoring natural carbon sinks (e.g., forests and peat bogs) and by facilitating increased carbon storage in human-dominated landscapes (e.g., by returning biomass to the soil in agricultural systems). Carbon can also be sequestered physically or chemically, for example, with biomass burial or subterranean injection.

INCREASE REFLECTION OF SOLAR RADIATION We can attempt to alter other aspects of the radiation balance on Earth, such as by increasing reflected solar radiation. There have been many proposals for how **geoengineering**—the intentional manipulation of large-scale environmental processes—might be applied to altering how much of the sun's energy reaches the Earth. Ideas like "cloud seeding" to increase cloud cover, spraying aerosols to reflect sunlight, and installing space mirrors as illustrated in Box Figure 11.1 remain controversial (Schneider 2008). Not only are these possibilities astronomically expensive, but the risk of unintended consequences is unpalatable to most.

MITIGATION VERSUS ADAPTATION Climate change mitigation efforts are essential, but even if we reduce carbon emissions, it will be a long time before the effects of anthropogenic climate change are stabilized or reversed. Therefore, **climate change adaptation** is an important counterpart to climate change mitigation. Climate change adaptation refers to actions to reduce the vulnerability of biological systems to anthropogenic climate change. For example, planting trees in urban areas can help moderate local temperatures, building levees can protect shores from extreme weather events, and breeding new heat-tolerant crops are all examples of climate change adaptation strategies. Initially, climate adaptation was seen as a distraction from climate mitigation, but it is now recognized as an essential part of climate change readiness.

Of course, many climate change mitigation and adaptation strategies are complicated and have trade-offs. For one example, wind turbines are an important technology in the renewable energy portfolio, but they have also been critiqued for causing bird and bat mortalities (Johnson et al. 2016; Thaxter et al. 2017). However, as shown in Box Figure 11.1, wind turbines affect many fewer individuals than other anthropogenic stressors: collisions with buildings and predation by domestic cats impact more than a billion birds per year in the United States alone (Loss et al. 2014, 2015). Ultimately, close coordination between scientists, managers, policymakers, and ethicists is essential for devising and implementing effective and responsible strategies for climate mitigation and adaptation.

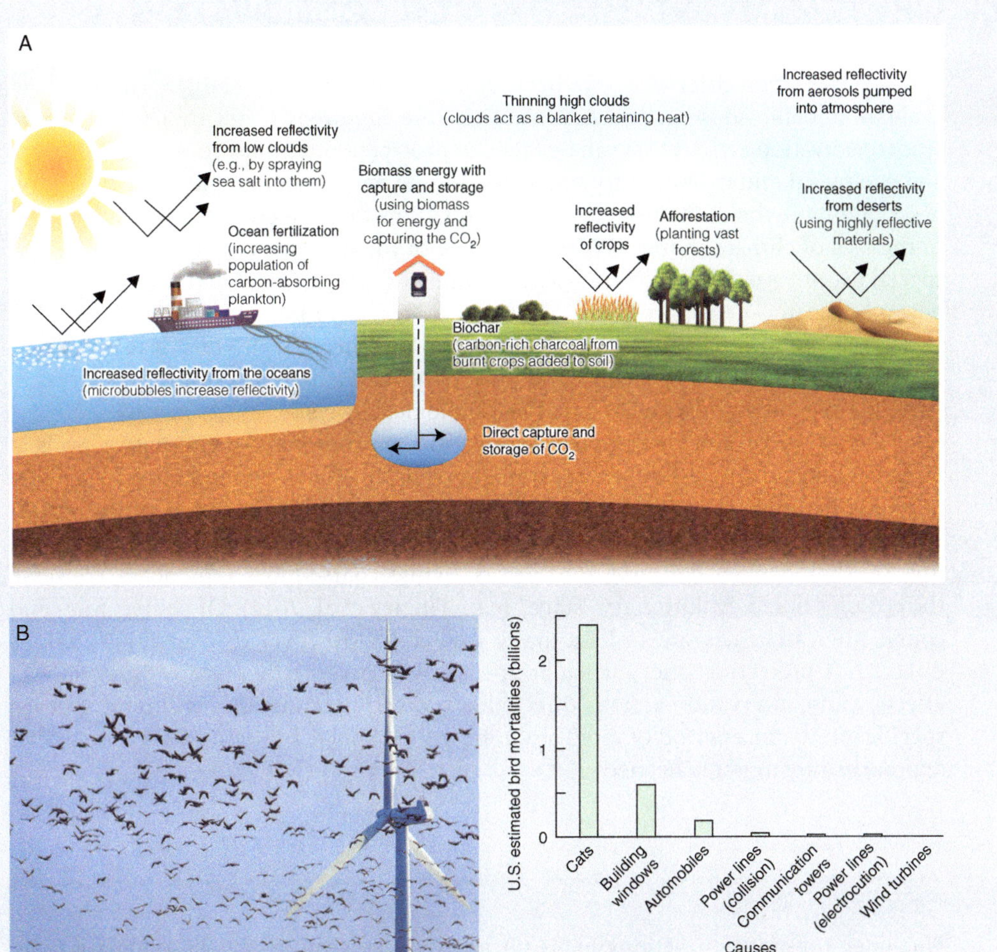

BOX FIGURE 11.1 Climate mitigation strategies often raise complexities and trade-offs. (A) Many different geoengineering approaches have been proposed to mitigate climate change through manipulation of large-scale environmental processes. Some of these focus on reducing incoming solar radiation and others on increasing carbon sequestration. Most are controversial given potential unintended side effects. (B) Alternative energy sources are powerful for reducing greenhouse gas emissions. However, these, too, can sometimes be in tension with conservation goals. For example, wind turbines have received some criticism because their construction and use can impact wildlife. However, other anthropogenic threats are much greater, as shown in this example of different causes of bird mortalities.

Reflection: If you were invited to a debate on the promise and risks of geoengineering, which side of the debate would you want to join? What would be the core of your argument for or against geoengineering efforts?

Source: (A) Piers Forster/University of Leeds; (B) Natalia Paklina/Shutterstock; (C) "Direct Mortality of Birds from Anthropogenic Causes. Scott R. Loss, Tom Will, Peter P. Marra Annual Review of Ecology, Evolution, and Systematics 2015 46:1, 99–120"

WHY IS IT IMPORTANT TO MATCH CONSERVATION ACTIONS TO PARTICULAR BIOLOGICAL LEVELS?

Once conservation priorities have been defined, conservation planners can begin to evaluate possible conservation actions. There are hundreds if not thousands of specific conservation actions that can be taken to protect populations, species, landscapes, and ecological and evolutionary processes. In one review, Heller and Zavaleta (2009) described more than 100 distinct recommendations for protecting biological diversity in the face of climate change. We review a few of these in this chapter's *Core Concepts* special feature. Another database—which addresses a variety of global change stressors beyond climate—lists more than 1,200 specific conservation actions that have been undertaken in different systems (Conservation Evidence 2017). Thus, it is important to determine which possible conservation actions have a high probability of success in particular circumstances and at particular biological levels.

Here we will evaluate examples of conservation actions at two alternative levels of biological organization. **Fine-filter strategies** typically focus on protecting individual populations or species. **Coarse-filter strategies** generally focus more at the ecosystem or landscape level. An analogy commonly used in conservation planning is that fine-filter approaches conserve the "actors" while coarse-filter approaches conserve the ecological and evolutionary "stage" (e.g., Tingley et al. 2014). Of course, fine- and course-filter strategies are not mutually exclusive. Moreover, conservation actions directed at protecting one particular species will necessarily have ecosystem-level effects, and conservation actions directed at protecting landscapes will have species-specific effects. In a period of rapid environmental change, both fine- and coarse-filter approaches are urgently needed.

WHAT ARE EXAMPLES OF FINE-FILTER CONSERVATION STRATEGIES?

Fine-filter conservation actions focus narrowly on protecting specific populations or species from specific threats. Actions focused at the population and species level are particularly appropriate when species are imperiled by global change stressors that can be addressed directly. Fine-filter strategies can have positive cascading effects on ecosystems. For instance, by protecting individual species, the ecosystems upon which they depend are often explicitly valued. Of course, fine-filter approaches can lead to costly efforts to bring relatively few species back from the brink of extinction. They also do not guarantee even distribution of conservation dollars across species or ecosystems. However, fine-filter approaches remain essential. Let us look at a few examples of effective approaches.

Reducing Overharvest

Sometimes it is possible to reduce individual threats on particular species. An obvious example is when hunting or harvesting is the primary cause of endangerment and can be easily regulated. Take illegal poaching of African elephants (*Loxodonta africana* and *Loxodonta cyclones*). Millions of elephants roamed the forest and savanna habitats of Africa less than a hundred years ago. But hunting and habitat loss decimated elephant populations around the continent (e.g., Maisels et al. 2013; Wittemyer et al. 2014).

The single greatest threat has been the killing of elephants to harvest their tusks for the ivory trade. To combat this threat, the Convention on International Trade in Endangered Species (CITES) banned international trade in ivory in 1989. This was a clear example of a fine-filter conservation action targeting a single threat for an iconic species.

Just because the cause of a population decline is easy to identify, it does not mean it is easy to reverse. Figure 11.2 shows that even after the ban in ivory trade, illegal killing has continued, even exceeding the natural mortality rate in some years. At a continental scale, ~100 elephants were killed per day in the peak poaching year. Moreover, the illegal killing rate is tightly correlated with the price of ivory, indicating that ivory trade is still driving elephant poaching (Wittemyer et al. 2014). Thus, reducing

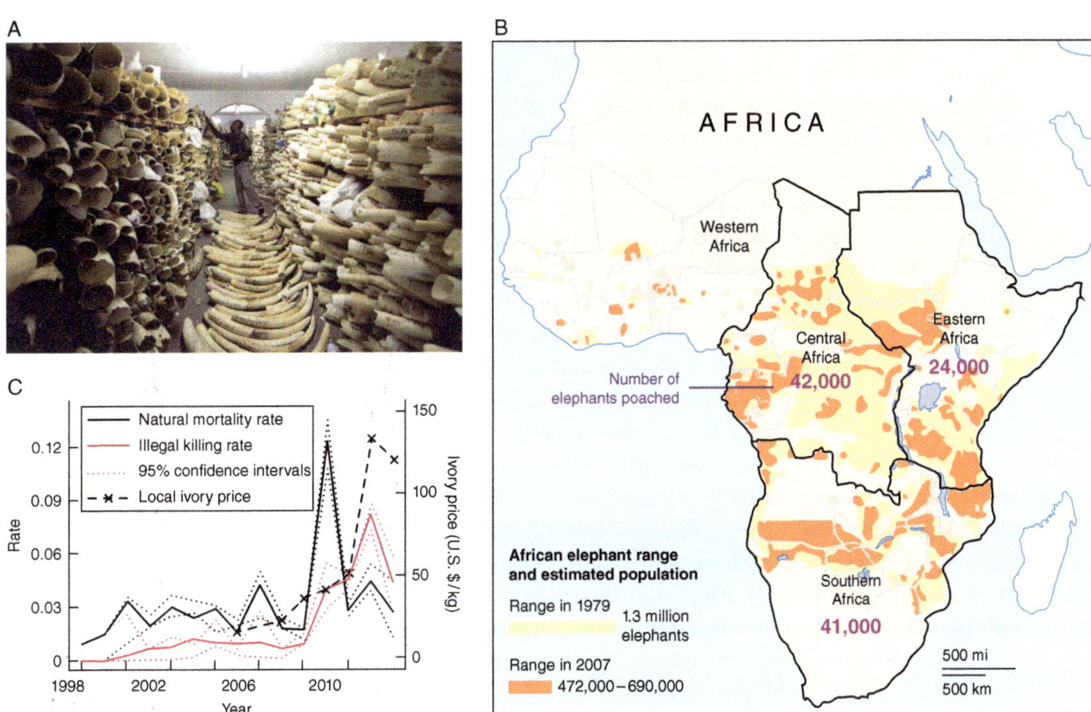

FIGURE 11.2 The need to reduce overharvest. (A) Confiscated tusks from illegally poached elephants. (B) Over the last 40 years, the population size and geographic range of the African elephant have shrunk dramatically. Continental scale data suggest that ~100,000 elephants were illegally killed from 2010 to 2012 alone. (C) Illegal poaching of elephants in one region of Kenya. Illegal killing rate for African elephants (red line) is contrasted with natural mortality rate (black line) and local price for ivory (dashed line).

Reflection: What are some reasons the natural mortality rate of elephants might fluctuate year to year? What are some factors other than local ivory price that might contribute to patterns of illegal elephant killing?

Source: (A) AP Photo/Tsvangirayi Mukwazhi, File; SOURCES: COLORADO STATE UNIVERSITY; SAVE THE ELEPHANTS; MONITORING THE ILLEGAL KILLING OF ELEPHANTS (MIKE); DEPARTMENT OF ZOOLOGY, UNIVERSITY OF OXFORD; KENYA WILDLIFE SERVICE; DIANE SKINNER, AFRICAN ELEPHANT SPECIALIST GROUP, IUCN; (B) (C) "Ivory poaching drives decline in African elephants George Wittemyer, Joseph M. Northrup, Julian Blanc, Iain Douglas-Hamilton, Patrick Omondi, Kenneth P. Burnham Proceedings of the National Academy of Sciences Sep 2014, 111 (36) 13117–13121"

even a single threat for a single species can require an incredible investment and coordination from local to international scales.

Reducing overharvest is not only important for large terrestrial mammals. Myriad species in both aquatic and terrestrial habitats are threatened by unsustainable collection practices. For example, many fish and aquatic invertebrates are taken for the food market, many amphibians and reptiles are collected for the pet trade, and many plant species are harvested for medicinal use (e.g., Carpenter et al. 2014; Purcell et al. 2014; van Wyk & Prinsloo 2018). Conservation actions that monitor and limit take to sustainable levels are therefore essential for the survival of these species.

Captive Breeding

Ideally, conservation actions can be used to protect organisms in their natural habitat, referred to as **in situ conservation**. However, in some cases, a species is protected by removing it from the area under threat—referred to as **ex situ conservation**. Hundreds of research facilities, zoos, aquaria, and botanical gardens around the world maintain captive—or cultivated—populations of plants and animals. Cryogenic preservation of reproductive cells is also increasingly common.

Perhaps the most high-profile examples of ex situ conservation are those in which an endangered species is bred in captivity and then rereleased to the wild. Take, for example, the California condor (*Gymnogyps californianus*), which declined precipitously in the 20th century due to habitat loss, hunting, and lead poisoning (Alagona 2004). The last California condor was brought into captivity in 1987, and the species was extinct from the wild for 5 years before the first individuals were reintroduced. The reintroduction of the California condor is often touted as a captive breeding success story: the wild population has grown consistently over the last 30 years (Ralls & Ballou 2004).

Captive breeding approaches are becoming increasingly sophisticated as researchers search for ways to ensure that released species will thrive under altered environmental conditions. For example, we evaluated the threats facing coral reefs in Chapter 7. In the last several years, coral biologists began to consider whether captively reared corals can be "preadapted" to cope with global warming and other anthropogenic stressors they will be exposed to upon release (e.g., van Oppen et al. 2015). As illustrated in Figure 11.3, a range of interventions is possible. One simple approach would be to induce an acclimation response by pre-exposing captive corals to elevated temperatures. A more aggressive approach would be to induce—and select for—mutations thought to confer heat tolerance. Similar **assisted evolution** approaches—where the process of adaptation is accelerated in the lab to enhance species persistence—are also being considered in other systems. The *Taking a Closer Look* feature in this chapter will further evaluate additional ways that genetic manipulation can be used in conservation and the associated ethical considerations.

Many practitioners consider captive breeding as the only option to avoid extinction for valuable species. However, captive breeding programs are expensive and necessarily focus on few species. Thus, they can be controversial. Take this quote from Pounds et al. (2006): *"Protecting populations in centers for captive breeding may evoke Noah's ark. In reality, these centers would be high-tech lifeboats, costly and of uncertain design, afloat indefinitely on perilous seas. Of the species that would obtain the inevitably limited seats, how many would make it home again, or have a home worth returning to?"*

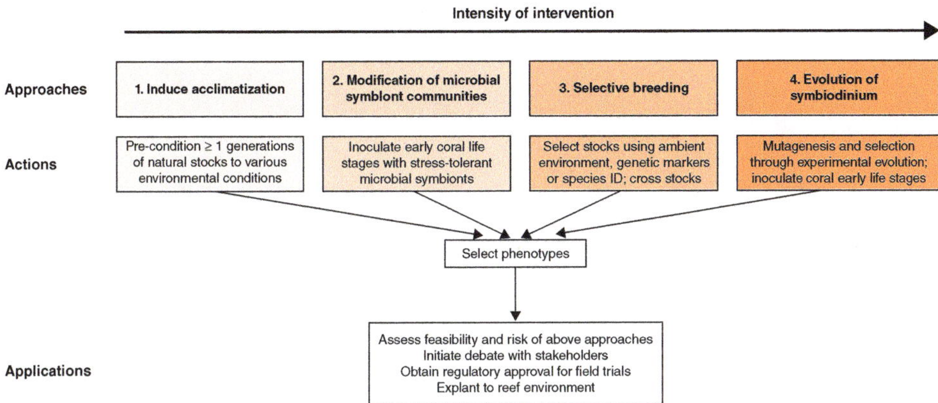

FIGURE 11.3 Possible approaches for enhancing the ability of corals to adapt or adjust to anthropogenic stressors.

Reflection: If you were a scientist working to save a local coral reef, where on the "intensity of intervention" spectrum would you be comfortable working?

Source: "Assisted evolution in coral Madeleine J. H. van Oppen, James K. Oliver, Hollie M. Putnam, Ruth D. Gates Proceedings of the National Academy of Sciences Feb 2015, 112 (8) 2307–2313"

Clearly, captive breeding efforts are most effective when coupled with efforts to reduce threats in the wild so that reintroduction efforts will ultimately be viable. Further, reintroduction efforts are more successful when breeding programs are guided by a nuanced understanding of species ecology and evolution. In this way, colonies can be explicitly managed–for example, to maintain genetic diversity, natural social behaviors, and key mutualisms.

Translocations

Many authors have suggested that the in situ–ex situ dichotomy is overly simplistic and that conservation actions in a changing world will need to draw on a variety of blurred approaches (e.g., Pritchard et al. 2011; Braverman 2014). One such approach is **translocation**: the transport and release of individuals from one location to another. Translocation can be used to recolonize the native range of a species—for example, if local populations were extirpated but the threat has been ameliorated. Translocations can also be used to move individuals to areas that were not previously occupied. In this case, translocation is sometimes referred to as **assisted migration**. Assisted migration is receiving increasing attention as a conservation strategy given that—as we saw in Chapters 5 and 8—current and future areas of suitable habitat will not overlap for many species, and dispersal limited species will be at particular risk of extinction.

One classic translocation success story comes from the black robin (*Petroica traversi*), a small songbird restricted to the Chatham Islands off the coast of New Zealand (shown in Figure 11.4). Having evolved on islands without large terrestrial predators, these birds were particularly vulnerable as introduced mammals such as rats invaded. By 1980, only five individuals remained on only one island. Moreover, the single breeding female—dubbed "Old Blue"—was under constant threat of predation. The last five individuals of

FIGURE 11.4 (A) A single reproductive female black robin (*Petroica traversi*) is credited with saving her species after being successfully translocated to an island without introduced predators off the coast of New Zealand. (B) The African oryx (*Oryx gazella*) is native to the arid plains of Southern Africa. In the 1970s, <100 these animals were introduced from the Kalahari Desert to the Chihuahuan Desert of New Mexico to provide big game hunting opportunities. Without natural predators, the population ballooned to >5,000.

Reflection: How would you define "success" for a translocation effort? Brainstorm what factors contribute to making a translocation successful.

Source: (A) Massaro M, Sainudiin R, Merton D, Briskie JV, Poole AM, Hale ML (2013) Human-Assisted Spread of a Maladaptive Behavior in a Critically Endangered Bird. PLoS ONE 8(12); (B) David Havel/Shutterstock

the species were translocated to a nearby island without introduced predators. Subsequent active management efforts, including cross-fostering chicks to boost reproductive output, led to a now stable population of more than 200 individuals. In this extreme case, a species was rescued from the brink of extinction by a single breeding individual.

However, there are many examples of unintended ecosystem-level consequences of translocations. Translocations have risks to both the focal species and the recipient ecosystem as food webs, energy, and nutrient flows are altered. One classic example is the introduction of African oryx (*Oryx gazella*) to the southwestern US desert (shown in Figure 11.4). In the early 1970s, 93 oryx were released into the Chihuahuan Desert of New Mexico with the intention of providing exotic large-game hunting experiences. However, with no natural predators, the oryx population ballooned. More than 5,000 oryx have been killed by hunters, but the population is still estimated at more than 5,000 strong. Millions of dollars have been spent trying to keep oryx population expansion in check and trying to protect fragile native plants and animals from their effects. Similar unintended consequences have been observed in aquatic systems where fish translocations have led to declines in other species and disrupted trophic interactions (e.g., Minckley 1995). Thus, translocations of any sort need to explicitly consider potential effects on ecological interactions in both donor and recipient communities.

One leading edge for translocations is the potential of moving individuals with favorable traits. This type of targeted **gene flow** could be used to introduce resistance to a particular global change threat more broadly in a species range (Kelly & Philips 2015, 2018). For example, some amphibian species exhibit variation in susceptibility to the emerging infectious disease chytridiomycosis, which we discussed in Chapter 8. If individuals from more resistant populations could be introduced to less resistant populations, it is possible that resistance could more quickly spread throughout the range of a species.

Although targeted gene flow refers to moving *individuals* with favorable alleles, it is also now possible in some cases to move the *alleles* themselves. Various types of **genetic engineering** raise the specter of directly modifying the genomes of endangered species. Of course, this approach raises numerous ecological and ethical concerns. We return to the ethical considerations of applying genetic engineering to conservation questions in the *Taking a Closer Look* feature at the end of this chapter.

WHAT ARE EXAMPLES OF COARSE-FILTER CONSERVATION STRATEGIES?

As we have just seen, fine-filter strategies focus on conserving specific *products* of evolution. But what about the *processes* that generated present-day biological diversity? Can conservation measures preserve species interactions, ecosystem functions, and evolutionary potential in a changing world? Coarse-filter conservation actions attempt to do this by focusing more generally on landscapes, ecosystems, and processes. Actions focused at the landscape level can have critical trickle-down effects to species currently inhabiting those areas. They can also provide a backdrop for ecological and evolutionary processes over larger spatial—and longer temporal—scales. Of course, landscape-level conservation is not inexpensive, but it is of fundamental importance in the face of global change. Let us look at a few examples of coarse-filter approaches.

Creating Reserves

From small local parks to large regional reserves, setting aside land for biodiversity has long been a staple of the conservation portfolio, as illustrated in Figure 11.5. Evidence that humans even created wildlife sanctuaries dates back thousands of years. In modern times, conservation planning tools for reserve selection are increasingly sophisticated (reviewed in Sarkar et al. 2006; Kukkala & Moilanen 2013). Computer algorithms can be used to apply optimization rules and select reserves that help protect particular biodiversity or landscape features.

For example, new reserves can be selected based on their complementarity to existing reserves (e.g., adding areas which contribute the most new species or attributes) or irreplaceability (e.g., areas with unique species or attributes). Moreover, reserve selection can explicitly integrate conservation and economic priorities, for example by maximizing environmental benefits while minimizing economic costs. Using a systematic conservation planning approach thus allows managers to locate and design reserves based on explicit criteria (Margules & Pressey 2000).

In one case study, researchers took a systematic reserve selection approach for Shanxi Province in China (Zhang et al. 2016), a region under threat from intensive coal mining. Since 1980, numerous small nature reserves have been established in this region, and researchers set out to assess the effectiveness of these reserves in protecting Shanxi's rich endemic flora. They evaluated geographic occurrence and habitat suitability data for more than 50 threatened plant species and used a reserve selection algorithm to identify high-priority areas for conservation. They found that 100% of the threatened plant species could be protected with only ~5% of the land area protected.

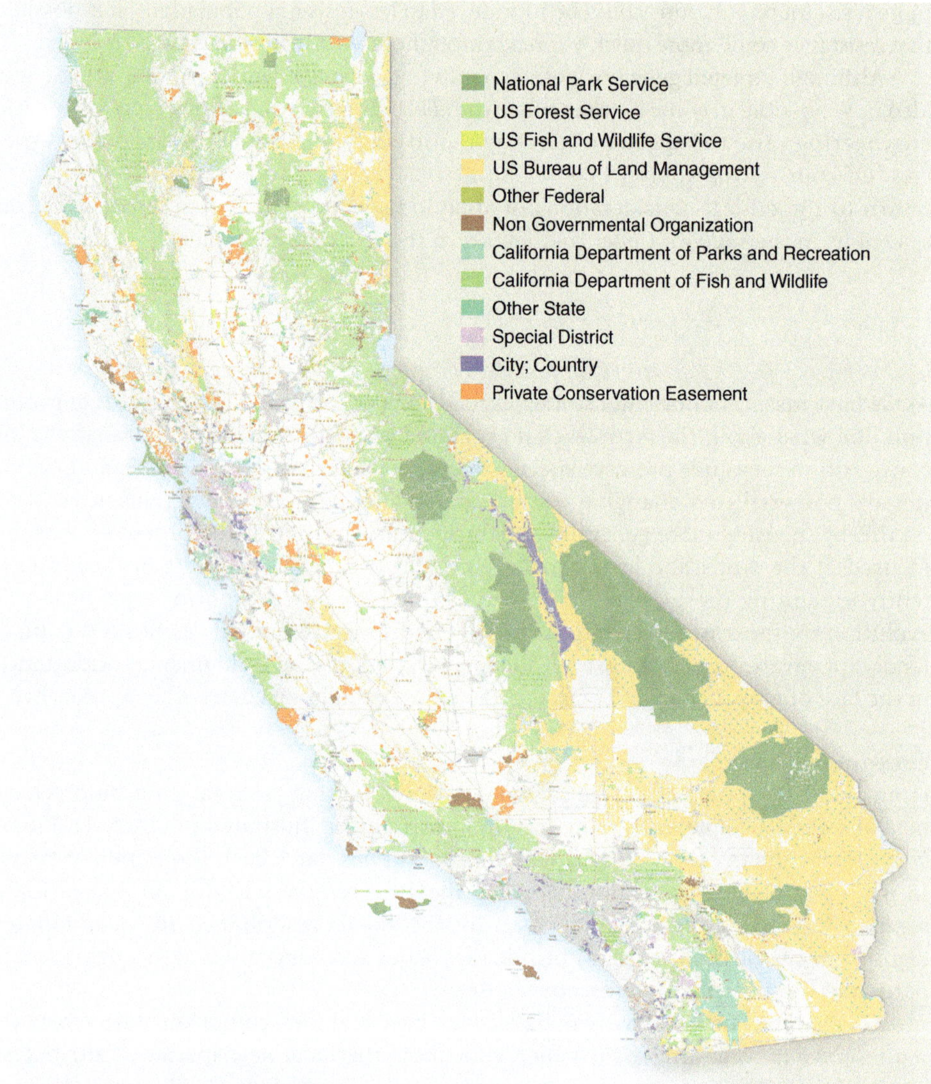

FIGURE 11.5 Networks of protected areas can be composed of lands owned and managed by diverse stakeholders. In this example from California, lands protected by federal, state, county, and city government are complemented by lands protected by nongovernmental organizations and private land holders.

Reflection: Compare and contrast how small local parks and large wilderness areas contribute to conservation goals and objectives.

Source: California protected areas database (www.calands.org)

Moreover, they identified several areas that would be highest priority for complementing the existing reserve system.

Examples such as this demonstrate that a systematic approach to reserve design can maximize benefits for biodiversity. However, important challenges remain. Identifying priority regions for conservation is easier than procuring, financing, and managing land. Obtaining land for reserves opens challenging questions about whether and what kinds of human activities will be permitted in these areas—from strict wilderness protection to sustainable use by indigenous populations. Moreover, even relatively remote reserves will change over time due to climate change and other human impacts, with no guarantee that they will continue to be suitable habitat for priority species. Finally, reserves are expensive, and current global expenditures on reserves may average ~6 billion US dollars per year (James et al. 2001). However, even these large expenditures represent a tiny fraction (perhaps <0.1%) of the annual value of biological diversity in ecosystem services (Pimentel et al. 1997).

Ultimately, robust reserve systems are essential for protecting biological diversity on our planet. In fact, many scientists have called for a rapid increase in protected lands. In one such proposal, the preeminent biologist E. O. Wilson advocates for fully 50% of Earth to be set aside in natural reserves (Wilson 2016). As Figure 11.5 illustrates, governments, nongovernmental organizations, private landowners (foundations, corporations, and individuals), and local communities are all important in establishing nature reserves (e.g., Pasquini et al. 2011).

It is important to draw particular attention to the central role **indigenous people** play in land conservation policy and practice. Indigenous communities have deep connection to place-based natural resources and to the stewardship of those resources. They have practiced complex, nuanced, and adaptive strategies of natural resource management for millennia (e.g., Eckert et al. 2018). Despite a brutal history of colonization, oppression, genocide, and forcible stripping of ancestral lands, indigenous people currently manage or have rights over more than 37 million km^2 of land across all the globe (Garnett et al. 2018). Thus, the importance of affirming—and returning—sovereignty over lands to indigenous communities cannot be overstated. Ecosystem stewardship and recovery efforts must be aligned with honoring indigenous cultural practices and actively championing indigenous rights. We will return to the question of how diverse groups of stakeholders can contribute to protecting-and financing—land for conservation in Chapter 12.

Re-establishing Corridors

As we just saw, identifying large tracts of intact habitat is typically the focus of reserve planning. However, sometimes the issue is not lack of core habitat, but lack of *connectivity* between core habitat patches. Connectivity is an important issue at several temporal scales. First, many species undergo annual migrations. Although key feeding and breeding grounds may be protected, anthropogenic activities may compromise movement corridors and stop-over sites along migratory routes. Second, populations of many species experience natural periodic cycles of extirpation and recolonization. In these cases, connectivity among habitat patches is essential for recolonization and recovery in these species. Third, many species will experience changes over longer timescales in suitable habitat, for example as climate change drives species poleward. When current and future suitable habitat patches do not overlap, connectivity is especially important for species that lack long-distance dispersal capacity.

One case study comes from Tanzania, a country that contains multiple biodiversity hotspots and numerous internationally recognized sites for plant, bird, and mammal conservation. Some local species undergo incredible long-distance migrations; for example, as shown in Figure 11.6, more than 1 million wildebeest (*Connochaetes taurinus*) migrate through the Serengeti-Mara ecosystem each year. In recent years, anthropogenic activities, including deforestation, land conversion for agriculture, and road construction have severed several key corridors to wildlife movement. One recent study evaluated existing and severed corridors using interviews with local communities, anthropogenic land conversion data, and computer models looking for least-cost corridors to wildlife movement (Riggio & Caro 2017). As illustrated in Figure 11.6,

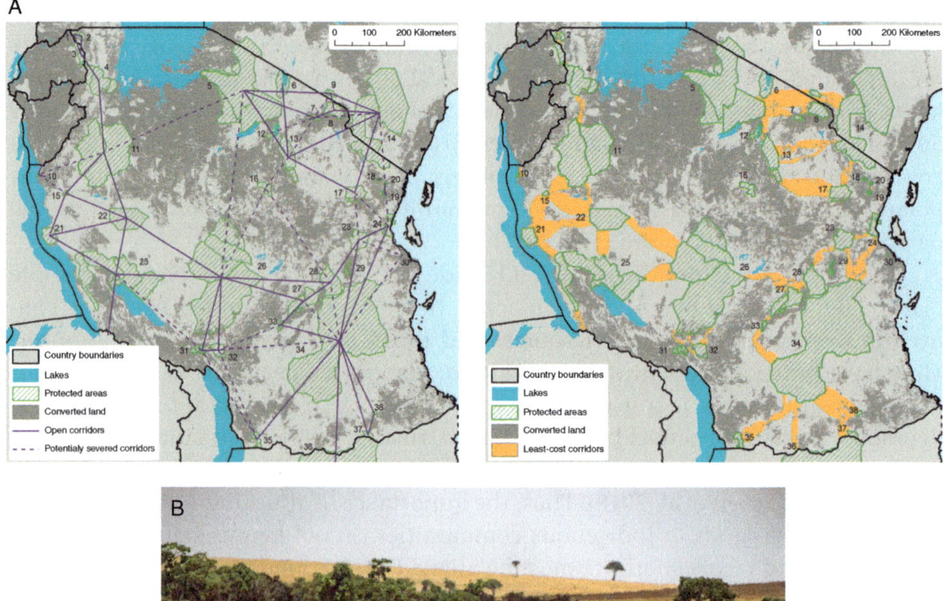

FIGURE 11.6 (A) Protected areas in Tanzania. Left panel shows open (purple lines) and potentially severed (dashed purple lines) corridors for wildlife movement and migration between protected areas. Right panel shows areas identified as "least-cost corridors" (in orange). These corridors are geographic areas that—if protected—would provide the greatest protection to migration routes and population connectivity. (B) Preserving connectivity at the landscape level is essential for migrating species—like the million wildebeest (*Connochaetes taurinus*) that traverse the Serengeti-Mara ecosystem each year.

Reflection: What are some similarities and differences in conservation strategies for protecting migratory routes for terrestrial, aquatic, and aerial species?

Source: (A) Riggio J, Caro T (2017) Structural connectivity at a national scale: Wildlife corridors in Tanzania. PLoS ONE 12(11); (B) Ricardo Matos Camarinha/Shutterstock

their analysis identified remaining open corridors and proposed key stepping-stones between National Parks that could be safeguarded to preserve migratory routes.

Although wildlife corridors are important, they are not a global solution to conservation in a changing world. Establishing narrow corridors between static reserve patches does not address the larger issue of habitat destruction or the concern that areas of suitable habitat are rapidly shifting. Moreover, terrestrial corridors for large migratory species tend to garner the most attention, but other types of corridors are necessary for other species. For example, one recent study revealed the importance of dark corridors for commuting bats, especially in a world increasingly lit at night by artificial light (Zeale et al. 2018). In addition, the importance of corridors in the marine realm is increasingly recognized (e.g., Pendoley et al. 2014; Krost et al. 2018), although tracking animal movement and creating corridors under water presents additional challenges. Ultimately, creating corridors between larger habitat patches is important, but the utility of these corridors for diverse organism under a range of possible future conditions must be considered.

Re-establishing Natural Disturbance Regimes

In addition to establishing reserves and rehabilitating corridors, it can also be important to re-establish natural disturbance regimes. While human activities have increased the frequency and severity of disturbance in many ecosystems, human societies have also reduced environmental fluctuations in other ecosystems. For instance, practices like levee building and fire suppression have reduced the prevalence and severity of floods and fires. Many species are actually adapted to high-disturbance ecological regimes. For example, the iconic giant sequoia (*Sequoiadendron giganteum*) relies on periodic wildfires for its seeds to germinate and has been negatively impacted by practices of fire suppression (Parsons & DeBenedetti 1979).

Re-establishing natural disturbance regimes can thus benefit native species adapted to volatile environments. For example, as Figure 11.7 shows, several highly endangered butterfly species have begun to recover with the reintroduction of disturbance—like fire and grazing—to their ecosystems (e.g., Schultz & Crone 1998; Haddad 2018). Of course, reintroduction of these natural processes comes with its own risks (e.g., Backer et al. 2004), but considering ways to re-establish natural regimes at a landscape level is an important coarse-filter conservation approach.

Enhancing Habitat in Highly Modified Settings

Setting aside land for biodiversity conservation is a hallmark of modern conservation. Traditional conservation efforts have for decades focused on preserving and restoring wild lands. However, anthropogenic impacts on the planet are now so widespread that conservation efforts no longer can focus solely on pristine habitats. As we have discussed, approximately 70% of the terrestrial surface of the planet is impacted by human activities. Even in areas with native vegetation, habitat patches are increasingly fragmented by urban or agricultural development. Thus, it is essential to consider how even highly modified landscapes can be enhanced to support local biological diversity. The practice of promoting conservation in human-dominated landscapes is sometimes referred to as **reconciliation ecology.** Reconciliation ecology recognizes that all landscapes can be enhanced to better support biological diversity.

There are many opportunities for conservation action outside of formally protected areas on both public and private lands. Take, for example, enhancing habitat for

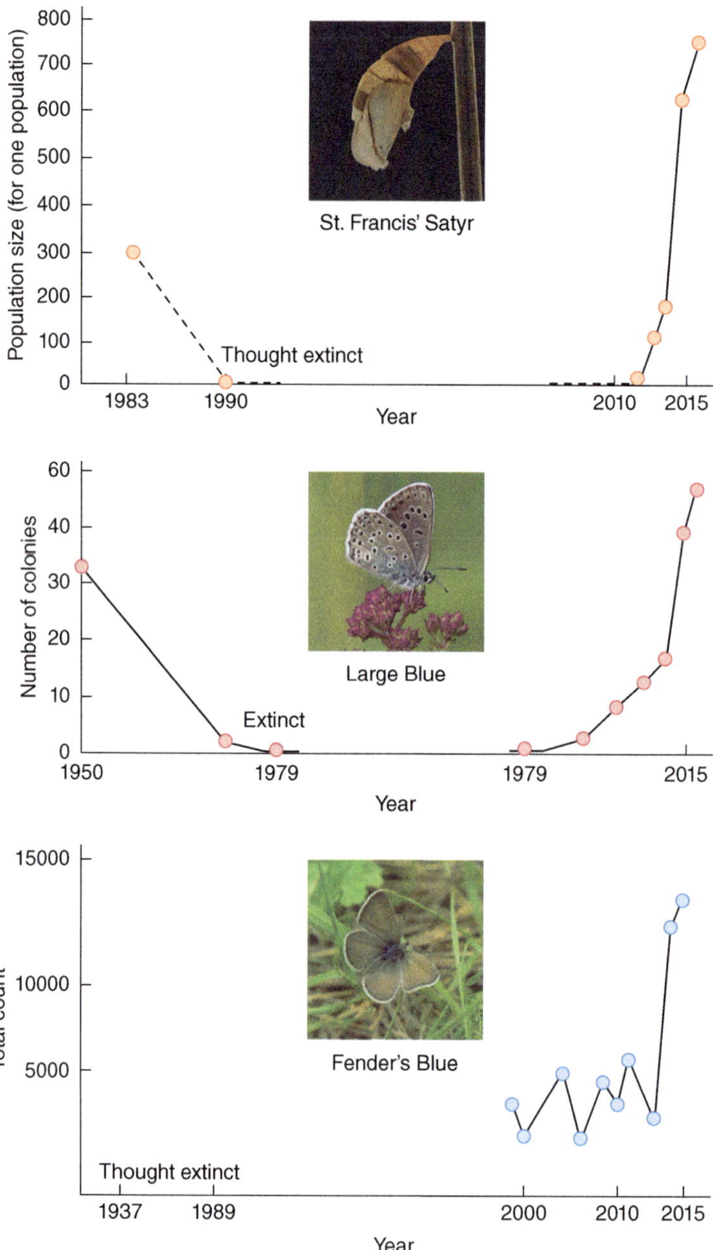

FIGURE 11.7 The recovery of three rare species of butterfly once natural disturbance regimes were re-established.

Reflection: How would you weigh the potential benefits of re-establishing natural fire regimes for an endangered species with the potential concerns of nearby human communities?

Source: Nick M. Haddad, Marcel Holyoak, Tawny M. Mata, Kendi F. Davies, Brett A. Melbourne, Kim Preston, Species' traits predict the effects of disturbance and productivity on diversity, Ecology Letters, Volume 11, Issue 4, April 2008, Pages 348–356

pollinators in agricultural settings. Pollinators—such as bees, butterflies, and birds—provide essential ecosystem services to agricultural lands (e.g., Klein et al. 2007). However, the same valuable pollinators are negatively impacted by mainstream agricultural practices, including pesticide use (which can contribute directly to mortality), monoculture planting (which can reduce foraging opportunities), and introduced pollinators (which can outcompete native species).

One clear example of enhancing habitat for native pollinators in agricultural settings without removing land from production is planting native vegetation at crop edges. These hedgerows can provide a more continuous resource for foraging and refuge from stressors like pesticides and soil tilling. Many studies have demonstrated that adding hedgerows significantly increases the number of species—and the total number—of native bees on agricultural lands (e.g., Morandin & Kremen 2013; M'Gonigle et al. 2015). As Figure 11.8 shows, native pollinators increase in density and diversity not only in the hedgerows but also in the adjacent fields.

Similar efforts to enhance habitat are increasingly common in urban settings. One example is an ambitious implementation of the rooftop garden taking root in some cities. As illustrated in Figure 11.9, these "vertical forests" create habitat on large buildings and can include diverse species and thousands and thousands of individual plants. Rooftop gardens and other urban green spaces provide myriad benefits, from reducing the urban heat island effect, to increasing local biodiversity, to improving human standards of living, to helping combat global climate change (e.g., Sisco et al. 2017; Sun et al. 2019). Although not all species will be able to thrive in highly modified environments, these efforts to increase biodiversity in even our most impacted landscapes will be critical in the 21st century and beyond.

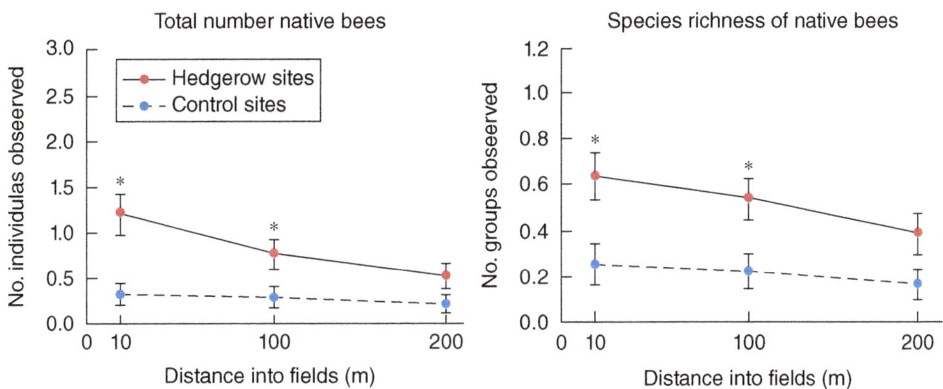

FIGURE 11.8 Enhancing biodiversity in agricultural settings. Bees and other native insects play important roles as pollinators not only in undisturbed ecosystems but also in highly modified agricultural environments. Restoring native flowering plants as hedgerows near crops provides opportunities for native bees to refuge, nest, and forage. Agricultural sites with hedgerows have a have a greater total number - and a greater diversity - of native bees than those without.

Reflection: Why do you think that the effect of hedgerows on native pollinator abundance and diversity extends hundreds of meters into the fields beyond the hedgerows themselves?

Source: © 2013 by the Ecological Society of America

FIGURE 11.9 Enhancing biodiversity in cities. A number of city areas around the world are investing in "vertical forests." Incredible numbers of plants can be grown on skyscrapers and can transform the urban landscape and the ecosystem services it provides.

Reflection: We have looked at numerous different conservation strategies, from protecting large, undeveloped wilderness areas to enhancing habitat in highly modified landscapes. How might these two approaches complement each other and/or be in conflict with each other? Where do your own views fall on the spectrum from considering conservation as an activity that should be focused in set-asides (like preserves) versus modified habitats (like cities)?

Source: Boeri Studio

In Sum

Of course, multiple different fine- and course-filter approaches can be applied simultaneously to support the recovery of a species or ecosystem. Ultimately, a portfolio of approaches will be needed: species- and ecosystem-level efforts in both wild and modified environments. We will return to the topic of integrative conservation strategies that engage diverse interest groups in the next chapter.

WHAT IS ADAPTIVE MANAGEMENT?

Regardless of which conservation approaches are implemented, conservation planning in an era of rapid global change must be iterative. Given complex interactions and feedbacks in ecosystems, conservation actions could have expected—and unexpected—consequences over many different timescales. Whether initial interventions are successful or unsuccessful, management plans need to be updated to reflect current conditions. Moreover, natural systems are dynamic entities and will change not only because of conservation actions but also because of natural temporal cycles and shifting anthropogenic stressors. Thus, monitoring, evaluation, and redesign are equally as important as initial design and implementation.

The holistic process from identifying conservation goals and assessing alternative actions, to planning and implementing conservation interventions, to tracking

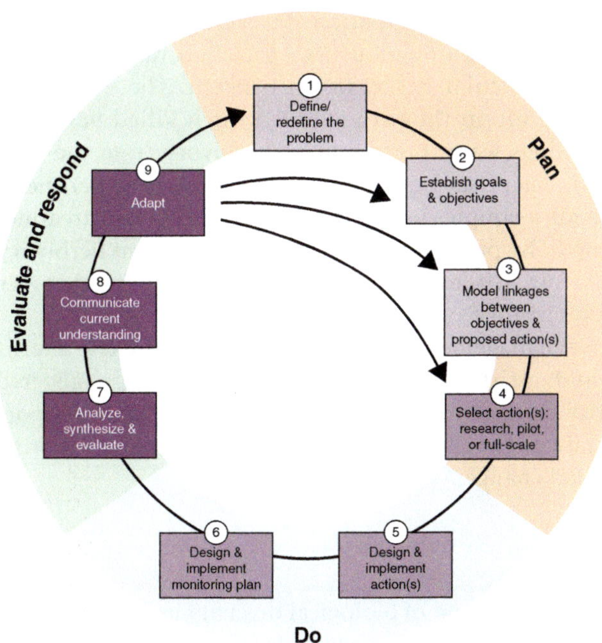

FIGURE 11.10 Adaptive management is an iterative, holistic approach to conservation planning, implementation, and evaluation. Ongoing monitoring and data collection allow a continuous improvement of conservation actions. The adapt phase could be a simple fine-tuning based on annual monitoring results or could be radical redesign based on new data about the system. Note that the use of "adapt" in this context refers not to organismal adaptation but to iteratively altering conservation approaches.

Reflection: Brainstorm at least three challenges you expect would confront adaptive management approaches.

Source: Delta Stewardship Council

effectiveness and ecological responses, to reassessing and modifying future goals is referred to as **adaptive management**. As shown in Figure 11.10, adaptive management promotes flexible decision making with the explicit intention that conservation approaches will be adjusted as conditions change and as outcomes from initial interventions are understood. Adaptive management also allows conservation planners to be explicit about key uncertainties and to interact dynamically with complex living systems.

Iterative adaptive management approaches require strong interaction and communication among many different stakeholders from scientists, to conservation managers, to policymakers, to local communities. In addition, the most powerful adaptive management approaches involve experimental replication and contrasting management regimes to truly assess effectiveness of alternative approaches. This is a high bar, and some authors argue that very few projects have truly implemented a comprehensive adaptive management approach (e.g., Rist et al. 2013; Westgate et al. 2013). However, there are increasingly explicit databases and software programs designed to support integrative and iterative conservation planning (e.g., www.conservationevidence.com; www.miradi.org; cmp-openstandards.org).

There are also some clear examples of adaptive management in action. Take, for example, the effort to rescue an endangered population of the black cypress pine (*Callitris endlicheri*) from extirpation (Mackenzie & Keith 2009). Although this species is not endangered throughout its Australian range, the Woronora Plateau population was of particular conservation concern. The population was initially decimated by a large fire in the early 2000s, which killed 98% of all individuals. After the fire, seedlings were threatened by herbivory from the invasive rusa deer (*Cervus timorensis*). Scientists implemented a controlled experiment whereby they tested the effects of a simple, low-cost herbivore exclusion treatment on seedling survival and growth. By protecting exposed seedlings from herbivory, they boosted seedling survival from 31% to 84%, assisting with recruitment at a critical juncture in this population's history.

Thus, adaptive management approaches can explicitly test alternative conservation strategies and assist in long-term persistence of endangered populations. Moreover, adaptive management strategies, which explicitly incorporate uncertainty and focus on iterative implementation, are especially important during this time of rapid environmental change.

CONCLUSION

Conservation and management of biological diversity in an era of global change present particular challenges. Biological systems face rapid rates of environmental change and synergies among anthropogenic stressors. However, there are a number of effective conservation strategies that can be applied at the population and landscape levels. These efforts are most successful when conservation priorities and goals are explicitly defined, when long-term resources are available for implementation, when ongoing data collection guides iterative planning, and when the sovereignty of indigenous people is deeply honored.

For those skeptical that conservation dollars spent have real impact, a recent study evaluated the relationship between conservation spending and global biodiversity loss (Waldron et al. 2017). Evaluating a decade of data from more than 100 countries, the researchers found that countries with higher conservation spending achieved reduced biodiversity loss. Moreover, they discovered that financial investment in conservation reduced biodiversity loss by an average of 29% per country. Thus, investing in conservation has measurable gains even over short time horizons.

However, to truly meet global conservation priorities, higher levels of conservation investment are required. An influential study that estimated the cost of meeting global conservation targets suggested that funding for conservation projects needs to increase by an order of magnitude (McCarthy et al. 2012). Although this may sound daunting, the total sums are not large relative to other societal priorities. In fact: *"less than 0.01% of global gross domestic product would be enough to all but stop species extinctions"* (Possingham & Gerber 2017). Ultimately, a global investment in biodiversity conservation could be highly effective if there is motivation for change, a topic we turn to in the next chapter.

MEET THE DATA MAXIMIZING EVOLUTIONARY DIVERSITY

Who Are the Scientists and What Did They Set Out To Do?

Here we will evaluate a study entitled "Comparing Strategies to Preserve Evolutionary Diversity," published by Magnuson-Ford et al. (2010) in the *Journal of Theoretical Biology*. This study was part of an international collaboration between scientists in Canada and New Zealand to understand how different conservation strategies could affect lemur diversity (see Box Figure 11.2).

As we explored earlier in this chapter, there are many different ways to define conservation priorities. For example, conservation planners can take an "egalitarian" approach by treating all species and regions as equally worthy of protection. Alternatively, conservation planners can take a "targeted" approach, choosing to prioritize particular species or communities. Targeted approaches are varied and can attempt to maximize many alternative aspects of biological diversity. For example, conservation planners can focus on maximizing endemism, species richness, phylogenetic diversity, or functional diversity. Applying alternative metrics of diversity may or may not lead to similar conservation outcomes.

The lemurs of Madagascar are an excellent case study for understanding the impact of different criteria on conservation outcomes. Lemurs are endemic and geographically restricted primates found only on the island of Madagascar. The lemur lineage is at least 50 million years old (Godinot 2006), and dozens of morphologically and behaviorally diverse species have evolved during that time. In modern times, lemurs have come under threat from habitat loss, habitat fragmentation, hunting for bushmeat, and capture for the pet trade. Of the more than 100 lemur species evaluated by the International Union for Conservation of Nature (IUCN) Red List, more than 90% are considered vulnerable to extinction (~20% listed as critically endangered, ~50% as endangered, and ~20% as threatened). Given their endemism, endangerment, occurrence across different climatic regimes, and important trophic role in Malagasy ecosystems, many different criteria could be appropriate for prioritizing lemur species for conservation.

What Are *Your* Predictions?

Before you read on, take a few minutes to think about how to apply what you have learned in this chapter about conservation priorities to this case study. Start by answering the following questions:

- *What are some conservation strategies that might be considered egalitarian versus targeted in this system?*
- *Do you expect that egalitarian and targeted conservation approaches would lead to different total amounts of future lemur diversity?*

BOX FIGURE 11.2 Lead author Karen Magnuson-Ford (A) and senior author Dr. Mike Steel (B) were part of an international collaboration between scientists in Canada and New Zealand to understand how different conservation strategies could affect lemur (C) diversity.

Source: (A) Karen Gordon; (B) Mike Steel; (C) Duke Lemur Center

(Continued)

- *For targeted approaches, do you think it makes more sense to protect the most endangered species or to conserve the maximum overall lemur diversity?*

Now write a summary statement about whether—and how—you expect different conservation strategies to affect lemur conservation priorities and biodiversity.

What Were the Scientists' Predictions?

The authors did not have formalized predictions, but they had clear questions. First, Magnuson-Ford et al. (2010) wanted to know whether egalitarian and targeted approaches are more effective at conserving phylogenetic diversity. Second, they wanted to identify the subset of lemur species that—if protected from extinction—would preserve the most overall phylogenetic diversity in this group. In their own words: *"Which 10 species if they were to become extinct would lead to the greater loss in phylogenetic diversity compared to any other combination of species?"*

What Data Were Collected?

The authors used mathematical models of extinction to compare the utility of alternative conservation strategies for preserving evolutionary diversity. The authors first developed an algorithm to calculate how much phylogenetic diversity would be lost if a species were to go extinct. They then constructed two conservation strategies in their model. First, they developed an "egalitarian" strategy, analogous to a conservation approach of helping all species a little. In practice this means reducing the extinction rate of all species by a small amount in the model. Second, they developed a "targeted" strategy, analogous to focusing conservation efforts on endangered species. They applied this approach to a case study of lemurs.

For the lemur case study, the authors constructed a phylogeny of 62 lemur species using standard methods and mitochondrial DNA sequence data. Next, they collected IUCN endangerment data to assess the extinction probability of each lemur species in the phylogeny. They used computational simulations to predict future lemur phylogenetic diversity under the two conservation strategies. They also used the phylogeny to identify the species whose extinction would lead to the greatest loss of phylogenetic diversity.

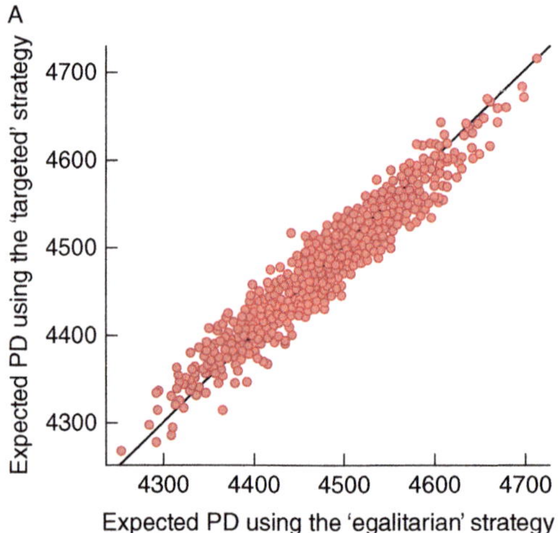

BOX FIGURE 11.3 (A) Predicted future phylogenetic diversity (PD) under two different conservation strategies. Each circle represents one iteration of the simulation, and the diagonal line shows when the strategies are equivalent. (B) Phylogenetic tree of the 62 lemur species included in this study. The red branches indicate the 10 lemur species whose extinction would lead to the greatest loss of phylogenetic diversity. The red arrows indicate those species listed as critically endangered by the IUCN.

Source: Karen Magnuson-Ford, Arne Mooers, Sébastien Rioux Paquette, Mike Steel, Comparing strategies to preserve evolutionary diversity, Journal of Theoretical Biology Volume 266, Issue 1, 7 September 2010, Pages 107–116

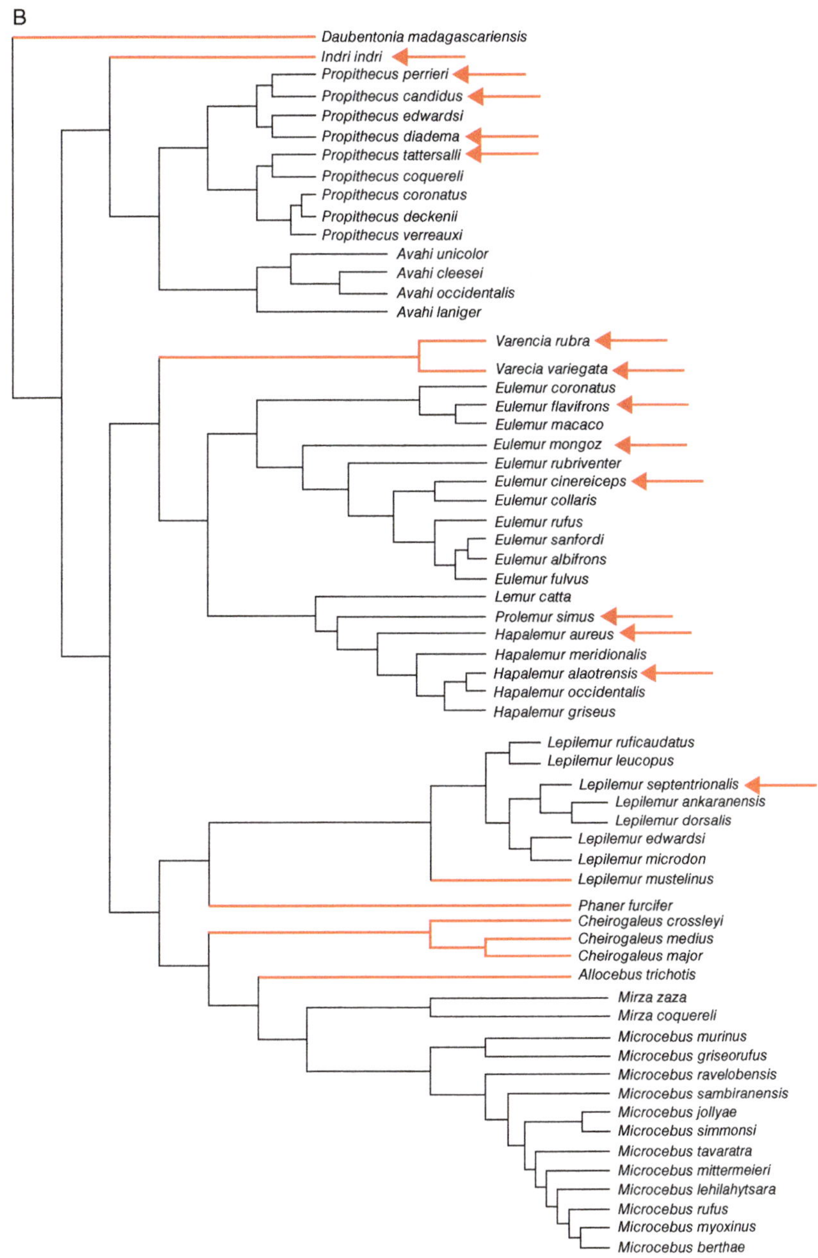

FIGURE 11.3 Continued.

What Is *Your* Interpretation of the Data?

Before you read on, take a few minutes to interpret the results in Box Figure 11.3. Consider the following questions as you look at the figures:

- *From the mathematical model, is there a large difference between expected phylogenetic diversity using a targeted versus an egalitarian approach?*
- *If a targeted approach was taken, would similar species be targeted if the goal was to conserve the most*

(Continued)

endangered species versus to conserve the species that contribute most to phylogenetic diversity?

- *To maximize phylogenetic diversity, what types of lineages should be protected?*

Write a sentence or two describing the *key findings* of this study and their *significance*.

What Was the Scientists' Interpretation of the Data?

From the simulation study, the mean expected future phylogenetic diversity using "egalitarian" and "targeted" approaches was nearly identical. The authors, however, caution against overinterpreting these results. Although the *"mean difference between strategies was negligible, outcomes for individual cases can be quite different."* Moreover, as we discussed in Chapter 1, mathematical modeling studies are highly sensitive to baseline assumptions and parameters used for the simulated dataset. Therefore, it is still an open question whether an egalitarian strategy (which spreads conservation efforts equally across species) or a targeted strategy (which focuses conservation efforts on particular species) would be most effective in this system.

From the phylogenetic analysis, the scientists identified 10 species that contribute most to evolutionary diversity in the lemur clade. If any one of these species went extinct at random, ~9% of phylogenetic diversity would be lost, but if all 10 species went extinct, ~33% of phylogenetic diversity in the clade would be lost. Most of these species exhibit relatively long branch lengths in the tree displayed in Box Figure 10.3, indicating their evolutionary distinctiveness. Therefore, a targeted approach focused on phylogenetic uniqueness could be a reasonable strategy for lemur conservation.

Preserving phylogenetic diversity can be a reasonable target for biodiversity conservation, and several other studies have shown that when phylogenetic diversity is maximized, other desirable traits (like economic or ecological importance) are not sacrificed (e.g., Forest et al. 2007). However, maximizing phylogenetic diversity is not necessarily an umbrella strategy. For example, in the Magnuson-Ford et al. study, prioritizing phylogenetic diversity did not lead to protecting the most highly endangered lemur species. In fact, there was only one species of lemur that would be a top-10 priority based on both the IUCN endangerment criteria and the phylogenetic diversity criteria. As the authors state: *"While lemurs provide an excellent example of how the mathematical models and algorithms presented here may be applied to real biological systems, we acknowledge that our model only considers the preservation of future phylogenetic diversity under one metric (PD) and does not include other factors such as cultural significance or ecological diversity."*

Thus, creating models for conservation prioritization that maximize across multiple attributes is an important area of ongoing work. One example is the Evolutionarily Distinct and Globally Endangered (EDGE) Initiative, which integrates multiple criteria by explicitly valuing both phylogenetic uniqueness and risk of extinction (e.g., Redding & Mooers 2015). This approach can be used not only for species selection but also for reserve design because geographic regions that add more unique evolutionary diversity can be prioritized for protection. Thus, mathematical models that explicitly integrate multiple criteria can help optimize conservation strategies for maximal benefit to biodiversity.

What Are *Your* Ideas for Future Research Directions?

Given the results from the Magnuson-Ford et al. study, what next steps do you envision for this research program? If you had been involved in the research, how would you follow up? Start by considering the following questions:

- *What hypotheses does this study generate that could be tested in other systems?*
- *Are there questions that could be asked using on-the-ground studies of natural lemur populations?*
- *What are some specific fine- or coarse-grain strategies for lemur conservation that could be studied using an adaptive management approach?*

Now write a few sentences about future directions on this research theme.

TAKING A CLOSER LOOK EMERGING TECHNOLOGIES AND CONSERVATION ETHICS

This is an exciting time for conservation in the context of global change. New tools and technologies allow scientists to assess population health, compare alternative management strategies, and monitor the effects of conservation actions in ever more sophisticated ways. Computational tools can be used for population viability analyses, predicting future suitable habitat, and optimizing the size, shape, and location of nature preserves. Physiological tools can be used to measure diverse aspects of population health and recovery—from stress responses to immunological health to ecotoxicology. Genetics and genomic tools can be used to assess levels of genetic variation, identify appropriate source populations for translocations, and monitor response to changing environmental conditions.

However, all novel technologies raise ethical questions about their appropriate application. This is particularly true for genetic approaches that now allow scientists unprecedented opportunity to manipulate the very code of life at the molecular level. Here we will explore several emerging approaches and the questions they raise for conservation in an era of global change. We will do this using a series of thought experiments about species reintroductions. For each thought experiment, take a moment to reflect on your opinion before continuing.

Thought Experiment 1

Imagine that an ecologically important species has very recently been extirpated from a local community, and the loss was clearly caused by human activities. All else being equal, should we reintroduce this recently extirpated species if it is still found nearby?

As we evaluated earlier in this chapter, translocations or reintroductions of species to areas where they have recently been lost is a commonly used approach in biodiversity management. Translocations and reintroductions can be data-driven to ensure a match between donor and recipient populations. For example, genetic data can be used to ensure that historical population structure is preserved and that populations have adequate genetic variation.

Even within species, translocations or reintroductions can be highly controversial. Take, for example, the reintroduction of wolves in the United States. Grey wolves (*Canis lupus*), shown in Box Figure 11.4, were hunted to near extirpation in the United States, and the removal of this top predator had catastrophic cascading effects. An intensive captive breeding program brought wolves back from the brink of extinction, and there are now hundreds of individuals throughout the western United States. However, controversy remains about having top predators—even

BOX FIGURE 11.4 Reintroducing ecologically important species or their analogs. (A) A grey wolf reintroduction to Yellowstone National Park. Grey wolves were hunted to near extinction in the United States, but captive breeding and release has re-established this top predator. (B) The last Pinta Island tortoise (Lonesome George) died in captivity in 2012. After this species was lost, closely related species were released on Pinta Island in hopes that they would perform a similar ecological role.

Reflection: Even if species appear to be ecologically similar in their native habitats, why might one species not perform an expected role when moved to a new environment?

Source: (A) Jim Peaco/National Park Service; (B) Discodollydiva/Shutterstock

(Continued)

those that are native to western ecosystems—in such close proximity to livestock and human populations (e.g., Bruskotter 2013).

Thought Experiment 2

Now imagine that an ecologically important species has recently become extinct globally, and the loss was clearly caused by human activities. All else being equal, should we reintroduce a close relative or ecological analog to fill its role in the community?

Although it is less common than translocations or reintroductions within species, introductions of ecological or evolutionary analogs occur. Sometimes referred to as taxon substitutions, these introductions are conducted to reduce secondary extinctions and other cascading effects. For example, following the extinction of Pinta tortoises (*Chelonoidis abingdonii*) in the Galapagos (shown in Box Figure 11.4), Saddleback and Domed tortoises were released on Pinta Island as ecological proxies. Although the species appeared ecologically similar, they colonized different habitats on the island and did not contribute equally to restoring local plant communities (Hunter & Gibbs 2014).

Thus, there is no guarantee that seemingly similar species will be ecologically interchangeable. The potential for unintended consequences is particularly important when considering higher risk interventions that aim to restore the ecological roles of species that went extinct long ago. For example, some conservationists have proposed the possibility of "rewilding" regions of the globe with ecological analogs to address historical megafaunal extinctions (e.g., Donlan et al. 2006). In the extreme, imagine reintroducing ungulates and carnivores from Africa to the plains of the United States, and it is easy to see why rewilding proposals remain controversial (e.g., Lorimer et al. 2015).

Thought Experiment 3

Again imagine that an ecologically important species has recently become globally extinct, and the loss was clearly caused by human activities. All else being equal, should we bring it back from extinction and reintroduce it to its original community?

Instead of introducing an ecological analog for an extinct species, what if we could just bring a species back from the grave? Referred to as **de-extinction**, technological advances now make it potentially possible to resurrect a species that has died out (or create a close functional analog). Of course, de-extinction is an incredibly controversial proposition both biologically and ethically. However, there are some scientists that are actively working to accomplish high-profile goals of returning extinct species to the wild. Let us look at several ways that de-extinction can potentially be accomplished.

IN-VITRO FERTILIZATION Perhaps the most straightforward method for de-extinction is **in-vitro fertilization**. If gametes were cryopreserved before the species was lost, it is technologically possible to produce embryos in the lab and use a close relative as a surrogate for gestation. One example is the 2018 loss of the last male northern white rhino (*Ceratotherium simum cottoni*), which is shown in Box Figure 11.5. Scientists saved genetic material from the last male and from the two remaining females of the subspecies. It remains to be seen whether in-vitro fertilization (using a southern white rhino as a surrogate) can bring this subspecies back from extinction (Ingledew 2018).

SELECTIVE BREEDING Another possibility is **selective breeding**, where living relatives of an extinct species are selectively mated to reproduce the traits of the lost species. Humans have used **artificial selection** for thousands of years to select desirable traits in domesticated plants and animals. A similar approach can be applied to reconstructing traits of extinct species. One example is the quagga (*Equus quagga quagga*), a South African zebra that went extinct in 1883. More than 20 years of selective breeding of a related subspecies has now retrieved many hallmark quagga traits (Harley et al. 2009; Heywood 2013).

CLONING **Cloning** is distinct from in-vitro fertilization because an individual is created from the genetic material of a single parent (not from the fertilization of an egg with a sperm). In brief, the nucleus of an adult cell is transferred into an unfertilized egg cell that has had the nucleus removed. This cell is then stimulated to divide and implanted into a surrogate carrier. The application of cloning to mammals was famously revealed by "Dolly" the sheep in 1996. Since then, several cloning projects for extinct species have been underway (Piña-Aguilar 2009), including a live birth of the previously extinct Pyrenean ibex (*Capra pyrenaica pyrenaica*; Folch 2009).

GENOME EDITING A final approach to de-extinction is **genome editing**, whereby scientists make targeted changes to the genome of a close relative. By deleting, inserting, or

BOX FIGURE 11.5 Different approaches to de-extinction. (A) In-vitro fertilization may be attempted to bring back the northern white rhino (shown here is the last male named Sudan, who died in 2018). (B) Generations of selective breeding have produced a quagga-like zebra (emulating the species that went extinct in 1883). (C) Cloning has led to the live birth of the previously extinct Pyrenean ibex. (D) Genome editing is underway in an attempt to resurrect the extinct woolly mammoth through making thousands of tiny changes in an Asian elephant genome.

Reflection: Based on the exploration in this *Taking a Closer Look*, where would you draw the line? Are you comfortable with reintroductions (a) between populations within a species, (b) between closely related species, (c) between species that are not close relatives but are ecologically similar, (d) by resurrecting a recently extinct species, (e) by resurrecting a species that went extinct a long time ago?

Source: (A) Steve Tum/Shutterstock; (B) LouisLotterPhotography/Shutterstock; (C) https://www.extinctanimals.org/wp-content/uploads/2015/04/Pyrenean-Ibex.jpg; (D) Reimar/Shutterstock

altering stretches of DNA sequence, it is possible to slowly reconstruct the genome sequence of an extinct species (Piaggio et al. 2017). The most high-profile example of this is the attempted resurrection of the woolly mammoth (*Mammuthus primigenius*), which has been extinct for thousands of years. The woolly mammoth genome has been sequenced from frozen DNA (e.g., Miller et al. 2008). By referencing this genomic data, scientists have slowly been modifying cells from Asian elephants to introduce key mammoth genetic variants. Whether we will ever see a live mammoth-like creature again is up for debate, but the technology exists to make this a possibility.

In Sum

Of course, all of these approaches raise complex biological, ethical, legal, political, social, and financial considerations (e.g., Carlin et al. 2013; Cohen 2014; Seddon et al. 2014; Lorimer et al. 2015; Peers et al. 2016; Bennett et al. 2017; Blockstein 2017; Shapiro 2017). These are costly and laborious processes that raise regulatory and animal welfare issues. Even if species are successfully resurrected, it is not clear that they would have adequate genetic diversity or a wild home to return to. Moreover, de-extinction efforts raise difficult questions about conservation priorities and whether funds are better spent on protecting species

(Continued)

that are endangered but not yet extinct. In fact, one recent analysis suggests that funding and attention focused on de-extinction could draw resources away from other conservation priorities and ultimately lead to a net biodiversity loss (Bennett et al. 2017).

The purpose here is not to advocate for or against the use of emerging technologies in biodiversity conservation, but to think critically about the unique opportunities and challenges that arise when technology can make possible what was previously restricted to the realm of science fiction.

KEY CONCEPTS

Why is it important to explicitly define conservation priorities?

- Conservation is the act of preserving, protecting, or restoring ecological systems.
- Given limited resources, conservation efforts can only focus on a subset of targets, so it is essential to define explicit criteria for conservation interventions.

Why is it important to match conservation actions to particular biological levels?

- Fine-filter and coarse-filter strategies address conservation goals at different biological levels.

What are examples of fine-filter conservation strategies?

- Fine-filter conservation strategies focus on protecting individual populations or species from specific threats, for example by reducing overharvest, investing in captive breeding programs, or designing effective translocations.

What are examples of coarse-filter conservation strategies?

- Coarse-filter conservation strategies focus on protecting landscapes, ecosystems, and evolutionary processes, for example by creating reserves, establishing corridors, re-establishing natural disturbance regimes, and enhancing habitat in highly modified landscapes.

What is adaptive management?

- Adaptive management is a holistic and iterative process of identifying conservation goals, assessing alternative actions, implementing conservation interventions, tracking effectiveness, and reassessing to modifying future goals.

Core concepts: What is a climate mitigation?

- Climate change mitigation refers to actions that reduce radiative forcing on the climate system by decreasing carbon emissions or enhancing carbon sinks. Climate mitigation addresses the root causes of climate change and is complemented by climate change adaptation, which aims to reduce the impact of anthropogenic climate change.

Meet the data: Maximizing evolutionary diversity

- Prioritizing phylogenetic diversity can be an effective strategy for targeted conservation efforts; however, many different species and ecosystem attributes are important to consider in conservation planning.

Taking a closer look: Emerging technologies and conservation ethics

- Novel genetic and genomic approaches are being applied to biodiversity conservation. Some of these applications are straightforward, but others (like in-vitro fertilization, selective breeding, cloning, and genome editing) raise complex biological and ethical questions about the opportunities and risks of bringing species back from extinction.

CONSOLIDATE YOUR KNOWLEDGE

Answer the following questions to assess your progress meeting the learning outcomes:

1. In your own words, write a one- or two-sentence synthesis of the big-picture take-away point of this chapter.
2. Revisit your answer to the *Blank Page* exercise in the beginning of this chapter. Would you refine your answer now based on knowledge you integrated from this chapter?
3. What are three examples of competing ways to define conservation priorities? What do you think are the most important criteria to use when choosing species or ecosystems for conservation efforts?
4. Compare and contrast fine- and coarse-filter conservation approaches. What are three specific examples for each?
5. Compare and contrast the conservation approaches of (a) setting aside pristine reserves for biodiversity versus (b) augmenting habitat in anthropogenically modified habitats. What are the relative merits and challenges of each approach?
6. What is adaptive management and why are iterative conservation approaches particularly important in an era of rapid global change?
7. What is an example of an unintended detrimental effect of a conservation action on a species or ecosystem? How can conservation practitioners minimize the risk of unintended consequences?
8. It seems counterintuitive to consider increasing disturbance as a conservation action. When might this be appropriate?
9. Compare and contrast climate adaptation and climate mitigation.
10. What is de-extinction? How can it be done? What concerns does it raise?
11. In your own words, define the bolded and italicized terms in this chapter.
12. What are some questions that *you* have about the content in this chapter? If you found some content particularly challenging or particularly interesting, identify these as areas for additional reflection or reading.

LITERATURE CITED

Alagona. 2004. "Biography of a 'Feathered Pig': The California Condor Conservation Controversy." *Journal of the History of Biology* 37: 557–83.

Backer, Jensen, & Mcpherson. 2004. "Impacts of Fire-Suppression Activities on Natural Communities." *Conservation Biology* 18 (4): 937–46.

Belote, Dietz, Jenkins, McKinley, Irwin, Fullman, Leppi, & Aplet. 2017. "Wild, Connected, and Diverse: Building a More Resilient System of Protected Areas." *Ecological Applications* 27 (4): 1050–56. doi.org/10.1002/eap.1527.

Bennett, Maloney, Steeves, Brazill-Boast, Possingham, & Seddon. 2017. "Spending Limited Resources on De-Extinction Could Lead to Net Biodiversity Loss." *Nature Ecology and Evolution* 1 (4): 1–4. doi.org/10.1038/s41559-016-0053.

Blockstein. 2017. "We Can't Bring Back the Passenger Pigeon: The Ethics of Deception Around De-Extinction." *Ethics, Policy and Environment* 20 (1): 33–37. doi.org/10.1080/21550085.2017.1291826.

Bower, Brownscombe, Birnie-Gauvin, Ford, Moraga, Pusiak, Turenne, Zolderdo, Cooke, & Bennett. 2018. "Making Tough Choices:

Picking the Appropriate Conservation Decision-Making Tool." *Conservation Letters* 11 (2): 1–7. doi.org/10.1111/conl.12418.

Braverman. 2014. "Conservation without Nature: The Trouble with in Situ Versus Ex Situ Conservation." *Geoforum* 51 (October): 47–57. doi.org/10.1016/j.geoforum.2013.09.018.

Bruskotter. 2013. "The Predator Pendulum Revisited: Social Conflict over Wolves and Their Management in the Western United States." *Wildlife Society Bulletin* 37 (3): 674–79. doi.org/10.1002/wsb.293.

Buckley. 2016. "Triage Approaches Send Adverse Political Signals for Conservation." *Frontiers in Ecology and Evolution* 4: 39. doi.org/10.3389/fevo.2016.00076.

Carlin, Wurman, & Zakim. 2013. "How to Permit Your Mammoth: Some Legal Implications of De-Extinction." *Stanford Environmental Law Journal* 33 (1): 3–57.

Carpenter, Andreone, Moore, & Griffiths. 2014. "A Review of the International Trade in Amphibians: The Types, Levels and Dynamics of Trade in CITES-Listed Species." *Oryx* 48 (4): 565–74. doi.org/10.1017/S0030605312001627.

Cohen. 2014. "The Ethics of De-Extinction." *NanoEthics* 8 (2): 165–78. doi.org/10.1007/s11569-014-0201-2.

Conservation Evidence. 2017. www.conservationevidence.com/.

Donlan, Berger, Bock, Bock, Burney, Estes, Foreman, et al. 2006. "Pleistocene Rewilding: An Optimistic Agenda for Twenty-First Century Conservation." *The American Naturalist* 168 (5): 660–81. doi.org/10.2307/3873461.

Eckert, Ban, Tallio, & Turner. 2018. "Linking Marine Conservation and Indigenous Cultural Revitalization: First Nations Free Themselves From Externally Imposed Social-Ecological Traps." *Ecology and Society* 23 (4): 23. doi/org/10.5751/.

Folch, Cocero, Chesné, Alabart, Domínguez, Cognié, Roche, et al. 2009. "First Birth of an Animal from an Extinct Subspecies (*Capra pyrenaica pyrenaica*) by Cloning." *Theriogenology* 71 (6): 1026–34. doi.org/10.1016/j.theriogenology.2008.11.005.

Forest, Grenyer, Rouget, Davies, Cowling, Faith, Balmford, et al. 2007. "Preserving the Evolutionary Potential of Floras in Biodiversity Hotspots." *Nature* 445: 757–60. doi.org/10.1038/nature05587.

Game, Kareiva, & Possingham. 2013. "Six Common Mistakes in Conservation Priority Setting." *Conservation Biology* 27 (3): 480–85. doi.org/10.1111/cobi.12051.

Garnett, Burgess, Fa, Fernández-Llamazares, Molnár, et. al. 2018. "A Spatial Overview of the Global Importance of Indigenous Lands for Conservation." *Nature Sustainability* 1: 369–74. doi.org/10.1038/s41893-018-0100-6

Godinot. 2006. "Lemuriform Origins as Viewed from the Fossil Record." *Folia Primatologica* 77 (6): 446–64. doi.org/10.1159/000095391.

Haddad. 2018. "Resurrection and Resilience of the Rarest Butterflies." *PLOS Biology* 16 (2): e2003488. doi.org/10.1371/journal.pbio.2003488.

Harley, Knight, Lardner, Wooding, & Gregor. 2009. "The Quagga Project: Progress Over 20 Years of Selective Breeding." *South African Journal of Wildlife Research* 39 (2): 155–63. doi.org/10.3957/056.039.0206.

Heller & Zavaleta. 2009. "Biodiversity Management in the Face of Climate Change: A Review of 22 Years of Recommendations." *Biological Conservation* 142 (1): 14–32. doi.org/10.1016/j.biocon.2008.10.006.

Heywood. 2013. "The Quagga and Science: What Does the Future Hold for This Extinct Zebra?" *Perspectives in Biology and Medicine* 56 (1): 53–64. doi.org/10.1353/pbm.2013.0008.

Hunter & Gibbs. 2014. "Densities of Ecological Replacement Herbivores Required to Restore Plant Communities: A Case Study of Giant Tortoises on Pinta Island, Galápagos." *Restoration Ecology* 22 (2): 248–56. doi.org/10.1111/rec.12055.

Ingledew. 2018. "Can IVF Save the Northern White Rhino from Extinction?" *The Veterinary Record* 182 (13): 366. http://www.ncbi.nlm.nih.gov/pubmed/29599255.

James, Gaston, & Balmford. 2001. "Can We Afford to Conserve Biodiversity?" *BioScience*

51 (1): 43–52. doi.org/10.1641/0006-3568 (2001)051[0043:cwatcb]2.0.co;2.

Johnson, Loss, Shawn Smallwood, & Erickson. 2016. "Avian Fatalities at Wind Energy Facilities in North America: A Comparison of Recent Approaches." *Human-Wildlife Interactions* 10 (1): 7–18.

Kelly & Phillips. 2015. "Targeted Gene Flow for Conservation." *Conservation Biology* 30 (2): 259–67. doi.org/10.1111/cobi.12623.

Kelly & Phillips. 2018. "Targeted Gene Flow and Rapid Adaptation in an Endangered Marsupial." *Conservation Biology* 33 (1): 112–21. doi.org/10.1111/cobi.13149.

Klein, Vaissière, Cane, Steffan-Dewenter, Cunningham, Kremen, & Tscharntke. 2007. "Importance of Pollinators in Changing Landscapes for World Crops." *Proceedings of the Royal Society B: Biological Sciences* 274 (1608): 303–13. doi.org/10.1098/rspb.2006.3721.

Krost, Goerres, & Sandow. 2018. "Wildlife Corridors Under Water: An Approach to Preserve Marine Biodiversity in Heavily Modified Water Bodies." *Journal of Coastal Conservation* 22: 87–104. doi.org/10.1007/s11852-017-0554-0.

Kukkala & Moilanen. 2013. "Core Concepts of Spatial Prioritisation in Systematic Conservation Planning." *Biological Reviews* 88 (2): 443–64. doi.org/10.1111/brv.12008.

Lorimer, Sandom, Jepson, Doughty, Barua, & Kirby. 2015. "Rewilding: Science, Practice, and Politics." *Annual Review of Environment and Resources* 40: 39–62. doi.org/10.1146/annurev-environ-102014-021406.

Loss, Will, Loss, & Marra. 2014. "Bird–Building Collisions in the United States: Estimates of Annual Mortality and Species Vulnerability." *The Condor* 116 (1): 8–23. doi.org/10.1650/condor-13-090.1.

Loss, Will, & Marra. 2015. "Direct Mortality of Birds from Anthropogenic Causes." *Annual Review of Ecology, Evolution, and Systematics* 46 (1): 99–120. doi.org/10.1146/annurev-ecolsys-112,414-054133.

M'Gonigle, Ponisio, Cutler, & Kremen. 2015. "Habitat Restoration Promotes Pollinator Persistence and Colonization in Intensively Managed Agriculture." *Ecological Applications* 25 (6): 1557–65. doi.org/10.1890/14-1863.1.

Mackenzie & Keith. 2009. "Adaptive Management in Practice: Conservation of a Threatened Plant Population." *Ecological Management and Restoration* 10 (Suppl. 1): 129–35. doi.org/10.1111/j.1442-8903.2009.00462.x.

Magnuson-Ford, Mooers, Paquette, & Steel. 2010. "Comparing Strategies to Preserve Evolutionary Diversity." *Journal of Theoretical Biology* 266 (1): 107–16. doi.org/10.1016/j.jtbi.2010.06.004.

Maisels, Strindberg, Blake, Wittemyer, Hart, Williamson, Aba'a, et al. 2013. "Devastating Decline of Forest Elephants in Central Africa." *PLoS ONE* 8 (3): e59469. doi.org/10.1371/journal.pone.0059469.

Margules & Pressey. 2000. "A Framework for Systematic Conservation Planning." *Nature* 405: 243–53. doi.org/10.1038/35012251.

McCarthy, Donald, Scharlemann, Buchanan, Balmford, Green, Bennun, et al. 2012. "Financial Costs of Meeting Global Current Spending and Unmet Needs." *Science* 338 (6109): 946–49.

Miller, Drautz, Ratan, Pusey, Qi, Lesk, Tomsho, et al. 2008. "Sequencing the Nuclear Genome of the Extinct Woolly Mammoth." *Nature* 456 (7220): 387–90. doi.org/10.1038/nature07446.

Minckley. 1995. "Translocation as a Tool for Conserving Imperiled Fishes: Experiences in Western United States." *Biological Conservation* 72: 297–09.

Morandin & Kremen. 2013. "Hedgerow Restoration Promotes Pollinator Populations and Exports Native Bees to Adjacent Fields." *Ecological Applications* 23 (4): 829–39. doi.org/10.1890/12-1051.1.

Myers, Mittermeier, Mittermeier, Fonseca, & Kent. 2000. "Biodiversity Hotspots for Conservation Priorities." *Nature* 403: 853–58. doi.org/10.1038/35002501.

Parsons & DeBenedetti. 1979. "Impact of Fire Suppression on a Mixed-Conifer Forest." *Forest Ecology and Management* 2: 21–33. doi.org/10.1016/0378-1127(79)90034-3.

Pasquini, Fitzsimons, Cowell, Brandon, & Wescott. 2011. "The Establishment of Large Private Nature Reserves by Conservation NGOs: Key Factors for Successful

Implementation." *Oryx* 45 (3): 373–80. doi .org/10.1017/S0030605310000876.

Peers, Thornton, Majchrzak, Bastille-Rousseau, & Murray. 2016. "De-extinction Potential Under Climate Change: Extensive Mismatch Between Historic and Future Habitat Suitability for Three Candidate Birds." *Biological Conservation* 197: 164–70. doi.org/10.1016/j .biocon.2016.03.003.

Pendoley, Schofield, Whittock, Ierodiaconou, & Hays. 2014. "Protected Species Use of a Coastal Marine Migratory Corridor Connecting Marine Protected Areas." *Marine Biology* 161: 1455–66. doi.org/10.1007/s00227 -014-2433-7.

Piaggio, Segelbacher, Seddon, Alphey, Bennett, Carlson, Friedman, et al. 2017. "Is It Time for Synthetic Biodiversity Conservation?" *Trends in Ecology and Evolution* 32 (2): 97–107. doi.org/10.1016/j .tree.2016.10.016.

Pimentel, Wilson, McCullum, Huang, Dwen, Flack, Tran, Saltman, & Cliff. 1997. "Economic and Environmental Benefits of Biodiversity." *BioScience* 47 (11): 747–57. doi .org/10.2307/1313097.

Piña-Aguilar, Lopez-Saucedo, Sheffield, Ruiz-Galaz, de J. Barroso-Padilla, & Gutiérrez-Gutiérrez. 2009. "Revival of Extinct Species Using Nuclear Transfer: Hope for the Mammoth, True for the Pyrenean Ibex, But Is It Time for 'Conservation Cloning'?" *Cloning and Stem Cells* 11 (3): 341–46. doi .org/10.1089/clo.2009.0026.

Possingham & Gerber. 2017. "Ecology: The Effect of Conservation Spending." *Nature* 551 (7680): 309–10. doi.org/10.1038/nature 24158.

Pounds, Carnaval, Puschendorf, Haddad, & Masters. 2006. "Responding to Amphibian Loss." *Science* 314 (5805): 1541–42. doi .org/10.1126/science.314.5805.1541.

Pressey & Bottrill. 2008. "Opportunism, Threats, and the Evolution of Systematic Conservation Planning." *Conservation Biology* 22 (5): 1340–45. doi.org/10.1111/j.1523-1739 .2008.01032.x.

Pritchard, Fa, Oldfield, & Harrop. 2011. "Bring the Captive Closer to the Wild: Redefining the Role of Ex Situ Conservation." *Oryx* 46 (1): 18–23. doi.org/10.1017 /S0030605310001766.

Purcell, Polidoro, Hamel, Gamboa, & Mercier. 2014. "The Cost of Being Valuable: Predictors of Extinction Risk in Marine Invertebrates Exploited as Luxury Seafood." *Proceedings of the Royal Society B: Biological Sciences* 281 (1781): 20133296. doi.org/10 .1098/rspb.2013.3296.

Ralls & Ballou. 2004. "Genetic Status and Management of California Condors." *The Condor* 106 (2): 215–28. doi.org/10.1650/7348.

Redding & Mooers. 2015. "Ranking Mammal Species for Conservation and the Loss of Both Phylogenetic and Trait Diversity." *PLoS ONE* 10 (12): e0141435. doi.org/10.1371 /journal.pone.0141435.

Riggio & Caro. 2017. "Structural Connectivity at a National Scale: Wildlife Corridors in Tanzania." *PLoS ONE* 12 (11): 1–16. doi .org/10.1371/journal.pone.0187407.

Rist, Campbell, & Frost. 2013. "Adaptive Management: Where Are We Now?" *Environmental Conservation* 40 (1): 5–18. doi .org/10.1017/S0376892912000240.

Sarkar, Pressey, Faith, Margules, Fuller, Stoms, Moffett, et al. 2006. "Biodiversity Conservation Planning Tools: Present Status and Challenges for the Future." *Annual Review of Environment and Resources* 31 (1): 123–59. doi.org/10.1146/annurev.energy.31.042606 .085844.

Schneider. 2008. "Geoengineering: Could We or Should We Make It Work?" *Philosophical Transactions of the Royal Society A: Mathematical, Physical and Engineering Sciences* 366 (1882): 3843–62. doi.org/10.1098/ rsta.2008.0145.

Schultz & Crone. 1998. "Burning Prairie to Restore Butterfly Habitat: A Modeling Approach to Management Tradeoffs for the Fender's Blue." *Restoration Ecology* 6 (3): 244–52. doi .org/10.1046/j.1526-100X.1998.00637.x.

Seddon, Moehrenschlager, & Ewen. 2014. "Reintroducing Resurrected Species: Selecting De-extinction Candidates." *Trends in Ecology and Evolution* 29 (3): 140–47. doi.org/10.1016/j.tree.2014.01.007.

Shapiro. 2017. "Pathways to De-extinction: How Close Can We Get to Resurrection of an Extinct Species?" *Functional Ecology* 31 (5): 996–1002. doi.org/10.1111/1365-2435.12705.

Sisco, Monzer, Farajalla, Bashour, & Saoud. 2017. "Roof Top Gardens as a Means to Use Recycled Waste and A/C Condensate and Reduce Temperature Variation in Buildings." *Building and Environment* 117: 127–34. www.sciencedirect.com/science/article/pii/S0360132317300847.

Sun, Xie, & Zhao. 2019. "Valuing Urban Green Spaces in Mitigating Climate Change: A City-Wide Estimate of Aboveground Carbon Stored in Urban Green Spaces of China's Capital." *Global Change Biology* 25 (5): 1717–32. doi.org/10.1111/gcb.14566.

Thaxter, Buchanan, Carr, Butchart, Newbold, Green, Tobias, Foden, O'Brien, & Pearce-Higgins. 2017. "Bird and Bat Species' Global Vulnerability to Collision Mortality at Wind Farms Revealed Through a Trait-Based Assessment." *Proceedings of the Royal Society B: Biological Sciences* 284 (1862): 20170829. doi.org/10.1098/rspb.2017.0829.

Tingley, Darling, & Wilcove. 2014. "Fine- and Coarse-Filter Conservation Strategies in a Time of Climate Change." *Annals of the New York Academy of Sciences* 1322 (1): 92–109. doi.org/10.1111/nyas.12484.

van Oppen, Oliver, Putnam, & Gates. 2015. "Building Coral Reef Resilience Through Assisted Evolution." *Proceedings of the National Academy of Sciences* 112 (8): 2307–13. doi.org/10.1073/pnas.1422301112.

van Wyk, & Prinsloo. 2018. "Medicinal Plant Harvesting, Sustainability and Cultivation in South Africa." *Biological Conservation* 227 (July): 335–42. doi.org/10.1016/j.biocon.2018.09.018.

Waldron, Miller, Redding, Mooers, Kuhn, Nibbelink, Roberts, Tobias, & Gittleman. 2017. "Reductions in Global Biodiversity Loss Predicted from Conservation Spending." *Nature* 552: 364–67. doi.org/10.1038/nature24295.

Westgate, Likens, & Lindenmayer. 2013. "Adaptive Management of Biological Systems: A Review." *Biological Conservation* 158: 128–39. doi.org/10.1016/j.biocon.2012.08.016.

Wilson. 2016. "Half-Earth: Our Planet's Fight for Life."

Wittemyer, Northrup, Blanc, Douglas-Hamilton, Omondi, & Burnham. 2014. "Illegal Killing for Ivory Drives Global Decline in African Elephants." *Proceedings of the National Academy of Sciences* 111 (36): 13117–21. doi.org/10.1073/pnas.1403984111.

Zeale, Stone, Zeale, Browne, Harris, & Jones. 2018. "Experimentally Manipulating Light Spectra Reveals the Importance of Dark Corridors for Commuting Bats." *Global Change Biology* 24 (12): 5909–18. doi.org/10.1111/gcb.14462.

Zhang, Liu, Fu, Phillips, Zhang, & Zhang. 2016. "Bridging the 'Gap' in Systematic Conservation Planning." *Journal for Nature Conservation* 31: 43–50. doi.org/10.1016/j.jnc.2016.03.003.

CHAPTER 12

Aligning the Interests of Biodiversity and Human Society

Learning Outcomes

After working with this chapter, you will be able to:

- Compare different approaches for realigning the interests of biodiversity and human society.
- Evaluate ways that individual, community, and policy actions can contribute to biodiversity conservation.
- Delineate the key factors influencing the extent of future environmental impacts.
- Articulate your own environmental worldview.
- Apply your knowledge to real-world case studies and interpret data from recent scientific studies.

THE BLANK PAGE

Often it seems that human needs are in conflict with the requirements of biological diversity. What are some specific ways that we could reconcile the needs of *Homo sapiens* and the millions of other species on the planet? Brainstorm at least one specific thing that you personally could do, that your community could do, and that your country could do. Given your own talents and interests, what role could you personally play in solutions at these different scales?

INTRODUCTION

In the last chapter, we evaluated approaches to biodiversity conservation and management. However, all of the conservation strategies we discussed require the support of individuals, communities, and governments. Moreover, on-the-ground conservation actions are deeply embedded in—and influenced by—broader social, cultural, and political contexts. Thus, successful biodiversity conservation in a changing world will require aligning the interests of individuals, societies, and the ecosystems on which humanity depends. The goal of this chapter is to explore the interdependency of biodiversity and human society and to evaluate different levers for protecting biodiversity in an era of rapid global change.

WHAT ARE COUPLED HUMAN–NATURAL SYSTEMS?

As we have seen throughout our explorations, living systems are complex and interactive with myriad biotic and abiotic feedbacks. The role of *Homo sapiens* in the biosphere is no exception, and ecological and societal systems are truly a single domain. Anthropogenic activities and the rest of the biosphere are directly linked through complex interactions and feedbacks. This interdependence is often referred to as **coupled human–natural systems** (or coupled social–ecological systems) (e.g., Liu et al. 2007; Carter et al. 2014;).

In previous chapters, we evaluated numerous examples of human enterprises impacting rates of environmental change and biological diversity. Reciprocally, changes to the natural world have dramatic repercussions on human society. Figure 12.1 reviews the concept of ecosystem services and highlights the links between social and ecological systems. Human survival depends on thriving ecosystems, and increasingly, ecosystem health depends on the choices humans make.

Analyses across spatial and temporal scales consistently show the interdependence of biodiversity and human well-being (e.g., Isbell et al. 2017). There are an overwhelming number of specific examples where loss of biodiversity impacts ecosystem

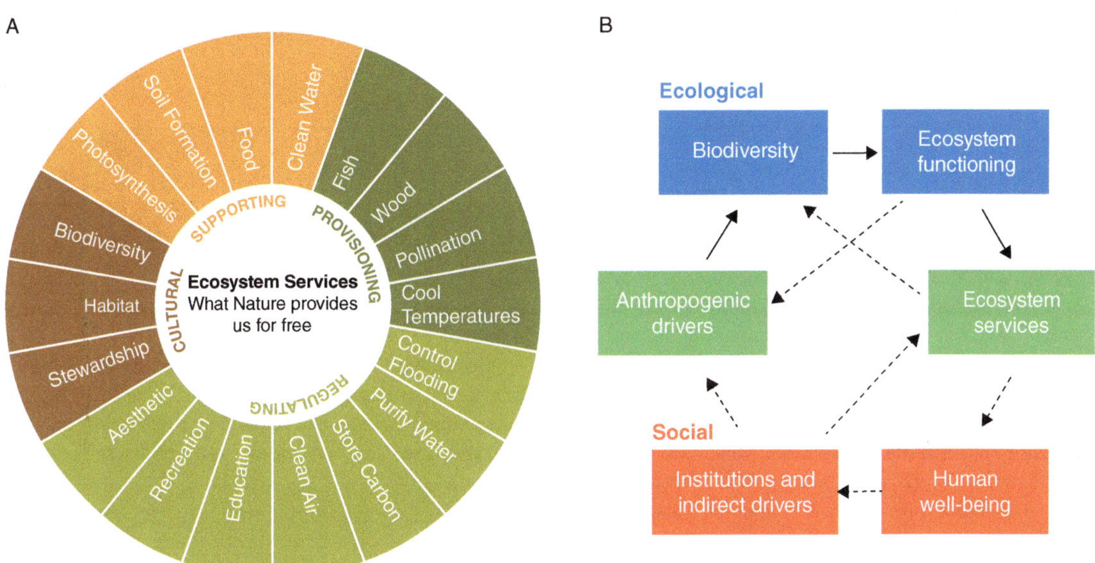

FIGURE 12.1 Coupled human–natural systems. (A) A review of key benefits human societies derive from nature. (B) Ways in which ecological and social systems are intimately tied.

Reflection: Using the ecological and social systems framework, pick one specific anthropogenic stressor and brainstorm some of its key social drivers. Next consider the impact of that stressor on biodiversity, ecosystem functioning, ecosystem services, and human well-being. Finally, speculate on what feedbacks might occur if changes in human well-being influence the drivers of anthropogenic change.

Source: (A) TEEB Europe; (B) Wikipedia, Pecl et al. 2017

processes, ecosystem services, and ultimately, human health (e.g., Pecl et al. 2017). Reciprocally, protecting biological diversity provides immense benefits to humanity, and clean water, food security, and sustainable livelihoods will only become more important in the coming decades.

Thus, long-term solutions to contemporary environmental problems will require systems-level thinking. We will need a renewed focus on the ways in which human societies and ecological systems share common interests. We will also need to understand complex feedbacks, not only within ecological systems but also between ecological and social systems. Ultimately, this will require a deeper look at the societal drivers of global change, the effects of environmental degradation on society, and the societal levers for change.

WHAT SOCIETAL LEVERS CAN BE USED TO SUPPORT BIODIVERSITY CONSERVATION?

Moving toward a future where *H. sapiens* live sustainably with the millions of other species on the planet will require changes at multiple levels of social organization. In the *Core Concepts* feature of this chapter, we look at the root social determinants of the magnitude of future anthropogenic impacts on our planet. Altering any of these key elements (e.g., population size, resource consumption, technological innovation) can change the trajectory of environmental stressors. Ultimately, the contribution of these factors is influenced by the choices we make as individuals, communities, nations, and a species as a whole.

CORE CONCEPTS

WHAT IS I=PAT?

Human activities have—and will continue to—dramatically influence the environmental conditions on Earth. What key factors govern the magnitude of future anthropogenic impacts? Many different frameworks have been proposed to summarize the elements that drive anthropogenic impacts on the global environment. One classic proposition is the **IPAT framework**, which was popularized in the early 1970s (Commoner 1972; Ehrlich & Holdren 1972) and argues that environmental impact (I) can be explained by three key determinants: population (P), affluence (A), and technology (T).

POPULATION The impact of population on the global environment is unequivocal. A larger human population means more land-use change, more resource extraction, and more waste production. Since I=PAT was proposed, the global population has nearly doubled. As Box Figure 12.1 shows, although the *rate* of population growth has slowed, the global population continues to increase and will likely exceed 11 billion people by 2100. Thus, population size remains a crux issue.

In addition to population *size*, there are many other important aspects of population *structure* that alter human impact on the environment. For example, human populations tend to be strongly clustered in space, and population density is a key modulator of environmental impact. For example, high-, medium-, and low-density settlements (such as urban, suburban, and pastoral) have different environmental footprints.

AFFLUENCE Affluence is often used as a proxy for resource consumption. Typically, as per-capita income increases, so does resource use. Consider, for example, the environmental impact of automobile ownership, a hallmark of affluence in many parts of the world. A typical personal vehicle emits nearly 5 tons of carbon dioxide

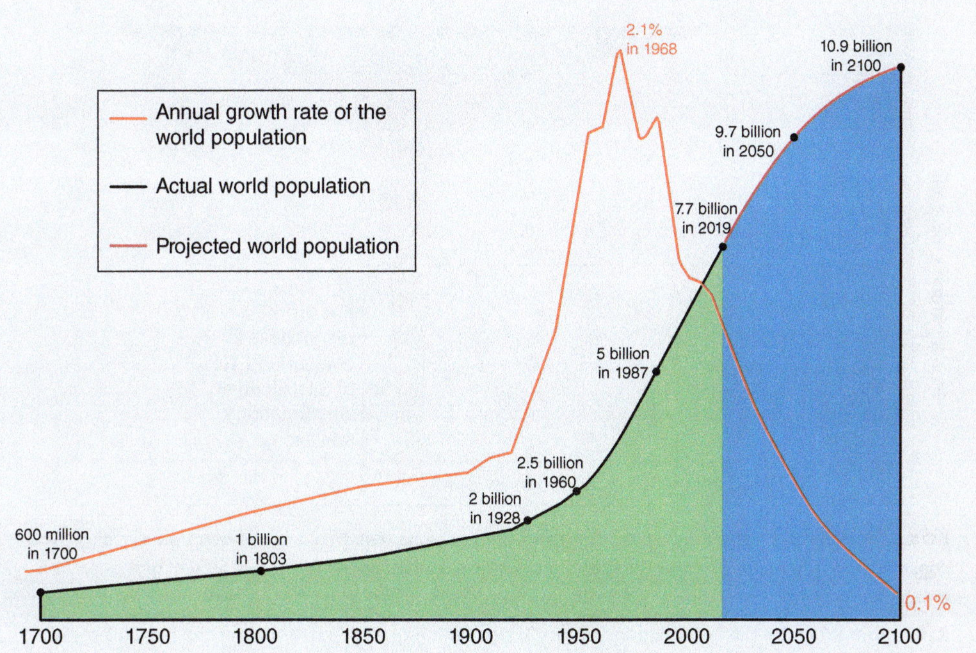

BOX FIGURE 12.1 Population growth rates and total global population size over time. Although population growth rates are slowing, they are still positive. Thus, the total number of people on planet Earth continues to increase.

Reflection: What are the key inflection points in population growth patterns over the last 300 years? What factors likely help explain these changes over time?

Source: Licensed under CC-BY-SA by the author Max Roser

per year. However, if we consider the entire life cycle of the car—which includes resource extraction for raw materials, water and energy for production, and toxic waste when discarded—the impacts are far greater.

Income is typically used to measure personal affluence, and the **gross domestic product** (GDP) is typically used to assess affluence at national or regional scales. Both wealth—and wealth disparities—among individuals and among countries drive environmental impacts, as shown in Box Figure 12.2. Not only do the most affluent individuals in a society typically contribute most dramatically to environmental degradation, but countries with the greatest wealth disparities contribute disproportionately to resource extraction, carbon emissions, and pollution (e.g., Boyce & Boyce 2007). The burden of the environmental impacts created by the affluent then fall disproportionately on the economically poor. Thus, there are both environmental and social benefits of economic equality, underscoring the need to confront wealth disparity.

TECHNOLOGY Technological innovation can reduce or increase human impacts on the environment. As we saw in Chapter 4, advances in production—and subsequent industrialization—were largely responsible for the birth of the Anthropocene. However, technological advances also provide hope for a more sustainable future. For example, as we explored in Chapter 11, improvements in energy efficiency and the development of alternative energy sources are important strategies for mitigating anthropogenic climate change.

Technological innovations can have complex tradeoffs, and there can be hidden costs associated with seemingly "green" technologies. Take the earlier example about personal transportation. As alternative technologies are integrated into the transportation sector, electric vehicles offer the promise of decreased greenhouse gas emissions. However, some electric vehicle supply chains actually increase resource extraction and environmental pollution (e.g., Hawkins et al. 2013). Therefore, the

(Continued)

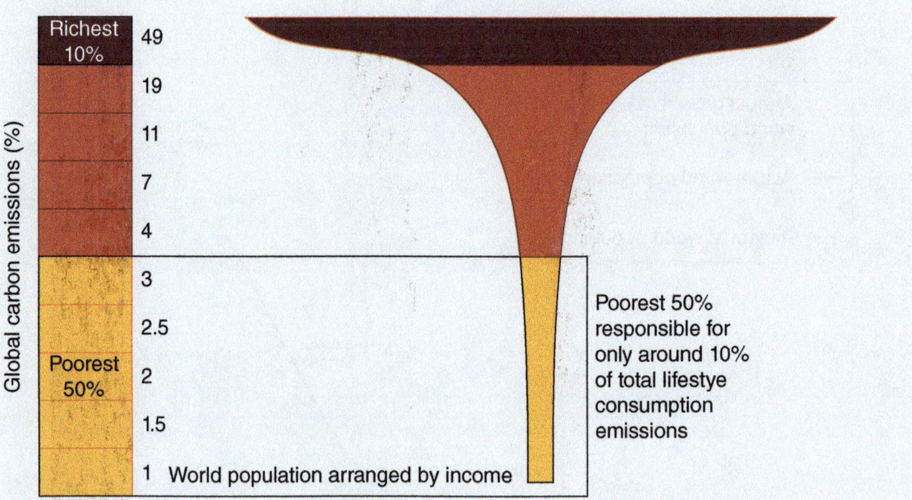

BOX FIGURE 12.2 Wealth disparity and carbon emissions. Personal income correlates strongly with consumption. The richest 10% of the global population is responsible for nearly 50% of global carbon emissions.

Reflection: What are the key inflection points in population growth patterns over the last 300 years? What factors likely help explain these changes over time?

Source: "The material https://www.theguardian.com/environment/2015/dec/02/worlds-richest-10-produce-half-of-global-carbon-emissions-says-oxfam - 'Percentage of CO_2 emissions by world population', credited to Oxfam'" is reproduced with the permission of Oxfam, Oxfam House, John Smith Drive, Cowley, Oxford OX4 2JY, UK www.oxfam.org.uk. Oxfam does not necessarily endorse any text or activities that accompany the materials

effects of technological innovation must be evaluated holistically, looking not only at supply chains and product life cycles but also at downstream, indirect, and long-term effects of new technologies on ecosystems.

BEHAVIOR There have been many debates about the I=PAT framework, how the terms are defined and measured, and whether additional terms should be added to the equation (e.g., Roca 2002; Waggoner & Ausubel 2002). One key example is the role of *behavior* in determining anthropogenic impacts on the environment. For example, individuals make choices as consumers, businesses make decisions as suppliers, nations make choices through legal and regulatory processes, and global organizations make decisions about international cooperation and governance (O'Rourke & Lollo 2015). Thus, the behavior of individuals, organizations, and governments can dramatically alter environmental impacts.

Of course, environmental impacts are also modulated by the biological system themselves. As we have evaluated throughout the book, ecosystems can exhibit different responses to anthropogenic stressors, and these responses can create feedback loops. Ultimately, the future of biological diversity on our planet will depend on the interaction between anthropogenic factors—like population, affluence, technology, and behavior—and the biological systems they affect. We are currently in a watershed moment where there is urgent need to authentically address the relationship between social and economic equity and environmental sustainability.

This chapter will evaluate examples of individual, community, and policy actions to support biodiversity conservation. Although this exploration is not as directly focused on the *biology* of Global Change Biology, it is essential. To conserve biological diversity, we must address the root causes of environmental degradation. Addressing these root causes requires us to confront the reality that a culture of extraction has not only impacted ecosystems but simultaneously marginalized and oppressed many people and cultures on the planet. Although it is easy to feel overwhelmed by the scale of the problem, this chapter highlights creative solutions with lasting impacts at multiple levels.

HOW CAN INDIVIDUALS SUPPORT BIODIVERSITY CONSERVATION?

Our individual values and choices impact other species on the planet. There are many daily decisions that can support biological diversity, from reducing use of pesticides and toxic household chemicals, to growing gardens with native plants, to reducing emissions by walking, bicycling, or carpooling. Although actions taken by any one person may not alter the fate of biological diversity on our planet, the combined force of individual choices can have a powerful influence on community priorities, commercial markets, and larger-scale political change (e.g., Ehrlich & Pringle 2008). Here we will look at several ways that individuals can influence social drivers of change, keeping in mind that our potential as individual actors goes far beyond any single category.

As Consumers

From an economic perspective, one strong lever individuals wield is as consumers in a marketplace. Increasingly, consumers have access to "greener" products, and many different tools have been developed to support sustainability choices. Of course, the cost of green products is prohibitive for many. In addition, providing sustainability information itself is not necessarily a panacea. Many consumers have ambivalent attitudes toward products marketed as "green" stemming from a perception that green claims may be exaggerated and that green products may be lower quality at higher price points (Chang 2011). In addition, sustainability information does not always affect purchasing choices for the average consumer—only for individuals that are directly seeking these data (O'Rourke & Ringer 2016). Thus, changing attitudes and behaviors of individuals in the marketplace will require more than simply providing green options. Education and incentivization are also key.

For those who are already motivated to seek sustainability information, there are many resources to support aligning purchasing with environmental values. Numerous certifications have been developed to signal a conservation ethos to consumers, for instance USDA Organic for food production, Energy Star for energy efficiency, Rainforest Alliance for farms that protect endangered species, Leadership in Energy and Environmental Design (LEED) for green building, and Fair Trade for food, coffee, and clothing.

In addition, there are larger searchable databases to assist consumers in finding products that meet their standards. For example, the GoodGuide (www.goodguide.com) has evaluated >75,000 personal care and household products for their chemical composition, helping consumers identify products that may be health hazards and contribute to environmental toxicity. Similarly, the Green Seal (www.greenseal.org) provides recommendations of products and services in more than 450 categories that meet their health and environmental sustainability criteria. There are specialized databases in nearly every sector to assist consumers, architects, scientists, and business owners in making purchasing decisions. Thus, there are many ways that consumers can drive market values toward sustainability.

As Funders

Individuals can also directly contribute to financing biodiversity conservation. In addition to traditional ways of donating time or money to conservation organizations, new platforms provide opportunities for individuals to make a demonstrable impact on the ground. For example, crowdfunding approaches are increasingly used to finance conservation projects. Gallo-Cajiao et al. (2018) conducted the first review of the impact of crowdfunding as a way to mobilize resources for conservation. The researchers found that more than 4.7 million US dollars had been raised over the last decade. These funds supported 577 different conservation projects in 80 different countries. As shown in Figure 12.2, the study demonstrated a global flow of funds, indicating that citizens are willing to contribute financially to meaningful conservation projects around the world.

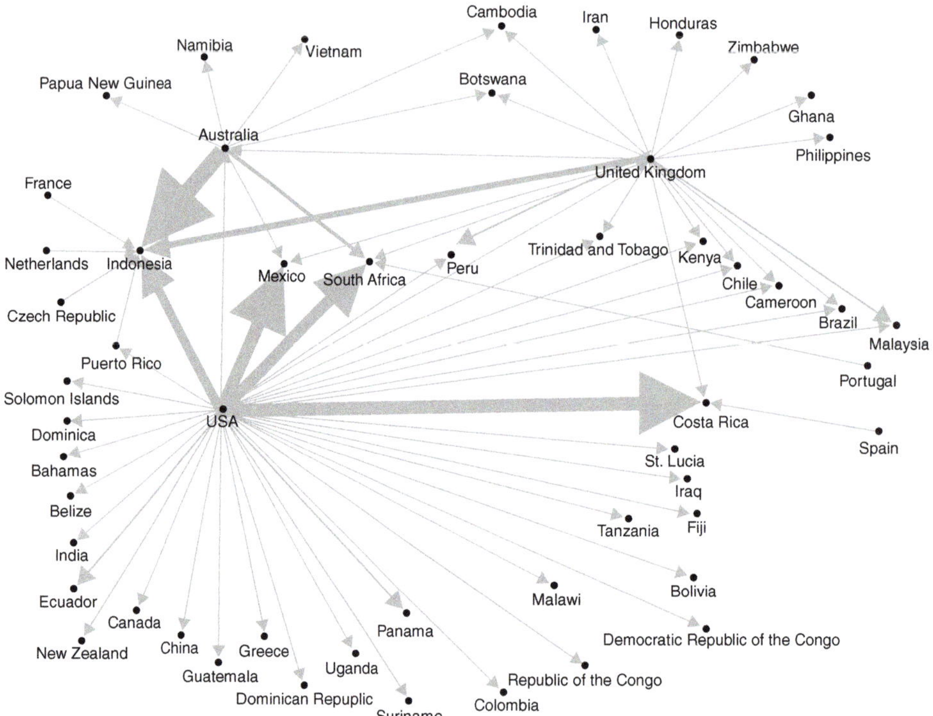

FIGURE 12.2 Global flows of crowdfunding for conservation. Shown here are size-weighted arrows of contributions from the three countries with the highest outflows. However, citizens from 16 different countries donated funds, and these funds supported projects in 80 different countries.

Reflection: If you were going to propose a conservation project of any scale for crowdfunding, what would it be?

Source: Gallo-Cajiao, E., Archibald, C., Friedman, R., Steven, R., Fuller, R.A., Game, E.T., Morrison, T.H. and Ritchie, E.G. (2018), Crowdfunding biodiversity conservation. Conservation Biology, 32: 1426–1435.

The crowdfunded conservation projects addressed both fine- and coarse-scale objectives for protecting individual species and ecosystems. Some projects focused on research (40%), some on action implementation (21%), and some on raising awareness (31%). Many of the leaders of these conservation projects came from nongovernmental organizations (35%) or universities (30%). However, a substantial fraction (26%) was composed of freelancers. Thus, citizens unaffiliated with a specific conservation organization are increasingly engaged in taking action to support biodiversity conservation. Moreover, in today's global economy where funds can be easily transferred person to person, it is possible for ordinary citizens to make direct and meaningful contributions to biodiversity conservation.

As Practitioners and Activists

When individuals choose to act in defense of the natural world, these actions can have dramatic impacts on both ecosystems and societies. Individuals are more than just consumers in a marketplace, and the most powerful individual actions for biological diversity typically come from the heart, not the wallet. Figure 12.3 shows a handful of inspirational individuals from the last decades who have poured their time and energy into reconciling the needs of humanity and biological diversity. Their stories are briefly summarized here.

Wangari Maathai (1940–2011) was the founder of the influential Green Belt movement. Maathai was the first woman in Central and Eastern Africa to earn a PhD, and she later played many roles in her professional life, from scientist to professor to activist to parliament member. Her pioneering work with the Green Belt movement empowered rural Kenyan women to plant trees to alleviate the impacts of deforestation and

FIGURE 12.3 The power of individual action. These individuals devoted their lives to raising awareness and protecting our natural heritage, with powerful community and ecosystem effects: (A) Dr. Wangari Maathai, (B) Chico Mendes, (C) Dr. Jane Goodall, (D) Dr. John Francis.

Reflection: Individuals can contribute to environmental problem solving in myriad small and large ways. Given your skills and interests, what do you think a natural role for you would be in this sphere?

Source: (A) Joseph Sohm/Shutterstock; (B) http://moralheroes.org/chico-mendes/; (C) Tinseltown/Shutterstock; (D) John Francis Institute

develop a source of income. She was awarded a Nobel Peace Prize for linking environmental restoration, sustainable livelihoods, and social equity.

Chico Mendes (1944–1988) was an important social activist, famous for linking rainforest protection and indigenous rights in Brazil. Beginning at age 9, Mendez worked as a rubber tapper, extracting latex from trees. Mendes helped found a union for rubber tappers that was influential in improving working conditions and protecting Amazonian forest stands for sustainable use. After campaigning to block a local deforestation effort, Mendes was murdered. The Chico Mendes Extractive Reserve is established in his memory to protect indigenous livelihoods while supporting sustainable use of forest resources.

Jane Goodall (1934–present) is a primatologist and conservation biologist most famous for her work linking chimpanzee conservation with local education and livelihood initiatives. Her early immersive field studies for her PhD on chimpanzees revolutionized the way we understand animal intelligence, emotions, and social organization. Her activism was initially focused on protecting the local population of chimpanzees in Gombe, Tanzania, but quickly grew to have a global reach. Her Roots & Shoots program now includes youth from 100 countries and aims to develop tomorrow's conservation leaders.

John Francis (1946–present) is an environmentalist most recognized for raising environmental awareness as a "planet walker." After seeing firsthand the impacts of a deadly oil spill, he spent 22 years without riding motorized vehicles, walking throughout the United States. He also spent 17 of these years in silence as a commitment to not defend his choices with argument. During this time, he obtained his bachelor's, master's, and PhD degrees. His Planetwalk organization aims to connect scientists, practitioners, and youth in a new vision for environmental responsibility and cooperation.

These four brief stories provide inspiring examples of how individuals can make an impact. However, there are myriad career paths that focus on environmental problem solving, whether as a scientist, conservation practitioner, educator, lawyer, policymaker, or activist. Opportunities to engage will only grow, and it is possible that the leaders of future sustainability efforts will work at jobs that do not yet exist in today's world. Moreover, there is no reason to wait until one has an established career to contribute. An increasing culture of youth engagement and activism is already shifting the needle on social priorities and demonstrating how individuals can create their own platforms to reach a broad audience. Many are increasingly galvanized to work at the intersection of environmental and social justice to address inequity and exploitation in all forms.

In Sum

Regardless of whether one operates as an informed consumer or an environmental practitioner, individuals engaged in environmental problem solving often confront feelings of overwhelming guilt, stress, grief, and/or fear about the future of our planet. Environmental engagement also requires evaluating our own worldview and beliefs about our role. We explore these topics in the *Taking a Closer Look* feature at the end of this chapter. The links between biodiversity conservation, human well-being, and individual action were expressed beautifully by Chico Mendes, when he said: *"At first*

I thought I was fighting to save rubber trees, then I thought I was fighting to save the Amazon rainforest. Now I realize I am fighting for humanity."

HOW CAN COLLECTIVES SUPPORT BIODIVERSITY CONSERVATION?

Individuals are embedded in collective structures, and these structures can have incredible influence on biodiversity conservation. **Collective actions** are those efforts undertaken cooperatively by a group of people toward a common goal. Collective action can range from small local grassroots efforts to large international partnerships. Most effective collective actions focus on aligning biodiversity conservation with human livelihoods, and there are numerous ways in which protecting ecosystem processes can support income, health, and food security for local communities. Here we will look at a just few examples, recognizing that there are a wealth of additional ways for communities, businesses, and institutions to support sustainability goals while honoring the sovereignty of local and indigenous peoples.

Ecotourism

Tourism is one sector where jobs can be created to enhance wildlife protection in vulnerable ecosystems. **Ecotourism** explicitly endeavors to link economic, social, and environmental benefits, as illustrated in Figure 12.4. Ecotourism has become a critical element of the economy of many countries, in some cases accounting for more than 10% of gross domestic product (Brandt & Buckley 2018).

There are hundreds of examples of community-based ecotourism around the world, and some of these have been extremely successful, creating local incentives to protect natural resources. For example, in Brazil, a collaboration between local communities, nongovernmental organizations (NGOs), and the government helped shift the local economy from exploitation to protection of endangered sea turtles. Members of the local community (including former egg poachers) are now employed to protect sea turtle nesting sites and lead educational and ecotourism programs (Marcovaldi & Dei Marcovaldi 1999).

However, not all ecotourism programs have positive outcomes for wildlife or local communities, especially when decision-making authority, land security, and fair distribution of economic benefits are not ensured for local and indigenous communities (Coria & Calfucura 2012; Das & Chatterjee 2015; Buckley et al. 2016). Moreover, ecotourism can have unexpected negative impacts on the very species it aims to protect. For example, one recent study evaluated the role of ecotourism on the osprey *Pandion haliaetus* in a marine protected area (Monti et al. 2018). As shown in Figure 12.4, the researchers found that osprey reproduction in the marine protected area was actually lower than outside the protected area. In this case, ecotourism led to increased boat traffic, and increased disturbance meant that adults spent more time flying off their nests and less time provisioning food resources. Thus, ecotourism is not a silver bullet, but it can be effective when any unexpected disturbances of flagship species are quickly addressed and when local communities are highly engaged and empowered (e.g., Mendoza-Ramos & Prideaux 2017).

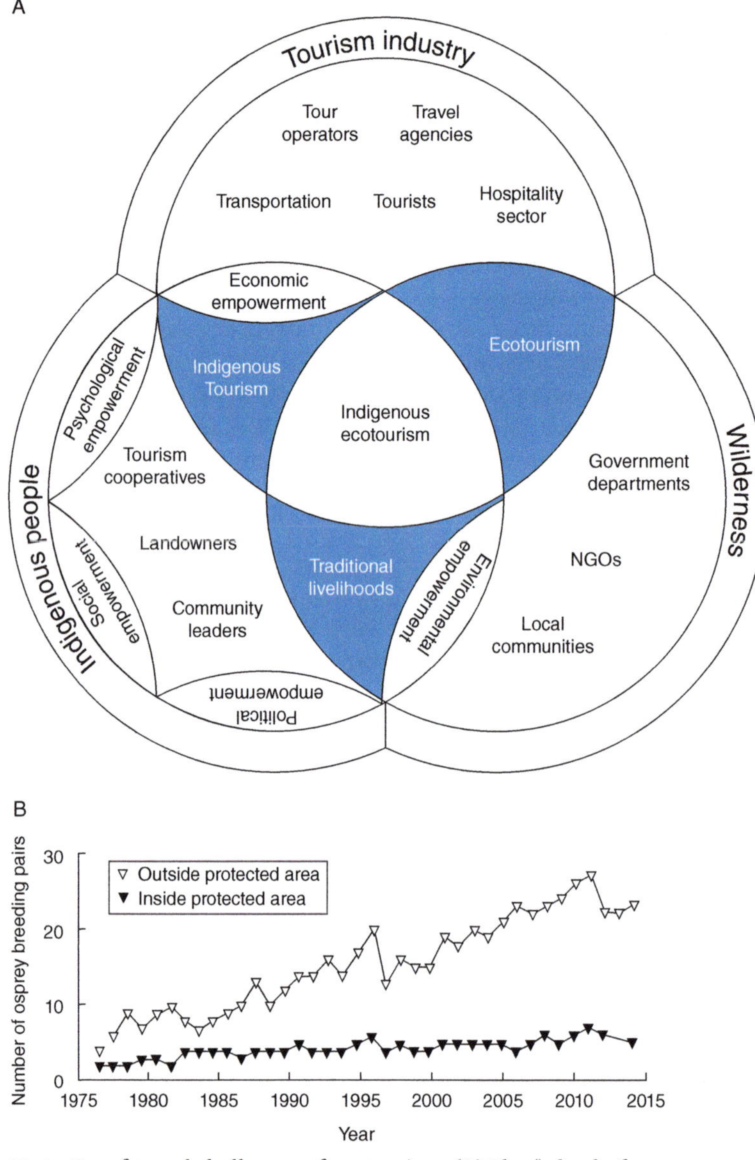

FIGURE 12.4 Benefits and challenges of ecotourism. (A) The "wheel of empowerment." A conceptual framework for how ecotourism provides social, environmental, and economic benefits to communities in the Mayan rainforest (Mexico, Belize, Guatemala, and Honduras). (B) Unexpected negative effects of ecotourism on ospreys on Corsica Island. Decreased number of osprey breeding pairs inside the marine protected area (black) compared to outside (white).

Reflection: If you were to take an adaptive management approach to addressing osprey conservation, what next action would you suggest based on the data?

Source: (A) Adrian Mendoza-Ramos & Bruce Prideaux (2018) Assessing ecotourism in an Indigenous community: using, testing and proving the wheel of empowerment framework as a measurement tool, Journal of Sustainable Tourism, 26:2, 277–291; (B) F. Monti O. Duriez J.-M. Dominici A. Sforzi A. Robert L. Fusani D. Grémillet, Animal Conservation, Volume 21, Issue 6, December 2018, Pages 448–458

Sustainable Food Production

Food production is another arena where collective action can be used to promote sustainability. From small community gardens to large commercial farms, the principles of **agroecology** can help align natural resource conservation, human livelihoods, and sustainable food production (Scherr & McNeely 2008; Thomich et al. 2011). Agroecology focuses on integrating modern agricultural science with indigenous knowledge and practices to generate resilient, productive, and biodiverse working landscapes (Altieri et al. 2012). As such, agroecology provides a new paradigm rather than just a "greening" of industrial agricultural systems (Ponisio & Ehrlich 2016).

One example is the rice–fish farming practice in China, recognized by the United Nations as a Globally Important Ingenious Agricultural Heritage System and shown in Figure 12.5. In these complex, biodiverse rice patties, fish eat natural insect pests (reducing pesticide use) and ducks eat aquatic weeds (preventing eutrophication). Thus, the diversified farming system offers significantly decreased pesticide use, increased nitrogen fixation in the soil, reduced methane emissions, and conservation of complex food webs.

The intimate ties between human and ecosystem health is also evident in other food provisioning settings. One example is local communities that rely on fishing for their nutrition and livelihoods. Fiorella et al. (2017) followed more than 300 households living near Lake Victoria in Kenya. The researchers found that when fishers were healthy, they were more likely to use sustainable fishing practices. In contrast, when fishers were ill, they did not reduce their fishing effort but instead shifted to more destructive—often illegal—nearshore fishing methods. Thus, there is a strong feedback between human health, subsistence, and sustainability. It is therefore key to address human health, livelihood, and food security, especially when communities have already been disproportionately impacted by climate change, biased power structures, and exploitation of their human and natural resources. Ultimately, aligning our food systems with biodiversity conservation and sustainable livelihoods would have a formidable impact on the biosphere.

Cradle-to-Cradle Manufacturing

Collective action can also alter practices in other sectors. For example, stakeholder pressure can shift manufacturing practices for commercial goods. A specific illustration comes from the apparel industry, as shown in Figure 12.6. Apparel has many environmental costs throughout its life cycle from production (such as pesticide use in cotton farming) to distribution (like resources for packaging and energy for transportation), to use (including water for washing), to disposal (like landfill waste). These environmental inputs—from chemical contamination to greenhouse gas emissions—have direct and indirect effects on biodiversity.

However, the increasing emphasis on **"cradle-to-cradle" life cycle analysis** can decrease environmental impacts. Individual companies and collective entities (like the Sustainable Apparel Coalition) are innovating with stricter environmental standards at all stages of the product life cycle, from extraction of raw materials to ultimate disposal or recycling. Incentives of all sorts can motivate more sustainable manufacturing (e.g., Fargani et al. 2016). For example, environmentally friendly business practices can lead to cost savings (like reduction of waste or energy consumption),

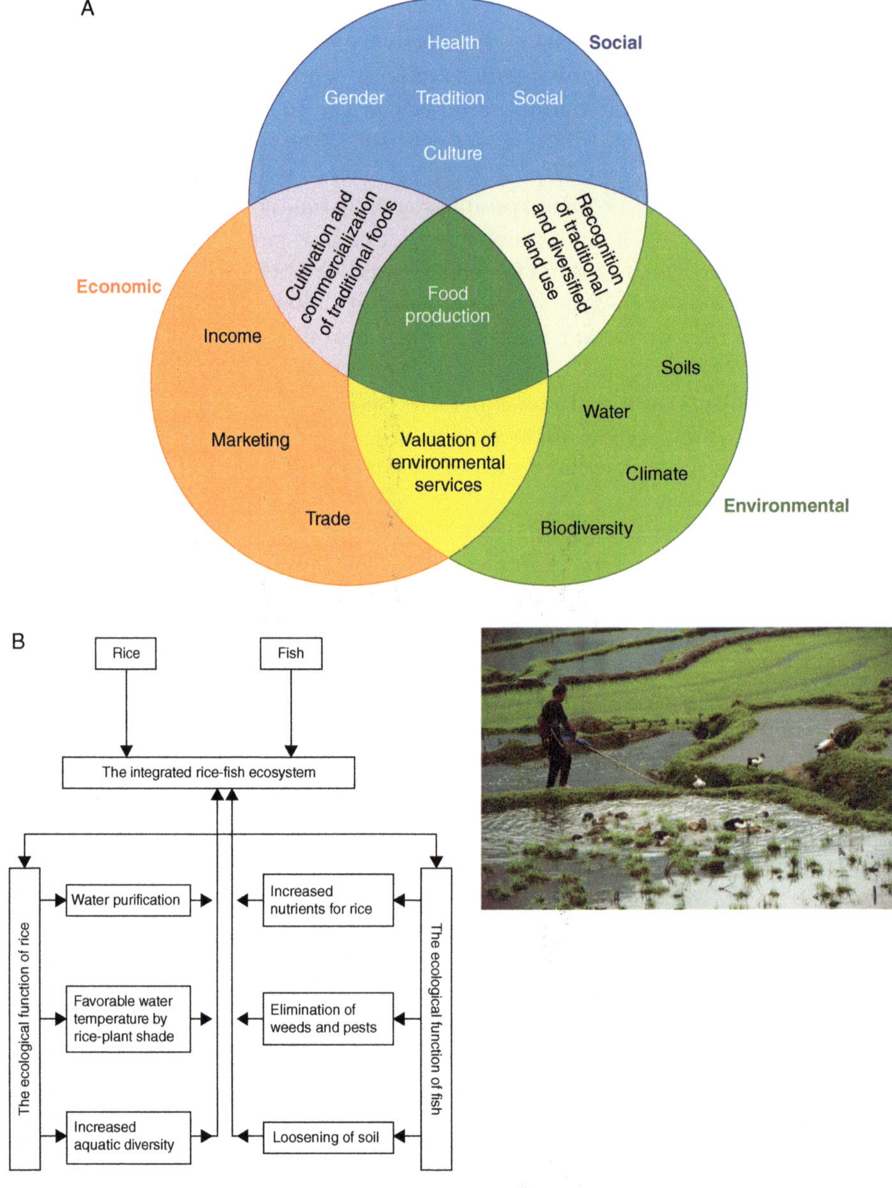

FIGURE 12.5 (A) Holistic approach of agroecology to food production. (B) The integrated rice–fish farming system in China, which has been in use for >1,500 years and maintains natural food webs even on food production lands.

Reflection: What are some examples of how human and ecosystem health are linked in the food production sector?

Source: (A) International Assessment of Agricultural Knowledge, Science and Technology for Development (IAASTD). 2009. IAASTD Global Report: Summary for Decision Makers. Island Press, Washington DC.; (B) Jianbo Lu, Xia Li, Review of rice–fish farming systems in China — One of the Globally Important Ingenious Agricultural Heritage Systems (GIAHS), Aquaculture, Volume 260, Issues 1–4, 29 September 2006, Pages 106–113; MOA/PRC

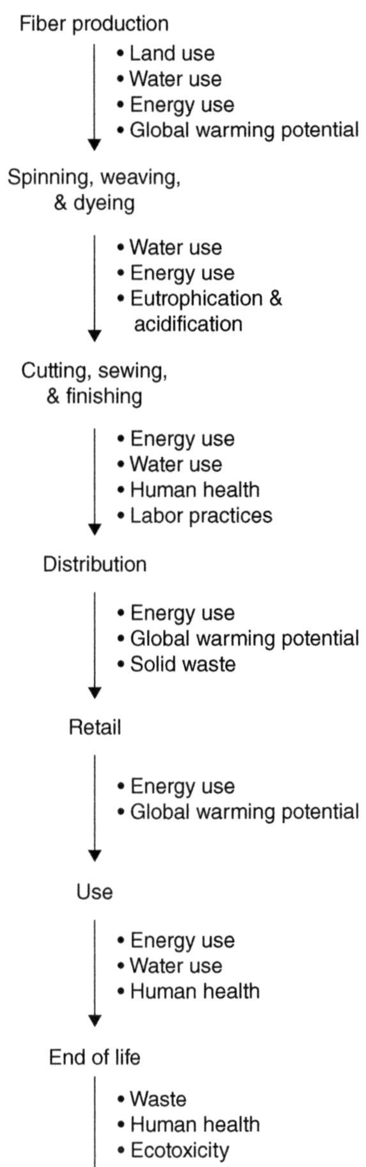

FIGURE 12.6 Environmental considerations during the life cycle of apparel products. At all stages of production, distribution, use, and disposal, choices can be made to decrease environmental impacts.

Reflection: In the three sectors discussed in this section (tourism, agriculture, and manufacturing), what general factors are likely to influence the success of initiatives to link biodiversity conservation with human needs?

Source: The science of sustainable supply chains. BY DARA O'ROURKE. SCIENCE 06 JUN 2014: 1124–1127

increased customer demand (due to desire for green products), and improved reputation (by altering public image). Thus, innovation in the manufacturing sector can dramatically reduce environmental impacts and set new standards for ecologically responsible business practices.

In Sum

Ultimately, anthropogenic stressors on ecosystems can provide an important trigger for collective action. As resources become stressed, communities are often motivated to find creative ways to protect the ecosystems on which they rely. Collective action can have incredible mutual benefits for biological diversity and human society but also requires high levels of trust, commitment, and community engagement (Kruijssen et al. 2009). In particular, when the sovereignty of local and indigenous communities is honored, collective action can simultaneously promote ecosystem recovery and address the human legacy of oppression and disenfranchisement. Collective action can also influence broader political and social change, a topic we will turn to next.

HOW CAN POLICY ACTION SUPPORT BIODIVERSITY CONSERVATION?

Individual and collective action can influence broader scale political change, and reciprocally, policy actions can influence individual and collective behavior. Local, national, and international governance bodies can enact policies to support biodiversity conservation. These policy actions can be in the form of *legislation* (enacting laws) and *regulation* (the process of monitoring and enforcing laws). Policy actions can also provide *motivation* in the form of incentives for individuals, businesses, and nations to make choices that support biological diversity and to coordinate across boundaries to increase the impact of these choices. Let us look at a few examples of how policy actions can align ecological, economic, and societal goals.

Fisheries Governance

It has long been recognized that shared resources can be subject to the "tragedy of the commons." If all individuals try to gain the greatest individual benefit, the shared resource can be rapidly depleted. Fisheries have been a classic test ground for ideas about resource use and exploitation. Pressure on individual fishers to maximize their resource yield has led to overfishing, high rates of by-catch, and environmentally destructive fishing practices (Fujita & Bonzon 2005). Ultimately, these practices have depleted fish stocks and impacted aquatic habitat in many parts of the world (e.g., Möllmann & Diekmann 2012).

However, in many traditional communities, the tragedy of the commons was avoided for centuries as individuals, families, collectives, or villages controlled specific fisheries and were invested in their health (Berkes 1985). Thus, fisheries managers have increasingly looked for ways to realign the interests of fishers and biodiversity. One promising strategy is **rights-based fishery management**, whereby stakeholders have more ownership in the fishery (e.g., Allison et al. 2012). If individuals or groups have exclusive rights in a fishing zone, it can reduce competitive depletion of fisheries resources (Viana et al. 2019). By ensuring that fishers have a more secure livelihood and more investment in the long-term health of the resource, stewardship is improved and overexploitation reduced.

A recent review attributes signs of recovery in many fisheries to this change in resource governance (Lubchenco et al. 2016). In the United States, now more than half of the fish caught each year operate under a rights-based approach, and at least 40 other countries globally have rights-based fisheries. Thus, altering incentives can align the interests of ecosystems and human societies, even in complex industries.

Endangered Species Legislation

A key example of legislation that attempts to directly address global change stressors on biological diversity is the US **Endangered Species Act** (ESA). The ESA was passed in 1973, and its purpose is to protect vulnerable species and the ecosystems upon which they depend. Species listed under the ESA receive legal protection and funding to support a recovery plan. The protection and restoration of *critical habitat* through **habitat conservation plans** is an important component of the ESA and helps link fine- and coarse-filter conservation strategies. As shown in Figure 12.7, there are currently more than 1,500 species listed as endangered or threatened in the United States.

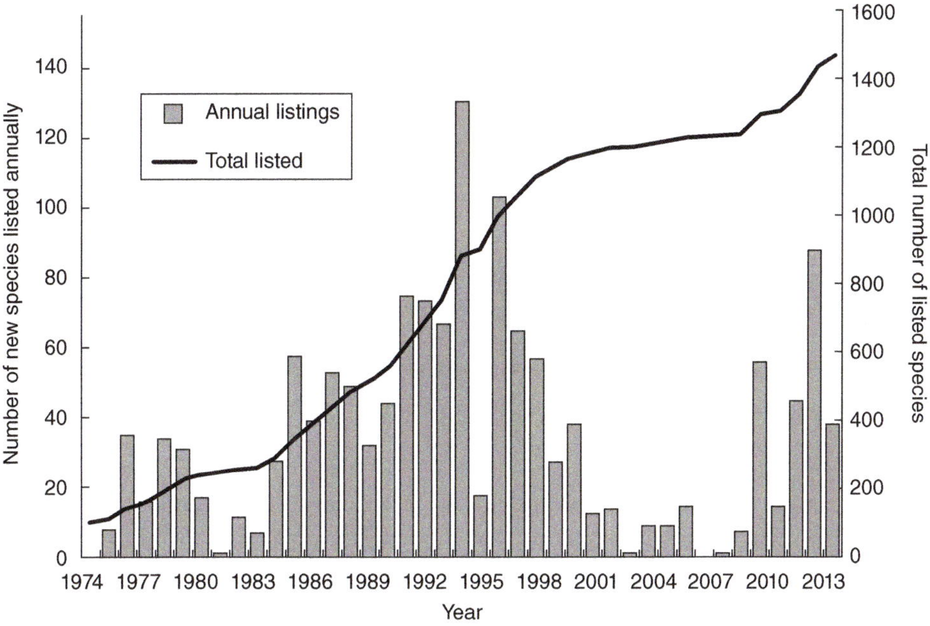

FIGURE 12.7 The Endangered Species Act (ESA) over time. Number of new species listed annually under the ESA (gray bars) and cumulative number of species listed (black line).

Reflection: The total number of species listed under the ESA has grown steadily over the last 50 years; however, there has been significant fluctuation. What factors do you think contribute to this interannual variation in number of species listed per year?

Source: "Emily E. Puckett, Dylan C. Kesler, D. Noah Greenwald, Taxa, petitioning agency, and lawsuits affect time spent awaiting listing under the US Endangered Species Act, Biological Conservation Volume 201, September 2016, Pages 220–229"

The ESA is a critically important legislative tool, but its implementation is also complex. Species can wait a long time to receive protection (one meta-analysis calculated an average of ~12 years), and the listing process can involve time-consuming and expensive lawsuits (Puckett et al. 2016). Moreover, it can be difficult to assess whether recovery tools (like species recovery plans, critical habitat designation, and species-specific funding) are effective. Many authors have argued this is because recovery plans often lack measurable and objective criteria (Gibbs & Currie 2012; Himes Boor 2014). Finally, even when recovery tools are effective, delisting can be risky because recovering species are then left with little regulatory protection (Doremus & Pagel 2001).

Scientists, managers, and policymakers continue to work to strengthen the ESA and other legislation contributing to biodiversity protection. For example, many of the analytical approaches (like population viability analysis) and management tools (like adaptive management) that we have discussed can be used to generate quantitative recovery criteria. Moreover, the ESA interfaces with multilateral treaties to protect endangered species around the world. For example, the **Convention on International Trade in Endangered Species** of Wild Fauna and Flora (CITES) is a 175-nation agreement that was enacted in 1975 to protect species that could be threatened by extinction due to international trade. There are currently more than 35,000 species protected by this multilateral agreement.

International Climate Treaties

Myriad environmental issues require coordination across national borders. Bilateral agreements are often made to negotiate the environmental regulation of a shared resource (like a lake spanning borders). However, multinational cooperation is required to tackle truly global issues. The United Nations Environment Programme (UNEP) is a key guiding body for international environmental legislature. One early UNEP win was the Montreal Protocol, which required the reduction of ozone-depleting substances. The Montreal Protocol was ratified in 1987 by 197 nations and led to a dramatic reversal in ozone degradation.

Now a critical arena for international cooperation is addressing anthropogenic climate change. In 1988, the **Intergovernmental Panel on Climate Change** (IPCC) was formed as a new branch of UNEP. The IPCC has produced definitive reports on the scientific, economic, political, and health implications of anthropogenic climate change. Informed by the IPCC's work, the United Nations Framework Convention on Climate Change (UNFCCC) created its first nonbinding treaty in 1992 establishing the need for countries to reduce their carbon emissions. The first legally binding commitment—the Kyoto Protocol—was adopted in 1997 with 192 participating parties (but not ratified by the United States). Additional agreements have followed, such as the Paris Agreement, that aim to curtail climate change with specific targets.

These multilateral agreements are exceedingly complex and raise tremendous equity, accountability, and enforcement issues (e.g., Duruji et al. 2018). One important issue has been creating a framework that allows for an equitable distribution of responsibility, whereby industrialized nations that have contributed the most to climate change also assume responsibility for making the most significant reductions in emissions (Dellink et al. 2009; Liu et al. 2017). However, this remains a thorny issue.

As shown in Figure 12.8, one recent study found that the majority of high-emissions countries are actually the least vulnerable to the impacts of climate change, while the majority of low-emissions countries are particularly vulnerable to climate change (Althor et al. 2016). Turning the tide on anthropogenic climate change will require addressing this climate change inequity. Ultimately, large industrialized countries will need to truly take responsibility for their historical and current contribution to atmospheric changes.

It is also essential to acknowledge the history of natural resource extraction and oppression of human populations that led to the current climate crisis and simultaneously to the social disparities between different human populations. Black,

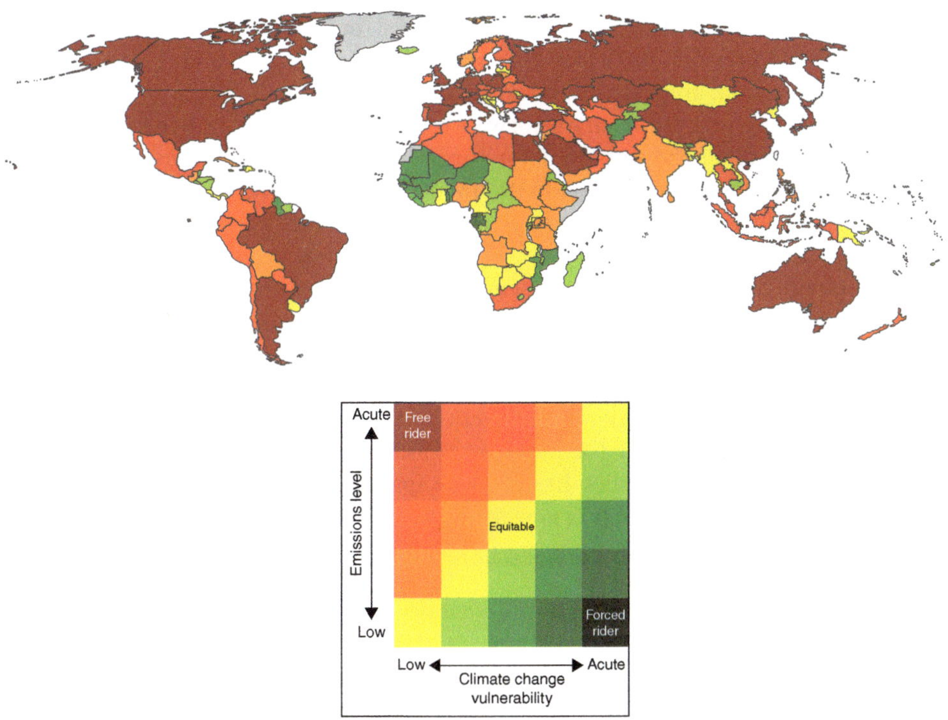

FIGURE 12.8 Climate change inequity. Not all countries contribute equally to climate change, and not all countries are equally vulnerable to the effects of climate change. Countries with the highest emissions and lowest vulnerability are darkest red and considered "free riders" while countries with the lowest emissions and highest vulnerability are darkest green and considered "forced riders". Very rarely are countries balanced in their contributions to—and risks from—climate change. Most industrialized countries contribute disproportionately to climate change, and most developing nations disproportionately bear the costs.

Reflection: In your opinion, what would it take for the countries with the largest impact to realign their emissions with international targets?

Source: Althor, G., Watson, J. & Fuller, R. Global mismatch between greenhouse gas emissions and the burden of climate change. Sci Rep 6, 20281 (2016)

indigenous, and people of color have been disproportionately impacted by the causes and consequences of environmental change, first by direct exploitation (including genocide, slavery, and exposure to toxic materials) and then by disproportionately bearing the burden of climate change (including drought, wildlife, and crop failure; Pierre 2020). Thus, it is not only that individuals, communities, and nations need to take responsibility for past emissions but also for how they have enabled, perpetuated, and benefited from systems of injustice that have wide-ranging social, economic, and ecological repercussions.

International Debt-for-Nature Swaps

A final example of creative international approaches for aligning economic and ecological needs is **debt-for-nature swaps**. In this approach, one country cancels the debt of another country in return for a commitment to habitat preservation (Hansen 1989). Initially, debt-for-nature swaps were commonly used for the protection of tropical rainforest stands (McFarland 2017). However, this voluntary mechanism has been used to create both terrestrial and marine preserves.

The first marine debt-for-nature swap recently occurred, and the case of the Seychelles highlights the potential for innovative partnerships between governments, NGOs, and private individuals. In this instance, the Leonardo DiCaprio Foundation secured enough private donors to assist the Nature Conservancy in purchasing debt from the Seychelles. With these funds, new marine protections were instated, increasing local protected waters from 1% to 30%. Moreover, an endowment was established to ensure the successful management of the protected area over the long term.

Of course, debt-for-nature approaches have potential pitfalls. For example, insufficient environmental protection, misdirection of funds, and lack of benefit to local communities can all hamper their success (Cassimon et al. 2011). However, innovative transnational approaches like debt swaps can turn debt into conservation financing. Ultimately, this can be a great boon both to local economies and ecosystem health provided that local and indigenous communities have agency in all stages of the process.

In Sum

There are many other tools by which local, national, and international governing bodies can mitigate global change stressors and positively impact biodiversity conservation. From ambitious international efforts (like Costa Rica's aim to become the first "carbon-neutral" country) to local endeavors (like incentivizing private landowners to contribute to conservation goals), policy action is a key element in biodiversity conservation. Ultimately, transparency, participatory governance, long-term vision, accountability, and new metrics for progress will be essential ingredients (O'Rourke & Lollo 2015). We are living in unprecedented times where new visions of equitable governance and new ways of partnering across sectors are emerging. We will evaluate one other creative example of public–private partnerships for aligning human and conservation interests in the *Meet the Data* feature of this chapter.

WHAT IS THE FORECAST FOR THE FUTURE?

As we have seen throughout this book, there are many analytical tools that can be used to project future trajectories for specific global change stressors or particular species or ecosystem responses. But is there a way to predict what the future holds more generally? Are we heading toward planetary ecological collapse? Or are we on the brink of a new sustainable relationship with our planet and its resources?

Environmental scholars have long warned that many Earth systems are approaching—or have crossed—critical thresholds. Even the popular media is fascinated with the possibility of environmental apocalypse. One recent study documented the increase over the last 50 years of high-grossing films with a theme of ecological collapse (Kareiva & Carranza 2018). Do these films capture something inevitable about where humanity is heading?

Ecological apocalypse scenarios hit especially close to home in this time when so many are impacted by disease outbreaks, drought, floods, famine, and wildfires. But at the same time, people around the world are mobilizing in new ways to advocate for sustainability and make lasting changes in their communities. There is increasing mobilization to confront unjust power structures and envision a more sustainable future for human society and the biosphere as a whole.

Moreover, even grave challenges can have unexpected impacts on our environmental footprint. For example, during the COVID-19 shutdown of 2020, carbon emissions decreased up to 26% in a set of countries studied, bringing emissions down to a level comparable to 2006 (Quéré et al. 2020). However, the longer-term emissions trajectory remains uncertain as we enter into a period of accelerated change in both social and ecological spheres.

Ultimately, we know that ecosystems can have tremendous resilience, and we know that humans can alter their beliefs and behaviors when needed. So the underlying question remains: Can we reorient to a more sustainable structure? Although there are no definitive answers to these questions, it is sometimes the questions without answers that motivate us into new ways of thinking. As anthropogenic changes to the planetary system intensify, our relationship to the environment has arguably never been more important. The *Taking a Closer Look* feature in this chapter provides an opportunity for further reflection on how we see this time point in history and our place in it.

CONCLUSION

As we have seen throughout our exploration, there is no one-size-fits-all solution to biodiversity conservation in an era of global change. There are millions of species on our planet, and each has a unique story. Species have different evolutionary trajectories, different roles in their ecosystems, and different stressors affecting their persistence. In addition, species coexist in intricate webs of interaction—not only with each other but also with short- and long-term biogeochemical cycles. Thus, actions to ameliorate global change stressors must consider the interrelated whole of biotic and abiotic processes on planet Earth.

Thus, we confront a grand challenge of conserving different aspects of biological diversity (from genetic diversity to species diversity to functional diversity to ecosystem diversity) across different spatial scales (from local to global) across different timescales (from now into the imaginable future) amid uncertainty, in a world that will continue to change.

Despite this challenge, there are a wealth of data and tools for biodiversity conservation and for aligning the needs of ecosystems and human societies. As we have seen, there is unprecedented opportunity to leverage new technologies for rapid assessments of anthropogenic stressors and species vulnerability. There are also unprecedented opportunities for collaboration and integration across disciplines, across stakeholder groups, and across nations.

Arguably, our species is at a turning point in its history. The path to a life-sustaining future will require new ways of thinking and new ways of organizing. As Albert Einstein famously said and as we discussed in this chapter's *Taking a Closer Look* feature: *"The significant problems we face cannot be solved at the same level of thinking we were at when we created them."* It is time to generate creative solutions that align the needs of human society and the rest of the natural world. This begins by reconciling the false separation between humans and "the environment." The restricted way of viewing the environment as something separate from humans has led to a limited perspective on both the past and the future. It is time to reimagine the dynamic and interdependent relationship among humans, all other life forms, and the planet. There is an opportunity to recognize our true creative power, individually and collectively, and generate an inspired vision for a new age. What might that vision look like and what roles could you play?

MEET THE DATA FINANCIAL INCENTIVES FOR DYNAMIC CONSERVATION

Who Are the Scientists and What Did They Set Out To Do?

Here we focus on a recent study entitled "Dynamic Conservation for Migratory Species," published by Reynolds and collaborators in 2017 in the journal *Science Advances.* The study was led by the Nature Conservancy, but it was a collaboration between biologists and economists at multiple universities and nonprofits (see Box Figure 12.3). The study evaluated the impact of a novel conservation program that was established to align the interests of local farmers and habitat needs of migratory shorebirds.

Financial incentives are increasingly used as a policy mechanism for decreasing emissions, reducing pollution, and supporting biodiversity conservation (reviewed in de Vries & Hanley 2016). In the agriculture sector, direct payment or tax credits have long been used to encourage landowners with environmentally sensitive cropland to contribute to conservation objectives. For example, landowners can receive monetary compensation for entering into land preservation agreements with government agencies or land trusts—termed **conservation easements**. Moreover, in some regions, landowners can receive an agglomeration bonus if they voluntarily coordinate with their neighbors to protect contiguous areas and reduce habitat fragmentation (Parkhurst et al. 2002).

Traditionally, conservation set-asides have focused on permanent land protection. However, donating land to conservation for perpetuity can be a significant risk for farmers, and acquiring large tracts of land for conservation can require huge up-front financial investments for land trusts. In addition, some conservation objectives can

BOX FIGURE 12.3 The Nature Conservancy team set out to incentivize local farmers to contribute to migratory bird conservation. (A) The Nature Conservancy team members Dr. Eric Hallstein (left) and Dr. Mark Reynolds (center, carrying a spotting scope and tripod) with one of the program's participating rice farmers (right). (B) Migratory stopovers for birds in rice farms of the Central Valley of California.

Source: Nature Conservancy

be met without permanent land protection. For example, a dynamic approach can be especially powerful for the conservation of migratory species. Migratory species are at particular risk from global change stressors, because changes at any one site along the migratory route can devastate the entire population. One recent meta-analysis showed that only 9% of migratory birds have protected habitat across their entire migratory route, compared to 45% of sedentary species with habitat in protected areas (Runge et al. 2015).

The Central Valley of California has historically provided critically important, seasonal habitat for millions of migratory and over-wintering ducks, geese, and shorebirds traveling the Pacific Flyway (Shuford et al. 1998). However, more than 90% of historical wetlands have been lost, and most remaining habitat lacks any official conservation protection (Stralberg et al. 2011). As illustrated in Box Figure 12.4, a large proportion of land conversion has been toward rice farming, which now covers more than 20% of the local land area. This backdrop inspired a novel attempt to incentivize local rice farmers to create short-term habitat patches for migratory waterbirds while allowing farmers to maintain their lands under cultivation for the rest of the year.

What Are *Your* Predictions?

Before you read on, take a few minutes to apply what you have learned in this chapter about aligning biodiversity conservation with human interests and livelihoods to the Central Valley system. Start by considering the following questions:

- *How specifically could rice farmers contribute to the conservation of migratory waterbirds?*
- *What incentives could be offered to encourage rice farmers to participate in a conservation program?*
- *How could the program outcomes be monitored and what metrics could be used to measure the success of the program?*

Now write a summary statement about how incentives might be used to encourage farmers to contribute to conservation outcomes and how the success of these incentives might be measured.

What Were the Scientists' Predictions?

The authors did not include a formal prediction but were explicit about the key objectives motivating their research: *"to be able to adaptively provision habitat when and where migratory species most need it, conservationists need to . . . be able to (i) predict where the species will be over the course of the year, (ii) identify areas that are suitable for the migrants or that can be modified to make them suitable, and (iii) create cost-effective mechanisms to ensure that the habitat will be there when the species arrive."*

What Data Were Collected?

Reynolds et al. used an integrative approach that combined citizen science, mathematical modeling, policy

(Continued)

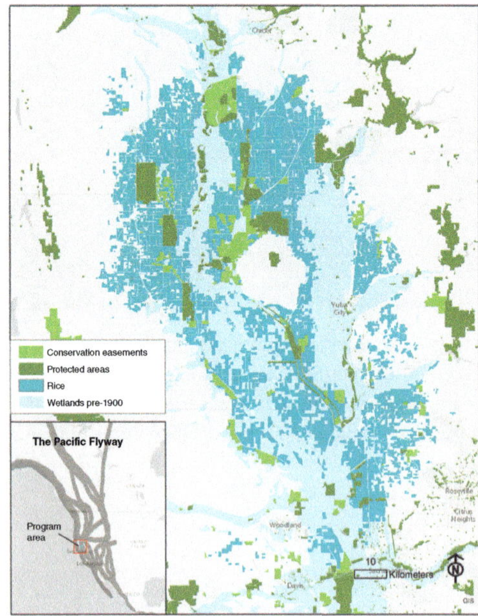

BOX FIGURE 12.4 There is very little protected habitat (shown in light and dark green) available for the millions of migratory and over-wintering waterbirds passing through California's Central Valley. Historical wetlands (shown in light blue) have largely been converted to urban and agricultural land uses, including extensive areas for rice farming (shown in dark blue).

Source: Dynamic conservation for migratory species. By Mark D. Reynolds, Brian L. Sullivan, Eric Hallstein, Sandra Matsumoto, Steve Kelling, Matthew Merrifield, Daniel Fink, Alison Johnston, Wesley M. Hochachka, Nicholas E. Bruns, Matthew E. Reiter, Sam Veloz, Catherine Hickey, Nathan Elliott, Leslie Martin, John W. Fitzpatrick, Paul Spraycar, Gregory H. Golet, Christopher McColl, Candace Low, Scott A. Morrison. Science Advances 23 Aug 2017: E1700707

incentives, and long-term monitoring. First, the authors obtained data on bird sightings from a citizen science project called eBird, where bird enthusiasts electronically record bird observations. They then created mathematical models to predict local bird abundances at weekly intervals and to generate a spatially explicit map of areas predicted to be of high conservation value. Next, the scientists used satellite imagery to assess the availability of suitable wetland habitat across the Central Valley. By finding times and places where bird abundance was predicted to be high—but surface water habitat was lacking—the scientists identified where and when birds would most desperately need stopover habitat. This approach is illustrated in Box Figure 12.5.

Next the Nature Conservancy implemented a **reverse auction** where they paid rice farmers to keep their fields flooded with irrigation water just for weeks where wetland habitat was critically needed. In the reverse auction model, farmers bid for these conservation leases (with the lowest bidders winning) and enter into a short-term agreement with the Nature Conservancy. Thus, farmers can set their own price for temporarily flooding their fields, and the Nature Conservancy can select habitat parcels holistically, considering the entire landscape and migratory season. Finally, the scientists compared shorebird responses in enrolled fields (those participating in temporary flooding) and unenrolled fields using on-the-ground visual surveys.

What Is *Your* Interpretation of the Data?

Before you read on, take a few minutes to interpret the results from the study shown in Box Figure 12.6. Consider the following questions as you look at the figure:

- *How do enrolled and unenrolled fields differ in shorebird density and species richness?*

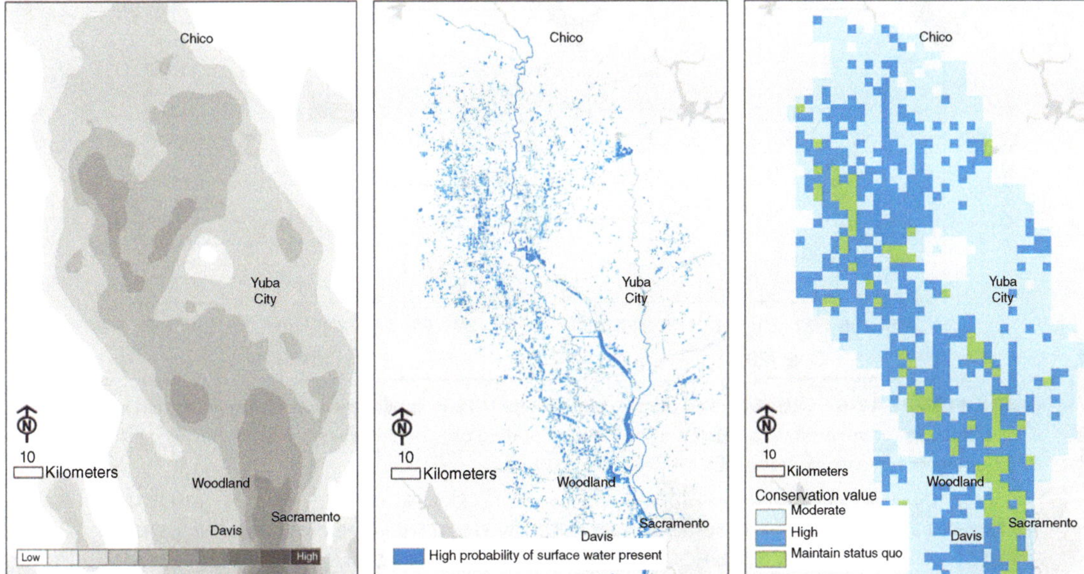

BOX FIGURE 12.5 Shorebird abundance data (left) and surface water availability data (center) were integrated to find areas of high conservation value (right). Areas of high conservation value (blue) are those where shorebird abundance is predicted to be high but water availability is expected to be low and thus represent opportunity areas for dynamic conservation.

Source: "Dynamic conservation for migratory species. BY MARK D. REYNOLDS, BRIAN L. SULLIVAN, ERIC HALLSTEIN, SANDRA MATSUMOTO, STEVE KELLING, MATTHEW MERRIFIELD, DANIEL FINK, ALISON JOHNSTON, WESLEY M. HOCHACHKA, NICHOLAS E. BRUNS, MATTHEW E. REITER, SAM VELOZ, CATHERINE HICKEY, NATHAN ELLIOTT, LESLIE MARTIN, JOHN W. FITZPATRICK, PAUL SPRAYCAR, GREGORY H. GOLET, CHRISTOPHER MCCOLL, CANDACE LOW, SCOTT A. MORRISON. SCIENCE ADVANCES 23 AUG 2017 : E1700707"

- *What temporal patterns are revealed by the data?*
- *What do the error bars tell you about the variation among replicates in this study?*

Write a sentence or two describing the *key findings* of this study and their *significance.*

What Was the Scientists' Interpretation of the Data?

By combining citizen science data on bird sightings and satellite data on surface water availability, the scientists found particular times and locations that were especially promising for habitat enhancement. In particular, wetland habitat was most desperately needed in late winter and early spring. The Nature Conservancy solicited bids from farmers willing to temporarily flood their fields for monetary renumeration during this peak bird season. They received 55 bids and accepted 80% of these bids.

The enrolled fields provided incredible benefits for migratory birds. In the authors' words: *"We observed high bird use of temporary wetlands and recorded more than 180,000 observations of birds representing 57 different species using the enrolled fields in spring of 2014. On average, shorebird species richness was more than three times greater and average shorebird density was five times greater on enrolled fields than unenrolled fields."*

The authors also highlight the economic advantages of dynamic short-term conservation strategies. They calculate the economic cost of their reverse auction approach and contrast that with the cost of buying, restoring, and managing a comparable amount of land in the Central Valley. The total cost of temporary habitat augmentation during the migratory season for the next 100 years would be approximately an order of magnitude lower than the cost of buying the same area of land. Thus, they suggest:

(Continued)

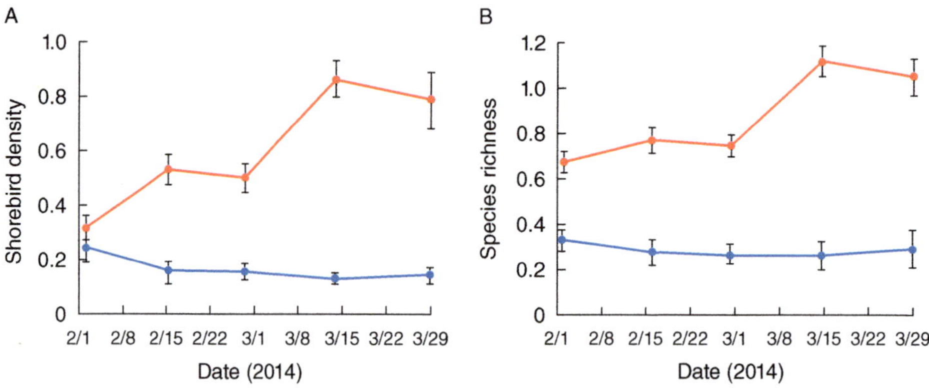

BOX FIGURE 12.6 Observed bird density per hectare (A) and species richness (B) in fields that were enrolled (red) or unenrolled (blue) in the short-term flooding program. Results are from the key over-wintering and migratory stopover time periods in 2014.

Source: "Dynamic conservation for migratory species. BY MARK D. REYNOLDS, BRIAN L. SULLIVAN, ERIC HALLSTEIN, SANDRA MATSUMOTO, STEVE KELLING, MATTHEW MERRIFIELD, DANIEL FINK, ALISON JOHNSTON, WESLEY M. HOCHACHKA, NICHOLAS E. BRUNS, MATTHEW E. REITER, SAM VELOZ, CATHERINE HICKEY, NATHAN ELLIOTT, LESLIE MARTIN, JOHN W. FITZPATRICK, PAUL SPRAYCAR, GREGORY H. GOLET, CHRISTOPHER MCCOLL, CANDACE LOW, SCOTT A. MORRISON. SCIENCE ADVANCES 23 AUG 2017 : E1700707"

"Dynamic conservation may offer important advantages over static protected area strategies, especially for migratory species. Potential advantages include the ability to adjust the timing, extent, and location of provisioned habitat to better match species' full life-cycle needs (for example, breeding, migration, stopover), and to adapt to climate change, droughts, habitat conversion, and other threats. Dynamic conservation strategies may also have greater scalability and cost-effectiveness because temporary habitat enhancements can be less expensive than permanent protection."

In summary, the authors conclude that dynamic conservation strategies *"can help extend the impact of scarce conservation dollars and, importantly, engage citizen scientists and private landowners."* However, they also *"caution that dynamic conservation is best considered as an additional tool and not a replacement for permanent protection, especially in areas of high, year-round conservation value."*

What Are *Your* Ideas for Future Research Directions?

Given the findings of Reynolds et al. about the positive effects of short-term habitat augmentation for migratory birds in the Central Valley of California, what next steps do you envision for this research program? If you had been involved in this study, how would you follow up? Start by considering the following questions:

- *Are there ways to extend the reach and impact of the habitat auction program?*
- *Are there other species or regions that would lend themselves particularly well to similar dynamic conservation efforts?*
- *Are there other complementary approaches that could be used to align biodiversity conservation with human livelihoods in agricultural settings or beyond?*

Now write a few sentences about future directions on this research theme.

TAKING A CLOSER LOOK ENVIRONMENTAL WORLDVIEWS

Our individual and collective behavior is shaped by our values, beliefs, and ultimately our worldview. As attributed to the philosopher Plato: *"Things are, for each person, the way he perceives them."* There are many different possible environmental worldviews and many different classification schemes for these viewpoints. Let us look at one formulation by the environmental scholar Dr. Joanna Macy (Macy & Brown 2014) as a way to illustrate the dramatic differences among environmental worldviews and to explore our own environmental perspective. The purpose of this feature is not to advocate for any particular worldview, but rather to draw attention to the fact that we all operate from conscious or subconscious environmental worldviews. Each of the following viewpoints has productive—and flawed—aspects. Which do you most relate to?

What Is the Industrial Growth Worldview?

Arguably one of the most influential cultural paradigms of the last millennium has been that of humans as a pinnacle species. As we saw in Chapter 2, Aristotle's *Scala Naturae* (Ladder of Life) depicted evolution as progress toward perfection with *Homo sapiens* on top. Even with a more modern understanding of evolutionary process and human history, the concept of humans as a rightfully dominant species is pervasive in many cultures. The attitude that the environment is essentially a resource for humans reached a peak in the Industrial Age in what Macy calls the "Industrial Growth" worldview.

An Industrial Growth perspective has fueled tremendous growth and contributed to the development of human society in many ways (such as increased living standards, advances in medicine, novel services and goods). However, industrial growth has also come at a great cost. The promise of health, wealth, and happiness for all has not been realized. Incredible inequities exist in different sectors of human society, for example in life span or income, as illustrated in Box Figure 12.7. Moreover, contemporary human impacts on ecosystems suggest that an unbridled Industrial Growth mentality is unsustainable both for human society and for the environment. Some contend that human ingenuity and technological advances will be able to solve our current environmental crisis. But many argue that nothing can grow endlessly without limits and that a different environmental narrative is needed.

What Is the Great Unraveling Worldview?

The most common alternative to the Industrial Growth worldview is what Macy refers to as the "Great Unraveling" worldview. As the magnitude of environmental challenges facing the planet has become increasingly clear, a different cultural paradigm has gained strength. Rather than seeing humans as approaching perfection, many have begun to see our own species as a cosmic problem precipitating an apocalypse on Earth. The fundamental message of the Great Unraveling worldview—illustrated in Box Figure 12.8—is that "we messed everything up" and the environment needs to be protected from intrinsic human greed.

The Great Unraveling worldview can be motivating for action and activism. However, this worldview may go beyond honest self-appraisal. It pits human beings against the environment, with little optimism for reconciliation and coexistence. Moreover, living from an inherently negative worldview can create additional problems. For example, at the societal level, opposition tends to entrench rather than alter destructive behavior. Similarly, at the personal level, guilt and anger are often depleting rather than motivating. Thus, it is not clear that the Great Unraveling worldview is providing a viable large-scale alternative solution to contemporary environmental issues.

What Is the Great Turning Worldview?

Many environmental thinkers have suggested that a new environmental worldview is necessary for this period of human and planetary history. These thinkers draw on a sentiment expressed by Albert Einstein when he said, *"The significant problems we face cannot be solved at the same level of thinking we were at when we created them."* The fundamental message of the "Great Turning" worldview is that new and creative human responses are needed to promote a society that lives in harmony with the rest of the biosphere. In addition to traditional environmental protection, a shift in our worldview and in the foundations of our culture may also be needed. This foundational shift involves a deep awareness of the inter-relatedness of all living things and a recognition that new social structures must support a vision of equality and sustainability for all.

The Great Turning worldview is expressed by Macy in this way: *"The most remarkable feature of this historical*

(Continued)

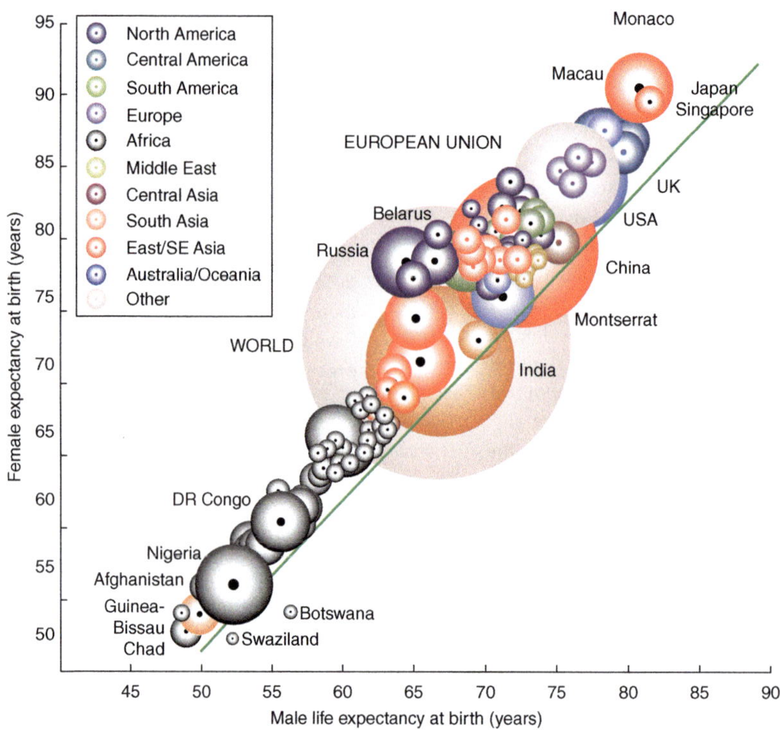

BOX FIGURE 12.7 Global variation in life expectancy. Size of circles is proportional to national population size.

Reflection: How are social inequities linked to environmental degradation? Are there underlying factors that contribute to both?

Source: Cmglee/Wikimedia Commons

moment on Earth is not that we are on the way to destroying the world—we've actually been on the way for quite a while. It is that we are beginning to wake up, as from a millennia-long sleep, to a whole new relationship to our world, to ourselves and each other." The idea of collective change in underlying worldview is appealing to many, but difficult to operationalize. What would it take for us to enter a new relationship with the biosphere?

This chapter has explored many ways of engaging in active environmental problem solving, and this special feature encourages a complementary approach. Shifting our way of relating to contemporary environmental issues may be as important as the actions we take. Many authors have suggested that a new sustainable era will be ushered in by those living in accordance with the principles of interconnectedness (e.g., Tolle 2005). As the sage Sri Nisargadatta Maharaj said: *"Everything affects everything. In this universe when one thing changes, everything changes. Hence the great power of man in changing the world, by changing himself."* This serves as a reminder not to underestimate the power of our own conscious creativity. As we change the way we relate to the millions of life forms on planet Earth, so too will our collective future begin to change.

If you are interested in further exploring your own environmental worldview, ask yourself the following questions: Who am I? Where am I? What is wrong in the world? What will make things better? Notice how the answers to these questions might change over time or with different life experiences. Try answering these questions from the perspective of one of your grandparents, or from your own perspective at an earlier age. Try answering these questions from the perspective that you think will be most productive for ushering humanity into a new era of sustainable coexistence with the biological heritage of planet Earth.

BOX FIGURE 12.8 (A) Cartoon representing the Great Unraveling worldview. (B) Dr. Joanna Macy, teacher of the Great Turning worldview.

Reflection: How does each of these worldviews (Great Unraveling vs Great Turning) affect our individual and collective environmental choices?

Source: Photo by Adam Loften

KEY CONCEPTS

What are coupled human–natural systems?

- Ecological and social domains are truly one integrated system with complex interactions and feedbacks.
- Successful biodiversity conservation will require an alignment of conservation values and the needs of human society.

What societal levers can be used to support biodiversity conservation?

- Human impact on the biosphere depends on the choices made by individuals, collectives, nations, and the global community.

How can individuals support biodiversity conservation?

- Individuals can contribute to biodiversity conservation in many ways–for example, by conserving natural resources, purchasing sustainably produced goods, and building environmental knowledge and awareness through research, activism, and education.

How can collectives support biodiversity conservation?

- Communities can contribute to biodiversity conservation in varied ways, including creating sustainable livelihood opportunities in ecotourism, agroecology, and manufacturing.

How can policy action support biodiversity conservation?

- Local, national, and international governance can contribute to biodiversity conservation through legislation, regulation, and motivation. Some examples include rights-based fisheries, endangered species legislation, international climate treaties, and debt-for-nature swaps.

What is the forecast for the future?

- There are no definitive forecasts for our complex coupled human–natural system, but there are many creative approaches that could help human society orient toward a more sustainable future.

Core concepts: What is I=PAT?

- The extent of future environmental impacts will depend strongly on patterns of human population, affluence, technological innovation, and the behavior of individuals, communities, and governments.

Meet the data: Financial incentives for dynamic conservation

- By integrating citizen science, mathematical modeling, incentive programs for local farmers, and biological monitoring efforts, the Nature Conservancy is creating opportunities for dynamic, short-term habitat augmentation for migratory waterbirds.

Taking a closer look: Environmental worldviews

- Our worldview affects the way we engage in environmental problem solving. Our personal views can change over time, as can cultural beliefs about the relationship between humans and the rest of the biosphere.

CONSOLIDATE YOUR KNOWLEDGE

Answer the following questions to assess your progress meeting the learning outcomes:

1. In your own words, write a one- or two-sentence synthesis of the big-picture takeaway point of this chapter.
2. Revisit your answer to the *Blank Page* exercise in the beginning of this chapter. Would you refine your answer now based on knowledge you integrated from this chapter?
3. How would you define a coupled human–natural system? What is a specific example of a feedback between human activities and ecological systems?
4. What are three specific ways that you as an individual can best contribute to biodiversity conservation?
5. Pick one example of community efforts to support sustainability and critique its benefits and risks.
6. What is the importance of the Endangered Species Act and what have been some challenges to its implementation?
7. What are other policy actions that can support biodiversity conservation at local, national, or international levels?
8. What is an example of dynamic conservation in space and/or time? Why are dynamic conservation approaches so important in this time of rapid global change?

9. From your perspective what are the ultimate root causes of anthropogenic stressors on biodiversity and what would it take to address them?
10. If you were writing an epilogue for this textbook to express what you hope students take away from reading it, what would it say?
11. In your own words, define the bolded terms in this chapter.
12. What are some questions that *you* have about the content in this chapter? If you found some content particularly challenging or particularly interesting, identify these as areas for additional reflection or reading.

LITERATURE CITED

Allison, Ratner, Åsgård, Willmann, Pomeroy, & Kurien. 2012. "Rights-Based Fisheries Governance: From Fishing Rights to Human Rights." *Fish and Fisheries* 13 (1): 14–29. doi.org/10.1111/j.1467-2979.2011.00405.x.

Althor, Watson, & Fuller. 2016. "Global Mismatch Between Greenhouse Gas Emissions and the Burden of Climate Change." *Scientific Reports* 6: 1–6. doi.org/10.1038/srep20281.

Altieri, Funes-Monzote, & Petersen. 2012. "Agroecologically Efficient Agricultural Systems for Smallholder Farmers: Contributions to Food Sovereignty." *Agronomy for Sustainable Development* 32 (1): 1–13. doi.org/10.1007/s13593-011-0065-6.

Berkes. 1985. "Fishermen and 'The Tragedy of the Commons.'" *Environmental Conservation* 12 (3): 199–206.

Boyce & Boyce. 2007. "Is Inequality Bad for the Environment?" Political Economy Research Institute Working Papers, no. 135. scholarworks.umass.edu/peri_workingpapers/121.

Brandt & Buckley. 2018. "A Global Systematic Review of Empirical Evidence of Ecotourism Impacts on Forests in Biodiversity Hotspots." *Current Opinion in Environmental Sustainability* 32: 112–18. doi.org/10.1016/j.cosust.2018.04.004.

Buckley, Morrison, & Castley. 2016. "Net Effects of Ecotourism on Threatened Species Survival." *PLoS ONE* 11 (2): 23–25. doi.org/10.1371/journal.pone.0147988.

Carter, Viña, Hull, McConnell, Axinn, Ghimire, & Liu. 2014. "Coupled Human and Natural Systems Approach to Wildlife Research and Conservation." *Ecology and Society* 19 (3): 43

Cassimon, Prowse, & Essers. 2011. "The Pitfalls and Potential of Debt-for-Nature Swaps: A US-Indonesian Case Study." *Global Environmental Change* 21 (1): 93–102. doi.org/10.1016/j.gloenvcha.2010.10.001.

Chang. 2011. "Feeling Ambivalent About Going Green: Implications for Green Advertising Processing." *Journal of Advertising* 40 (4): 19–31. doi.org/10.2753/JOA0091-3367400402.

Commoner. 1972. *The Closing Circle; Nature, Man, and Technology*. New York: Knopf.

Coria & Calfucura. 2012. "Ecotourism and the Development of Indigenous Communities: The Good, the Bad, and the Ugly." *Ecological Economics* 73: 47–55. doi.org/10.1016/j.ecolecon.2011.10.024.

Das & Chatterjee. 2015. "Ecotourism: A Panacea or a Predicament?" *Tourism Management Perspectives* 14: 3–16. doi.org/10.1016/j.tmp.2015.01.002.

de Vries & Hanley. 2016. "Incentive-Based Policy Design for Pollution Control and Biodiversity Conservation: A Review." *Environmental Resource Economics* 63: 687–702.

Dellink, den Elzen, Aiking, Bergsma, Berkhout, Dekker, & Gupta. 2009. "Sharing the Burden of Financing Adaptation to Climate Change." *Global Environmental Change* 19 (4): 411–21. doi.org/10.1016/j.gloenvcha.2009.07.009.

Doremus & Pagel. 2001. "Why Listing May Be Forever: Perspectives on Delisting Under the U.S. Endangered Species Act." *Conservation Biology* 15 (5): 1258–68. doi.org/10.1046/j.1523-1739.2001.00178.x.

Duruji, Olanrewaju, & Duruji-Moses. 2018. "From Kyoto to Paris: An Analysis of the

Politics of Multilateralism on Climate Change" In *Promoting Global Environmental Sustainability and Cooperation*. Hershey, PA: IGI Global.

Ehrlich & Holdren. 1972. "Critique." *Bulletin of the Atomic Scientists* 28 (5): 16–27. doi.org/10.1080/00963402.1972.11457930.

Ehrlich & Pringle. 2008. "Where Does Biodiversity Go from Here? A Grim Business-as-Usual Forecast and a Hopeful Portfolio of Partial Solutions." *Proceedings of the National Academy of Sciences* 105 (suppl. 1): 11579–86. doi.org/10.1073/pnas.0801911105.

Fargani, Cheung, & Hasan. 2016. "An Empirical Analysis of the Factors that Support the Drivers of Sustainable Manufacturing." *Procedia CIRP* 56: 491–95. doi.org/10.1016/j.procir.2016.10.096.

Fiorella, Milner, Salmen, Hickey, Omollo, Odhiambo, Mattah, Bukusi, Fernald, & Brashares. 2017. "Human Health Alters the Sustainability of Fishing Practices in East Africa." *Proceedings of the National Academy of Sciences* 114 (16): 4171–76. doi.org/10.1073/pnas.1613260114.

Fujita & Bonzon. 2005. "Rights-Based Fisheries Managements an Environmentalist Perspective." *Reviews in Fish Biology and Fisheries* 15 (3): 309–12. doi.org/10.1007/s11160-005-4867-y.

Gallo-Cajiao, Archibald, Friedman, Steven, Fuller, Game, Morrison, & Ritchie. 2018. "Crowdfunding Biodiversity Conservation." *Conservation Biology* 32 (6): 1426–35. doi.org/10.1111/cobi.13144.

Gibbs & Currie. 2012. "Protecting Endangered Species: Do the Main Legislative Tools Work?" *PLoS ONE* 7 (5): e35730. doi.org/10.1371/journal.pone.0035730.

Hansen. 1989. "Debt for Nature Swaps—Overview and Discussion of Key Issues." *Ecological Economics* 1 (1): 77–93. doi.org/10.1016/0921-8009(89)90025-6.

Hawkins, Singh, Majeau-Bettez, & Strømman. 2013. "Comparative Environmental Life Cycle Assessment of Conventional and Electric Vehicles." *Journal of Industrial Ecology* 17 (1): 53–64. doi.org/10.1111/j.1530-9290.2012.00532.x.

Himes Boor. 2014. "A Framework for Developing Objective and Measurable Recovery Criteria for Threatened and Endangered Species." *Conservation Biology* 28 (1): 33–43. doi.org/10.1111/cobi.12155.

Isbell, Gonzalez, Loreau, Cowles, Díaz, Hector, Mace, et al. 2017. "Linking the Influence and Dependence of People on Biodiversity Across Scales." *Nature* 546 (7656): 65–80. doi.org/10.1038/nature22899.

Kareiva & Carranza. 2018. "Existential Risk Due to Ecosystem Collapse: Nature Strikes Back." *Futures* 102 (January): 39–50. doi.org/10.1016/j.futures.2018.01.001.

Kruijssen, Keizer, & Giuliani. 2009. "Collective Action for Small-Scale Producers of Agricultural Biodiversity Products." *Food Policy* 34 (1): 46–52. doi.org/10.1016/j.foodpol.2008.10.008.

Liu, Taylor, Dietz, Carpenter, Alberti, Folke, Moran, et al. 2007. "Complexity of Coupled Human and Natural Systems." *Science* 317 (5844): 1513–16. doi.org/10.1126/science.1144004.

Liu, Wu, & Huang. 2017. "An Equity-Based Framework for Defining National Responsibilities in Global Climate Change Mitigation." *Climate and Development* 9 (2): 152–63. doi.org/10.1080/17565529.2015.1085358.

Lu & Li. 2006. "Review of Rice–Fish Farming Systems in China—One of the Globally Important Ingenious Agricultural Heritage Systems (GIAHS)." *Aquaculture* 260: 106–13. doi.org/10.1016/j.aquaculture.2006.05.059.

Lubchenco, Cerny-Chipman, Reimer, & Levin. 2016. "The Right Incentives Enable Ocean Sustainability Successes and Provide Hope for the Future." *Proceedings of the National Academy of Sciences* 113 (51): 14507–14. doi.org/10.1073/pnas.1604982113.

Macy & Brown. 2014. *Coming Back to Life: The Guide to the Work that Reconnects*. Gabriola Island, Canada: New Society Publishers.

Marcovaldi & Dei Marcovaldi. 1999. "Marine Turtles of Brazil: The History and Structure of Projeto TAMAR-IBAMA." *Biological Conservation* 91 (1): 35–41. doi.org/10.1016/S0006-3207(99)00043-9.

McFarland. 2017. *Conservation of Tropical Rainforests*. Cham, Switzerland: Palgrave MacMillan. doi.org/10.1007/978-3-319-63236-0.

Mendoza-Ramos & Prideaux. 2018. "Assessing Ecotourism in an Indigenous Community: Using, Testing and Proving the Wheel of Empowerment Framework as a Measurement Tool." *Journal of Sustainable Tourism* 26: 277–91. doi.org/10.1080/09669582.2017.1347176.

Möllmann & Diekmann. 2012. "Marine Ecosystem Regime Shifts Induced by Climate and Overfishing." *Advances in Ecological Research* 47: 303–47. doi.org/10.1016/b978-0-12-398315-2.00004-1.

Monti, Duriez, Dominici, Sforzi, Robert, Fusani, & Grémillet. 2018. "The Price of Success: Integrative Long-Term Study Reveals Ecotourism Impacts on a Flagship Species at a UNESCO Site." *Animal Conservation* 21 (6): 448–58. doi.org/10.1111/acv.12407.

O'Rourke. 2014. "The Science of Sustainable Supply Chains." *Science* 344 (6188): 1124–27. doi.org/10.1126/science.1248526.

O'Rourke & Lollo. 2015. "Transforming Consumption: From Decoupling, to Behavior Change, to System Changes for Sustainable Consumption." *Annual Review of Environment and Resources* 40 (1): 233–59. doi.org/10.1146/annurev-environ-102014-021224.

O'Rourke & Ringer. 2016. "The Impact of Sustainability Information on Consumer Decision Making." *Journal of Industrial Ecology* 20 (4): 882–92. doi.org/10.1111/jiec.12310.

Parkhurst, Shogren, Bastian, Kivi, Donner, & Smith. 2002. "Agglomeration Bonus: An Incentive Mechanism to Reunite Fragmented Habitat for Biodiversity Conservation." *Ecological Economics* 41 (2): 305–28. doi.org/10.1016/S0921-8009(02)00036-8.

Pecl, Araújo, Bell, Blanchard, Bonebrake, Chen, Clark, et al. 2017. "Biodiversity Redistribution Under Climate Change: Impacts on Ecosystems and Human Well-Being." *Science* 355 (6332): eaai9214. doi.org/10.1126/science.aai9214.

Pierre. 2020. "Temporal Impacts of Ecological Harms Across Human Populations." Figure at doi.org/10.6084/m9.figshare.12670916.v1.

Ponisio & Ehrlich. 2016. "Diversification, Yield and a New Agricultural Revolution: Problems and Prospects." *Sustainability* 8 (11): 1–15. doi.org/10.3390/su8111118.

Puckett, Kesler, & Greenwald. 2016. "Taxa, Petitioning Agency, and Lawsuits Affect Time Spent Awaiting Listing Under the US Endangered Species Act." *Biological Conservation* 201: 220–29. doi.org/10.1016/j.biocon.2016.07.005.

Quéré, Jackson, Jones, Smith, Abernethy, et al. 2020. Temporary reduction in daily global CO_2 emissions during the COVID-19 forced confinement. *Nature Climate Change* 10: 647–53.

Reynolds, Sullivan, Hallstein, Matsumoto, Merrifield, Spraycar, Golet, et al. 2017. "Dynamic Conservation for Migratory Species." *Science Advances* 3 (8): 1–9. doi.org/10.1126/sciadv.1700707.

Roca. 2002. "The IPAT Formula and Its Limitations." *Ecological Economics* 42: 1–2.

Runge, Watson, Butchart, Hanson, Possingham, & Fuller. 2015. "Protected Areas and Global Conservation of Migratory Birds." *Science* 350 (6265): 1255–58. doi.org/10.1126/science.aac9180.

Scherr & McNeely. 2008. "Biodiversity Conservation and Agricultural Sustainability: Towards a New Paradigm of 'Ecoagriculture' Landscapes." *Philosophical Transactions of the Royal Society B: Biological Sciences* 363 (1491): 477–94. doi.org/10.1098/rstb.2007.2165.

Shuford, Page, & Kjelmyr. 1998. "Patterns and Dynamics of Shorebird Use of California's Central Valley." *The Condor* 100 (2): 227–44. doi.org/10.2307/1370264.

Stralberg, Cameron, Reynolds, Hickey, Klausmeyer, Busby, Stenzel, Shuford, & Page. 2011. "Identifying Habitat Conservation Priorities and Gaps for Migratory Shorebirds and Waterfowl in California." *Biodiversity and Conservation* 20 (1): 19–40. doi.org/10.1007/s10531-010-9943-5.

Tolle. 2005. *A New Earth: Awakening to Your Life's Purpose*. New York: Penguin.

Tomich, Brodt, Ferris, Galt, Horwath, Kebreab, Leveau, et al. 2011. "Agroecology: A Review from a Global-Change Perspective." *Annual Review of Environment and Resources* 36: 193–222. doi.org/10.1146/annurev-environ-012110-121302.

Viana, Gelcich, Aceves-Bueno, Twohey, & Gaines. 2019. "Design Trade-Offs in Rights-Based Management of Small-Scale Fisheries." *Conservation Biology* 33 (2): 361–68. doi.org/10.1111/cobi.13208.

Waggoner & Ausubel. 2002. "A Framework for Sustainability Science: A Renovated IPAT Identity." *Proceedings of the National Academy of Sciences* 99 (12): 7860–65. doi.org/10.1073/pnas.122235999.

GLOSSARY

ACCLIMATION The process of an individual adjusting to changing conditions in a laboratory setting.

ACCLIMATIZATION The process of an individual adjusting to changing conditions in a natural setting.

ADAPTATION The process by which populations become better suited to their environment via heritable genetic change across generations.

ADAPTIVE MANAGEMENT A holistic process of identifying conservation goals, implementing management interventions, studying outcomes, and adjusting future goals.

ADAPTIVE POTENTIAL The likelihood that a species will exhibit an adaptive response to changing conditions.

AGROECOLOGY The integration of ecological principles and study into agricultural systems.

ALBEDO The proportion of heat and light reflected by a surface.

ALLEE EFFECT When decreased population size or density disrupts key aspects of survival or reproduction.

ALLELE A particular variant of a gene.

ANTHROPOCENE The geological time period when human activities dominate key Earth processes.

ANTHROPOGENICALLY INDUCED ADAPTATION TO INVADE HYPOTHESIS Populations adapted to human habitats in their native range are more likely to be transported and survive in other human-altered environments outside their native range (abbreviated as AIAI).

ARCHAEA An ancient domain of single-celled organisms without nuclei, originally classified with bacteria but now recognized as a distinct lineage.

ARDIPITHECUS An early diverging group of hominids present ~4–6 million years ago.

ARTIFICIAL SELECTION The process whereby humans intentionally breed particular individuals with desirable traits over many generations to alter the characteristics of a variety, breed, or species.

ASSISTED EVOLUTION Intentional acceleration of adaptation to enhance desirable traits for species confronted with rapidly changing environments.

ASSISTED MIGRATION Intentional movement of species to areas of suitable habitat that they are unlikely to disperse to on their own.

AUSTRALOPITHECUS An early species of hominid present ~2–4 million years ago.

BACTERIA An ancient domain of single-celled organisms without nuclei, likely the first cellular life on Earth (sometimes referred to as "eubacteria" or "true bacteria").

BENEFICIAL MUTATION A genetic change that increases fitness and is selected for by natural selection.

BIODIVERSITY HOTSPOTS Biodiverse geographic regions that contain many endemic species and are especially vulnerable to global change stressors.

BIOGEOCHEMICAL CYCLES Interlinked cycles that govern the flow of energy and matter through biotic and abiotic components of the environment.

BIOSPHERE The sum total of all regions of the Earth inhabited by living things.

BIOTIC HOMOGENIZATION The increased similarity (genetically, taxonomically, and functionally) of ecological communities across regions of the globe.

BIOTIC INTERACTION CHAINS The direct and indirect effects of species across trophic levels.

BOLIDE A celestial object (like a meteor or meteorite) that hits the Earth's atmosphere.

BOTTOM-UP EFFECTS The ways in which nutrients, resources, and lower trophic levels (such as producers) impact higher trophic levels in an ecosystem.

BROWN FOOD WEB The portion of the food web anchored by decomposers, occurring primarily below ground.

CARBON SEQUESTRATION The process of storing carbon, which occurs naturally (e.g., in plants, soils, and the ocean) and can also be augmented technologically (e.g., through subterranean injection).

CASCADING EFFECTS Rippling impacts of one change on other components of an ecosystem.

CHROMOSOMAL INVERSION Large-scale DNA rearrangement where a chromosomal segment is reversed.

CHROMOSOME Structural unit of the genome.

CHYTRIDIOMYCOSIS A disease caused by the fungus *Batrachochytrium dendrobatidis* that is impacting amphibians around the world.

CLIMATE CHANGE ADAPTATION Actions to reduce the vulnerability of biological systems to anthropogenic climate change.

CLIMATE CHANGE COMMITMENT Amount of future global warming that is inevitable even if all emissions were curtailed due to inertia in the climate system.

CLIMATE CHANGE MITIGATION Actions to reduce radiative forcing on the climate system.

CLONING The process of creating a genetic copy of an individual.

COARSE-FILTER STRATEGIES In the conservation arena, those efforts to preserve, protect, or restore biodiversity with a focus on landscape or ecosystem level.

COLLECTIVE ACTION Efforts undertaken cooperatively by a group of people toward a common goal.

COMMON GARDEN EXPERIMENT When different genotypes are reared in a shared environment; often to assess whether traits are heritable.

COMMUNITY Set of interacting species in a particular area.

CONNECTIVITY Ability of organisms and resources to flow among habitat patches.

CONSERVATION The act of preserving, protecting, or restoring ecological systems.

CONSERVATION EASEMENT When power over a particular land area is vested in a land trust or government agencies to maximize conservation benefits.

CONSERVATION TRIAGE Decision making about which conservation targets will be given resources and which will not; typically some species are abandoned to focus on others with a higher chance of recovery.

CONSUMERS Organisms with an ecological role of obtaining energy by feeding on producers or other consumers.

CONTROL A standard of comparison that is not exposed to a treatment and provides a benchmark for data interpretation.

CONVENTION ON INTERNATIONAL TRADE OF ENDANGERED SPECIES Multilateral treaty to regulate international trade of endangered species (abbreviated as CITES).

CORAL BLEACHING When corals eject their symbionts (and lose associated pigment); often caused by high temperature and resulting in coral death.

COUPLED HUMAN–NATURAL SYSTEMS Dynamic interrelationship between social, economic, ecological, and biogeochemical systems.

CRADLE-TO-CRADLE LIFE CYCLE ANALYSIS Considering environmental effects of all stages of the product life cycle, from raw materials to recycling.

CRITICAL THERMAL MAXIMUM The temperature at which animals lose key motor function and above which they likely cannot survive.

DE-EXTINCTION The process of resurrecting a species that has died out (or creating a close functional analog).

DEBT-FOR-NATURE SWAPS When debt between countries is canceled in return for a commitment to habitat preservation.

DECOMPOSERS Organisms with an ecological role of breaking down organic matter.

DEFAUNATION The loss of animal populations from a community.

DELETERIOUS MUTATION A genetic change that decreases fitness and is selected against by natural selection.

DEOXYRIBONUCLEIC ACID Double-stranded macromolecule that is self-replicating and serves as the source of genetic inheritance for most forms of life on Earth (abbreviated as DNA).

DEPENDENT VARIABLE A factor whose value depends on the independent variable (also called response variable).

DIRECT EFFECT The impact of one factor (or individual or species) on another without any mediation through additional factors (contrast with indirect effect).

DISPERSAL CAPABILITY Organismal ability to move, with dispersal typically referring to one-way movement away from natal or breeding territory.

ECOSYSTEM All interacting biotic and abiotic components in a given area.

ECOSYSTEM COLLAPSE Rapid, substantial, and lasting loss of ecosystem structure and function.

ECOSYSTEM FUNCTION Physical, chemical, and biological processes that occur in a given ecosystem.

ECOSYSTEM RESILIENCE Ability of a system to absorb—or recover from—a perturbation while maintaining its overall structure and function.

ECOSYSTEM SERVICE Ecosystem functions that directly benefit humans.

ECOSYSTEM STRUCTURE Biotic and abiotic components of an ecosystem.

ECOTOURISM Commercial activities focused on visiting and supporting conservation areas, typically focused on linking economic, social, and environmental benefits.

ENDANGERED SPECIES ACT US legislation passed in 1973 to protect vulnerable species and the ecosystems upon which they depend (abbreviated as ESA).

ENDOSYMBIOSIS An ecological relationship where one organism lives inside the other and both interacting partners benefit.

ENVIRONMENTAL JUSTICE Equitable distribution of environmental costs and benefits across human society without excluding or burdening any group.

ENVIRONMENTAL STOCHASTICITY Unpredictable changes in environmental conditions over space or time.

EPIGENETIC INHERITANCE Heritable changes in gene expression that do not involve changes at the DNA sequence level.

EUKARYOTES All organisms whose cells contain a membrane-enclosed nucleus.

EVOLUTIONARY RADIATION The rapid accumulation of taxonomic diversity in a lineage, typically due to increased speciation rates (referred to as adaptive radiation if the role of natural selection is clear).

EVOLUTIONARY RESCUE Genetic adaptation that allows a population to recover from an environmental change that would otherwise cause extirpation.

EX SITU CONSERVATION "Off-site" efforts to protect vulnerable species by removing individuals from the area under threat.

EXPERIMENTAL EVOLUTION The use of controlled selection regimes to study evolutionary processes in real time across generations in laboratory experiments or field manipulations.

EXTANT Lineages that are still alive today.

EXTINCT Lineages that have died out with no surviving members.

EXTINCTION The loss of an entire species (or higher taxonomic group).

EXTINCTION DEBT Future extinctions caused by past or current events.

EXTINCTION RISK The probability of species loss (typically defined over a particular timescale).

EXTINCTION VORTEX Factors impacting small populations that create a positive feedback on population size and increase the likelihood of extinction.

EXTIRPATION The loss of an entire population.

FACULTATIVE Optional, particularly in regard to species that are not restricted to a particular ecological role or can interact with ecological partners depending on conditions.

FEEDBACK A process or system modified by its own effects.

FINE-FILTER STRATEGIES In the conservation arena, those efforts to preserve, protect, or restore biodiversity with a focus on the population or species level.

FITNESS The ability of an organism to survive and reproduce under particular environmental conditions.

FORCING A factor that alters the radiation balance of the climate system.

FULLY ADDITIVE EFFECT When the total impact of multiple factors is exactly the sum of their individual impacts.

FUNCTIONAL REDUNDANCY When multiple species share similar ecological roles and support similar ecological processes.

FUNDAMENTAL NICHE The conditions and resources that a population could use under ideal conditions (contrast with realized niche).

GENE Segment of DNA that codes for a protein.

GENE EXPRESSION The process by which a segment of DNA is actively transcribed into RNA.

GENE FLOW The movement of genetic variation between populations.

GENE REGULATION The process by which cells increase or decrease the production of RNA and/or proteins from particular genes.

GENETIC ASSIMILATION The process by which a trait that was initially induced by environmental conditions evolves decreased plasticity and becomes genetically encoded.

GENETIC DRIFT The process by which the genetic composition of populations changes over time due to random sampling of organisms and alleles.

GENETIC ENGINEERING Making deliberate, targeted changes to the genome of an organism.

GENETIC STOCHASTICITY Random changes in allele frequency, particularly pronounced in small populations (see genetic drift).

GENOME The sum total of genetic material in a cell or organism.

GENOME EDITING Making targeted changes to the genome of an organism by deleting, inserting, or altering stretches of DNA.

GENOTYPE The genetic makeup on an organism (can be used broadly to describe strains or narrowly to describe genetic variation influencing a particular trait).

GENOTYPE BY ENVIRONMENT INTERACTION When environmental conditions affect different genotypes in different ways.

GEOENGINEERING The intentional manipulation of large-scale environmental processes.

GEOGRAPHIC DISTRIBUTION The arrangement of species on Earth; all points where individuals of a particular species occur.

GEOGRAPHIC RANGE The total spatial area where a species is found.

GLOBAL WARMING COMMITMENT Amount of future global warming that is inevitable even if all emissions were curtailed due to inertia in the climate system.

GLOBALIZATION The interaction and increasing interdependence of people, organizations, and governments worldwide.

GREEN FOOD WEB The portion of the food web anchored by photosynthetic organisms, occurring primarily above ground.

GREENHOUSE EFFECT Natural process whereby the Earth's atmosphere traps incoming solar radiation.

GROSS DOMESTIC PRODUCT Total value of goods and services, typically reported per nation on an annual basis (abbreviated GDP).

HABITAT CONSERVATION PLAN Agreement under the Endangered Species Act about how lands will be managed and endangered species will be impacted.

HABITAT SUITABILITY MAP A representation of the area where a species is expected to survive based on its ecological needs.

HEREDITY Means by which traits (and the genetic information that encodes them) are passed from parent to offspring.

HERITABLE Traits that are genetically passed from parents to offspring.

HETEROZYGOSITY A specific metric of genetic diversity where a single individual has multiple alleles at a specified locus.

HOMINIDAE Family designation for all of the great apes (orangutan, gorilla, chimpanzee, bonobo, humans); also referred to as hominids.

HOMO Genus designation for all archaic and modern humans with *Homo sapiens* as the only surviving member.

HORIZONTAL GENE TRANSFER Movement of genetic information between individuals that do not have a vertical (parent–offspring or ancestor–descendant) relationship.

HOT HOUSE Geological periods during the Earth's history that were relatively warm and supported less glaciation.

HYBRIDIZATION The interbreeding of distinct species.

ICE AGE Any glacial episode; most often used to refer to the glaciation during the Pleistocene.

ICE HOUSE Geological periods during the Earth's history that were relatively cool and supported more glaciation.

IMMIGRATION Movement of organisms into a population.

IMPERVIOUS SURFACE COVER Hard groundcovers (like pavement) that do not allow water infiltration to the soil.

IN SITU CONSERVATION "On-site" efforts to protect vulnerable species by addressing threats to organisms in their natural habitat.

IN-VITRO FERTILIZATION Fertilization of egg with sperm outside of the body to produce an embryo in the lab.

INBREEDING The mating of close genetic relatives.

INCOMING SOLAR RADIATION Energy from the sun, which enters the Earth's atmosphere as shortwave radiation.

INDEPENDENT VARIABLE A factor whose value does not depend on other factor(s) studied (also called explanatory variable).

INDIGENOUS PEOPLE Ethnic or cultural groups that inhabited an area prior to occupation or colonization.

INDIRECT EFFECT The impact of one factor (or individual or species) on another that is mediated through an additional factor (contrast with direct effect).

INDUSTRIAL REVOLUTION Period in the late 1700s and early 1800s in which a number of manufacturing innovations first appeared.

INTERACTION EFFECTS The joint effect of two or more independent variables.

INTERGOVERNMENTAL PANEL ON CLIMATE CHANGE An international body convened by the United Nations to analyze and communicate contemporary patterns of climate change (abbreviated as IPCC).

INTERSPECIFIC Occurring between species, as in interspecific interactions or interspecific competition.

INTRASPECIFIC Occurring within species, as in intraspecific interactions or intraspecific competition.

INTROGRESSION The movement of genetic variation between species.

INVASIVE SPECIES A non-native species that has negative economic or ecological impacts in its expanded range.

IPAT FRAMEWORK The formulation that environmental impact (I) can be explained by three key determinants: population (P), affluence (A), and technology (T).

KEY BIODIVERSITY AREAS Geographic regions contributing significantly to the global persistence of biodiversity.

KEYSTONE SPECIES A species with a disproportionate impact on its ecosystem, whose loss has dramatic effects.

LAG EFFECTS Time delay between cause and consequence.

LINEAGE An ancestor and all of its descendants.

LINEAGE SPECIES CONCEPT A formulation that defines species as independently evolving segments of the tree of life.

LOCUS Position of a gene on a chromosome (plural: loci).

LUXURY EFFECT The observation that human wealth is often positively correlated with organismal diversity in human-modified environments.

MAIN EFFECTS The effect of one independent variable, ignoring the effects of any other independent variables.

MASS EXTINCTIONS Periods of time in the Earth's history where a disproportionate number of lineages around the world have been lost.

MEAN Average; sum of all data points divided by the total number of points.

META-ANALYSIS Explicit reanalysis of data from numerous individual studies in a unified framework.

METAPOPULATION A group of populations interacting through migration of individuals, governed by dynamics of extinction and recolonization.

MINIMUM VIABLE POPULATION A threshold size below which a population is likely to be lost.

MODEL ORGANISM Widely studied species (like yeast, fruit flies, and mice) that are easy to keep in the laboratory.

MUTATION A genetic change at the DNA level.

MUTUALISM An ecological relationship between two or more species that is beneficial to all interacting partners.

NATIVE RANGE Geographic distribution of a species prior to human influence.

NATURAL SELECTION Differential survival or reproduction of organisms based on their fit to their environment; the evolutionary process leading to adaptation.

NEGATIVE FEEDBACK LOOP A self-regulatory process where the response decreases the stimulus.

NEOLITHIC REVOLUTION The period in history when human societies began to transition toward larger settlements and agricultural cultivation.

NEW MUTATION A genetic change that occurs after an environmental change.

NICHE The relationship between a species and its environment, specifically the species' environmental requirements and the species' impacts on its environment.

NICHE CONSERVATISM Tendency for species to retain similar ecological roles and requirements as their ancestors.

NONINVASIVE SAMPLING Collecting samples without harming the study organisms, for example by using feathers, fur, or scat.

NUCLEAR GENOME Sum total of genetic material in the nucleus of the cell (for eukaryotes).

NULL HYPOTHESIS A statistical concept to capture the default position, for example, no significant relationship between variables.

OBLIGATE Required, particularly in regard to species that are restricted to a particular ecological role or set of interacting partners.

OUT OF AFRICA HYPOTHESIS The now well-accepted formulation that modern *Homo sapiens* initially evolved in Africa.

OUTGOING LONGWAVE RADIATION Energy emitted from Earth and its atmosphere as thermal radiation.

P-VALUE A statistical concept to capture the probability that results have not occurred by chance (typically a p-value of <0.5 is used to indicate statistical significance).

PARANTHROPUS An early species of hominid present ~1–3 million years ago.

PARTIALLY ADDITIVE EFFECT When the total impact of multiple factors is less than the sum of their individual impacts.

PHENOLOGICAL MISMATCH When interacting species are no longer temporally or spatially aligned (also called trophic asynchrony).

PHENOLOGY The study of cyclic or seasonal natural events.

PHENOTYPE An observable organismal trait, resulting from the interaction between genotype and environment.

PHENOTYPIC PLASTICITY The capacity of a single genotype to exhibit different phenotypes in different environments.

PHYLOGENETIC TREE A diagram of evolutionary relationships, typically reconstructed based on genetic and/or phenotypic data.

PLASTICITY-FIRST HYPOTHESIS The premise that phenotypic plasticity might not only precede but also promote adaptive genetic change.

PLEISTOCENE The first geological epoch of the Quaternary and the most recent period of extensive glaciation.

POINT SOURCE A localized origin, often of pollution.

POPULATION VIABILITY ANALYSIS Quantitative modeling methods used to evaluate the likelihood that a population will persist for a given number of years (abbreviated as PVA).

POSITIVE FEEDBACK LOOP An amplifying process where the response increases the stimulus.

PRODUCERS Organisms with an ecological role of producing energy by converting solar energy into biomass via photosynthesis.

PROJECTED CLIMATE CHANGE Amount of global warming yet to come as predicted by mathematical climate models.

PROKARYOTES Single-celled organisms without a membrane-enclosed nucleus are often referred to as prokaryotes, although the term does not describe any unified lineage of the tree of life.

PROPAGULE BANK Reservoir of seeds, eggs, cysts, or spores that is stably preserved in sediments, soil, or ice (which can often be revived later).

PROPAGULE SIZE The number of individuals or gametes released, particularly when referring to initial release of a non-native species in a new area.

PROTEINS Macromolecules composed of amino acids.

QUATERNARY MEGAFAUNAL EXTINCTIONS Global loss of many large-bodied species, largely due to human hunting and climate shifts in the geological period prior to the Anthropocene.

RANDOMIZATION Avoiding systematic bias in experiments by using chance methods to assign individuals or samples to treatment groups.

RANGE CONTRACTION The decrease of a species' geographic distribution such that the species occupies a subset of its historical range.

RANGE EXPANSION The increase of a species' geographic distribution such that the species occupies a larger area than it did previously.

RANGE MARCH The shift of a species' geographic distribution due to a contraction in one part of the range and an expansion in another.

REALIZED CLIMATE CHANGE Amount of global warming that has already occurred (in contrast with projected climate change).

REALIZED NICHE The conditions and resources that a population actually uses given constraints such as those imposed by predators and competitors (contrast with fundamental niche).

RECIPROCAL TRANSPLANT EXPERIMENTS When different genotypes are reared reciprocally in both "home" and "away" conditions, often to disentangle genetic and environmental effects.

RECOMBINATION Rearrangement of genetic material, typically via chromosomal crossing over.

RECONCILIATION ECOLOGY The study of ways to enhance human-dominated landscapes to better support biological diversity.

RED LIST OF ECOSYSTEMS A standard for assessing the status of ecosystems compiled by the International Union for Conservation of Nature (IUCN).

RED LIST OF THREATENED SPECIES A database defining conservation status and threat levels for species around the world compiled by the International Union for Conservation of Nature (IUCN).

REFLECTED SOLAR RADIATION The fraction of incoming solar radiation that is reflected back to space as shortwave radiation by clouds, atmospheric gases, or the surface of the Earth.

REFUGIA An area where environmental conditions have allowed isolated relict populations to persist.

RELAXATION TIME The amount of time required for a system to reach a new equilibrium.

REPLICATION Repetition of an experiment in multiple independent individuals, samples, groups, or locations.

RESURRECTION ECOLOGY The study of organisms "frozen in time," for example by hatching or germinating preserved eggs or seed.

REVERSE AUCTION An exchange where buyer and seller roles are reversed: Sellers bid for the price they are willing to sell their services to one buyer.

RIBONUCLEIC ACID Single-stranded macromolecule that can both store and process information; involved in many cellular functions such as gene expression and protein synthesis and is the sole basis of genetic inheritance in many viruses (abbreviated as RNA).

RIGHTS-BASED FISHERIES MANAGEMENT A governance system where stakeholders have more secure rights and ownership over resources, typically to address pressures of overfishing.

SELECTIVE BREEDING See artificial selection.

SESSILE Immobile, not free to move.

SINGLE NUCLEOTIDE POLYMORPHISM A variant at a single DNA base pair (abbreviated as SNP).

SIXTH MASS EXTINCTION Contemporary global extinction spasm caused by human activities on the planet.

SPECIATION The evolutionary process of diversification by which new lineages are formed.

SPECIES DISTRIBUTION MODELING Mathematical models that relate information about where species are currently found to environmental data to predict future areas of suitable habitat or future community composition.

SPECIES DIVERSITY Number of species in an ecological community.

SPECIES RICHNESS The total number of different species found in an ecological community.

STANDING GENETIC VARIATION The presence of multiple alleles at a locus in a population.

STATISTICAL SIGNIFICANCE When the relationship between variables cannot be explained simply by chance.

SYNERGISTIC EFFECT When the combined impact of multiple factors is greater than the sum of their individual impacts.

TIME LAG A gap in time between cause and effect.

TIPPING POINT An abrupt and significant change in a system or process, often referring to a threshold effect from one state to another.

TOP-DOWN EFFECTS The ways in which higher trophic levels (such as predators) impact lower trophic levels, nutrients, and resources in an ecosystem.

TRANSLOCATION The intentional transport and release of individuals from one location to another for conservation.

TROPHIC CASCADE Impact of changes in one species on other species multiple trophic levels away.

TROPHIC LEVELS The functional roles of organisms in a food web (e.g., producers, primary consumers, secondary consumers).

TROPHIC MISMATCH The lack of synchrony between consumers and their resources; increasingly observed with climate change.

TROPHIC STRUCTURE The organization of organisms in an ecosystem into functional groups based on their role in a food web.

URBAN HEAT ISLAND EFFECT The increased temperature in cities due primarily to lower albedo, lower rates of evapotranspiration, and increased heat trapping.

VARIANCE A statistical concept to capture the variation or spread of values around a mean.

VULNERABILITY The degree to which a species or system is susceptible to environmental change (influenced by sensitivity, exposure, and capacity to respond).

ZOONOSES Diseases transmitted between humans and animals.

ZOOXANTHELLAE Photosynthetic algae that serve as key symbionts to many species of corals.

INDEX

abiotic factors, 11, 13, 14, 17
  adaptation and, 182, 187
  climate influenced by, 83, 179
  ecosystems and, 266, 267, 268, 279, 294
  evolution and, 36–37, 38, 40
  extinction rates and, 45
  fitness and, 173
  geographic range and, 111, 115, 123
  Homo evolution and, 60
  move response and, 106
  niches and, 112
  speciation and, 42
  terrestrial systems and, 274
  urbanization and, 160, 164
above- and below-ground food webs, 248, *249bf*, *250bf*
"Accelerated Modern Human-Induced Species Losses: Entering the Sixth Mass Extinction" (Ceballos et al.), 222–25
acclimation, 139, 142–43, 308
acclimatization, 139, 142–43
acidification, 146, 185, 195, *196bf*
acorn ants *(Temnothorax curvispinosus)*, 155
active locomotion, 107
activists, 341–42
adaptation, 107, 139, 165, 171–98, 209, 237, 242
  climate change, 304
  conditions required for, 172–74
  defined, 171
  extinction prevention and, 187–89
  global change pressures and, 177–81
  identifying, 181–87
adaptive management, 318–20, *319f*, 350
adaptive potential, 187, 207
adaptive radiations, 42
adjust response, 93, 123, 139–65, 171, 209, 237, 242
    See also phenotypic plasticity
aerosols, 100, 274, 304
affluence, 336–37, *338bf*
Africa, 60, 63
African elephants *(Loxodonta africana* and *Loxodonta cyclones)*, 206, 295, 306–8, *307f*
African oil palm *(Elaeis guineensis)*, 116
African oryxes *(Oryx gazella)*, 310, *310f*
Agricultural Revolution, 68
agriculture, 4, 68, 274
  centers of origin, 65, *66f*
  early environmental impact of, 65–66
  habitat enhancement and, 317, *317f*
agroecology, 345, *346f*
Alaska, 244, 260
albedo, 99, 281
algae, 239–40, *239f* *See also Symbiodinium*
algal blooms, 17, 277
Algerian mice *(Mus spretus)*, 175
Allee effect, 206, 221
alleles, 140, 141, 206, 311
alligator weed *(Alternanthera philloxeoides)*, 152, *153f*
allis shad *(Alosa alosa)*, 121–23
Amazon rainforest, 342
amensalism, *238t*
American bison, 116
American pika *(Ochotona princeps)*, 109, *109f*, 119
amphibian declines, 226–31
amplification, 241
Animalia, 42
Animal Life in the Yosemite (Grinnell & Storer), 125

ant(s)
 acorn *(Temnothorax curvispinosus)*, 155
 Argentine *(Linepithema humile)*, 129–30, 130*bf*, 241–42, 242*f*
 leaf-cutter *(Atta sexdens)*, 161–63, 163*bf*
Antarctic, 278
Antarctic emerald rockcod *(Trematomus bernacchii)*, 146
Anthropocene, 80–102
 debate over beginning of, 81*f*, 82
 defined, 80
 population growth and, 85–86
 proposal to declare a formal epoch, 81
 types of environmental pressures, 86–91
anthropogenically induced adaptation to invade (AIAI) hypothesis, 132
anthropogenic stressors. *See also* global change pressures
 of contemporary civilizations, 86–91
 defined, 2
 development of key activities over time, 81*f*
 of early civilizations, 63–68
 geographic range changes in absence of, 113–15
 geographic range changes in response to, 115–20
 interactions of, 91–92, 91*f*
 IPAT framework on, 336–38
antibacterial soap, 171
Anura, 226, 227*bf*
apex predators, 251, 252*f*, 284
apple maggot flies *(Rhagoletis pomonella)*, 179
aquatic systems. *See also* marine systems
 biodiversity hotspots in, 281
 biological interactions in, 242
 biotic homogenization in, 132
 climate influenced by, 82
 collapse of, 285
 contamination of, 228
 global change pressures and, 271, 274, 277
 harvesting in, 308
 land-use change and, 88
 phenotypic plasticity and, 146, 152, 153*f*
 translocations in, 310
Archaea, 36–37, 42, 46, 47, 287
Arctic, 4, 97, 270, 278, 279*f*
Arctic foxes *(Vulpes lagopus)*, 112
Arctic geese, 242–43
Ardipithecus, 58, 59, 59*f*
Argentina, 253–56
Argentine ants *(Linepithema humile)*, 129–30, 130*bf*, 241–42, 242*f*
Aristotle, 48, 49*bf*, 359
armyworms *(Spodoptera frugiperdam)*, 178, 178*f*, 179
Arrhenius, Svante, 3
arthropods, 38
artificial selection, 73–74, 73–74*bf*, 326
ASPM genes, 71
assisted evolution, 308
assisted migration, 309
Atlantic cod *(Gadus morhua)*, 285
Atlantic Forest of Brazil, 271, 272*f*
Atlantic salt marshes, 284, 284*f*
atmospheric systems, 17, 89
 climate influenced by, 82
 global change pressures and, 271, 274–77
 sea otters and, 258, 259
atoms, 33
ATP-binding gene, 178
atrazine, 228
Australia, 113, 117, 187–89
Australopithecus, 58, 59, 59*f*
Australopithecus afarensis, 59
automobile ownership, 336–37
avian influenza, 229

Bacillus thuringiensis (Bt), 177–78, 178*f*
background extinction rates, 223, 224, 225*bf*, 226*bf*
Bacteria, 36–37, 42, 46, 47, 287
Bahamas, 182
bald eagles *(Haliaeetus leucocephalus)*, 259
banding patterns, 82
bark beetles, 273*f*
Barnosky, Anthony, 82
barred owls, 122–23*f*
base pairs, 141*bf*
Batrachochytrium dendrobatidis (Bd), 229, 230, 230*bf*, 241
Batrachochytrium salamandrivorans (Bsal), 231
bats, 315
bat white nose syndrome, 229
bears
 brown *(Ursus arctos)*, 244–45, 244*f*
 polar *(Ursus maritimus)*, 146, 147*f*, 243
bees, 72, 94–96, 96*bf*, 317, 317*f*
 bumblebee *(Bombus terrestris)*, 95, 96*bf*
 honeybee *(Apis mellifera)*, 95, 96*bf*, 149, 150*f*
 solitary *(Osmia bicornis)*, 95, 96*bf*
beetles
 bark, 273*f*
 mountain pine *(Dendroctonus ponderosae)*, 145, 145*f*, 271
 Western Pine *(Dendroctonus brevicomis)*, 121
behavior
 IPAT framework and, 338
 phenotypic plasticity and shifts in, 146–47

beneficial mutations, 174
Bering Land Bridge, 61, 114*f*, 115
bi-directional learning, 4
Big Bang, 33, 34*f*
"Big Five" mass extinctions, 45–46, 45*t*, 222
binomial naming system, 42
biodiversity
  anthropogenic stressors on *See* (anthropogenic stressors)
  conservation of *See* (conservation)
  evolutionary processes shaping, 39–41
  urbanization and, 161
  value of, 25–26
biodiversity databases, 216–19
biodiversity hotspots, 274, 302, 313
  defined and explained, 279–81
  global distribution of, 280*bf*
biogeochemical cycles, 279
  biome classification and, 269*f*
  defined, 267
  types of, 267–68
  urbanization and, 160
biological interactions, 238–45
  direct *See* (direct effects)
  facultative, 239
  global change pressures and, 240–45
  indirect *See* (indirect effects)
  interspecific, 238
  intraspecific, 238
  key types of, 238–40, 238*t*
  multispecies, 239–40
  obligate, 238–39
biological levels of change, 48–51
biomass burial, 304
biopiracy, 26
bioprospecting, 26, 26*bf*
biosphere
  biological organization level, 50*bf*, 51
  defined, 51, 237
  global change influences at, 294
biotic factors, 11, 13, 14
  adaptation and, 179, 182, 183, 187
  climate influenced by, 83, 179
  ecosystems and, 266, 267, 268, 279
  evolution and, 36–37, 38, 40
  extinction rates and, 45
  fitness and, 173–74
  geographic range and, 111, 115, 123, 124*f*
  Homo evolution and, 60
  move response and, 106
  niches and, 112
  speciation and, 42
  terrestrial systems and, 274
  urbanization and, 165
biotic homogenization, 132
biotic interaction chains, 251
bipedalism, 71
birdsong, 147, 149, 155, 161
bison, American, 116
black cypress pine *(Callitris endlicheri)*, 319–20
black robins *(Petroica traversi)*, 309–10, 310*f*
blue whales *(Balaenoptera musculus)*, 19*f*, 109, 109*f*
bolides, 45, 99
bottlenecks, 39
bottom-up effects, 248, 250*bf*, 293
BP oil spill, 88
brain evolution, 71
Brazil, 342, 343
breeding experiments, 182
brown anole lizards *(Anolis sagrei)*, 179, 180*f*, 182–83
brown bears *(Ursus arctos)*, 244–45, 244*f*
brown food web, 248
bullfrogs *(Lithobates catesbeianus)*, 241
bumblebees *(Bombus terrestris)*, 95, 96*bf*
butterflies, 145, 155, 315
  Edith's checkerspot *(Euphydryas editha)*, 119
  Glanville fritillary *(Melitaea cinxia)*, 208, 208*f*
  monarch *(Danaus plexippus)*, 107, 111

caecilians, 226
calcification, 185
California, 116, 117, 354–58
California condors *(Gymnogyps californianus)*, 308
Callendar, Guy, 3
CAM (crassulacean acid metabolism), 146
Cambrian Explosion, 38
capacity to respond, 92*f*, 93
captive breeding, 308–9
carbon, 267, 270, 287, 288–89, 288*bf*
carbonaria, 176–77, 176*f*
carbon cycle, 267
carbon deficit, 271
carbon dioxide (CO2), 3, 34, 65, 89–90, 89*f*, 100, 146, 185, 267, 274, 275–77, 289, 290*bf*, 291*bf*
carbon emissions, 350, 353
carbon sequestration, 271, 304
Caribbean islands, 179
Carson, Rachel, 4
cascading effects, 251, 252*f* *See also* trophic cascades
categorical variables, 20
caterpillars *(Nemoria arizonaria)*, 142, 143*f*
Caudata, 226, 227*bf*

causation *vs.* correlation, 63
cavity-nesting birds, 161, 163*bf*
Ceballos, Gerardo, 222–25, 223*bf*
cellular level, 49, 50*bf*
cellular life, evolution of, 36–38
cellular membranes, 35
Cenozoic era, 81*f*
Central Valley of California, 355–58
cerebral cortex, 71
cerebrum, 71
change, biological levels of, 48–51
charismatic poison frogs *(Mantella)*, 110, 111*f*
Chatham Islands, 309
Chico Mendes Extractive Reserve, 342
Chihuahuan Desert, 310
chimpanzees *(Pan)*, 57, 58, 342
China, 311–12, 345
Chinese giant salamanders, 226*bf*
chlorofluorocarbon, 274
chloroplast genomes, 37, 140
chromatin marking, 141
Chromista, 42
chromosomal inversion, 180
chromosomes, 140
chytrid fungus. *See Batrachochytrium dendrobatidis*
chytridiomycosis, 229–30, 310–11
CITES. *See* Convention on International Trade in Endangered Species
cities. *See* urbanization
citizen science, 15
Civic Laboratory for Environmental Action Research (CLEAR), 6, 7*f*
classification systems, 42, 48, 58, 58*bf*
classism, 10, 11, 164
CLEAR. *See* Civic Laboratory for Environmental Action Research
cliff-nesting birds, 161, 163*bf*
climate
  defined, 82
  feedbacks in, 277, 281–82
  measurement of, 82–83, 83*bf*, 84*bf*
  urbanization and, 161
climate change, 3–6, 13, 14, 91, 91*f*, 219 *See also* global warming
  adaptation to, 179–80, 181*f*
  amphibian decline and, 228–9
  aquatic systems and, 277
  atmospheric systems and, 274–77
  cryospheric systems and, 278–79
  geographic range and, 116, 124*f*
  historical and contemporary, 97–101
  inequity in contributions to, 350–51, 351*f*
  kelp forests and, 260
  projected, 276, 276*f*
  Quaternary extinctions and, 63–65
  realized, 275
climate change adaptation, 304
climate change commitment, 221
climate change migration, 304, 305*bf*
climate treaties, 350–52
cloning, 326, 327*bf*
cloud seeding, 304
coal, 3, 89, 176–77, 304
coal mining, 311
coarse-filter strategies, 341, 349
  defined, 306
  examples of, 311–18
coextinction, 245–50, 251
  defined, 245
  of mutualists, 245–46
  of parasites, 246–47, 247*f*
  of predators and herbivores, 247–48
collective actions, 343–48
commensalism, 238*t*
common garden experiment, 152, 153*f*
communication, animal, 161
communities, 237–62
  biological organizational level, 50*bf*, 51
  cascading effects in, 251, 252*f*
  defined, 51, 237
  extinction effects on, 245–50
  global change influences at, 293–94
  species interactions in *See* (biological interactions)
"Comparing Strategies to Preserve Evolutionary Diversity" (Magnuson-Ford et al.), 321–24
competition, 35, 112, 174, 238, 238*t*
complete migration, 107
computational tools, 41
  adaptation and, 187
  defined and explained, 19–20
condors, California *(Gymnogyps californianus)*, 308
confounding variables, 8*t*
connectivity, 286, 303*f*, 313
Connell, Joseph, 113*bf*
conservation, 301–29, 334–63
  adaptive management and, 318–20, 319*f*, 350
  coarse-filter strategies *See* (coarse-filter strategies)
  collective actions for, 343–48
  defined, 301
  egalitarian approach to, 321, 323
  emerging technology and ethics of, 325–29
  ex situ, 308–9

financial incentives for, 354–58
fine-filter strategies *See* (fine-filter strategies)
forecast for, 353
individual support for, 339–43
in situ, 308
policy actions supporting, 348–53
prioritization in, 302, 303*f*
societal levers used to support, 336–39
targeted approach to, 321, 323–24
in urban areas, 163
Conservation Biology (journal), 2
conservation easements, 354
conservation set-asides, 354
conservation triage, 302
consumers, 248, 251, 252*f*, 339–40
contaminants, 4 *See also* pollution
adaptation and, 177–78
amphibian decline and, 228
coral reefs and, 195
geographic range and, 121
kelp forests and, 260
continuous variables, 20
control, 10
control group, 8*t*
Convention on International Trade in Endangered Species (CITES), 307, 350
copper mining, 143–44
coral bleaching, 195, 196*bf*
coral reefs, 23, 193–97, 193*bf*, 217, 219, 281, 308, 309*f*
core responses. *See* adaptation; adjust response; die response; move response
correlation *vs.* causation, 63
correlative species distribution models (SDMs), 121, 123
corridors, 313–15, 314*f*
Costanza, Robert, 22–23, 22*bf*
Costa Rica, 352
cougars *(Puma concolor)*, 241, 251
coupled human-natural systems, 335–36, 335*f*
COVID-19, 229, 353
crabs, 251, 252*f*, 284*f*
cradle-to-cradle life cycle analysis, 345–48
crassulacean acid metabolism (CAM), 146
Cretaceous period, 45, 45*t*, 81*f*, 99
Critical Ecology Lab, 6, 7*f*
critical habitat designation, 349–50
critical thermal maximum, 192
crowdfunding, 340–41, 340*f*
cryopreservation (of reproductive cells), 308, 326
cryospheric systems, 82, 277–79, 279*f*
cultural value of biodiversity, 25
culture (of early humans), 72–73
curly-tail lizards *(Leiocephalus carinatus)*, 179, 180*f*, 182–83
cuttlefish *(Sepia officinalis)*, 147, 147*f*
cyanobacteria, 37, 99–100

daily movements, 107
damselflies *(Coenagrion scitulum)*, 186
Daphnia, 142, 143*f*, 151, 155, 191–93, 192*bf*
Darwin, Charles, 39, 49, 172, 174, 182, 245, 257
data display, 20–21
Dawkins, Richard, 35
dead zones, 277
debt-for-nature swaps, 352
decomposers, 248
deer
mule *(Odocoileus hemionus)*, 241
rusa *(Cervus timorensis)*, 320
de-extinction, 325–27, 326*bf*
defaunation, 271, 272*f*
deforestation, 89, 210, 251, 267, 274, 314
de Groot, Rudolf, 22–24, 22*bf*, 194
deleterious mutations, 174
De Meester, Luc, 190*bf*
Denovian period, 45*t*
deoxyribonucleic acid. *See* DNA
depauperate communities, 132, 133*bf*
dependent variables, 8*t*, 9, 20
development shifts, 145
devil facial tumor disease, 207
Devil's Hole pupfish *(Cyprinodon diabolis)*, 109*f*, 110
die response, 94, 190, 205–33, 240 *See also* extinction; extirpation
across multiple biological levels, 206–9
as a nonresponse, 206
differential survival, 173–74
dinosaurs, 44, 44*f*, 45, 99
direct effects, 240, 255, 256*bf*
disease. *See also* specific diseases
amphibian, 228
emerging infectious, 229, 229*bf*
dispersal capability, 107, 123, 131, 132–33, 241, 291–92, 313
defined and explained, 110–11
phenotypic plasticity for, 293
diversity, 6 *See also* biodiversity; evolutionary diversity
DNA, 35, 36, 149, 150–51, 172, 173
environmental, 19
"junk," 140
Neanderthal, in modern human genome, 60, 68–70, 71*bf*
structure of, 141*bf*

DNA methylation, 151
DNA sequencing, 18–19, 19*f*, 20
dogs, 73, 73*bf*
Dolly (cloned sheep), 326
dolphins, 71, 210*f*, 211
Domed tortoises, 326
domestication of plants and animals, 65–66, 73, 73–74*bf*
drought, 146, 149
"Dynamic Conservation for Migratory Species" (Reynolds et al.), 354–58

eagles, bald *(Haliaeetus leucocephalus)*, 259
Earth, 32–53
  development of a crust and hydrosphere, 34
  formation of, 34
  key transitions leading to emergence of life on, 33–35
  origin of heredity, 35
  prebiotic chemistry, 34–35
  timeline of early events in the history of life on, 33*f*
  timeline of major transitions on, 36*f*
Earth Day, 4
earthworms *(Lumbricus terrestris)*, 251, 252*f*
"Ebers Papyrus," 3*f*
Ebola, 229
eButterfly, 15
eccentricity, 99, 100*bf*
echidna, 41*f*
ecological integrity, 303*f*
ecology
  reconciliation, 315
  resurrection, 191, 192, 192*bf*, 193
Ecology (journal), 2
economic value
  of biodiversity, 25
  of nature, 22–24
ecosystem(s), 266–96
  biogeochemical cycles and *See* (biogeochemical cycles)
  biological organization level, 50*bf*, 51
  characteristics of invaded, 131–32
  conservation and, 302
  defined, 51, 237, 267
  global change influences at, 294
  global change pressures and, 268–71
ecosystem collapse, 283–85
ecosystem functions, 22, 268–71
ecosystem representation, 303*f*
ecosystem resilience, 286, 286*f*
ecosystem services, 193, 270–71
  defined, 22, 270
  examples of, 270*t*
  value of, 22–23, 24*bf*
Ecosystem Services (journal), 22
ecosystem structure, 268, 270–71
ecotourism, 343, 344*f*
EDGE (Evolutionarily Distinct and Globally Endangered) Initiative, 322
Edith's checkerspot butterflies *(Euphydryas editha)*, 119
egalitarian approach to conservation, 321, 322–23
Einstein, Albert, 354, 359
elaenia birds, white-crested *(Elaenia albiceps)*, 253–56, 255*bf*
electrons, 33
elephants, African *(Loxodonta africana* and *Loxodonta cyclones)*, 206, 295, 306–8, 307*f*
emerging infectious diseases (EIDs), 229, 229*bf*
emperor penguins *(Aptenodytes forsteri)*, 116, 117*f*
E/MSY, 224, 225*bf*
Endangered Species Act (ESA), 216, 349–50, 349*f*
endemism, 279–81
endosymbiosis, 37, 48
energy, 267
Energy Star, 339
energy use and sources, 304
environmental DNA (e-DNA), 19
environmental justice, 12
environmental monitoring tools, 17
Environmental Protection Agency, 12
environmental stochasticity, 206
environmental tolerance, 161
environmental worldviews, 359–61
Eocene epoch, 81*f*
epigenetic inheritance, 141, 150–51
error bars, 20, 21*bf*
ESA. *See* Endangered Species Act
Escherichia coli, 183–85
ethics
  biodiversity value in, 25
  of technology and conservation, 325–28
Eucalyptus, 247
Eucarya, 42
eukaryotes, 37, 46–48, 47*bf*, 140
European house mice *(Mus musculus)*, 175
European rabbits, 187–89, 188*f*
European salamanders, 231
eutrophication, 277, 345
evergreen shrub *(Aristotelia chilensis)*, 253–56, 255*bf*, 256*bf*, 257*bf*
evolution. *See also* natural selection
  of Archaea, 36–37
  assisted, 308

of Bacteria, 36–37
biodiversity shaped by, 39–41
brain, 71
of cellular life, 36–38
continuing, 73–74
of early hominids, 56–60
of eukaryotes, 37
experimental, 183–85, 184*f*, 191, 192, 192*bf*
gene, 186
geographic range and, 110, 115, 123, 124*f*
of the Homo group, 60
of multicellularity, 37–38, 38*f*
success of in humans, 71–74
Evolution (journal), 2
Evolutionarily Distinct and Globally Endangered (EDGE) Initiative, 322
evolutionary diversity, maximizing, 321–24
evolutionary radiations, 42–44, 44*f*
evolutionary rescue, 187–89, 188*f*
experimental approaches, 12, 14, 15
defined and explained, 13
phenotypic plasticity and, 152–55
experimental evolution, 183–85, 184*f*, 191, 192, 192*bf*
exploitation, 228, 238*t*
exposure, 92*f*, 93
ex situ conservation, 308–9
extant organisms, 42
extinction, 39, 205, 293, 294
adaptation and, 187–89
background rates, 223, 224, 225*bf*, 226*bf*
co- *See* (coextinction)
community effects of, 245–50
de-, 326–28, 327*bf*
defined, 40, 208
explained, 40–41
global, 208
global change pressures and, 209–11
high rates of, 41–46
local, 208
mass *See* (mass extinctions)
in metapopulations, 119
Quaternary megafaunal, 63–65, 211, 246
range contractions and, 116
reintroducing close relatives/analogs of affected species, 326
extinction debt, 220–21, 221*bf*
extinction risk, 292
estimating, 211–16
summarizing global patterns of, 216–20
extinction vortex, 206, 207, 207*f*, 231
extinct organisms, 42
extirpation, 208–9, 240, 313
adaptation and, 187, 189
defined, 208
reintroducing species affected by, 325–26

facultative interactions, 239
Fair Trade, 339
falcons, peregrine *(Falco peregrinus)*, 161
feedbacks, 277, 281–83, 293
defined, 281
negative, 281
non-climate, 283
positive, 281, 282*bf*, 283, 284, 290
Fertile Crescent, 65
field studies, 13, 182–83
file-drawer problem, 15
finches, Galapagos ground *(Geospiza fortis)*, 182, 183*f*
fine-filter strategies, 341, 349
defined, 306
examples of, 306–11
"Fingerprints of Global Warming on Wild Animals and Plants" (Root), 157–59
fire
early use of, 65, 72
suppression of, 315
fire poppies *(Papaver californicum)*, 209
fisheries governance, 348–49
fishing, 260, 284, 285, 345
fitness, 173–74, 173*f*, 181–82
flies
apple maggot *(Rhagoletis pomonella)*, 179
damselflies *(Coenagrion scitulum)*, 186
fruit flies *(Drosophila subobscura)*, 179–80, 181*f*
flu virus, 292
food webs, 248, 249*bf*, 250*bf*
forced riders, 351*f*
forcing, 97, 100, 101*bf*, 277
forest clearance, 65, 274
fossil fuels, 89, 267, 274, 304
fossils, 82
founder effect, 39
foxes, Arctic *(Vulpes lagopus)*, 112
Francis, John, 341*f*, 342
free riders, 351*f*
frogs, 226, 228
bullfrogs *(Lithobates catesbeianus)*, 241
charismatic poison *(Mantella)*, 110, 111*f*
microhylid, 227*bf*
mountain yellow-legged, 230
fruit flies *(Drosophila subobscura)*, 179–80, 181*f*

Fu, Qiaomei, 68–70, 69*bf*
fully additive effects, 91*f*, 92
functional redundancy, 286
fundamental niche, 112, 113*bf*
Fungi, 42

Galapagos ground finches *(Geospiza fortis)*, 182, 183*f*
Galapagos Islands, 182, 209, 326
galaxies, 33
Geerts, Aurora, 191–93, 191*bf*
geese, Arctic, 242–43
gene(s)
  ASPM, 71
  ATP-binding, 178
  defined, 140
  RARS, 187
gene evolution, 186
gene expression, 149, 186
gene flow, 70, 107, 175, 310–11
gene regulation, 149
genetic assimilation, 155, 190
genetic diversity, 292, 292*bf*
genetic drift, 39–40, 40*f*, 70, 181, 206, 293
genetic engineering, 311
"Genetic History of Ice Age Europe, The" (Fu et al.), 68–70
genetic stochasticity, 206
genetic variation, 172, 175*bf*, 292
  natural selection and, 189, 189*f*
  reduced, 206–7
  source of, 174–75
  standing, 174, 176
genome(s)
  chloroplast, 37, 140
  defined, 140
  eukaryotic, 46–48
  human, 140
  mitochondrial, 37
  Neanderthal, in modern humans, 60, 68–70, 71*bf*
  nuclear, 140
genome editing, 326–27, 327*bf*
genomic sequencing, 185, 186, 186*f*
genotype(s), 140–41, 143–44, 144*f*
genotype by environment interaction, 144
Genus species naming system, 42
geoengineering, 304
geographic distribution, 109, 110
geographic range, 109–24, 293, 294*bf*
  adaptation and, 186
  changes in absence of anthropogenic stressors, 113–15
  defined, 109
  factors determining, 110–12
  prediction of, 120–24
  in response to anthropogenic stressors, 115–20
giant ground sloths, 63, 64*f*
giant sequoia tree *(Sequoiadendron giganteum)*, 112, 295, 315
gibbons, 58
glacial maxima, 63, 113, 115
glaciers, 82, 115, 277–78, 279*f*
Glanville fritillary butterflies *(Melitaea cinxia)*, 208, 208*f*
glaucous-winged gulls *(Larus glaucescens)*, 259
Global Change Biology
  defined, 1
  development of, 2–6
  key research approaches, 12–17
  key tools used in, 17–20, 18*f*
  study design, 7–12
Global Change Biology (journal), 2, 287
global change pressures. *See also* anthropogenic stressors
  adaptation in response to, 177–81
  biological interactions affected by, 240–45
  ecosystems impacted by, 268–71
  extinction in response to, 209–11
  factors influencing response to, 292–95
  influences on vulnerability to, 92–93, 92*f*
  large-scale earth systems impacted by, 271–79
  phenotypic plasticity in response to, 144–48
"Global Estimates of the Value of Ecosystems and Their Services in Monetary Units" (de Groot et al.), 22–24
global extinction, 208
globalization, 90–91
  defined, 90
  invasive species and, 128–33
"Globally Coherent Fingerprint of Climate Change Impacts Across Natural System, A" (Parmesan), 157–59
Globally Important Ingenious Agricultural Heritage System, 345
global warming, 3, 13, 90, 100, 145, 146, 147*f*, 271 *See also* climate change
  adaptation to, 180, 191–93
  amphibian decline and, 228
  aquatic systems and, 277
  atmospheric systems and, 275–77
  biological interactions and, 242
  cryospheric systems and, 278–79

feedbacks and, 281, 282*bf*, 283
geographic range changes and, 117–19, 118*f*, 121, 186
move response and, 106
oceans affected by, 195
phenology and, 157–59
soil impact of, 287–91
in Yosemite National Park, 125–28
global warming commitment, 277
golden-mantled ground squirrels *(Callospermophilus lateralis)*, 120*f*
gold mining, 87*f*, 88
Goodall, Jane, 341*f*, 342
GoodGuide, 339
gorillas *(Gorilla)*, 57, 58
Grant, Peter, 183*f*
Grant, Rosemary, 183*f*
graphs, 20
grass *(Andropogon gerandii)*, 149
grazers, 251
Great Accerleration, 82
great apes, 56–57, 58
greater glider *(Petauroides volans)*, 215, 215*f*
Great Turning worldview, 359–60
Great Unraveling worldview, 359, 360*bf*
Greek philosophers, 2
Green Belt movement, 341–42
green food web, 248
greenhouse effect, 97
greenhouse gases, 100, 101*bf*, 277
climate change migration and, 304, 305*bf*
in the soil, 287–91
Green Seal, 339
grey wolves *(Canis lupus)*, 215, 325, 325*bf*
Grinnell, Joseph, 125–27, 126*bf*
"Grinnell Resurvey Project," 125–27
gross domestic product (GDP), 23, 337
Gulf of Mexico, 88
Gulf War, 88
gulls, glaucous-winged *(Larus glaucescens)*, 259
Gymnophiona, 226, 227*bf*

habitat conservation plans, 349
habitat enhancement (in highly modified settings), 315–17, 318*f*
habitat loss, 91, 91*f*, 227, 270, 283, 306, 308
habitat suitability maps, 121
habitat use, 160
Hallstein, Eric, 355*bf*
harvesting, 65, 91*f*, 92
of coral reefs, 195
geographic range and, 116
kelp forests and, 260
reducing excessive, 306–8, 307*f*
health value of biodiversity, 25
heat shock proteins, 151, 197
heavy metals, 9–10, 13
hedgerows, 317, 317*f*
herbivores, coextinction of predators and, 247–48
heredity, 172, 173*f*
defined, 35
mechanisms of, 140–41
origin of, 35
heritable traits, 172, 181–82
Hetch Hetchy Valley, 87*f*, 88
heterozygosity, 206
hierarchical classification system, 42
HIV, 229
Holocene epoch, 81*f*, 82
Hominidae, 56, 58
hominids, 56–60
Hominina, 58
Homininae, 58
Hominini, 58
Homo, 58, 59, 59*f*
cranium capacity of, 72*bf*
evolution of, 60
Homo erectus, 60
Homo floresiensis, 60
Homo habilis (handy man), 60
Homo heidelbergensis, 60
Homo neanderthalensis. See Neanderthals
Homo sapiens, 38, 46, 56, 57, 58, 60, 335, 336
early impact on environment, 63–68
evolutionary success of, 71–74
origin date for, 61
spread of around the world, 61–62
honeybees *(Apis mellifera)*, 95, 96*bf*, 149, 150*f*
horizontal gene transfer, 42, 46, 51, 175
Hot House periods, 97
Hubble Space Telescope, 34*f*
Hudson Bay, 147*f*
humans, 56–76 *See also* anthropogenic stressors
conservation aligned with interests of, 334–63
coupled human-natural systems, 335–36
defined, 58
evolutionary success of, 71–74
evolution of early hominids, 56–60
impact of contemporary civilizations on environment, 86–91
impact of early civilizations on environment, 63–68

humans (*Continued*)
- overlapping lineages of early, 57–60
- social systems in cities, 164
- spread of modern around the world, 61–62

hummingbirds *(Sephanoides sephaniodes)*, 253–56, 255*bf*, 256*bf*, 257*bf*

hunting, 91, 92, 246, 306, 308, 310
- earliest evidence for, 63
- ecosystems and, 270, 271
- feedbacks and, 283
- kelp forests and, 260
- Quaternary megafaunal extinctions and, 63–65

hybridization, 51, 175
hydrogen sulfide, 34
hydrosphere, 34
hypoxia, 277

ibexes, Pyrenean *(Capra pyrenaica pyrenaica)*, 326, 327*bf*
Ice Age, 68–70, 82
ice and sediment cores, 82
Ice House periods, 97
ice plant *(Carpobrotus edulis)*, 116–17
immigration, 175 *See also* migration
impervious surface cover, 161, 162*bf*
inbreeding, 206
incoming solar radiation, 98, 99*bf*, 100*bf*
independent variables, 8*t*, 9, 20
indigenous people, 4, 313, 343, 345, 352
indirect effects, 240, 256
individuals
- biological organization level, 50*bf*
- loss of, 206–8

Indonesia, 60
induction, 151
Indus Basin, 67, 160
Industrial Growth worldview, 359
Industrial Revolution, 82, 85, 85*t*, 176
inequity, 10–12, 350–52
insecticides. *See* pesticides
in situ conservation, 308
instantaneous responses, 149, 150*f*
integrative approaches, 7–8
interaction effects, 8*t*, 10
interbreeding, 51, 60, 70, 175
Intergovernmental Panel on Climate Change (IPCC), 3, 350
international climate treaties, 350–52
International Geological Congress, 81
International Union for Conservation of Nature (IUCN), 216–17, 222, 223, 227, 281, 321
interspecific competition, 112
interspecific interactions, 238
intertidal barnacle experiment, 113*bf*
intraspecific competition, 112
intraspecific interactions, 238
introduced/non-native species, 128 *See also* invasive species
- adaptation to, 179
- amphibian decline and, 227–29
- increase of in Europe, 129*bf*
- mutualisms disrupted by, 251, 253–57
- terrestrial systems and, 274

introgression, 175
Inuit, 4, 5*f*
invasive species. *See also* introduced/non-native species
- adaptation to, 179
- biological interactions and, 241–42, 242*f*
- biotic homogenization and, 132–33
- characteristics defining, 131–32
- defined, 128
- feedbacks and, 283
- geographic range and, 117, 119, 121
- globalization and, 128–33
- impacts of, 128–29
- mutualism collapse and, 253–57
- stages of invasion, 129, 131*bf*

in-vitro fertilization, 326, 327*bf*
IPAT framework, 336–38
IPCC. *See* Intergovernmental Panel on Climate Change
iron mining, 87*f*, 88
IUCN. *See* International Union for Conservation of Nature
ivory trade, 307–8, 307*f*

jaguars, 251
Journal of Theoretical Biology, 321
"junk" DNA, 140
Jurassic period, 81*f*

kelp forests, 251, 257–60, 258*bf*, 259*bf*, 260*bf*
Kenya, 341, 345
key biodiversity areas, 281
keystone species, 251, 258, 271
killifish *(Rundulus heteroclitus)*, 154–55, 154*f*
koalas *(Phascolarctos cinereus)*, 247
Kodiak Island, 244
Krause, Johannes, 69*bf*
Kyoto Protocol, 350

laboratory studies, 13, 183–85
Ladder of Life. *See Scala Naturae*

lag effects, 285
Lake, James, 46–48
Lake Victoria, 345
Landsat 8 satellite, 17–18
land-use change, 4, 23, 87*f*, 274
  ecosystems affected by, 268
  geographic range and, 115, 121
  impacts of, 86–88
  move response and, 106
  urbanization and, 160
large-scale earth systems, 271–79
Leadership in Energy and Environmental Design (LEED), 339
lead poisoning, 308
leaf-cutter ants *(Atta sexdens)*, 161, 163*bf*
legislation, 348, 349–50
lemurs, 321–22, 321*bf*, 323–24*bf*
Lenski, Richard, 183–85
Leonardo DiCaprio Foundation, 352
Liboiron, Max, 7*f*
life expectancy, 359, 360*bf*
limpets, 65
lineages
  defined, 37, 42
  early human, 57–60
lineage species concept, 51
Linnaeus, Carl, 42, 48
lionfish *(Pterois)*, 117
living fossils, 41, 41*f*
lizards, 216
  brown anole *(Anolis sagrei)*, 179, 180*f*, 182–83
  curly-tail *(Leiocephalus carinatus)*, 179, 180*f*, 182–83
  monitor, 63, 64*f*
  Phrynosoma, 129
  Puerto Rican crested anole *(Anolis cristatellus)*, 186–87
local extinction, 208
loci, 140, 141
lodgepole pine *(Pinus contorta)*, 271
Lonesome George (tortoise), 209, 325*bf*
Lucy (fossil), 59
Luxury Effect, 164
Lyell, Charles, 48–49

Maathai, Wangari, 341–42, 341*f*
macromolecules, 35
Macy, Joanna, 359–61, 361*bf*
Madagascar, 110, 245, 321–23
Magnuson-Ford, Karen, 321–22, 321*bf*
Mahara, Sri Nisargadatta, 360
main effects, 8*t*, 10
maize *(Zea mays)*, 66, 66*f*
malaria, 61
manufacturing, 345–48, 347*f*
marine systems, 6, 17, 258, 267, 352 *See also* aquatic systems
  adaptation and, 177, 185
  biogeochemistry classifications for, 269*f*
  cascading effects in, 251
  corridors in, 315
  extinction and, 219
  food webs in, 248, 250*bf*
  geographic range and, 116, 117, 118, 119, 120*f*
  global change pressures and, 274, 277
  nitrogen-fixing microbes and, 268
  phenotypic plasticity and, 146, 152
mark-recapture studies, 182, 183
mass extinctions
  "Big Five," 45–46, 45*t*, 222
  defined, 45
  sixth, 205, 220, 222–25
matter, 267
mean, 20
mechanistic species distribution models (SDMs), 121–23
megafaunal extinctions. *See* Quaternary megafaunal extinctions
Mendes, Chico, 341*f*, 342–43
mesocosms, 13, 14*f*
Mesopotamia, 67, 160
Mesozoic era, 81*f*
meta-analyses, 14–17, 15*f*, 22
  of adaptation, 187
  defined, 14
  of extinction risk, 219
  of geographic range changes, 118
  of phenology and global warming, 157–59
metapopulations, 119, 208, 208*f*
methane (CH4), 34, 100, 274, 289–90, 290*bf*, 291*bf*
mice
  Algerian *(Mus spretus)*, 175
  European house *(Mus musculus)*, 175
  piñon *(Peromyscus truei)*, 119, 120*f*
microbes, 287
microhylid frogs, 227*bf*
Middleton, Beth Rose, 4, 5*f*
migration
  assisted, 309
  climate change, 304, 305*bf*
  complete, 107
  partial, 107
migratory shorebirds, 354–58, 355*bf*, 356*bf*, 357*bf*, 358*bf*

Milankovitch cycles, 99, 100*bf*
Miller, Stanley, 35
Miller-Urey experiment, 35
minimum viable population (MVP) size, 212–15
mining, 9–10, 9*f*, 87*f*, 88, 143–44, 311
Miocene epoch, 81*f*
mistletoe *(Tristerix corymbosus)*, 253–56, 255*bf*, 256*bf*, 257*bf*
mitochondrial genomes, 37
moa birds, 246
modeling approaches, 12, 15 *See also* species distribution modeling
  adaptation and, 187
  climate measurement and, 82, 84*bf*
  defined and explained, 13–14
  evolutionary diversity and, 322–24
  extinction risk and, 212–16
  geographic range changes and, 120
model organisms, 185
molecular approaches
  adaptation and, 185–87
  defined and explained, 18–19
  phenotypic plasticity and, 154–55
molecular cloud, 33, 34*f*
molecular level
  in biological organization, 49, 50*bf*
  global change influences at, 292
molecules, 35, 49, 50*f*
mollusks, 38
monarch butterflies *(Danaus plexippus)*, 107, 111
monitor lizards, 63, 64*f*
Montreal Protocol, 350
Moritz, Craig, 125–27, 125*bf*
morning glory *(Ipomoea purpurea)*, 141, 142*f*
morphology shifts, 148
mosquitoes, pitcher-plant *(Wyeomyia smithii)*, 180
moths
  peppered *(Biston betularia)*, 176–77, 176*f*
  sloth (*Cryposes* spp.), 239–40, 239*f*
  sphinx *(Xanthopan morganii)*, 245, 246*f*
motivation, 348
mountain pine beetles *(Dendroctonus ponderosae)*, 145, 145*f*, 271
mountain yellow-legged frogs, 230
move response, 102, 106–35, 173, 209, 237
  geographic range and *See* (geographic range)
  how and why of, 107
mule deer *(Odocoileus hemionus)*, 241
multicellularity, evolution of, 37–38, 38*f*
multispecies interactions, 239–40
mussels, 128, 251
mutations, 35, 173, 175*bf*, 292
  beneficial, 174
  deleterious, 174
  new, 174–75
  point, 174
mutualisms, 238*t*, 241, 242*f*, 251
  coextinction and, 245–46
  collapse of, 253–57
MVP. *See* minimum viable population
myxoma virus, 187–88, 188*f*

National Ecological Observatory Network, 20
National Institutes of Health (NIH), 20
National Oceanic and Atmospheric Administration (NOAA), 20
National Park System, 4
National Science Foundation (NSF), 20
native range, 128
natural disturbance regimes, 315, 316*f*
natural gas, 89, 304
natural selection, 35, 39–40, 40*f*, 42, 70, 172–74, 189, 189*f*, 190, 207, 208
  defined, 39, 172
  example of evolution by, 176–77
  explained, 39
  mathematical formulation for predicting trait change by, 174
  molecular approaches to study of, 185–86
nature (economic value of), 22–24
Nature (journal), 2, 22, 46, 68, 94
Nature Climate Change (journal), 191
Nature Conservancy, 352, 354–58
NatureServe, 216
Neanderthals, 60, 68–70, 71*bf*
negative feedback loops, 281
Neolithic Revolution, 65, 67, 68, 74, 85
neonicotinoids, 94–96, 96*bf*
nested classification system, 42, 43*bf*, 49
neutralism, 238*t*
neutrons, 33
New Guinea highlands, 65
new mutations, 174–75
newts, 226
New York Times, 157
New Zealand, 117
niche(s), 115, 123
  characteristics of, 112
  defined, 111
  fundamental, 112, 113*bf*

realized, 112, 113*bf*
niche adaptation, 112
niche conservatism, 112
Nile Valley, 67, 160
nitrogen cycle, 267–68
nitrogen-fixing microbes, 268
nitrous oxide (N2O), 100, 274, 289–91, 290*bf*, 291*bf*
"Node-by-Node Disassembly of a Mutualistic Interaction Web Driven by Species Introductions" (Rodriguez-Cabal et al.), 253–55
noise, 146–47, 147*f*, 149, 150*f*, 155, 161
noninvasive sampling, 19
non-native species. *See* introduced/non-native species
North Africa, 283
northern white rhinos *(Ceratotherium simum cottoni)*, 326, 327*bf*
nuclear genome, 140
nuclei, 33, 37
nucleotides, 141*bf*
null hypothesis, 158

O'ahu, Hawaii, 210
oak trees, 206
obligate interactions, 238–39
observational approaches, 12, 14, 15, 17
  defined and explained, 13
  geographic range changes and, 120
oceans
  acidification of, 185, 195, 196*bf*
  global map of human impact in, 278*f*
  warming of, 195, 277
oil, 89
oil spills, 8, 17, 88, 342
Old Blue (black robin), 309
Oligocene epoch, 81*f*
One Health, 25
opossums *(Dromiciops gliroides)*, 253–56, 255*bf*
opposable thumbs, 71
orangutans *(Pongo)*, 57, 58
orchids *(Angræcum sesquipedale)*, 245, 246*f*
Ordovician period, 45*t*
organismal level
  in biological organization, 49, 50*bf*
  global change influences at, 292
organismal monitoring tools, 17–18
Origin of Species, The (Darwin), 49, 172
orthologs, 47
oryxes, African *(Oryx gazella)*, 310, 310*f*
ospreys *(Pandion haliaetus)*, 343, 344*f*
otters. *See* sea otters
outgoing longwave radiation, 97, 99–100, 99*bf*
Out of Africa hypothesis, 61–62, 62*f*
overexploitation, 228
overharvesting. *See* harvesting
owls
  barred, 122–23*f*
  powerful *(Ninox sterna)*, 215
  spotted, 122–23*f*
oxygen cycle, 268
oxytocin, 73
ozone, 274, 350

Pääbo, Svante, 69*bf*
Pacific Flyway, 355
Paleocene epoch, 81*f*
Paleozoic era, 81*f*
paralogs, 47
Paranthropus, 59, 59*f*
parasites, coextinction of, 246–47, 247*f*
Paris Agreement, 350
Parmesan, Camille, 157, 158, 157*bf*
partially additive effects, 91*f*, 92
partial migration, 107
participatory approaches, 15, 17
passenger pigeons *(Ectopistes migratorius)*, 210, 210*f*
passive locomotion, 107
penguins, emperor *(Aptenodytes forsteri)*, 116, 117*f*
peppered moths *(Biston betularia)*, 176–77, 176*f*
peregrine falcons *(Falco peregrinus)*, 161
performance assays, 182
permafrost, 278, 288, 289, 290
Permian period, 45*t*
pesticides, 4, 317, 345
  adaptation and, 177–78, 178*f*
  pollinators and, 94–97, 95*bf*
phenological mismatch, 158
phenology
  defined, 157
  global warming and, 157–59, 180
  shifts in, 242–45, 243*f*, 244*f*
phenotypes, 140–41
phenotypic plasticity, 140–56, 156–57, 181, 250, 293
  assessment and prediction of, 152–55
  capacity for across traits and species, 143–44
  defined, 142
  global change pressures and, 144–48
  long-term persistence and, 155, 156*f*
  underlying mechanisms of, 148–51
phenotypic variation, 172–73
phi-X virus, 140
phosphorus cycle, 268
photosynthesis, 36–37, 45, 146, 267, 270

phyla, 38
phylogenetic tree, 41, 43*bf*
  defined and explained, 42
  of humans and living primate relatives, 57*f*
physiology shifts, 146
Phytophthora ramorum, 206
phytoplankton *(Emiliania huxleyi)*, 185
Pierre, Suzanne, 7*f*
pigeons, passenger *(Ectopistes migratorius)*, 210, 210*f*
pika, American *(Ochotona princeps)*, 109, 109*f*, 119
"Pillars of Creation," 34*f*
piñon mice *(Peromyscus truei)*, 119
Pinta Island tortoises *(Chelonoidis nigra abingdonii)*, 209, 210*f*, 325*bf*, 326
pitcher-plant mosquitoes *(Wyeomyia smithii)*, 180
Planetwalk, 342
Plantae, 42
plasmids, 140
Plasmodium falciparum, 61
plasticity-first hypothesis, 155
Plato, 359
Pleistocene epoch, 68, 81*f*, 82, 97, 278
Pliocene epoch, 81*f*
point mutations, 174
point source pollution, 88, 89*f*
polar bears *(Ursus maritimus)*, 146, 147*f*, 243
polar ice melting, 5*f*, 116, 117*f*, 277, 279
pollination crisis, 94
pollinators, 94–96, 95*bf*, 240, 245–46, 254–56, 317
pollution, 88–90, 89*f*, 277 *See also* contaminants
Polynesian tree snails *(Partula nodosa)*, 210, 210*f*
population(s)
  biological organization level, 49, 50*f*
  defined, 51
  global change influences at, 293, 294*bf*
  loss of, 208–9 *See also* (extirpation)
  meta-, 119, 208, 208*f*
  minimum viable, 212–15
population declines, 13
population growth, 4, 23, 336, 337*bf*
  contemporary patterns of, 85–86, 85*t*, 86*f*
  in early humans, 67–68, 67*f*
  measurements of, 182
  urban, 160, 160*bf*
population size, 20, 65, 67*f*, 110, 174, 175, 188*f*, 189, 206, 207*f*, 208, 215, 221, 238, 293, 294*bf*, 336, 337*bf*
population viability analysis (PVA), 212–16, 214*f*, 215*f*, 350
positive feedback loops, 281, 282*bf*, 283, 284, 290
potato blight, 229
powerful owls *(Ninox sterna)*, 215
practitioners, 341–42
prairie dogs, 72
prebiotic chemistry, 34–35
predators
  apex, 251, 252*f*, 284
  coextinction of herbivores and, 247–48
  kelp forests and, 257–60
  keystone *See* (keystone species)
primary consumers, 251, 252*f*
Proceedings of the National Academy of Sciences (journal), 253
producers, 248
projected climate change, 276, 276*f*
prokaryotes, 36, 37, 46, 140
propagule banks, 191
propagule size, 132
proteins, 35
Proterozoic era, 81*f*
protons, 33
Protozoa, 42
Puerto Rican crested anole lizards *(Anolis cristatellus)*, 186–87
PVA. *See* population viability analysis
p-values, 20, 21*bf*
Pyrenean ibexes *(Capra pyrenaica pyrenaica)*, 326, 327*bf*

quaggas *(Equus quagga quagga)*, 326, 327*bf*
Quaternary period, 81*f*
Quaternary megafaunal extinctions, 63–65, 211, 246

rabbit hemorrhagic disease (RHD), 189
rabbits, 187–89, 188*f*
racism, 6, 10, 11, 164, 351–52
racoons, 111
Rainforest Alliance, 339
rainforests, 113, 342, 352
randomization, 8*t*, 10
range contractions, 113, 115, 116, 116*f*
range expansions, 113, 115, 116–17, 116*f*
range marches, 115, 116*f*, 117–19, 118*f*
"Rapid Evolution of Thermal Tolerance in the Water Flea *Daphnia*" (Geerts et al.), 191–93
RARS gene, 187
realized climate change, 275
realized niche, 112, 113*bf*
reciprocal transplant experiments, 152, 153*f*, 196
recolonization, 119, 208, 209, 309, 313
recombination, 174
reconciliation ecology, 315
red elderberry, 244*f*, 245

Red List of Ecosystems, 281
Red List of Threatened Species, 216–19, 217*f*, 218*f*, 228*bf*, 281, 321
red sea urchins *(Strongylocentrotus franciscanus)*, 258–60
red squirrels *(Tamiasciurus hudsonicus)*, 180
reed buntings *(Emberiza schoeniclus)*, 149, 150*f*
reflected solar radiation, 97, 99, 99*bf*
refugia, 113
regulation, 348
Reich, David, 69*bf*
relaxation time, 220, 221*bf*
renewable energy sources, 304
replication, 8*t*, 10, 35
reproduction, 173–74
reserves, 311–13, 312*f*
resource extraction, 88
response magnitude, 151
response permanence, 151
response time, 151
response window, 151
resurrection ecology, 191, 192, 192*bf*, 193
reversal, 151
reverse auctions, 356
Reynolds, Mark, 354–58, 355*bf*
rhinos, northern white *(Ceratotherium simum cottoni)*, 326, 327*bf*
ribonucleic acid. *See* RNA
rice-fish farming system, 345, 346*f*
rights-based fishery management, 348–49
"Ring of Life Provides Evidence for a Genome Fusion Origin of Eukaryotes, The" (Rivera and Lake), 46–48
Rivera, Maria, 46–48, 47*bf*
RNA, 35, 149, 172
robins, black *(Petroica traversi)*, 309–10, 310*f*
Rocky Mountains, 145
Rodriguez-Cabal, Mariano, 253–57, 254*bf*
Rome, 67, 85
Root, Terry, 157–59, 157*bf*
Roots & Shoots, 342
royal jelly, 149, 150*f*
Rundlöf, Maj, 94–97, 94*bf*
rusa deer *(Cervus timorensis)*, 320

saber-tooth tigers, 63, 64*f*
Saccharomyces cerevisiae, 37
Saddleback tortoises, 326
Sahal region, 283
Sahara, green to desert, 283
Sakhalin fir *(Abies sachalinensis)*, 153–54
salamanders, 226
  Chinese giant, 227*bf*
  European, 231
  slender *(Batrachoseps)*, 108*f*
salmon, 244*f*, 245
San Francisco Bay, 131, 131*bf*
Scala Naturae (Ladder of Life), 48, 49*bf*, 359
Schell, Chris, 11
Science (journal), 2
Science Advances (journal), 222, 354
scientific method, 8
Scientific Revolution, 2
scientific value of biodiversity, 25
SDM. *See* species distribution modeling
sea cucumbers, 294, 295*bf*
sea ice melting, 116, 117*f*, 270, 278, 279*f*, 283
sea level rise, 277, 279
sea lions, stellar, 212
sea otters *(Enhydra lutris)*, 258–60, 258*bf*, 259*bf*, 260*bf*
sea star wasting disease, 229
sea turtles, 343
sea urchins *(Strongylocentrotus)*, 258–60, 259*bf*, 260*bf*
secondary extinction. *See* coextinction
Second Industrial Revolution, 85
"Seed Coating with a Neonicotinoid Insecticide Negatively Affects Wild Bees" (Rundlöf et al.), 94–97
seed dispersal, 245–46, 253–57, 270–71, 272*f*, 283
selective breeding, 326, 327*bf*
sensitivity, 92*f*, 93
Serengeti-Mara ecosystem, 314
serotonin, 73
sessile organisms, 107
"sex reversal," 148, 228
Seychelles, 352
Shanxi Province, China, 311–12
shipping routes, 90, 90*f*
Sierra Nevada Mountains, 119, 120*f*
Silent Spring (Carson), 4
single nucleotide polymorphisms (SNPs), 69
sixth mass extinction, 205, 220, 222–26
slave trade, 91
slender salamanders *(Batrachoseps)*, 108*f*
sloth moths (*Cryposes* spp.), 239–40, 239*f*
sloths
  giant ground, 63, 64*f*
  three-toed, 239–40, 239*f*
smallpox virus, 210*f*, 211
snails
  Euglandina rosea, 210
  Polynesian tree *(Partula nodosa)*, 210, 210*f*

snakes, timber rattlesnake *(Crotalus horridus)*, 106
soap, 171
social structure (of early humans), 72–73
soil, greenhouse gases in, 287–91
solar radiation reflection, 304
solar systems, 33
solitary bees *(Osmia bicornis)*, 95, 96*bf*
song rate (animal), 147 *See also* birdsong
South Africa, 116
space mirrors, 304
sparrows, white-crowned *(Zonotrichia leucophrys nuttalli)*, 155
speciation
  defined and explained, 39–40
  high rates of, 41–46
species
  conservation and, 302
  defined, 51
  effects of change in, 242–45
  gain in, 241–42
  global change influences at, 293
  interactions in *See* (biological interactions)
  loss of, 209, 240–41 *See also* (extinction)
  phenotypic plasticity capacity and, 143–44
species distribution modeling (SDM)
  correlative, 121, 123
  defined, 120–21
  extinction risk and, 212, 213*f*, 219
  geographic range changes and, 120–23, 122–23*f*, 124*f*
  mechanistic, 121–23
species diversity, 286
species recovery plans, 350
species richness, 132, 132*bf*, 220
species-specific funding, 350
sperm whales, 71
sphinx moths *(Xanthopan morganii)*, 245, 246*f*
spillover, 241
spiritual value of biodiversity, 25
spotted owls, 122–23*f*
squirrels
  golden-mantled ground *(Callospermophilus lateralis)*, 120*f*
  red *(Tamiasciurus hudsonicus)*, 180
standing genetic variation, 174, 176
statistical significance, 20, 21*bf*
Steel, Mike, 321*bf*
stellar sea lions, 212
STEM fields, 6
struggle for existence, 174
subarctic tundra warming, 287–91
subfamilies, 58, 58*bf*
subterranean injection, 304
subtribes, 58
Sudan (northern white rhino), 327*bf*
sudden oak death, 206, 229
sulfur cycle, 268
sun, 33, 98
superb cyanea *(Cyanea superba)*, 210–11, 210*f*
super colonies, 129
superfamilies, 58, 58*bf*
"survival of the friendliest" hypothesis, 72–73
Sustainable Apparel Coalition, 345
sustainable food production, 345
swallows, tree *(Tachycineta bicolor)*, 148
Symbiodinium, 194*bf*, 195, 196*bf*, 197
symbionts, 51, 193, 195, 197
synergistic effects, 91*f*, 92
synthesis approaches, 12, 14–17
  adaptation and, 187
  defined and explained, 14–15
  geographic range changes and, 120

tabletop corals *(Acropora hyacinthus)*, 196–97
Tahiti, 210
tan oak trees, 206
Tanzania, 313–15, 314*f*, 342
targeted approach to conservation, 321, 322–24
Tasmanian devils *(Sarcophilus harrisii)*, 207
tax credits, 354
taxonomy, 42
taxon substitutions, 326
Technological Revolution, 85
technology, 325–28, 336
teosinte, 66, 66*f*
terrestrial systems, 17, 258, 267, 270, 352
  adaptation and, 177
  biodiversity hotspots in, 279–80
  biogeochemistry classifications for, 269*f*
  biological interactions in, 242
  biotic homogenization in, 132–33
  cascading effects in, 251
  climate influenced by, 82
  contemporary alterations in, 87, 87*f*, 88
  extinction and, 219
  food webs in, 248, 249*bf*
  geographic range and, 118, 119, 120*f*
  global change pressures and, 271, 274, 275*f*
  harvesting in, 308
  nitrogen-fixing microbes and, 268

phenotypic plasticity and, 146, 152, 153*f*
Tertiary period, 81*f*, 97
Thales of Miletus, 3*f*
thermoconformers, 216
thermoregulators, 216
Thoreau, Henry David, 49
three-toed sloths, 239–40, 239*f*
tigers, saber-tooth, 63, 64*f*
timber rattlesnakes *(Crotalus horridus)*, 106
time lag, 220
tipping points, 283–84, 285*f*
toads, 226, 227*bf*
tool use, 60, 71
top-down effects, 248, 250*bf*, 293
tortoises
Domed, 326
Pinta Island *(Chelonoidis nigra abingdonii)*, 209, 210*f*, 325*bf*, 326
Saddleback, 326
Toxoplasma gondii, 91
tragedy of the commons, 348
translocations, 309–11
transportation networks, 106, 115, 229, 337
treaties, climate, 350–52
treatment group, 8*t*
tree of life, 18, 27, 36, 38, 42, 49, 49*bf*, 51, 161, 205
adaptation and, 172, 175, 177
amphibians in, 226, 227, 229
development shifts in, 145
extinction in, 209, 211, 216, 217, 219, 220
geographic range and, 110
hereditary systems in, 172
key processes shaping, 39
mutation rates in, 174, 292
niche shifts and, 112
tree swallows *(Tachycineta bicolor)*, 148
Triassic period, 45*t*, 81*f*
tribes, 58, 58*bf*
trophic cascades, 284
defined, 240
kelp forests and, 257–60
trophic generalists, 239
trophic levels, 248, 249*bf*, 251, 267, 284
trophic mismatch, 242–45
trophic specialists, 238, 247–48
trophic structure, 248, 249*bf*
tropical forests, 270–71
tuatara, 41*f*
turtles, sea, 343
United Nations, 345
United Nations Environment Programme (UNEP), 350
United Nations Framework Convention on Climate Change (UNFCCC), 350
universe, origin of, 33
urban heat island effect, 161, 162*bf*, 164, 317
urbanization, 4, 159–64
adaptation and, 186–87
early humans and, 67–68
habitat enhancement and, 317, 318*f*
Urey, Harold, 35
USDA Organic, 339

"Value of the World's Ecosystem Services and Natural Capital, The" (Costanza et al.), 22–23
variables
categorical, 20
confounding, 8*t*
continuous, 20
dependent, 8*t*, 9, 20
independent, 8*t*, 9, 20
variance, 20
variation, 35, 173*f*
genetic *See* (genetic variation)
phenotypic, 172–73
vertical flow of genetic information, 42, 46
"vertical forests," 317, 318*f*
Voigt, Carolina, 287–91, 287*bf*
volcanic activity, 8, 45, 99, 255
vulnerability, 92–93, 92*f*

warfarin, 175
"Warming of the Subarctic Tundra Increases Emissions of All Three Important Greenhouse Gases" (Voigt et al.), 287–91
wasps *(Vespula germanica)*, 253–56, 256*bf*, 257*bf*
water cycle, 267
Wells, Emily, 7*f*
Western Pine beetles *(Dendroctonus brevicomis)*, 121
Wet Tropics of Australia, 113, 114*f*
whales
blue *(Balaenoptera musculus)*, 19*f*, 109, 109*f*
sperm, 71
wheel of empowerment, 344*f*
whelk, 251, 252*f*
white campion *(Silene latifolia)*, 132*bf*
white-crested elaenia birds *(Elaenia albiceps)*, 253–56, 255*bf*
white-crowned sparrows *(Zonotrichia leucophrys nuttalli)*, 155

wildebeest *(Connochaetes taurinus)*, 314, 314*f*
wildfires, 315
Wilson, E. O., 205, 313
wind turbines, 304, 305*bf*
wolves, 73, 73*bf*, 215, 325, 325*bf*
woolly mammoths *(Mammuthus primigenius)*, 63, 64*f*, 327, 327*bf*
worms
  armyworms *(Spodoptera frugiperdam)*, 178, 178*f*, 179
  earthworms *(Lumbricus terrestris)*, 251, 252*f*
Woronora Plateau, 320

x-axis (graph), 20

Yangtze River Basin, 65
Yangtze River dolphins *(Lipotes vexillifer)*, 210*f*, 211
y-axis (graph), 20, 21*bf*
Yellow River Basin, 65
Yorok Tribe, 4
Yosemite National Park, 87*f*, 88, 125–28, 125*bf*, 126*bf*, 127*bf*, 128*bf*

zebra mussels, 128
Zion National Park, 241, 251
zoonoses, 229
zooxanthellae, 195